T0212793

Lecture Notes in Computer Science 9079

Commenced Publication in 1973
Founding and Former Series Editors:
Gerhard Goos, Juris Hartmanis, and Jan van Leeuwen

Editorial Board

David Hutchison
Lancaster University, Lancaster, UK
Takeo Kanade
Carnegie Mellon University, Pittsburgh, PA, USA
Josef Kittler
University of Surrey, Guildford, UK
Jon M. Kleinberg
Cornell University, Ithaca, NY, USA
Friedemann Mattern
ETH Zürich, Zürich, Switzerland
John C. Mitchell
Stanford University, Stanford, CA, USA
Moni Naor
Weizmann Institute of Science, Rehovot, Israel
C. Pandu Rangan
Indian Institute of Technology, Madras, India
Bernhard Steffen
TU Dortmund University, Dortmund, Germany
Demetri Terzopoulos
University of California, Los Angeles, CA, USA
Doug Tygar
University of California, Berkeley, CA, USA
Gerhard Weikum
Max Planck Institute for Informatics, Saarbrücken, Germany

More information about this series at http://www.springer.com/series/7407

Vangelis Th. Paschos · Peter Widmayer (Eds.)

Algorithms and Complexity

9th International Conference, CIAC 2015
Paris, France, May 20–22, 2015
Proceedings

 Springer

Editors
Vangelis Th. Paschos
LAMSADE
Université Paris-Dauphine
Paris Cedex 16
France

Peter Widmayer
Inst. of Theoretical Computer Science
ETH Zürich
Zurich
Switzerland

ISSN 0302-9743
Lecture Notes in Computer Science
ISBN 978-3-319-18172-1
DOI 10.1007/978-3-319-18173-8

ISSN 1611-3349   (electronic)

ISBN 978-3-319-18173-8   (eBook)

Library of Congress Control Number: 2015936678

LNCS Sublibrary: SL1 – Theoretical Computer Science and General Issues

Printed on acid-free paper

Springer International Publishing AG Switzerland is part of Springer Science+Business Media (www.springer.com)

Preface

This volume contains the papers presented at CIAC 2015: 9th International Conference on Algorithms and Complexity held during May 20–22, 2015 at the Université Paris-Dauphine, Paris. This series of conferences presents original research contributions in the theory and applications of algorithms and computational complexity.

The volume begins with invited papers and continues with contributed papers alphabetically arranged by the last names of their authors. There were 93 submissions for this CIAC edition. Each submission was reviewed by at least three Program Committee members. The committee decided to accept 30 papers. The program also includes four invited talks.

We thank all the authors who submitted papers, the members of the Program Committee, and the external reviewers. We are also grateful to the four invited speakers, Bernard Chazelle (Princeton University), Haim Kaplan (Tel Aviv University), Elias Koutsoupias (Oxford University), and Mikkel Thorup (University of Copenhagen), who kindly accepted our invitation to give plenary lectures at the conference.

We gratefully acknowledge the support from Université Paris-Dauphine, its Laboratory LAMSADE, the AGaPe Research Group of the LAMSADE, and the EATCS.

We would also like to thank the Local Organizing Committee and in particular Katerina Kinta, Cécile Murat, Olivier Rouyer, and Juliette de Roquefeuil for their active participation in several organization tasks. Many thanks also to Mireille Le Barbier for her valuable help in allocating all the necessary rooms in the University, for all the conference main and side activities.

We finally would like to thank EasyChair for providing us a very friendly environment for handling the contributions and editing the proceedings of CIAC 2015.

March 2015

Vangelis Th. Paschos
Peter Widmayer

Organization

Program Committee

Hannah Bast	University of Freiburg, Germany
Vincenzo Bonifaci	IASI-CNR, Rome, Italy
Jérémie Chalopin	CNRS and Université d'Aix-Marseille, France
Victor Chepoi	Université d'Aix-Marseille, France
Marek Chrobak	University of California, Riverside, USA
Pierluigi Crescenzi	University of Florence, Italy
Jurek Czyzowicz	Université du Québec en Outaouais, Canada
Yann Disser	Technische Universität Berlin, Germany
Thomas Erlebach	University of Leicester, UK
Bruno Escoffier	Université Pierre-et-Marie-Curie, France
Irene Finocchi	Sapienza University of Rome, Italy
Fedor Fomin	University of Bergen, Norway
Pierre Fraigniaud	CNRS and University Paris Diderot, France
Herman Haverkort	Technische Universiteit Eindhoven, The Netherlands
Matúš Mihalák	ETH Zürich, Switzerland
Luca Moscardelli	University of Chieti-Pescara, Italy
Yoshio Okamoto	University of Electro-Communications, Japan
Vangelis Th. Paschos	Université Paris-Dauphine, France
David Peleg	Weizmann Institute of Science, Israel
Marie-France Sagot	INRIA and Université Lyon 1, France
Piotr Sankowski	University of Warsaw, Poland
Maria Serna	Polytechnic University of Catalonia, Spain
Paul Spirakis	University of Liverpool, UK and CTI, Greece
Dimitrios Thilikos	CNRS, LIRMM, France and UoA, Greece
Roger Wattenhofer	ETH Zürich, Switzerland
Peter Widmayer	ETH Zürich, Switzerland
Gerhard J. Woeginger	Technische Universiteit Eindhoven, The Netherlands

Organizing Committee

Additional Reviewers

Adamczyk, Marek	Barbay, Jérémy
Albers, Susanne	Bein, Wolfgang
Angelini, Patrizio	Belmonte, Rémy
Bampas, Evangelos	Bilò, Vittorio

Blum, Christian
Bonnet, Edouard
Bonsma, Paul
Borassi, Michele
Boria, Nicolas
Brandes, Philipp
Brandstadt, Andreas
Brandt, Sebastian
Brunetti, Sara
Bulteau, Laurent
Calamoneri, Tiziana
Campelo, Manoel
Chikhi, Rayan
Chuzhoy, Julia
Couëtoux, Basile
Crochemore, Maxime
D'Angelo, Gianlorenzo
Dabrowski, Konrad Kazimierz
Das, Shantanu
Demange, Marc
Dürr, Christoph
Elbassioni, Khaled
Emek, Yuval
Estellon, Bertrand
Fanelli, Angelo
Feige, Uri
Fernau, Henning
Foerster, Klaus-Tycho
Garg, Naveen
Garnero, Valentin
Gaspers, Serge
Gavoille, Cyril
Gawrychowski, Pawel
Giroudeau, Rodolphe
Godard, Emmanuel
Goldwasser, Michael
Golovach, Petr
Gu, Qianping
Habib, Michel
Hackfeld, Jan
Jansen, Klaus
Keller, Barbara
Kellerer, Hans
Kim, Eun-Jung
Klimm, Max
Knauer, Kolja

Kowalik, Lukasz
König, Michael
Labourel, Arnaud
Langner, Tobias
Lauria, Massimo
Lazarus, Francis
Liotta, Giuseppe
Lv, Yuezhou
M.S., Ramanujan
MacQuarrie, Fraser
Mamageishvili, Akaki
Markakis, Evangelos
Marques-Silva, Joao
Mary, Arnaud
Meissner, Julie
Melideo, Giovanna
Messeguer, Arnau
Mnich, Matthias
Monaco, Gianpiero
Montanari, Sandro
Mäkinen, Veli
Naves, Guyslain
Nicolas, Bourgeois
Nisse, Nicolas
Nutov, Zeev
Otachi, Yota
Pacheco, Eduardo
Pajak, Dominik
Pascual, Fanny
Pasquale, Francesco
Préa, Pascal
Radoszewski, Jakub
Raffinot, Mathieu
Ravelomanana, Vlady
Rigaill, Guillem
Rossi, Gianluca
Russo, Luis M.S.
Sacomoto, Gustavo
Sau, Ignasi
Saurabh, Saket
Schaefer, Marcus
Schewior, Kevin
Seidel, Jochen
Sgouritsa, Alkmini
Shioura, Akiyoshi
Sikdar, Somnath

Sikora, Florian
Sinaimeri, Blerina
Sirén, Jouni
Spieksma, Frits
Stachowiak, Grzegorz
Stamoulis, Georgios
Stolz, David
Suchý, Ondřej
Telelis, Orestis
Uitto, Jara

Uno, Yushi
Uznański, Przemysław
van Stee, Rob
Vaxès, Yann
Viglietta, Giovanni
von Niederhäusern, Leonard
Wenk, Carola
Wiese, Andreas
Winkler, Peter
Young, Neal

Fast and Powerful Hashing Using Tabulation (Extended Abstract)

Mikkel Thorup*

University of Copenhagen, Department of Computer Science, Universitetsparken 5,
2100 Copenhagen East, Denmark

Abstract. Randomized algorithms are often enjoyed for their simplicity, but the hash functions employed to yield the desired probabilistic guarantees are often too complicated to be practical. Here we survey recent results on how simple hashing schemes based on tabulation provide unexpectedly strong guarantees.

Simple tabulation hashing dates back to Zobrist [1970]. Keys are viewed as consisting of c characters and we have precomputed character tables $h_1, ..., h_q$ mapping characters to random hash values. A key $x = (x_1, ..., x_c)$ is hashed to $h_1[x_1] \oplus h_2[x_2].....\oplus h_c[x_c]$. This schemes is very fast with character tables in cache. While simple tabulation is not even 4-independent, it does provide many of the guarantees that are normally obtained via higher independence, e.g., linear probing and Cuckoo hashing.

Next we consider *twisted tabulation* where one character is "twisted" with some simple operations. The resulting hash function has powerful distributional properties: Chernoff-Hoeffding type tail bounds and a very small bias for min-wise hashing.

Finally, we consider *double tabulation* where we compose two simple tabulation functions, applying one to the output of the other, and show that this yields very high independence in the classic framework of Carter and Wegman [1977]. In fact, w.h.p., for a given set of size proportional to that of the space consumed, double tabulation gives fully-random hashing.

While these tabulation schemes are all easy to implement and use, their analysis is not.

This invited talk surveys result from the references below.

References

1. Dahlgaard, S., Knudsen, M.B.T., Rotenberg, E., Thorup, M.: Hashing for statistics over k-partitions. CoRR abs/1411.7191 (2014), http://arxiv.org/abs/1411.7191
2. Dahlgaard, S., Thorup, M.: Approximately minwise independence with twisted tabulation. In: Proc. 14th Scandinavian Workshop on Algorithm Theory (SWAT). pp. 134–145 (2014)
3. Pătraşcu, M., Thorup, M.: The power of simple tabulation-based hashing. Journal of the ACM 59(3), Article 14 (2012), announced at STOC'11
4. Pătraşcu, M., Thorup, M.: Twisted tabulation hashing. In: Proc. 24th ACM/SIAM Symposium on Discrete Algorithms (SODA). pp. 209–228 (2013)
5. Thorup, M.: Simple tabulation, fast expanders, double tabulation, and high independence. In: Proc. 54th IEEE Symposium on Foundations of Computer Science (FOCS). pp. 90–99 (2013)

Research partly supported by an Advanced Grant from the Danish Council for Independent Research under the Sapere Aude research carrier programme.

Contents

Communication, Dynamics, and Renormalization

Bernard Chazelle[(✉)]

Department of Computer Science,
Princeton University Princeton, Princeton, NJ 08540, USA
chazelle@cs.princeton.edu

Abstract. This paper explores a general strategy for analyzing network-based dynamical systems. The starting point is a method for parsing an arbitrary sequence of graphs over a given set of nodes. The technique is then harnessed to carry out a dynamic form of renormalization for averaging-based systems. This analytical framework allows us to formulate new criteria for ensuring the asymptotic periodicity of diffusive influence systems.

Keywords: Dynamical systems · Networks · Renormalization · Influence systems

1 Introduction

There is by now a wide, well-established body of techniques for decomposing networks into smaller pieces [3, 8]. These include methods based on spectral partitioning, SDP relaxation, diffusion, coarsening, flows, metric embeddings, local search, etc. By comparison, the cupboard of decomposition tools for dynamic networks looks bare. Allowing edges to come and go puts basic connectivity questions in a new light and calls for a novel set of tools [11]. This need, of course, hinges on the relevance of dynamic graphs in the first place. It is easy to argue that, in practice, networks rarely come with a fixed set of edges. On the Web, for example, hyperlinks are added and deleted all the time. The same is true of social networks and virtually any large graph subject to failure. The present motivation for investigating dynamic networks emanates from a specific concern, however: the dynamics of multiagent systems and, more ambitiously, the emergence of collective behavior in living systems.

Think of how fireflies synchronize their flashes, birds form V-shaped flocks, ants find shortest paths, and bacteria perform quorum sensing. The standard approach to modeling such systems is to look at the individual organisms as *agents* endowed with two kinds of rules: *communication* rules to specify which agents "listen" to which ones under what circumstances; and *action* rules to

This work was supported in part by NSF grants CCF-0963825 and CCF-1420112. Part of it was done when the author was an Addie and Harold Broitman Member at the Institute for Advanced Study, Princeton, NJ.

© Springer International Publishing Switzerland 2015
V.Th. Paschos and P. Widmayer (Eds.): CIAC 2015, LNCS 9079, pp. 1–32, 2015.
DOI: 10.1007/978-3-319-18173-8_1

instruct the agents on what to do with the information they acquire. Communication between the agents is channeled through a dynamic network whose topology changes endogenously as the system evolves over time. Before we describe our approach to analyzing such systems, we provide a few words of intuition.

1.1   Parsing Dynamic Networks

The analysis relies on specific methods for tracking the propagation of information at all scales. This means monitoring how often any two groups of agents of *any size* communicate with each other. This, in turn, opens the door to divide-and-conquer. To take an extreme example, suppose that the agents can be partitioned into two subsets that never exchange information. If so, each subgroup is decoupled from the other and can be analyzed separately. Somewhat trickier is the case of two groups A and B with no edges pointing from B to A. Since information runs in the reverse direction of the edges,[1] the group B is decoupled and can be analyzed on its own. The same is not true of A, however, since the group relies on information from B for its dynamics. What to do then? The key observation is that, should the system B eventually settle around a fixed point or a limit cycle, its predictable behavior will allow us to treat the group B, after a suitable period of time, as a single *block-agent*: think of it as some sort of super-agent with static or periodic behavior. In this way, the original system can now be analyzed by first working on B and then turning our attention to a *block-directional* system of $|A| + 1$ agents ($+1$ because of the block-agent). Think of an Ancient Régime monarchy where the royal court B has its own internal dynamics, one that is vital to the dynamics of A yet entirely oblivious to it.

Breaking down a system into its decoupled and block-directional parts is a process that can be iterated recursively. One difficulty is that the decomposition scheme (ie, the choice of A and B) cannot be fixed once and for all but must adapt to the changes in the communication network. This is unusual. In statistical physics, for example, this kind of decomposition hierarchy is known as *real-space renormalization* (more on which below). Unlike here, the decomposition scheme is fixed and known a priori. In the planar Ising model, for example, the communication graph is a grid, so block-spin renormalization naturally results in a hierarchy of subgrids. The process can be specified without knowing anything about the dynamical system itself. By contrast, we have here a feedback loop between the dynamics of an influence system and the topology of its communication channels, so that it is impossible to build any hierarchy before we know which graphs will crop up when. This is the reason we call this brand of renormalization *algorithmic*, to distinguish it from its *closed-form* version from statistical physics and quantum mechanics.

How do we choose the partition of the agent set into A and B? There are many considerations at play but one of them stands out: updating A and B should be as infrequent as possible. To take a fanciful example, consider a basketball

---

[1] An edge pointing from me to you indicates that I am listening to you and therefore that the information flows from you to me.

game. It might be sensible to choose one team to be A and the other one B on the grounds that the ball stays within a given team more often than not. On the other hand, this might not be true when the action is near the basket and possession switches rapidly between shooters, blockers, and rebounders. One could then imagine changing the choice of the groups A and B every now and then in order to keep the interactions between the two groups to a minimum. Yet to find the absolute minimum is not an option. The choice must not only mirror the flow of information across the networks but also proceed on-line: in particular, one should not have to look ahead into the future to decide how to split the agents into groups A and B.[2]

The dynamic assignment of A and B partitions the timeline $t = 0, 1, 2, \ldots$ into a sequence of consecutive intervals within which the assignment is time-invariant. Within each such interval, the intra-communication among the agents of A (or B) could be itself quite complex, thus prompting us to partition A and B. Proceeding in this way recursively produces a hierarchical decomposition of the timeline, ie, a *parse tree* such as $\big((01)(234)\big)\big((5)(67)(8\cdots)\big)$. In this example, the timeline forms the root of the tree. The root has two children associated with the time intervals 01234 and $5678\cdots$, which themselves have respectively two and three children, with intervals 01 and 234 for one and 5, 67, and $8\cdots$ for the other. As we show below, this allows us to view the transmission of information across the agents at all relevant timescales and "renormalize" the system accordingly. The parsing procedure is a *message passing* algorithm—as are, we should point out, most spectral methods and belief propagation algorithms used for fixed networks (a word we use interchangeably with "graphs").

1.2   Temporal Renormalization

The systems are deterministic, so a given starting configuration of the agents yields a single orbit. The question is under which conditions is the orbit attracted to a limit cycle. Whereas attraction to a fixed point can often be analyzed by looking at the orbit of interest and setting up a suitable Lyapunov function for it, asymptotic periodicity does not lend itself to this kind of investigation. It is often necessary to reason about the space of all orbits, which is, of course, a source of complication: as though trying to understand the behavior of an infinite sequence of networks was not hard enough, we must consider the space of all possible such sequences at once. How do we even encode such a structure so we can reason about it? Each orbit gives rise to a parse tree, so the challenge is how to organize the set of all possible parse trees into a single structure: this is the role of the *coding tree*. By way of analogy, consider the set of all English sentences. Via lexicographic ordering, we can organize this set as an infinite coding tree whose paths are in bijection with the sentences. Such data structures are known in computer science as prefix, digital, or radix trees. Being

---

[2] At the risk of belaboring the obvious, we mention that this concerns only the *analysis* of the dynamical system: the choice of A and B has no incidence on the dynamics itself.

associated with a sentence, each path can be parsed in accordance with the rules of English grammar. The key insight is that sentences whose paths share long prefixes will have parse trees with big overlaps: this in turn allows us to infer a hierarchical decomposition of the coding tree itself.

In the case of influence systems, the coding tree is infinite and each path in it corresponds to a (chronological) sequence of communication networks. The trick is to infer from the parse trees associated with the paths a recursive decomposition of the entire coding tree itself. This can be thought of as a form of temporal renormalization not carried out in closed form but algorithmically (a remark the discussion below should help clarify). What makes the coding tree useful for the analysis is that it is not merely a combinatorial structure but a geometric one: indeed, a path of the coding tree corresponds to a whole set of nearby orbits that share the same sequence of networks. These look like "tubes" in spacetime $\mathbb{R}^n \times \mathbb{N}$. For example, an influence system with 2 agents would produce a coding tree in three dimensions that might look a bit like a real tree of the sort we encounter in nature. Paths are branches whose cross-sections have an area (or volume in higher dimensions). For reasons we discuss below, bounding the rate at which new branches are formed (entropic growth) and how thin they get as we move further from the root (dissipation) is the key to the analysis. Roughly speaking, bushy trees with thick branches correspond to chaotic systems. We have proven in a suitable model that a tiny random perturbation of the input sends the agents into a limit cycle almost surely [6]. We follow a different tack here and establish a more general result under plausible heuristic assumptions. To replace these assumptions by established facts appears to be a major mathematical challenge which is left as an open problem. As a side-benefit, our approach gives us a platform for working out the renormalization scheme in full, something that was not needed in [6].

We define the model of diffusive influence formally in the next section (§2) and discuss specific constructions as a warmup. In §3, we show how to parse an arbitrary sequence of networks. The section is of independent interest. By generalizing the idea of breadth-first search, this gives us a principled way of tracking the propagation of information (eg, rumors or contagion) in dynamic graphs. We sketched the idea earlier [6] but we give it the full treatment here. In §4, we lay down the foundations of algorithmic renormalization and explain in §5 how to use the framework to mediate between the two "forces" driving the system: entropy and dissipation. Finally, we show how to analyze a diffusive influence system in §6 by setting up the relevant recurrence relations. It bears mentioning that the first three sections are highly general. It is only in §4 that the piecewise-linearity of the system is used and in §5,6 that the "averaging" nature of diffusive influence systems is exploited. In particular, the algorithmic renormalization scheme applies to any discrete-time network-based dynamics.

2   The Model

Our model draws from the classic *Hegselmann-Krause* model of opinion dynamics [9]. Part of its appeal is the simplicity of its definition: Fix a real parameter

$r > 0$ and initialize n agents on the real line $\mathbb{R}$. At each time step, each agent moves to the mass center of its neighbors, in this case, any agent at distance r or less. In other words, the position $x_i \in \mathbb{R}$ of agent i becomes $x_i \leftarrow |N_i|^{-1} \sum_{j \in N_i} x_j$ at the next step, where $N_i = \{ j : |x_i - x_j| \leq r \}$. The updates are carried out synchronously and repeated ad infinitum. Numerical simulations suggest fast convergence. Although the typical relaxation time has yet to be pinned down, it is known that, within polynomial time, the agents end up frozen in single-point clusters at least r away from each other [1, 5, 10, 12–15].

2.1   Generalized HK-Systems

There are three natural ways to extend the original Hegselmann-Krause (HK) model. One of them is to lift the agents into higher dimension instead of confining them to the real line. Regardless of the dimension, the agents will still converge to a fixed configuration in polynomial time [1]. Another modification is to replace the update rule with a weighted mass center. Assuming nonzero self-weights, convergence is still guaranteed but it might become asymptotic. To see why, consider two agents at distance less than r moving toward each other one third of the way at each step: $x_1 \leftarrow \frac{1}{3}(2x_1 + x_2)$ and $x_2 \leftarrow \frac{1}{3}(x_1 + 2x_2)$. In the phase space $\mathbb{R}^2$, any orbit is attracted exponentially fast to the line $x_1 = x_2$. The third type of extension is to redefine what it means to be a "neighbor." Despite massive empirical evidence that the system should converge to a fixed point, a change as simple as allowing a different threshold r_i for each agent i produces a dynamics that is still unresolved. On the other hand, certain minor variants are known to produce periodicity and even chaos [6]. To grasp the subtleties at play, we need a more expressive palette to work with. We begin with a slight generalization of *HK* systems (lin-DNF) and then push the generalization to its natural limit (diffusive influence systems). Though looking vastly different to the naked eye, this is an illusion: the two formulations are in fact equivalent (modulo an adjustment in the number of agents).

A *lin-DNF* is a set of linear constraints expressed as a disjunction of conjunctions, ie, $P_1 \vee P_2 \vee \cdots$, where each P_l is of the form $Q_1 \wedge Q_2 \wedge \cdots$ and each Q_k is a halfspace $u^T x \leq v$ (or $u^T x < v$), where $u, x \in \mathbb{R}^n$. We define the *communication graph* $G(x)$ by associating a node to each agent. The edges of the n-node graph $G(x)$ depend on $x = (x_1, \ldots, x_n)$: for each pair $i \neq j$, we choose a lin-DNF ϕ_{ij} and we declare (i, j) to be an edge of $G(x)$ if $\phi_{ij}(x)$ is true. A natural extension of *HK* systems is provided by the update rule: for $i = 1, \ldots, n$,

$$x_i \leftarrow \frac{1}{|N_i|} \sum_{j \in N_i} x_j \quad \text{and} \quad N_i = \{ j \mid (i, j) \in G(x) \}. \tag{1}$$

Note that the original *HK* system is put in lin-DNF form very simply by setting $\phi_{ij}(x)$ as

$$\left(x_j - x_i \leq 0 \wedge x_i - x_j \leq r \right) \vee \left(x_i - x_j \leq 0 \wedge x_j - x_i \leq r \right). \tag{2}$$

2.2 Diffusive Influence Systems

The definition of a diffusive influence system is identical to that of an *HK* system, with the only difference coming from the communication network, specifically the criterion used to include a pair (i, j) as an edge of $G(x)$. We equip the pair with its own first-order predicate $\phi_{ij}(x)$ and make (i, j) an edge of $G(x)$ if and only if $\phi_{ij}(x)$ is true. Recall that a first-order predicate over the reals is a logical sentence consisting of universal and existential quantifiers bound to real variables $y_1, \ldots, y_m$, along with polynomial inequalities from $\mathbb{Q}[x_1, \ldots, x_n, y_1, \ldots, y_m]$ tied together by Boolean connectives. Formulation (2) is a particularly simple instance of a first-order predicate, as it lacks quantifiers and bound variables, and uses only linear polynomials.

Do we really need the full first-order theory of the reals? The answer is yes. Here is a simple example taken from the field of robotics [2]. The agents are represented by points in a room full of obstacles and $G(x)$ is defined as the "constrained Delaunay graph": this means that (i, j) is an edge of $G(x)$ if there exists a sphere passing through the agents i and j with no agent or obstacle protruding inside. Instead of single real numbers, a variable x_i is now a point in $\mathbb{R}^3$ and an edge (i, j) is characterized by the truth-condition of a first-order formula with both existential and universal quantifiers. In plain English, the formula reads as follows: "*There exist a center and a radius such that, for all points p on an obstacle or at an agent, p does not lie in the corresponding ball.*" This is formally expressed as a first-order predicate over the reals: the formula contains the symbols $\exists, \forall, \vee, \wedge$, a number of bound variables, and a set of polynomial inequalities with rational coefficients. We see that, even for a simple communication network from robotics, alternating quantifiers are already necessary.

Whereas the update rule itself (1) is kept deliberately simple, it is the high expressiveness of the language in which the communication network is specified that gives diffusive influence systems their distinctive feature. Because virtually any edge selection rule is acceptable, the definition meets the primary objective of the model, which is to allow the agents to have their own, distinct communication rules.

2.3 Equivalence of the Models

Phrased in the language of first-order logic, diffusive influence systems seem far removed from the lin-DNF formulation of generalized *HK* systems. The two definitions are in fact equivalent. Indeed, any diffusive influence system can be put in lin-DNF form after suitable transformation. This involves a number of steps including quantifier elimination, linearization, tensoring, addition of new agents, etc. (See [6] for details.) It might seem surprising that polynomials of arbitrary degree can thus be replaced by linear ones, but this is made possible by the piecewise linearity of the underlying dynamics.

We can go further and rid the lin-DNF formulation of all its Boolean connectives. To do that, we consider the set $\mathcal{D}$ of hyperplanes formed by the linear constraints in the formulas ϕ_{ij}. By adding a variable if necessary,[3] we can always assume that these hyperplanes, called the *discontinuities*, are of the form $u^T x = 1$. They form an arrangement whose full-dimensional cells c, the *continuity pieces* of f, are each assigned a directed n-node graph G_c. We extend the labeling to all of $\mathbb{R}^n$ by defining $G(x)$ as the graph with no edges if x lies on a discontinuity and $G(x) = G_c$ if x lies in cell c.

For convenience (specifically, to avoid cluttering the notation), we assume that the number of hyperplanes in $\mathcal{D}$ is at most polynomial in n, so that the number of graphs is bounded by $n^{O(n)}$. After a translation along the all-one vector $\mathbf{1}$ and rescaling if necessary, we can always choose the unit cube $X \triangleq [0,1]^n$ as the phase space, so the continuity pieces are bounded open n-dimensional polyhedra. To summarize the discussion,

DEFINITION 1. *A diffusive influence system (f, X) is a piecewise-linear system specified by a map f from X to itself, $x \mapsto P(x)x$, where $P(x)$ is the incidence matrix of the communication graph $G(x)$ augmented with self-loops, with each row rescaled so as to sum up to 1. The matrix $P(x)$ is constant over the cells of a hyperplane arrangement in $\mathbb{R}^n$.*

We assume a positive diagonal to avoid spurious periodicities of no mathematical interest.[4] We define $P(x)$ as the identity matrix for any x on a discontinuity or on the boundary of the unit cube. Our discussion generalizes easily to update rules based on weighted mass centers, and we use uniform weights across each matrix row only for notational convenience.

A more substantive simplification in this paper is our assumption that the systems are *locally coupled*, meaning that each $\phi_{i,j}$ depends only on x_i, x_j and not on the other agents. Local coupling means that the presence of the edge (i, j) in $G(x)$ depends only on the sign-condition of x with respect to discontinuities $u^T x = 1$ for which only the coefficients u_i and u_j may be nonzero. To summarize, a diffusive influence system (f, X) is specified by an arrangement of hyperplanes, where each continuity piece c (an open n-cell) is labeled by a directed graph whose edges (i, j) depend only the projection of c onto the (x_i, x_j) plane. Figure 1 illustrates the case of two agents.

2.4   A New Brand of Renormalization

Most dynamic regimes can be observed in low dimensions (eg, fixed-point attraction, periodicity, quasi-periodicity, chaos, strange attractors), which is why research in dynamics has had a tendency to focus on systems with few degrees of freedom. At the opposite extreme, statistical mechanics prefers infinite-dimensional systems for the mathematical benefits it can draw from the thermodynamic limit. Influence

---

[3] For example, $x_1 = 0$ might becomes $x_1 + x_0 = 1$, with x_0 set to 1. Perturbation now involves both x_0 and x_1.

[4] For example, consider the two-node cycle $G(x)$ with $P(x) = \begin{pmatrix} 0 & 1 \\ 1 & 0 \end{pmatrix}$.

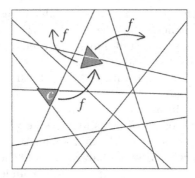

Fig. 1. A map f for $n = 2$. Note that the continuity piece c is mapped continuously but $f(c)$ is not. It is this sort of fragmentation that makes the dynamics difficult to analyze.

systems sit in the "mesoscopic" middle: many agents, but still too few, too diverse, and too autonomous for any classical treatment. These network-based systems seek to model the emergence of collective behavior through the multiplicity of individual interactions. But how do we go about analyzing the endogenous interaction of many diverse agents?

The physical analogue would be to allow each particle in a gas to follow its own laws of physics. While this feature alone may push influence systems beyond the purview of statistical mechanics, the concept of *renormalization* will play a key role (just as it does in the study of one-dimensional systems like the logistic map). The basic idea is to rescale a system while retaining its dynamics. In our case, scaling is with respect to both time and the number of agents. Renormalizing an influence system is to produce another one with fewer agents and a behavior resembling a coarse-grained version of the original one. If the communication graph was fixed then graph clustering techniques might suggest a way to do that: agents interacting with their neighbors in a grid, for example, can be clustered into subgrids, so that a $\sqrt{n} \times \sqrt{n}$ grid of n agents can be renormalized as a $\sqrt{n/k} \times \sqrt{n/k}$ grid of "block-agents" consisting of k agents each. Naturally, this decomposition can be repeated recursively.

This form of block-spin renormalization, famously introduced by Kadanoff, works only if the interaction among the block-agents can be understood reasonably well without having to track the precise behavior of their constituent agents. In other words, a block-agent needs to be able to "hide" the internal role of its agents from the outside. The other requirement is that the coarse-grained system should look like a "blurred" version of the larger one. To achieve this, it is customary to view the coarse-graining process itself as a dynamical system mapping an influence system to another, simpler one. The goal is then to adjust the coarse-graining parameters to home in on a fixed-point attractor and thus allow the basic behavior to be retained throughout the renormalization steps.

What is the role of time in all of this? Presumably, a block-agent has a characteristic timescale: the time it takes its own agents to settle in their long-term mode. If this time is the same for all the block-agents, then we can rescale the time axis by redefining its basic unit at every step in the coarse-graining hierarchy. To carry out this plan, of course, one needs to cope with the ever-changing topology of the communication graph. Worse, the changes being endogenous, one cannot postulate a prior on the graph distribution. Determinism is not necessarily an impediment to using tools from statistical mechanics [4] because the bifurcation analysis needs to focus on the "edge of chaos," a region replete with pseudorandomness.

In [6] we outlined an approach to network-based renormalization and derived sufficient conditions for the asymptotic periodicity of diffusive influence systems. We revisit the method here and show how to replace a certain timing assumption by a general-position heuristic under which the critical region can be shown to be of measure zero—the details are given below in §6. This is an opportunity to develop the renormalization framework to its fullest, something our earlier timing assumption [6] allowed us to bypass. The algorithmic renormalization of dynamic networks is a general, powerful idea and one, we believe, likely to be useful elsewhere.

2.5 Mixed Timescales

Diffusive influence systems have been shown to span the entire range of dynamic modes, from fixed-point attraction to chaos. Predicting their behavior can be undecidable [6]. It is no surprise therefore that they should exhibit periodic orbits of any length. Remarkably, long limit cycles with large basins of attraction can be manufactured as well: think of them as extremely long periodic orbits robust under perturbation. We give such a construction below: the period is a tower-of-twos of height proportional to the number of agents. This type of behavior may be pathological but it touches upon a phenomenon at the heart of influence systems: the *mixing* of timescales. If one regards attraction as a process of amnesia, then to be caught in a limit cycle means to forget one's origins. Indeed, an agent begins with an unbounded amount of information (encoded in its initial position) but, once trapped in a limit cycle, carries only a few positional bits from then on. This near-total loss of memory is what we term "amnesia."

In a diffusive influence system, some agents can "collude" to form memory devices that can hold information over periods of time much higher than their own characteristic times. These gadgets allow agents to recover their memory just as they are about to reach amnesia, which in turn allows them to extend their internal characteristic time and delay the eventual attraction to a limit cycle. This archival mechanism can be nested hierarchically to create bigger and bigger timescales. Remarkably, such constructions are robust to noise. The existence of hierarchical schemes that combine to create new dynamical regimes points to the need for time-based renormalization, which is the main theme of this work. The ability to create higher scales both in length and time seems a necessary component of any living system, and the relevance of the present investigation should be appreciated in this light.

The Very Slow Clock. We describe an n-agent diffusive influence system with a limit cycle of length roughly equal to a tower-of-twos of height $n/2$. This is "two to the two to the two..." repeated $n/2$ times. The design relies on a small gadget which converts an $(n-2)$-agent system with period $p_{n-2} = k$ into an n-agent system with period $p_n \approx 2^k$. All gadgets share two static agents a, b positioned at -1 and 1, respectively; for convenience, we do not count them in n. Each gadget also has its own pair of mobile agents. Applying the construction (roughly) $n/2$ times leads to the desired result. As an inessential but convenient exception, we allow zero entries in the matrix diagonals. The n (mobile) agents are at positions $x_1, \ldots, x_n$. The update rules will automatically ensure that, for all i: $x_i \leq 0$ precisely when the time t is $-1 \pmod{p_n}$; and $x_i = (-1)^i$ precisely when $t = 0 \pmod{p_n}$. For $n = 2$, we use a, b to (easily) engineer the system so that (x_1, x_2) follows the periodic orbit $(-1, 1), (1, 1), (1, -1), (-1, -1)$. In general, to form $G(x)$, we extend the communication graph of the $(n-2)$-agent system by adding to it the two nodes n and $n-1$, together with the edges specified below:

(i) If $x_i \geq 0$ for some $i < n - 1$, then add the self-loop (x_{n-1}, x_{n-1}) and the edge (x_{n-1}, x_n). Note: agent $n-1$ moves half way toward agent n; this case is the most frequent, as it occurs whenever $t \neq -1 \pmod{k}$.

(ii) Otherwise, add (x_{n-1}, a); if $x_n \leq 0$ then add (x_n, b) else add (x_n, x_{n-1}). Note: in the first (resp. second) case, agent n moves to the location of agent b (resp. agent $n-1$) and agent $n-1$ moves to -1; this occurs only when $t = -1 \pmod{k}$, which is typically very rare.

The mechanism of the clock is quite simple. The n-agent system consists of a subsystem of $n-2$ agents (the "subclock") and a gadget that interacts only with the static agents a, b. There is never any edge between the gadget and the subclock: their interaction is mediated entirely through the positional constraints.

At time 0, the odd-labeled agents are at -1 and the others at 1. Rule (i) kicks in for the next $k-1$ steps, pulling agent $n-1$ toward agent n, so they end up separated by $\approx 2^{-k}$. At time $k-1$, the subclock is entirely in negative territory, so rule (ii) kicks in. If agent n (which stays always to the right of agent $n-1$) is still at a positive location, then it slides left to the position of agent $n-1$ while the latter is repositioned at -1. This is the key moment: in k steps, the system has managed to move agent n to the left by as little as $\approx 2^{-k}$. As long as $x_n > 0$, we can repeat these k steps, which means the subclock can run through roughly 2^k cycles. (The leftward slide of agent n varies a little in length at each iteration but always remains on the order of 2^{-k}.) Note that we do not rush to reset the subclock as soon as $x_n \leq 0$ but, rather, allow it to complete its natural cycle. The initial position of the agents does not require fine calibration and the astronomical period is robust under perturbation.

The recursive construction of the Very Slow Clock (VSC) suggests how renormalization should proceed. The n-agent system VSC_n results from nonlocal-coupling between a clock VSC_{n-2} and a gadget consisting of two private agents and two shared ones. The characteristic timescale of the gadget is exponentially higher than that of the VSC with which it is coupled. Renormalizing the system is easy because it was essentially put in by hand. In general, to tease out the subparts of the dynamics that can be "factored out" is no easy task. In the end, renormalization is about dimension reduction. When the graph never changes, the dynamics can be expressed by the powers of a fixed matrix and the dimension reduction is done by breaking down the system into its eigenmodes. When the communication topology keeps changing, however, linear algebra is of little use. The dimension reduction cannot be carried out in "closed form." It would seem that it can only be performed as a step-by-step process that evolves alongside the dynamics of the system. This is what algorithmic renormalization tries to achieve.

THEOREM 1. *For any $n > 2$, there exists an n-agent diffusive influence system with a periodic orbit of length equal to a tower-of-twos of height proportional to n. The dynamics is robust under perturbation.*

3 Parsing Graph Sequences

The algorithmic renormalization of influence systems relies on a general procedure of independent interest: a method for *parsing* arbitrary graph sequences. Recall that the analysis of influence systems rests crucially on our ability to decompose an infinite sequence of communication graphs hierarchically. By building a tree on top of the network sequence, we are thus able to break down the original system into subsystems. This is similar to forming the parse tree of an ordinary English sentence. While this operation is driven by the rules of syntax, what sort of rules will we use then to parse a sequence of communication graphs? In a nutshell, the idea is to track the propagation of information across the networks (think of a virus spreading in a population) and use the pauses (for example, due to poor connectivity) as breakpoints to guide the construction. The parsing algorithm can be seen as a grand generalization of breadth-first search for dynamics graphs.

Given a (finite or infinite) sequence $\mathcal{G} = G_0 G_1 G_2 \cdots$ of directed graphs over the same set of n labeled nodes, make the lowest-labeled node "wet" and declare wet any node pointing to a wet node in G_0. Next, switch to G_1 and call wet any node that is either already wet or points to a wet node via an edge of G_1. Water flows only to immediate neighbors in a single step and wet nodes remain wet. We use water as an anology but we could substitute rumor propagation or viral contagion. Transmission runs in the reverse direction of the edges, following the principle that information is acquired only by pointing to it. As soon as all the nodes become wet (if ever), the subsequence of graphs considered up to this point is said to form a *wave*. When this occurs, we dry up all the nodes and repeat the entire process, restarting from the current point in the graph sequence. It might happen that some nodes never get wet or that wetness propagation is interrupted for long periods of time before resuming again. What to do in such cases is at the heart of the parsing procedure, which is called the *flow tracker*.

3.1   Making Waves

We treat the sequence $\mathcal{G}$ as a word in a language over an alphabet of 2^{n^2} letters, where each letter represents an n-node graph (or, equivalently, an n-by-n 0-1 matrix). Parsing the sequence means producing a *parse tree* $T(\mathcal{G})$: this is a rooted tree whose leaves are associated with the graphs $G_0, G_1, G_2, \ldots$ from left to right and whose internal nodes reflect important transitions during the flooding. As we shall see, even when the sequence $\mathcal{G}$ is infinite, the depth of the tree remains finite. For example, if all the graphs are strongly connected (or if the graphs are connected but undirected), then[5]

$$T(\mathcal{G}) = \big((G_0 \cdots G_{i_1})^{wave} (G_{i_1+1} \cdots G_{i_2})^{wave} \cdots \big)^{seq}, \qquad (3)$$

where i_k is the time at which the graph becomes entirely wet for the k-time. The tree $T(\mathcal{G})$ has depth only two because it is a simple case. The superscripts indicate the nature of the nodes: the node $(\cdots)^{seq}$ is the root of the parse tree of the original sequence; its children $(\cdots)^{wave}$ correspond each to a wave and their number is infinite; the leaves of the tree are given by G_0, G_1, etc. In this particular case where all the graphs are assumed to be strongly connected, the waves are less than n-long but in general there is no a priori bound on their lengths. The first two levels of any parse tree look like (3); most trees will have higher depth, however, and recursive rewriting rules will be needed to express them.

It is time to put this informal description on a more rigorous footing. Recall that $\mathcal{G}$ is an arbitrary (bounded or not) sequence of directed graphs over n labeled nodes. We denote by B the set of nodes initialized as wet. So far, we have considered only the case where B is the singleton consisting of the lowest-labeled node. A variant of the procedure will soon require wetting more than one node at the outset, hence the use of the set B. We put together a partial parse tree in the form of $(\texttt{wavemaker}\,(\mathcal{G}, B))^{seq}$, which we then proceed to refine. This is how $\texttt{wavemaker}$ works:

wavemaker $(\mathcal{G}, B)$ *" $\mathcal{G}$ is a graph sequence and $B \subset [n]$ "*

     $W_0 \leftarrow B$ and print '('
     for $t = 0, 1, \ldots$
         print 'G_t'
         $W_{t+1} \leftarrow W_t \cup \{ i \mid \exists j \in W_t \text{ such that } (i,j) \in G_t \}$
         if $|W_{t+1}| = n$ then $W_{t+1} = B$ and print ')^{wave}('
     print ')^{wave}'

We bend the usual conventions of programming language theory by allowing the last print statement to be executed even if the sequence $\mathcal{G}$, hence the for-loop, is infinite.[6] The output of $\texttt{wavemaker}$ looks like the right-hand side of (3).

---

[5] Throughout this paper, to save space, we represent trees not as stick figures but as nested parenthesis systems. Here is a tree whose root has two children, with the leftmost one having two children of its own: $(\,((\,)(\,))\,(\,)\,)$.

[6] In the odd chance that $\texttt{wavemaker}$ prints a single spurious $(\,)^{wave}$ at the end, we simply erase it.

A wave can be either complete or incomplete, depending on whether all the nodes get wet: a wave is *complete* if it is terminated by the successful outcome of the conditional "if $|W_{t+1}| = n$." For that reason, an infinite wave is always incomplete but the converse is not true: a finite sequence $\mathcal{G}$ might simply run out of graphs to flood all the nodes, hence leaving an incomplete final wave. If $\mathcal{G}$ is infinite, then either all the waves are finite but their number is not (as is the case with strongly connected graphs) or there are a finite number of waves but the last one is infinitely long. In this case, the partial parse tree $(\texttt{wavemaker}\,(\mathcal{G}, B))^{seq}$ is of the form:

$$\left((G_0 \cdots G_{i_1})^{wave} \cdots (G_{i_l+1} \cdots)^{wave} \right)^{seq}. \tag{4}$$

Why isn't this the end of the story? The reason is that waves themselves need to be parsed, especially when they are very long. For example, suppose that B is a singleton and its node is always disconnected from the other ones. In that case, water never flows anywhere and $\texttt{wavemaker}$ parses $\mathcal{G}$ as $((G_0 G_1 G_2 \cdots)^{wave})^{seq}$, which is as simple as it is useless. Clearly, parsing must be carried out even during flooding delays. We explain now how to do this.

3.2   Parsing a Wave

A flooding delay implies that the current "wet" set W_t stops growing: it could be momentary or permanent (there is no way to tell ahead of time). During any such delay, the graphs G_k have no edges pointing from the dry nodes to the wet ones, ie, from $[n] \setminus W_t$ to W_t. This motivates the definition of a *block-sequence* as any subsequence of $G_0 G_1 G_2 \cdots$ such that, for some partition $[n] = A \cup B$ of the nodes, no edge from the subsequence points from B to A: the block-sequence is said to be of *type $A \to B$*. Let $\mathcal{G} = G_0 G_1 G_2 \cdots$ be a wave. Recall that, in the case of a complete wave, this means that all the nodes are wet at the end but not earlier. As usual, B is set to the singleton consisting of the lowest-labeled node. By definition of a wave, $\texttt{wavemaker}\,(\mathcal{G}, B)$ terminates at or before the first time $|W_{t+1}| = n$. Let $t_1, t_2, \ldots$ be the times t at which $W_t \subset W_{t+1}$ (ie, strict inclusion). These *coupling times* indicate when the water propagates. They break down the wave $\mathcal{G}$ into block-sequences $\mathcal{G}_k$ via $(\texttt{blockseqmaker}\,(\mathcal{G}, B))^{wave}$. Using superscripts to distinguish the wave nodes from the block-sequence kind, the output is of the form:

$$\left((\mathcal{G}_{A_0 \to B_0})^{bseq} G_{t_1} (\mathcal{G}_{A_1 \to B_1})^{bseq} G_{t_2} (\mathcal{G}_{A_2 \to B_2})^{bseq} G_{t_3} \cdots \right)^{wave}. \tag{5}$$

$\texttt{blockseqmaker}\,(\mathcal{G}, B)$ "$\mathcal{G}$ *is a wave and* $B \subset [n]$ "

$\quad W_0 \leftarrow B$ and print '('
$\quad$ for $t = 0, 1, \ldots$
$\quad\quad\quad W_{t+1} \leftarrow W_t \cup \{ i \mid \exists j \in W_t \text{ such that } (i, j) \in G_t \}$
$\quad\quad\quad$ if $W_t = W_{t+1}$ then   print 'G_t'
$\quad\quad\quad\quad\quad\quad\quad\quad\quad\quad$ else   print ')^{bseq} G_t('
$\quad$ print ')^{bseq}'

As was the case with `wavemaker`, we remove empty parenthesis pairs $(\,)^{bseq}$; so, say, in the case of strongly connected graphs, the output of `blockseqmaker` for a wave will simply be $G_{t_1} G_{t_2} \cdots$ $(t_i = i - 1)$. The block-sequence $\mathcal{G}_{A_k \to B_k}$ is a maximal subsequence of the wave $\mathcal{G}$ that witnesses no water propagation: it is of type $A_k \to B_k$, where $A_k = W_{t_k+1}$, $B_k = [n] \setminus A_k$, and $B \subseteq A_k \subset A_{k+1}$. The initialization requires setting $t_0 = -1$ so that $A_0 = W_0 = B$. Note how the letters A and B switch roles (more on this below). Until now, B has been a singleton: parsing block-sequences, our next topic, will change all that.

3.3   Parsing a Block-Sequence

We have built the first three levels of the parse tree so far. The root is associated with the full sequence and its children with its constituent waves. The root's grandchildren denote either single-graphs (leaves of the tree) or block-sequences. Parsing the latter repeats the treatment of general sequences described above with two small but crucial differences. Let $\mathcal{G}_{A \to B}$ denote a block-sequence of type $A \to B$. First, we break it down into *block-waves* by applying `wavemaker`$(\mathcal{G}_{A \to B}, B)$. The output is of the form

$$\left((G_0 \cdots G_{i_1})^{bwave} (G_{i_1+1} \cdots G_{i_2})^{bwave} \cdots \right)^{bseq}. \tag{6}$$

We note that B is now wet at the outset. This contrasts with the circumstances behind the creation of a block-sequence of type $A \to B$, which feature wet A, dry B, and delay in water propagation. This radical change in wetness status explains the need for a recursive method that encapsulates wetness status by local variables hidden from the outside. It is in that sense that the renormalization is truly *algorithmic*.

The block-waves constitute the children of the node associated with the block-sequence $\mathcal{G}_{A \to B}$. Let $\mathcal{H}$ denote any one of the subsequences of the form $G_{i_k+1} \cdots G_{i_{k+1}}$ $(i_0 = -1)$. To extend the parse tree further, we replace in (6) each occurrence of the block-wave $(\mathcal{H})^{bwave}$ by $(\texttt{blockseqmaker}\,(\mathcal{H}, B))^{bwave}$. Each block-wave is thus parsed as

$$\left((\mathcal{G}_{A\|B})^{dec} G_{t_1} (\mathcal{G}_{A_1 \to B_1})^{bseq} G_{t_2} (\mathcal{G}_{A_2 \to B_2})^{bseq} G_{t_3} \cdots \right)^{bwave}. \tag{7}$$

The parsing of a block-wave differs from that of a wave (5) in two ways:

- First, we notice the difference in the first term $\mathcal{G}_{A\|B}$. As we scan through the sequence $\mathcal{H}$, we may have to wait for a while before some edge actually joins A to B. Since there is no edge from B to A, the two sets are decoupled until then (hence the superscript *dec*). In other words, A and B form a cut in all the graphs in the sequence $\mathcal{G}_{A\|B}$: such "decoupled" sequences are parsed as a single leaf in the tree.
- The second difference is more subtle. When parsing an ordinary wave, recall that the block-sequence $\mathcal{G}_{A_k \to B_k}$ satisfies $A_k = W_{t_k+1}$. In the case of a block-wave, instead we have $A_k = W_{t_k+1} \setminus B$ (and, as usual, $B_k = [n] \setminus A_k$). Note that it is still the case that $A_k \subset A_{k+1}$. This assignment of A_k satisfies the main requirement of a block-sequence of type $A_k \to B_k$, namely the absence of any edge from B_k to A_k.

We had to amend the old invariant $A_k = W_{t_k+1}$ because of inductive soundness. Here is an example to illustrate why: Let $G_0 = a \to b \quad c$; $G_1 = a \to b \to c$; $G_2 = a \leftarrow b \to c$; $G_3 = a \quad b \to c$; and $G_4 = G_1$. In this notation, G_0 has a single edge (a, b). The block-wave $G_0 \cdots G_4$ of type $\{a, b\} \to \{c\}$ is parsed as

$$\left(\left(\mathcal{G}_{\{a,b\}\|\{c\}} \right)^{dec} G_1 \left(\mathcal{G}_{\{b\}\to\{a,c\}} \right)^{bseq} G_4 \right)^{bwave}, \tag{8}$$

where $\mathcal{G}_{\{b\}\to\{a,c\}} = G_2 G_3$. The reason why setting $A_k = W_{t_k+1}$ is a bad idea is that the block-sequence $G_2 G_3$ would then be interpreted as $\mathcal{G}_{\{b,c\}\to\{a\}}$, which would have the effect of making a block-sequence of type $\{b, c\} \to \{a\}$ a child of another one of type $\{a, b\} \to \{c\}$. To ensure the soundness of the recursion, we need the cardinality of the A-set to drop by at least by 1 as we go from from parent to child: this property is ensured by setting $A_k = W_{t_k+1} \setminus B$ since this implies that $A_k \subset A$.

There is a subtlety in the recursion. The arrival of G_4 brings in the edge (a, b), which is incompatible with a block-sequence of type $\{b\} \to \{a, c\}$. It is therefore no surprise to see $\mathcal{G}_{b \to ac}$ in (8) terminate and give way to the single-graph G_4. Actually this is not the true reason for the termination of the block-sequence. Indeed, if G_4 were of the form $b \to a \to c$, the block-sequence would still terminate regardless of its continuing compatibility with the type $\{b\} \to \{a, c\}$. The true cause of termination is the growth of W_t. This shows that a node can only be parsed if the relevant parameters of its ancestors are already known.

3.4 Rewriting Rules

It is convenient to think of the flow tracker (the name for the entire parsing algorithm) as the application of certain productions of the sort found in context-free grammars: *seq, wave, bseq, bwave* are used as nonterminal symbols and *sg, dec* as terminals; here, *sg* is shorthand for "single-graph." Via the formulas (3, 5, 6, 7), the flow tracker yields the following productions:

$$\left\{ \begin{array}{lll} seq & \Longrightarrow & (wave, wave, \dots) \\ wave & \Longrightarrow & (bseq, sg, bseq, sg, \dots) \\ bseq & \Longrightarrow & (bwave, bwave, \dots) \\ bwave & \Longrightarrow & (dec, sg, bseq, sg, bseq, sg, \dots) \end{array} \right. \tag{9}$$

How sensible is it to make $\mathcal{G}_{A\|B}$ a leaf of the parse tree? On the face of it, not very much. Indeed, suppose that $\mathcal{G}$ is a long, complicated sequence with a rich parse tree and V is its node set. Form a new sequence $\mathcal{G}'$ by adding an isolated node a (with no edge to or from V). Further, suppose that a is given the lowest label so that it defines the starting singleton B. The flow tracker will parse $\mathcal{G}'$ as the 6-node single-path tree: $T(\mathcal{G}') = \left(\left(\left(\mathcal{G}_{\{a\}\to V} \right)^{bseq} \right)^{wave} \right)^{seq}$, with $\mathcal{G}_{\{a\}\to V} = \left(\left(\mathcal{G}_{a\|V} \right)^{dec} \right)^{bwave}$. Adding a single node hides the richness of $T(\mathcal{G})$ by producing a completely trivial parse tree. Of course, it is quite possible that picking another node for B besides a would produce a more interesting parse tree. But optimizing the selection of the starting singleton is out of the scope of our discussion.

As for the parallel treatment of A and B in $\mathcal{G}_{A\|B}$, this is something for the renormalization of the dynamical system itself to handle. The reason we delegate this task is that A and B may not operate along the same time scale. Indeed, in a block-sequence of type $A \to B$, the set B, unlike A, can be handled separately over the entire length of the sequence. It follows from our previous discussion that the parse tree has depth at most $2n + O(1)$, as a typical downward path reads

$$seq, \; wave, \; bseq, \; bwave, \; bseq, \; bwave, \; \ldots, \; bseq, \; bwave, \; dec/sg.$$

The number of waves or block-waves that are the children of a given parent can be unbounded, however. In the case of diffusive influence systems, large degrees express structural properties of the dynamics and play a key role in the analysis. We conclude our discussion of graph-sequence parsing with a simple example of the flow tracker in action. Let $G_0 = G_1 = G_4 = G_6 = a \leftarrow b \leftarrow c$, $G_3 = G_5 = a \to b \to c$, and $G_2 = a \quad b \to c$. Setting $\mathcal{G} = G_0 \cdots G_6$ and choosing a to form the initial singleton B, we have

$$T(\mathcal{G}) = \left((G_0 G_1)^{wave} \left((((G_2)^{dec} G_3)^{bwave})^{bseq} G_4 ((G_5)^{bwave})^{bseq} G_6 \right)^{wave} \right)^{seq}.$$

4   Algorithmic Renormalization

Let (f, X) be a system as in Definition 1. The *coding tree* $\mathcal{T}_f(X)$ of (f, X) captures both its geometry and its symbolic dynamics [6]. We defined it informally as a combinatorial object tracing the graph sequences of the orbits and we made a passing remark about the geometric information it encodes. We formalize these ideas now. The levels of the coding tree correspond to the times $t = 0, 1, 2$, etc. Combinatorially, its paths encode the entire symbolic dynamics of the system by listing in chronological order all the possible communication graph sequences formed by the orbits. In addition, nodes carry geometric information about the corresponding orbits. Each node v of the coding tree is associated with two cells U_v, V_v, where t_v is the depth of node v: the cell U_v is a continuity piece of f^{t_v}, ie, a maximal connected subset of $X = [0,1]^n$ over which f^{t_v} is continuous; and $V_v = f^{t_v}(U_v)$ (which is not necessarily n-dimensional). The coding tree is defined inductively by starting with the root, $U_{root} = V_{root} = X$, and attaching a child w to a node v for each of the cells c into which the discontinuities of f break V_v. We then set $V_w = f(c)$ and $U_w = U_v \cap f^{-t_v}(c)$.[7] We define $V_t = \bigcup_{t_v = t} V_v$ and note that V_t includes all the points reachable in t steps. Any point reachable in $t+1$ steps can also be reached in t steps (just start at the second step); therefore,

$$V_{t+1} \subseteq V_t. \tag{10}$$

The *nesting time* $\nu = \nu(\mathcal{T}_f)$ is the smallest t such that V_t intersects no discontinuity. This number cannot be bounded if the system is chaotic; on the other

---

[7] Recall that f is the identity along the discontinuities so there is no need to define children for them. This causes a node to become a leaf if V_v falls entirely within one of the discontinuities, which can be handled separately and hence assumed away.

hand, a finite nesting time implies asymptotic periodicity. To see why, observe that, for any v at depth ν, $f^k(f(U_v))$ does not intersect any discontinuity for any $k \leq \nu$; therefore, $f(U_v) \subseteq U_w$ for some node w ($t_w = \nu$). It follows that the symbolic dynamics can be modeled by paths in the functional graph whose nodes are labeled by v ($t_v = \nu$) and (v, w) is an edge if $f(U_v) \subseteq U_w$.[8] Asymptotic periodicity follows from classic Markov chain theory: the powers of a stochastic matrix with a positive diagonal always converge to a fixed matrix. It appears therefore that the key question is to pin down the conditions under which the nesting time remains bounded. A node v is called *shallow* if $t_v \leq \nu$: only those nodes can witness an intersection between their cell V_v and a discontinuity. Observe also that the period (plus preperiod) of any limit cycle is at most equal to the number of shallow nodes: the reason is that all eventually periodic orbits trace a path in the functional graph, whose number of nodes is itself bounded by the number of shallow nodes.

Being infinite, the coding tree has no "size" to speak of. It has a growth rate, however, which is captured by the *word-entropy* $h(\mathcal{T}_f)$: it is defined as the logarithm (to the base 2) of the number of shallow nodes. Since any node of the coding tree has at most $O(|\mathcal{D}|^n) = n^{O(n)}$ children, the word-entropy is at most $O(\nu n \log n)$: crucially, it can be much smaller. Intuitively, attraction to a limit cycle hinges on keeping the word-entropy low relative to the dissipation rate of the system (the evolution rate of its contracting parts). In physical terms, this means that the loss of energy must outpace the increase in entropy. Although the system is deterministic, its behavior in the critical region (between periodic and chaotic) is essentially "random" and lends itself naturally to the language of statistical physics [7]: this is the concept of deterministic chaos familiar to dynamicists.

4.1 Parsing a Dynamical System

Every path of the coding tree corresponds to an infinite string $\mathcal{G} = G_0 G_1 G_2 \cdots$ of n-node directed graphs, where G_i is the graph associated with the continuity piece c_i. The set of paths is in bijection with the refinement of the iterated pullback, ie, the language $\{c_0 c_1 c_2 \cdots \mid \bigcap_{t>0} f^{-t}(c_t) \neq \emptyset\}$, which we explained how to parse in §3. In this way, every infinite path of the coding tree can be parsed according to $T(\mathcal{G})$. Parsing "renormalizes" the orbits of the system with respect to time, but this is not what we want. The objective is to renormalize the system itself, not individual orbits; for this, we need to "parse" the coding tree itself. We do this by combining together the individual parse trees of the infinite paths to form a single parse tree for $\mathcal{T}_f(X)$.

In the language of compiler theory, the parsing algorithm is of type LR, meaning that it can be executed by reading the string from left to right without backtracking. For this reason, the parse trees of paths with long common prefixes must share large subtrees. The productions in (9) specify exactly one rewriting

---

[8] A graph is functional if exactly one edge points out of any node (possibly to itself). In any such graph, infinite paths end up in cycles.

rule for each left-hand side term, so all the parse trees are identical up to the number of terms inside the parentheses and the coupling times. This allows us to express the coding tree recursively via three tensor-like operations:

Direct Sum. When the context is clear, we use A either to refer to a set of m agents or to denote the corresponding phase space $[0,1]^m$; same with the set B of $n-m$ agents. The *direct sum* $\mathcal{T}_f(A) \oplus \mathcal{T}_g(B)$ models the evolution in $A \times B$ of two decoupled systems (f, A) and (g, B). A path in the direct sum is of the form $(u_0, v_0), (u_1, v_1), \ldots$, where u_i and v_i $(i \geq 0)$ form paths in $\mathcal{T}_f(A)$ and $\mathcal{T}_g(B)$, respectively. A node w is formed by a pair (u, v) of nodes from each tree and $U_w = U_u \times U_v$, $V_w = V_u \times V_v$.

Direct Product. Let (f, X) and (g, X) be two systems over the same phase space $X = [0,1]^n$. We choose an arbitrary set of nodes in $\mathcal{T}_f(X)$ and prune the subtrees rooted at them. This creates new leaves in the coding tree, which we call *absorbing*.[9] Next, we attach $\mathcal{T}_g(V_v)$ to the absorbing leaves v. The reason we use V_v (defined here with respect to f) and not X as argument for $\mathcal{T}_g$ is to make the glueing seamless. The resulting tree is called the *direct product* $\mathcal{T}_f(X) \otimes \mathcal{T}_g(X)$. The operation is not commutative. In fact the two operands play very different roles: on the left, the tree $\mathcal{T}_f(X)$ gets pruned; on the right, a copy of the tree $\mathcal{T}_g(X)$ gets glued to every absorbing node, each copy cropped in a different way.

Lift. A system is called *block-directional* of type $A \to B$ if no edges in any of the communication graphs point from B to A. Its coding tree is not a direct sum because, although B evolves independently, the agents of A are coupled with those of B. This one-way coupling is expressed by the *lift* $\mathcal{T}_f(A \nearrow B)$. The arrow highlights both the dependency of A on B and the fact that the coding tree of the whole system is a lift of $\mathcal{T}_f(B)$ into $\mathbb{R}^n \times \mathbb{N}$.

Nesting Time and Word-Entropy. The nesting time of a direct sum is at most the bigger of the two nesting times. The word-entropy of a direct sum/product is subadditive, even when it is infinite:

$$h(\mathcal{T}_f\{\oplus, \otimes\}\mathcal{T}_g) \leq h(\mathcal{T}_f) + h(\mathcal{T}_g). \tag{11}$$

4.2 The Renormalized Coding Tree

Translating (3, 5, 6, 7) in the tensor language of coding trees gives us:

$$\begin{cases} 1. & \mathcal{T}(X) & \Longrightarrow & \bigotimes_k \mathcal{T}(X)^{wave} \\ 2. & \mathcal{T}(X)^{wave} & \Longrightarrow & \bigotimes_{k<n} \left\{ \mathcal{T}(A_k \nearrow B_k)^{bseq} \otimes \mathcal{T}(G_{t_{k+1}})^{sg} \right\} \\ 3. & \mathcal{T}(A \nearrow B)^{bseq} & \Longrightarrow & \bigotimes_k \mathcal{T}(A_k \nearrow B_k)^{bwave} \\ 4. & \mathcal{T}(A \nearrow B)^{bwave} \Longrightarrow & (\mathcal{T}(A) \oplus \mathcal{T}(B))^{dec} \otimes \mathcal{T}(G_{t_1})^{sg} \otimes \\ & & \bigotimes_{k<n} \left\{ \mathcal{T}(A_k \nearrow B_k)^{bseq} \otimes \mathcal{T}(G_{t_{k+1}})^{sg} \right\}. \end{cases} \tag{12}$$

---

[9] The terminology "absorbing" is by analogy with the absorbing states of a Markov chain.

The notation borrows from the theory of programming languages, which makes it concise but nonstandard, so a few words of explanation are in order. Recall that water starts flowing from agent 1 to the nodes of G_0 that point to it, and then proceeds in this manner in G_1, G_2, etc. The propagation of the water determines how to break up the graph sequence into waves as shown in (3). In other words, any path of $T_f(X)$ from the root forms a sequence of waves. Let us call "absorbing" the node corresponding to the end of the first wave along each such path. Pruning the subtrees rooted at the absorbing nodes leaves us with a coding tree which we denote by $T_f(X)^{wave\text{-}1}$ as a reminder that it encodes all first waves. In this way, $T_f(X) = T_f(X)^{wave\text{-}1} \otimes T_f(X)$. Repeating this process for each attached tree yields:

$$T_f(X) = \bigotimes_{k=1}^{\ell} T_f(X)^{wave\text{-}k}.$$

Note that ℓ is a variable and not a fixed parameter: it counts the number of waves along each path, meaning that ℓ can be finite as well as infinite. Dropping all the subscripts that can be inferred from the context leads us to 12.1, which we express as a rewriting rule. In rule 12.2, both A_k and B_k involve fewer than n agents, so k ranges from 0 to $n-2$ or less. The arguments X, A_k, B_k, etc, are used as a reminder of the sets of agents involved. In rule 12.4, the index k extends from 1 to less than $|A|$. The induction is sound because $|A_k| < |A| < n$. We write $k < n$ not to specify the precise range but to indicate whether there is an a priori bound on k or not.

The paths of a coding tree can be infinite but they can all be parsed by trees of linear depth. The rewriting rules (12) give us a quick-and-dirty way to bound the nesting time in terms of the number N of waves and block-waves. Let $\nu(n)$ and $\nu(m, n)$ be upper bounds on the nesting time for, respectively, a general n-agent system and an n-agent block-directional system of type $A \to B$ with $|A| \leq m$. We derive from (12) the recurrence

$$\begin{cases} \nu(n) \leq N(n(\nu(n-1, n) + 1)) \\ \nu(n-1, n) \leq N(\nu(n-1) + n\nu(n-2, n)). \end{cases}$$

It follows that $\nu(n) \leq (Nn)^{O(n)}$. To bound N and the word-entropy requires a closer look at the branching structure of the coding trees.

5   Pseudorandomness and Dissipation

The discussion so far applies to any piecewise-linear network-based system. For that reason, our emphasis has been on purely *syntactic* considerations such as the flow of information across the networks. We now turn our attention to diffusive systems and exploit the averaging nature of the updates. In other words, we expand our investigation from the communication of information to its actual processing. The dynamics of a diffusive influence system features a clash between two "forces." One of them is *entropic* and a source of pseudorandomness: when a cell V_v bumps into a discontinuity, it is broken apart and its pieces are mapped to random-like locations. This is not always the case, however, and since chaos hinges on this entropic explosion, the process bears close examination. To counter this entropic effect, we have the dissipation of energy provided by the stochastic matrices. These linear maps are contractive along all the eigendirections except the principal ones (with eigenvalue 1). To appreciate how this complicates matters, an illustration might help.

Picture a balloon bouncing on a lined, corrugated surface. Imagine that the balloon has a tiny hole and deflates slowly at each bounce. The probability of bouncing right across a line will decrease over time. This illustrates the case of a cell V_v decreasing in volume as v goes down a path of the coding tree. Hitting a discontinuity (the balloon falling across a line) results in the splitting of V_v, which can often be described as a (pseudorandom) branching process. A high enough deflation rate might be able to overcome the splitting rate, with the production of new pieces slowing down over time and the branching process dying out: this is how limit cycles are produced. The difficulty is that the balloon does *not* contract along the principal direction(s); furthermore these directions can change and span spaces of varying dimension.

Because of the non-commutativity of the matrices, the system does not have coherent eigenmodes. The true picture, therefore, is not that a round balloon deflating over time but, rather, of a balloon turning into a football and then into a sausage; then back into a football, etc. While our earlier intuition had no trouble with a shrinking balloon hitting a line with decreasing frequency, this new picture of footballs and sausages is more difficult to grasp. If you throw a sausage on the ground, its thickness has little effect on its probability of crossing a fixed line (think of Buffon's needle): only the length matters. But if this length does not decrease, then how can the branching process die out? To answer this question, which is at the heart of the renormalization process, we need to introduce an important classification of the agents.

5.1 Dominance Structure

We begin with the simple case of a fixed communication graph G. The standard decomposition of the corresponding Markov chain partitions the nodes into essential and inessential classes. For completeness, we describe this process. The strongly connected components of G partition the node set into subsets, which, if contracted into single nodes, are seen to be joined together by edges so as to form an acyclic graph.[10] The sinks of this graph form the classes of *dominant* agents: no path in G can exit from a dominant class. The other agents are called *subdominant*.

For example, Figure 2a features a 9-node graph G with three dominant classes: $\{1, 2, 4\}, \{5, 7, 9\}, \{6\}$. With a single graph, the system evolves as $P^t x$,

---

[10] The property that two nodes are joined by paths in both directions forms an equivalence relation. The strongly connected components of the graph are the equivalence classes.

where P is the stochastic matrix associated with G. Because of the positive diagonal, each dominant class, being strongly connected, is attracted to a fixed point (more on which below) and the number of such classes represent the long-term rank of the system (ie, its dimensionality as time goes to infinity). The subdominant agents 3 and 8 are attracted to some convex combination of the dominant agents (Fig. 2b). The system is block-directional of type *subdominant →* *dominant*, which in this running example is $\{3, 8\} \to \{1, 2, 4, 5, 6, 7, 9\}$. In our discussion, the agents lie on the real line. In the figure, however, we have placed them in the plane for visual convenience (the same ideas work in higher dimension, anyway).

Suppose now that the communication graph changes with time but that the dominance structure does not. Furthermore, suppose that none of the three dominant classes can communicate with one another. The dominant agents end up frozen in place and only the two subdominant agents can move. Asymptotic periodicity means that agents 3 and 8 go around a cyclic trajectory, not necessarily with the same period. Crucially, the subdominant agents can never leave the convex hull of the dominant ones. This remains true even if we allow the dominance structure to change. In the case of a limit cycle, the system finally settles on a fixed dominance pattern with the dominant agents converging to fixed-point attractors and the others forever gyrating around inside their convex hull or perhaps settling toward fixed points.

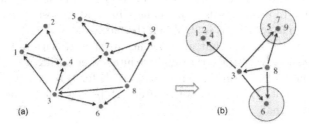

Fig. 2. A one-graph system evolving to a fixed-point attractor, with its three dominant classes highlighted

5.2   Attraction Rate

We go back to the case of a fixed graph and show how to bound the convergence rate. In anticipation of our treatment of dynamic networks, however, we do it by using our water propagation mechanism, instead. Let $B_1, \ldots, B_r$ be the dominant classes, and let $n_i = |B_i|$ and $m = n - (n_1 + \cdots + n_r)$. The system is block-directional of type $A \to B$, where $B = B_1 \cup \cdots \cup B_r$ and $A = [n] \setminus B$. By abuse of terminology, the stochastic matrix P is of the form

$$P = \begin{pmatrix} A & C \\ \mathbf{0} & B \end{pmatrix}.$$

Limit of A^t. Our next observation provides the key link between water propagation and dissipation: it is simple and crucial. Every time water propagates to new agents, something shrinks: in the case of B, it is the length of the smallest interval enclosing the wet agents; in the case of A, it is memory about itself. We explain. Initialize the agents of A at 1 and those of B at 0. Since G models a block-directional system of type $A \rightarrow B$, the agents of B will never be able to leave 0. We assume that $m > 0$. Because G is fixed, there is no flooding delay, so the *dec* and *bseq* sequences in (7) are empty, and formulas (6, 7) give us

$$\big(\underbrace{A \cdots A}_{\leq m} \big)^{bwave} \big(\underbrace{A \cdots A}_{\leq m} \big)^{bwave} \cdots$$

Consider a single block-wave and let λ_t be the length of the smallest interval enclosing the wet agents (which includes those in B) at time t, with $\lambda_0 = 0$. We denote by R the complement of this interval in $[0, 1]$. Note that a dry agent becomes wet as soon as it leaves 1. For λ_t to increase, a wet agent must move into R: it could be one currently wet that slides to the right into R or a dry agent at 1 moving left into R and becoming wet. In both cases, the agent moves to the mass center of its neighbors (including itself), so R can shrink by at most a factor of n; in other words

$$1 - \lambda_{t+1} \geq \tfrac{1}{n}\big(1 - \lambda_t\big).$$

Since water propagates at each step, it follows that $\lambda_t \leq 1 - n^{-n}$ at the end of the first block-wave. In other words, the smallest interval enclosing wet agents can grow but only up to length $1 - n^{-n}$. When the block-wave terminates, the agents are then wet, which means that all n agents fit strictly within $[0, 1]$. In other words, each block-wave shrinks the smallest interval enclosing the whole system by at least a factor of $1 - n^{-n}$; hence $\lambda_t \leq 2^{-\gamma t}$, for some fixed $\gamma \geq \Omega(n^{-n})$. Since the placement of the agents of A after t steps is given by the coordinates of $A^t \mathbf{1}$, it follows that

$$\|A^t \mathbf{1}\|_\infty \leq 2^{-\gamma t}. \tag{13}$$

The previous argument relies on the absence of any flooding delay. Indeed, a delay of θ steps might allow R to shrink by a factor exponential in θ (if wet agents point to dry ones but not the other way around), which can lead to the sort of crawling behavior in evidence in the *Very Slow Clock* of §2.5.

Limit of B^t. To bound the convergence rate of B^t proceeds along similar lines. The matrix B is the block diagonal matrix $(B_1, \ldots, B_r)$. Each B_i can be treated separately, so we might as well assume that $r = 1$. Since B is strongly connected, formulas (3, 5) simplify into[11]

$$\big(\underbrace{B \cdots B}_{< n-m} \big)^{wave} \big(\underbrace{B \cdots B}_{< n-m} \big)^{wave} \cdots$$

---

[11] By abuse of notation, we write B for its induced subgraph.

Initialize all the agents B at 0, except for the k-th one, which is placed at 1. As before, we define λ_t as the length of the smallest interval enclosing the wet agents. Removing this interval from $[0, 1]$ leaves us with two intervals L, R. The previous argument shows that neither L nor R can shrink by more than a factor of n in a single step. Since B is strongly connected, water propagates at each time step, so $\lambda_t \leq 1 - n^{-n}$ at the end of the first wave $(t < n)$. The placement of the B-agents is given by the k-th column of B^t. It follows that, for some fixed probability distribution vector $z \in \mathbb{R}^n$,

$$\left\| B^t - \mathbf{1}z^T \right\|_{\max} \leq 2^{-\gamma t}. \tag{14}$$

Limit of P^t. With the estimates on the powers of A and B given by (13, 14), it is now routine to bound

$$P^t = \begin{pmatrix} A^t & C_t \\ \mathbf{0} & B^t \end{pmatrix},$$

where $C_1 = C$ and $C_{t+1} = AC_t + CB^t$. By (13), the matrix $I - A$ is nonsingular, hence the elementary identity

$$C_t = (I - A^t)(I - A)^{-1}C\,\mathbf{1}z^T + \sum_{k=0}^{t-1} A^{t-k-1}C(B^k - \mathbf{1}z^T).$$

By (13, 14), any matrix entry in the k-th summand is bounded by $2^{-\gamma(t-k-1)-\gamma k}$ in absolute value. The max-norm of the matrix sum itself is therefore at most $t2^{-\gamma(t-1)}$ entry-wise. The entries of $(I - A)^{-1}$ are at most $\sum_t 2^{-\gamma t} = O(n^n)$. By rescaling γ, we reach the following conclusion: there exists $\gamma \geq \Omega(n^{-n})$ such that, for any $t > 1/\gamma$, $\| P^t - \mathbf{P} \|_{\max} \leq 2^{-\gamma t}$, where

$$\mathbf{P} = \begin{pmatrix} \mathbf{0} & (I - A)^{-1}C\,\mathbf{1}z^T \\ \mathbf{0} & \mathbf{1}z^T \end{pmatrix}. \tag{15}$$

If the number r of dominant classes is larger than 1, then $\mathbf{1}z^T$ needs to be replaced by a rank-r stochastic matrix of the form $\text{diag}(\mathbf{1}z_1^T, \ldots, \mathbf{1}z_r^T)$, where z_i is a probability distribution vector in $\mathbb{R}^{|B_i|}$. The limit cycles of directional systems of type $A \rightarrow B$ are described by (time-invariant) matrices such as P, with one difference: all the agents in A are subdominant but not all of B needs to be dominant. As long as the powers of B converge at the rate given above, however, P^t will also converge accordingly.

6 The Analysis

As in [6], we assume a *snap rule*: the status (in or out) of (i, j) as an edge of $G(x)$ is constant over the slab $|x_i - x_j| < \varepsilon_0$; in other words, the edge cannot vanish and reappear incessantly as the agents i and j get infinitesimally close to each other.[12] The idea of renormalization is to prove that a certain dynamical behavior is recursively preserved as we move up the hierarchy of subsystems. In this case we prove inductively that each one of the systems on the right-hand side of the rules in (12) satisfies the following properties:

- There exists a region E, called the *exclusion zone*, such that the coding tree $T(X \setminus E)$ has a nesting time and a word-entropy bounded by ν and h respectively. The exclusion zone is the union of a set of δ-*slabs*, which are regions of $\mathbb{R}^n$ of the form $\{x : |u^T x - 1| \leq \delta\}$, for $\|u\|_2 = n^{O(1)}$ and δ at most a small positive constant. (These bounds can be adjusted liberally.) Only shallow nodes contribute δ-slabs and at most $|\mathcal{D}|$ each. Given a node v of the coding tree, let $P_{\leq v}$ denote the stochastic matrix encoding the linear restriction of f^{t_v} to U_v. The node is said to contribute to the exclusion zone if the latter includes one (or several) δ-slabs of the form $\{x : |u^T P_{\leq v} x - 1| \leq \delta\}$, where $u^T x = 1$ is a discontinuity of $\mathcal{D}$. Because $\|P_{\leq v}^T u\|_2 = n^{O(1)}$, the polynomiality condition on the coefficients of δ-slabs is preserved under any of the tensor operations, since these only require updating $P_{\leq v}$. The total number of δ-slabs in the exclusion zone is bounded by $|\mathcal{D}|2^{h(T)}$. Because $P_{\leq v}^T u$ cannot be too short for a δ-slab to intersect the unit cube X, its width is $2\delta/\|P_{\leq v}^T u\|_2 = O(\delta\sqrt{n})$, so it spans a volume of $O(\delta n^n)$ within X; hence

$$\text{Vol}(X \setminus E) \geq 1 - n^{O(n)}\delta 2^{h(T)}. \tag{16}$$

- Every orbit starting in $X \setminus E$ (and not hitting an absorbing node) is a limit cycle (possibly of period 1). There is a single infinite path descending from any nonshallow node v. If p (resp. q) is its corresponding period (preperiod), then $p + q \leq 2^h$ and $P_{\leq v} = P_{\leq w}P^{(t_v - t_w)/p}$, where w is an ancestor of v of depth between q and $p + q$ (assuming v deep enough); there are p matrices P associated with the given path, each one the product of p of the original stochastic matrices associated with the linear restrictions of f. We assume a uniform lower bound γ for the values of $\gamma = \gamma(P)$ in (15) over all subsystems of any type with a given number of agents.

We analyze the effect of the four rules (12) on the coding tree. By convention, ν, h (resp. ν', h') denote the parameters for the right-hand (resp. left-hand) side of the rule under consideration.

6.1   Sequence to Block-Sequence

Rule 12.1 rewrites $T(X)^{seq}$ as the iterated direct product $\bigotimes_k T(X)^{wave}$. We show that, upon completion of the s-th wave (should it exist), all the agents are covered by an interval of length $2^{-s/n^{O(1)}}$. The argument is a variant of the one we used in §5.1 to prove (14). The number of direct products in rule 12.2 is less

---

[12] We can choose ε_0 to be arbitrarily small but, for convenience, we set $\varepsilon_0 = 2^{-O(n)}$. See [6] for an explanation why this is needed. We note that the snap rule is automatically implied by *metrical systems*, where discontinuities are of the form $x_i - x_j = u$.

than n and every single-graph $G_{t_{k+1}}$ signifies the propagation of water to dry agents. Let λ_k be the length of the smallest interval enclosing A_k at time t_k, ie, at the formation of the block-directional sytem of type $A_k \rightarrow B_k$. One difference with §5.1 is that, by the time A_k gives way to A_{k+1} at time t_{k+1}, its enclosing interval might have grown to cover almost all of $[0,1]$. This might happen if B_k has agents at 0 and 1, for example, and A_k has edges pointing to them.[13] Obviously, the worst case features all the agents B_k located at 0 or 1. Neither L nor R can shrink by more than a factor of n in a single step, so the length of neither one can fall below ε_0/n prior to t_{k+1} (note that it can be smaller than ε_0/n to begin with: it just cannot become so). It follows that

$$1 - \lambda_{k+1} \geq \tfrac{1}{n} \min\{1 - \lambda_k, \varepsilon_0\} \geq \varepsilon_0 n^{-n} \geq n^{-O(n)},$$

which proves that all the agents are covered by an interval of length $2^{-s/n^{O(n)}}$ after s waves. Once all the agents lie within an interval of length ε_0, the snap rule freezes the communication graph and, by (15), the system is attracted to a fixed point at a rate of $2^{-t/n^{O(n)}}$. The system is then of rank 1 but, of course, the high likelihood of incomplete waves can increase the rank (and the period) by creating several dominant classes. By repeated applications of subadditivity (11), $h' \leq (\#\texttt{waves})h$; hence, by $\varepsilon_0 \geq 2^{-O(n)}$, $h' \leq n^{O(n)}h$. Rule 12.2 expresses $T(X)^{wave}$ by fewer than n direct products whose factors are themselves products with a single-graph coding tree, so using primes to denote the trees formed by application of the two rules 12.1 and 12.2:

$$\nu' \leq n^{O(n)}\nu \quad \text{and} \quad h' \leq n^{O(n)}h. \tag{17}$$

Bidirectional Systems. Before we move on to the analysis of the last two rules, it is helpful to build some intuition by resolving the *bidirectional* case. This is the version of the model where the communication graph $G(x)$ is undirected: every edge (i,j) comes with (j,i). Such systems are known to converge [10,12,14]. We show how the renormalization framework leads to a bound on the relaxation time. The parsing rules in (12) reduce to these two:

$$\begin{cases} T(X) & \Longrightarrow & \bigotimes_k T(X)^{wave} \\ T(X)^{wave} & \Longrightarrow & \bigotimes_{k<n} \left\{ (T(A_k) \oplus T(B_k)) \otimes T(G_{t_{k+1}})^{sg} \right\}. \end{cases}$$

Let ν, h denote upper bounds on the nesting time and word-entropy of A_k and B_k. We can show inductively that the period is 1 (fixed-point attraction) so that, along any given path of the coding tree of the direct sum, a node v of depth $t_v > \nu$ is such that $P_{\leq v} = P_{\leq w} P^{t_v - \nu}$ for some node w of depth ν. The matrix P is of the form

$$P = \begin{pmatrix} A_w & \mathbf{0} \\ \mathbf{0} & B_w \end{pmatrix},$$

---

[13] The *Very Slow Clock* builds on this idea.

with both of A_w and B_w playing the role of B in (15).[14] This implies the existence of an idempotent matrix $\mathbf{P}$ such that $\|P^l - \mathbf{P}\|_{\max} \leq 2^{-\gamma l}$, for $\gamma \geq 1/n^{O(n)}$ and any $l \geq 0$. (Better bounds can be found but they are not needed here.) How much higher than ν the nesting time of the direct sum can be depends on how deep a node v can be such that V_v intersects a discontinuity $u^T x = 1$ involving agents from both A_k and B_k. This occurrence implies that $u^T P_{\leq v} x = 1$ for some $x \in U_v$; hence,

$$|u^T P_{\leq w} \mathbf{P} x - 1| \leq \|u\|_1 2^{-\gamma(t_v - \nu)} \leq 2^{-\gamma(t_v - \nu) + O(\log n)}.$$

To make this into a δ-slab, we set $t_v \geq \nu + |\log \delta| n^{bn}$ for constant b large enough. The slab does not depend on v but on its path, so adding it to the exclusion zone guarantees the absence of absorptions deeper than t_v. Accounting for all the direct sums sets an upper bound of $n\nu + n^{O(n)} |\log \delta|$ on the nesting time of $\mathcal{T}(X)^{wave}$. Using subscripts to indicate the number of agents, by (17),

$$\nu(\mathcal{T}_n) \leq n^{O(n)} (\nu(\mathcal{T}_{n-1}) + |\log \delta|) \leq n^{O(n^2)} |\log \delta|. \tag{18}$$

By subadditivity (11), a conservative upper bound on the word-entropy of $\mathcal{T}(X)^{wave}$ is $n(2h(\mathcal{T}_{n-1}) + \log \nu(\mathcal{T}_n))$; hence

$$h(\mathcal{T}_n) \leq n^{O(n)} \big(h(\mathcal{T}_{n-1}) + \log |\log \delta| \big) \leq n^{O(n^2)} \log |\log \delta|.$$

By (16),

$$\mathrm{Vol}\,(X \backslash E) \geq 1 - n^{O(n)} \delta 2^{h(\mathcal{T}_n)} \geq 1 - \delta |\log \delta|^{n^{O(n^2)}} > 1 - \sqrt{\delta},$$

for $\delta > 0$ small enough. This proves that the tiniest random perturbation of the starting configuration—obtained by, say, shifting each agent randomly left or right by a constant, but arbitrarily small amount—will take the system to a fixed-point attractor with probability close to 1. By (18) and the convergence rate of single-graph systems, the system is at a distance ε away from its attractor after a number of steps equal to $n^{O(n^2)} |\log \delta\varepsilon|$. The dependency on δ (and, of course, ε) cannot be avoided. Indeed, there is no uniform bound on the convergence rate over the entire phase space $[0,1]^n$. The following result requires only the snap rule stated at the beginning of §6:

THEOREM 2. *With probability arbitrarily close to one, a perturbed bidirectional diffusive influence system is attracted to a fixed point. Specifically, with probability at least* $1 - \delta$, *the system is at distance* ε *of the fixed point after* $c_n |\log \delta\varepsilon|$ *steps, where* c_n *depends on the number* n *of agents.*[15]

---

[14] Both A_w and B_w can have several dominant classes.

[15] We showed in [6] how to improve the time bound via the s-energy [5].

Entropy vs. Energy. We return to the case of general diffusive influence systems (ie, with no bidirectionality assumption). The nesting time ν tells us how deep we have to go down the coding tree for the dynamics to stabilize. The average degree μ of a shallow node can be defined by $2^h = \mu^\nu$, so that, as δ goes to 0, one would expect of a periodic system $(f, X \backslash E)$ that $\mu = 2^{h/\nu}$ should tend to 1. The average degree measures the tension between the entropic forces captured by h and the energy dissipation expressed by ν via the water propagation. The coding tree can branch at a maximum rate of roughly $|\mathcal{D}|^n$ per node. For the system to be attracting, the rate must be asymptotically equal to 1. It was fairly easy to achieve this without heuristic assumptions in the bidirectional case. We have shown that it is possible in the general case [6] provided that we assume a certain timing mechanism to prevent the reentry of long-vanished edges. In the absence of such conditions, the critical region in parameter space between chaos attraction remains mysterious. In the next section, we sketch minimal heuristic assumptions to ensure asymptotic periodicity.

6.2   Block-Sequence to Block-Sequence

Rule 12.3 rewrites $T(A \nearrow B)^{bseq}$ as $\bigotimes_k T(A_k \nearrow B_k)^{bwave}$. This is our main focus in this section, with a brief mention of rule 12.4 at the end. Fix a path in the coding tree and let $v = v(s)$ the first node (if at all) after the first s block-waves. By definition,

$$P_{\leq v} = \begin{pmatrix} A_s & C_s \\ 0 & B_s \end{pmatrix}.$$

Repeating the argument from §5.1 used for the limit of A^t, we find that

$$\|A_s \mathbf{1}\|_\infty \leq 2^{-\gamma' s}, \tag{19}$$

for some $\gamma' > 0$. Note that we need not assume that B consists only of dominant agents. We used a fairly technical argument to bound γ' in [6] under some mechanism to control the reappearance of long-absent edges. We pursue a simpler, more general approach here.

The Intuition. Recall that the word-entropy measures how likely a typical shallow node v sees its cell V_v intersect a discontinuity and split accordingly. The best way to bound the growth of the word-entropy, therefore, is to show that the cells V_v shrink and hence split with diminishing frequency as t_v grows. The system is not globally contractive, however, and the diameter of V_v, far from shrinking, might actually grow. Indeed, consider the two-agent system with the single graph $a \rightarrow b$: the iterates of the cell $[0, 0.1] \times [0, 1]$ converge to the segment $[(0, 0), (1, 1)]$, so that the area vanishes while the diameter grows by roughly a factor of $\sqrt{2}$. In this example, the cell thins out in the horizontal direction but stays unchanged along the vertical (ie, the dominant) one. The solution is first to factor out the dominant agents and then restore their dynamics in a neighborhood of the periodic points via coarse-graining.

For a mechanical analogy, think of the B-agents as forming the frame of a box-spring. First, we consider a fixed frame and study the vibrations of the springs subject to an impulse:[16] the network of springs (the A-agents) may see its topology change over time but the frame itself remains rigid. In a second stage, we allow the frame to be deformed under its own internal forces. The dynamics of the frame itself is decoupled from the springs (just as B is decoupled from A in a block-directional sytem of type $A \rightarrow B$). This sort of quotient operation is precisely what algorithmic renormalization aims for. We flesh out this intuition below.

Freezing the B-Agents. We begin with the case of a fixed "box-spring frame." The phase space becomes $[0, 1]^m \times \{x_B\}$, where x_B is now viewed as a parameter in $[0, 1]^{n-m}$. Let $\mathcal{T}_{\leq s}$ denote the coding tree of the first s block-waves in rule 12.3 and let v be a node deep enough that at least s block-waves occur before time t_v. Because of the approximation on A_s in (19), the projection of V_v onto $[0, 1]^m$ is contained in an m-dimensional cube of side-length at most $2^{-\gamma's}$. If V_v intersects a discontinuity $u_A^T x_A + u_B^T x_B = 1$, it then follows that

$$\left| (u_A^T C_s + u_B^T B_s)x_B - 1 \right| \leq 2^{-\gamma's+O(\log n)}, \tag{20}$$

for some $x = (x_A, x_B)$. Fix an arbitrarily large threshold s_0 and observe that $\sigma_0 \triangleq 2^{-\gamma's_0}$ is an upper bound on the side-length of the cube enclosing V_v for any v deeper than the s_0-th block-wave. We model the children of v as the outgrowth of a branching process whose reproduction rate (the average node degree) is at most $n^2\sigma_0|\mathcal{D}|$.

Here is a quick heuristic justification. We begin with our earlier observation (10) that the union V_t of the cells V_v for a given depth $t_v = t$ forms a nested sequence as t grows. In the absence of any process biasing the orbits towards the discontinuities, a random point from V_t should not be significantly closer to a discontinuity than if it were random within X itself. Thus, if a typical cell $f(V_v)$ ends up being thrown randomly within V_t, one would expect it to intersect a discontinuity with probability that depends on the size of its enclosing cube. We show how to derive the reproduction rate when V_t is roughly X. (The argument is scalable so it can be extended to the case where V_t is much smaller than X.) If a point is at distance $\sqrt{n}\,\sigma_0$ from a hyperplane, it is possible to move it to the other side by changing a suitable coordinate by at most $n\sigma_0$ (easy proof omitted). The estimate of $n^2\sigma_0|\mathcal{D}|$ follows from a union bound on the n coordinates and the discontinuities. This heuristic validation does not hold in the chaotic construction given in [6], which is why we need the snap rule. As is shown there, when the B-agents are frozen, however, the reproduction rate can be provably bounded.

The coding tree $\mathcal{T}_{\leq s}$ is renormalized by rule 12.3 as a tree of block-wave trees: the latter's absorbing nodes are at most ν away from the root. Past the s_0-th block-wave, the probability that a given node v has its cell V_v is split by

---

[16] For the analogy to be accurate, one must think of the springs as being one-way—in flagrant violation of Newtonian mechanics...

a discontinuity is at most $n^2\sigma_0|\mathcal{D}|$. This creates a reproduction rate of $\mu_v = 1 + \sigma_0 n^{O(n)}$ for a given node v and $\mu \leq \mu_v^\nu$ for an entire whole block-wave tree. Assuming that σ_0 is sufficiently smaller than $1/\nu n^{O(n)}$,

$$\mu \leq 1 + \sigma_0 \nu n^{O(n)}. \tag{21}$$

We enforce nesting after s block-waves by adding to the exclusion zone a number of δ-slabs no greater than $|\mathcal{D}|2^{h(\mathcal{T}_{\leq s})}$: for this, we need to ensure that $s \geq s_0 + (|\log \delta| + b \log n)/\gamma'$, for constant b large enough, so that δ dominates the right-hand side in (20). Note that x_B is considered a parameter here, so the slabs do not split the cells V_v per se: they simply exclude certain positions of x_B, ie, certain configurations of the fixed box-spring frame. By subadditivity,

$$h(\mathcal{T}_{\leq s}) \leq (s_0 + 1)h + (s - s_0) \log \mu.$$

We artificially added 1 to s_0 to account for the fact that each of the coding trees for $\mathcal{T}(A \nearrow B)$ between block-waves s_0 and s needs its own exclusion zone: this slight overestimate of the word-entropy has the benefit of keeping $|\mathcal{D}|2^{h(\mathcal{T}_{\leq s})}$ as a valid upper bound on the number of slabs needed for the exclusion zone. Using primes to refer to the left-hand side of rule 12.3, the word-entropy can be bounded as follows:

$$h' \leq (s_0 + 1)h + \frac{1}{\gamma'}\big(|\log \delta| + O(\log n)\big) \log \mu. \tag{22}$$

Coarse-Graining. We turn to the case of the dynamic "box-spring frame." The previous analysis was premised on the assumption that the B-agents were frozen once and for all. Treating them as variables in $\mathbb{R}^{n-m}$ may violate the assumption that deep nodes rarely witness branching. By Crofton's Lemma (Buffon's needle), a random positioning of a cell V_v will hit a discontinuity with high probability if the diameter of the cell is large. All we can argue is that the volume decreases as the depth of v grows, but it is the diameter that matters, not the volume! The solution is to coarse-grain the phase space for B by subdividing it into tiny cubes and then treating each cube as a single point.

We subdivide $[0,1]^{n-m}$ into cubes of side-length σ_0 and restrict x_B to one of them, denoted c_B. Consider a path of $\mathcal{T}(B)$ and let p denote its period (taking multiples if necessary to ensure that $\nu + p$ is at least the preperiod). We denote by $B_{\leq v}$ the stochastic matrix encoding the linear restriction of f^{t_v} to the space of B-agents. If $t_v > \nu + p$ then, by our induction hypothesis (since $|B| < n$),

$$B_{\leq v} = B_{\leq w} B_w^{(t_v - t_w)/p},$$

where $t_w \leq \nu + p$ and B_w is one of p matrices associated with the periodic orbit; furthermore, there exists an idempotent matrix $\mathbf{B}_w$ such that $\|B_w^l - \mathbf{B}_w\|_{\max} \leq 2^{-\gamma l}$. Of course, this still holds if we switch our point of view and consider a node v of $\mathcal{T}(X_{|c_B})$ of depth $t_v > \nu + p$, where the notation $X_{|c_B}$ indicates that the phase space is still X but $U_{\text{root}} = [0,1]^m \times c_B$. If v is the first node after s block-waves then, by (19), for any $x \in U_v$,

$$\left\| f^{t_v}(x) - \begin{pmatrix} C_s \\ B_{\leq w}\mathbf{B}_w \end{pmatrix} x_B \right\|_\infty \leq 2^{-\gamma' s} + n2^{-(t_v - \nu - p)\gamma/p}. \tag{23}$$

By our assumption that x_B can vary by at most σ_0 in each coordinate and $x = (x_A, x_B)$, the cell V_v is enclosed within a cube of side-length σ_1, where

$$\sigma_1 \leq \sigma_0 + 2^{1-\gamma's} + n2^{1-(t_v-\nu-p)\gamma/p}. \tag{24}$$

We update (21) to estimate the new reproduction rate on the assumption that $s > s_0$ and σ_1 is sufficiently smaller than $1/\nu n^{O(n)}$:

$$\mu \leq 1 + \sigma_1 \nu n^{O(n)}. \tag{25}$$

If V_v intersects the discontinuity $u_A^T x_A + u_B^T x_B = 1$, then, by (23) and $\|u\|_2 = n^{O(1)}$,

$$\left| (u_A^T C_s + u_B^T B_{\leq w} \mathbf{B}_w) x_B - 1 \right| \leq 2^{-\gamma's+O(\log n)} + 2^{-(t_v-\nu-p)\gamma/p+O(\log n)} \leq \delta, \tag{26}$$

with the last inequality ensuring that the constraints fit within δ-slabs. Observe that the characteristic timescale is $1/\gamma'$ (measured in block-waves) for $T(A \nearrow B)$ and p/γ for $T(B)$. After a suitably large number s of block-waves ($s > s_0$), we add to the exclusion zone the relevant δ-slabs for each node at the depth corresponding to the end of the s-th block-wave.

To see why this causes nesting, we examine the coding tree for B first. The added slabs are cylinders, with their bases in c_B, which carve the cells U_v into subcells that are "essentially" invariant for all times in the relevant residue class modulo the corresponding period. The qualifier refers to the fact that the orbit converges toward a fixed point at a rate of $2^{-\gamma}$ per cycle. For t large enough so that the second exponential term in (26) is sufficiently smaller than δ, the orbits of the B-agents might still hit the slabs but not cross their mid-halfplane. From that point on, we can thus factor out the B-agents by pretending that they are fixed and, from there, infer nesting. Adding the contribution to the word-entropy of the grid decomposition of the space of B-agents, we update (22) as

$$h' \leq (n - m)|\log \sigma_0| + (s_0 + 1)h + (s - s_0) \log \mu.$$

We set $s = s_0 + \frac{b}{\gamma'} \log \frac{n}{\delta}$ and $t_v = \nu + \frac{bp}{\gamma} \log \frac{n}{\delta}$, for a constant b large enough (reused generically to alleviate the notation). These assignments satisfy (26). We will always choose δ smaller than σ_0. By (24) and the definition of σ_0 as $2^{-\gamma's_0}$, this implies that $\sigma_1 \leq 2^{2-\gamma's_0}$ and, by (25), $\log \mu \leq 2^{-\gamma's_0}\nu n^{O(n)}$. This upper bound is much less than 1 if we set

$$s_0 = \frac{1}{\gamma'} \log \left(\frac{n^{bn}\nu}{\gamma'} |\log \delta| \right).$$

It follows that $\nu' \leq s\nu \leq (s_0 + \frac{b}{\gamma'} \log \frac{n}{\delta})\nu$ and

$$h' \leq (s_0 + 1)(h + n\gamma') + 1.$$

By subadditivity, rule 12.4 adds factors of at most n to these bounds. Using primes to refer to the parameters of $T(A \nearrow B)^{bseq}$ and unprimed notation to refer to any of the trees in the right-hand side of rule 12.4, we use the inequalities $\delta < \sigma_0$ and $\nu \leq 2^h$ to derive (conservatively):

$$\begin{cases} \nu' \leq O(n/\gamma')^3 |\log \delta| \nu \\ h' \leq O(n/\gamma')^3 (\log |\log \delta|) h^2. \end{cases} \tag{27}$$

6.3   Putting It All Together

We are now in a position to bound the volume of the exclusion zone E. We denote by h_n the maximum word-entropy of $T(X \setminus E)$ for any n-agent system. We reserve the notation $h_{m,n}$ for the biggest of h_m, h_{n-m}, and the maximum word-entropy of any n-agent bidirectional system of type $A \to B$ with $|A| \leq m$. By (17, 27), $h_n \leq n^{O(n)} h_{n-1,n}$, and, for $0 < m < n$,

$$h_{m,n} \leq \max\Big\{ h_m, h_{n-m}, O(n/\gamma')^3 (\log |\log \delta|) h_{m-1,n}^2 \Big\}.$$

It follows that $h_n \leq \big(\gamma'^{-1} \log |\log \delta|\big)^{2^{O(n)}}$. Because δ appears as a double (and not single) logarithm in the upper bound, by (16),

$$\mathrm{Vol}\,(X \setminus E) \geq 1 - n^{O(n)} \delta 2^{h_n} < 1 - \sqrt{\delta},$$

for δ small enough. We conclude:

THEOREM 3. *Under the heuristic assumptions above, with probability arbitrarily close to one, perturbing the initial state of a diffusive influence system produces an orbit that is attracted to a limit cycle.*

This result makes no bidirectionality assumption. It comes with strings attached to it, however, notably pseudorandom discontinuity splitting and uniform bounds on the convergence rates. Our intuition that such heuristic assumptions are essentially correct is backed up by abundant empirical evidence. That said, one should not underestimate the mathematical difficulty in overcoming them rigorously. We could do it in [6] with a single enforceable (ie, non-heuristic) assumption in the model and this already required quite a bit of technical work. To remove all assumptions seems a formidable endeavor.

An avenue of research that seems more accessible is to develop a notion of fractional wetness. In our framework, an agent node is dry or wet depending on the information transmitted to it. But it is a binary predicate which does not measure the amount of information being transmitted. Since the dissipation rate depends on it, it might be useful to delay the completion of the waves until the wetness of each agent has reached a certain threshold. How to develop a "belief propagation" method of message passing for parsing the graph sequences using a variant of the flow tracker is a fascinating open problem.

References

1. Bhattacharyya, A., Braverman, M., Chazelle, B., Nguyen, H.L.: On the convergence of the Hegselmann-Krause system. In: Proc. 4th ITCS, pp. 61–66 (2013)
2. Bullo, F., Cortés, J., Martinez, S.: Distributed Control of Robotic Networks, Applied Mathematics Series. Princeton University Press (2009)
3. Buluc, A., Meyerhenke, H., Safro, I., Sanders, P., Schulz, C.: Recent advances in graph partitioning. arXiv:1311.3144 (preprint)
4. Castellano, C., Fortunato, S., Loreto, V.: Statistical physics of social dynamics. Rev. Mod. Phys. **81**, 591–646 (2009)
5. Chazelle, B.: The total s-energy of a multiagent system. SIAM J. Control Optim. **49**, 1680–1706 (2011)
6. Chazelle, B.: The dynamics of influence systems. In: Proc. 53rd IEEE FOCS, 311–320 (2012); To appear in J. SIAM Comput. (2014)
7. Chazelle, B.: An Algorithmic Approach to Collective Behavior. Journal of Statistical Physics **158**, 514–548 (2015)
8. Fortunato, S.: Community detection in graphs. Physics Reports **486**, 75–174 (2010)
9. Hegselmann, R., Krause, U.: Opinion dynamics and bounded confidence models, analysis, and simulation. J. Artificial Societies and Social Simulation **5**, 3 (2002)
10. Hendrickx, J.M., Blondel, V.D.: Convergence of different linear and non-linear Vicsek models. In: Proc. 17th International Symposium on Mathematical Theory of Networks and Systems (MTNS2006), Kyoto, Japan, pp. 1229–1240, July 2006
11. Kempe, D., Kleinberg, J.M., Kumar, A.: Connectivity and inference problems for temporal networks. Journal of Computer and System Sciences **64**, 820–842 (2002)
12. Lorenz, J.: A stabilization theorem for dynamics of continuous opinions. Physica A: Statistical Mechanics and its Applications **355**, 217–223 (2005)
13. Martínez, S., Bullo, F., Cortés, J., Frazzoli, E.: On synchronous robotic networks Part ii: Time complexity of rendezvous and deployment algorithms. IEEE Transactions on Automatic Control 52, 2214–2226 (2007)
14. Moreau, L.: Stability of multiagent systems with time-dependent communication links. IEEE Transactions on Automatic Control **50**, 169–182 (2005)
15. Touri, B., Nedić, A.: Discrete-time opinion dynamics. In: Proc. 45th IEEE Asilomar Conference on Signals, Systems, and Computers, pp. 1172–1176 (2011)

"Green" Barrier Coverage with Mobile Sensors

Amotz Bar-Noy[1], Dror Rawitz[2], and Peter Terlecky[1(✉)]

[1] The Graduate Center of the City, University of New York,
New York, NY 10016, USA
`amotz@sci.brooklyn.cuny.edu, pterlecky@gc.cuny.edu`
[2] Faculty of Engineering, Bar-Ilan University, Ramat Gan 52900, Israel
`dror.rawitz@biu.ac.il`

Abstract. Mobile sensors are located on a barrier represented by a line segment. Each sensor has a single energy source that can be used for both moving and sensing. A sensor consumes energy in movement in proportion to distance traveled, and it expends energy per time unit for sensing in direct proportion to its radius raised to a constant exponent. We address the problem of energy efficient coverage. The input consists of the initial locations of the sensors and a coverage time requirement t. A feasible solution consists of an assignment of destinations and coverage radii to all sensors such that the barrier is covered. We consider two variants of the problem that are distinguished by whether the radii are given as part of the input. In the *fixed* radii case, we are also given a radii vector ρ, and the radii assignment r must satisfy $r_i \in \{0, \rho_i\}$, for every i, while in the *variable* radii case the radii assignment is unrestricted. We consider two objective functions. In the first the goal is to minimize the sum of the energy spent by all sensors and in the second the goal is to minimize the maximum energy used by any sensor.

We present FPTASs for the problem of minimizing the energy sum with variable radii and for the problem of minimizing the maximum energy with variable radii. We also show that the latter can be approximated within any additive constant $\varepsilon > 0$. We show that the problem of minimizing the energy sum with fixed radii cannot be approximated within a factor of $O(n^c)$, for any constant c, unless P=NP. The problem of minimizing the maximum energy with fixed radii is shown to be strongly NP-hard. Additional results are given for three special cases: (i) sensors are stationary, (ii) free movement, and (iii) uniform fixed radii.

1 Introduction

Battery lifetime is a significant bottleneck on wireless sensor network performance. Thus, one of the fundamental problems in sensor networks is optimizing battery usage when accomplishing tasks such as covering, monitoring, tracking and communicating. We study the problem of covering a boundary or a barrier by mobile sensors, e.g., covering borders, coastlines, railroads, etc. Also, often

D. Rawitz—Supported by the Israel Science Foundation (grant no. 497/14).

V.Th. Paschos and P. Widmayer (Eds.): CIAC 2015, LNCS 9079, pp. 33–46, 2015.
DOI: 10.1007/978-3-319-18173-8_2

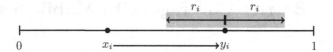

Fig. 1. Sensor i moves from x_i to y_i and covers the interval $[y_i - r_i, y_i + r_i]$

covering region boundaries is the cost efficient way of protecting the interior. The focus of this paper is to determine what is the most energy efficient way of covering a straight-line barrier for a predetermined amount of time with mobile sensors given some initial arrangement of these sensors on the barrier. Prior work tried to optimize either covering costs or mobility costs but not a combination of both costs. We consider a model where energy is consumed by sensing and movement from a single battery source as is most commonly the architecture [2].

Model. We consider a setting where there are n mobile sensors initially located on a barrier represented by the interval $[0, 1]$. (It is convenient, but not essential, to assume that sensors are located on the barrier.) Let $x = (x_1, \ldots, x_n)$ be the initial position vector, where x_i is initial position of sensor i. We consider the *set-up and sense* model [7,10–12,19], where sensors first move to their desired destinations and then begin sensing. Let $y = (y_1, \ldots, y_n)$ be the deployment vector, where y_i is the destination position of sensor i. The system works in two phases. In the *deployment phase*, sensor i moves from its initial position x_i to its destination y_i. This phase is said to occur at time 0. In the *covering phase*, sensor i is assigned a sensing radius r_i and covers the interval $[y_i - r_i, y_i + r_i]$; let $r = (r_1, \ldots, r_n)$ be the radii vector. We call this interval the *covering interval* of sensor i. An example of movement and coverage by one sensor is given in Fig. 1. It is required that the sensors collectively cover the unit interval, i.e. $[0, 1] \subseteq \cup_i [y_i - r_i, y_i + r_i]$. A pair (y, r) is called *feasible* if it covers $[0, 1]$.

Sensor i expends energy both in moving and sensing. Given a deployment point y_i, the energy sensor i spends in movement is proportional to the distance i has traveled, and given by $a|x_i - y_i|$, where a is the constant of proportionality, also referred to as the cost of *friction*. The energy sensor i uses for sensing *per time unit* is r_i^α, for some constant $\alpha \geq 1$. Given a radii assignment r, a sensor i is called *active* if $r_i > 0$, and otherwise it is called *inactive*. Given a deployment y, a radii assignment r, and a time t, sensor i needs at least $E_i^t(y, r) \stackrel{\text{def}}{=} a|y_i - x_i| + tr_i^\alpha$ energy in order to maintain coverage of the interval $[y_i - r_i, y_i + r_i]$ for t time. (We usually omit t and write $E_i(y, r)$, when t is clear from the context.)

Problems. Given an instance (x, t), we seek a feasible pair (y, r) that is "green" with energy expenditure or energy-efficient. We consider two objective functions: (i) minimizing the sum of the energy used, namely minimizing $\sum_i E_i(y, r)$; and, (ii) minimizing the maximum amount of energy expended, i.e., minimizing $\max_i E_i(y, r)$. We also consider two variants of the problem that are distinguished by whether the radii are given as part of the input. In the *variable radii* case the goal is to find a radii assignment r such that $r_i \geq 0$, for every i, while in the *fixed radii* case the input contains a radii vector ρ, and the goal is to find a radii assignment r, such that $r_i \in \{0, \rho_i\}$, for every i. Thus, we get four variants:

1. MIN SUM ENERGY WITH VARIABLE RADII (SUMVAR)
2. MIN SUM ENERGY WITH FIXED RADII (SUMFIX)
3. MIN MAX ENERGY WITH VARIABLE RADII (MAXVAR)
4. MIN MAX ENERGY WITH FIXED RADII (MAXFIX)

Sometimes when we consider a specific friction parameter a we add a subscript a to the problem name. For example, SUMVAR$_0$ stands for the problem of finding a pair (y, r), where r is variable, that minimizes $\sum_i E_i(y, r)$ for $a = 0$.

Given a SUMFIX or a MAXFIX instance (x, ρ, t), we say that the radii vector ρ is *uniform* if $\rho_i = \rho_j$, for every sensors i and j. Also, we assume that $\sum_i 2\rho_i \geq 1$ throughout the paper, since otherwise there is no feasible solution. A solution (y, r) (or a deployment y) is called *non-swapping* if $x_i < x_j$ implies $y_i \leq y_j$.

Related Work. Most previous research has implicitly assumed a two battery model, in which there is a separate battery for movement and a separate battery for sensing. These works attempt to optimize on only one of the parameters.

When only moving is optimized (covering energy is ignored) the problem is equivalent to having an infinite covering battery. In our model such problems can be described by setting $t = 0$. Czyzowicz et al. [11] addressed the problem of deploying sensors on a line barrier while minimizing the maximum distance traveled by any sensor, where radii are uniform. This is MAXFIX$_1$ with uniform radii and $t = 0$ in our model. (In this case we may assume w.l.o.g. that $a = 1$.) They provided a polynomial time algorithm for this problem. It follows that there is a polynomial time algorithm for MAXFIX with $t = 0$ and uniform radii, for any $a \in (0, \infty)$. They also gave an NP-hardness result for a variant of this problem with non-uniform radii in which one sensor is assigned a predetermined position. Chen et al. [10] gave a polynomial time algorithm for the more general case in which the sensing radii are non-uniform, namely for MAXFIX$_1$ with $t = 0$ and improved upon the running time for MAXFIX$_1$ with uniform radii and $t = 0$.

Czyzowicz et al. [12] studied the problem of covering a barrier with mobile sensors with the goal of minimizing the sum of distances traveled. This problem is a special case of SUMFIX$_1$ in which $t = 0$ (w.l.o.g., $a = 1$). They presented a polynomial time algorithm for SUMFIX$_1$ with uniform radii and $t = 0$ and showed that the non-uniform problem cannot be approximated within a factor of c, for any constant c.

There are other problems in which movement is optimized. We list several examples. Mehrandish et al. [18] considered the same model with the objective of minimizing the number of sensors which must move to cover the barrier. Dobrev et al. [14] studied the problem of covering a set of barriers attempting to optimize movement costs. Tan and Wu [20] presented improved algorithms for minimizing the max distance traveled and minimizing the sum of distances traveled when sensors must be positioned on a circle in regular n-gon position.

The problems were initially considered by Bhattacharya et al. [8]. Demaine et al. [13] studied minimizing movement of pebbles in a graph in order to obtain a property (such as connectivity, independence, and perfect matchability), and the goal is to minimize maximum movement, total movement, or number of movements.

In many papers it is assumed that sensors are static, and the goal is to minimize sensing energy. Li et al. [15] presented a polynomial time algorithm for SumFix$_\infty$ and an FPTAS for SumVar$_\infty$ with $\alpha = 1$. They also showed that SumVar$_\infty$ with $\alpha = 1$ is NP-hard. Agnetis et al. [1] considered an extension of SumVar$_\infty$ with $\alpha = 2$. They gave a closed form solution for this problem if the coverage set is given, and developed a branch-and-bound algorithm and heuristics. Some papers explored discrete coverage of points on the barrier by static sensors (see, e.g., [6,17]).

Another common research direction is to consider the dual problem which is to maximize the lifetime of the network where the battery sizes are given. See, e.g., [3–5,7,9,16,19].

To the best of our knowledge our earlier papers [7,19] and this work are the first to consider energy consumption from moving and sensing from a single battery source. In [19] we attempted to maximize the transmission lifetime of mobile battery-powered relays on a direct line from source to sink, and in [7] we considered maximizing barrier coverage lifetime of a network of mobile battery-powered sensors.

Our Results. SumVar is studied in Section 2, where we present an $O(n)$ time algorithm for SumVar$_0$ and an FPTAS for SumVar, for any a. The latter is based on the FPTAS for SumVar$_\infty$ with $\alpha = 1$ by Li et al. [15]. However, we introduce several new ideas in order to cope with sensor mobility and with $\alpha > 1$. In particular we show that there exists a non-swapping optimal solution and use the optimal value for $a = 0$ as a lower bound for the case where $a > 0$.

Section 3 deals with MaxVar. We present an FPTAS for MaxVar that is similar to the SumVar FPTAS. However, while the SumVar non-swapping property is reminiscent of previous non-swapping results for uniform radii (see, e.g., [11,12]), proving MaxVar non-swapping is more challenging and require rigorous case analysis. In the full version of the paper we present $O(n)$ time algorithms for MaxVar$_0$ and for MaxVar$_\infty$, and we also show that MaxVar can be approximated to within an additive approximation $\varepsilon > 0$, for any constant $\varepsilon > 0$. This result is based on the non-swapping property and on [7].

In Section 4 we study MaxFix. We provide an $O(n \log n)$ time algorithm for MaxFix$_0$ in the full version. We show that MaxFix is strongly NP-hard for every $a \in (0, \infty)$ and $\alpha \geq 1$. We also show that MaxFix is NP-hard, for every $a \in (0, \infty)$ and $\alpha \geq 1$, even if $x \in (\frac{1}{2})^n$, and this result implies that it is NP-hard to find an optimal ordering.

We study SumFix in Section 5. We show that SumFix cannot be approximated within a factor of $O(n^c)$, for any constant c, unless P=NP. In the full version we show that SumFix$_0$ is NP-hard for $\alpha = 1$, and provide an FPTAS for SumFix$_0$ for any α. We also prove that SumFix with uniform radii can be approximated to within an additive approximation $\varepsilon > 0$, for any constant $\varepsilon > 0$.

2   Minimum Sum Energy with Variable Radii

In this section we consider SumVar. We show that SumVar$_0$ can be solved in linear time, and the main result of the section is an FPTAS for the case where $a > 0$. Our FPTAS is based on the approach of Li et al. [15] who gave an FPTAS for SumVar$_\infty$ with $\alpha = 1$. We note that several new and non-trivial ideas were introduced in order to cope with mobility and with $\alpha > 1$.

We start with the case where $a = 0$.

Theorem 1. SumVar$_0$ *can be solved in* $O(n)$ *time with optimum* $nt \left(\frac{1}{2n}\right)^{\alpha}$.

Proof. Given a SumVar$_0$ assignment (x, t), let $r_i = \frac{1}{2n}$, for all i, and let $y_i = \sum_{j=1}^{i-1} 2r_j + r_i$, for every i. We show that (y, r) is an optimal solution. This solution assignment clearly covers $[0, 1]$. Consider any radii assignment $r' \neq r$ that covers the line. It follows that $\sum_i r'_i \geq \frac{1}{2} = \sum_i r_i$. Since sensors are free to move without energy consumption, by Jensen's Inequality we have that $\sum_i E_i(y, r) = nt \left(\frac{1}{2n}\right)^{\alpha} \leq \sum_i t(r'_i)^{\alpha} = \sum_i E_i(y', r')$. Thus, (y, r) is optimal as well. Finally, notice that (y, r) can be computed in linear time. $\square$

Observe that the optimal solution may only decrease as a decreases. Hence, $nt \left(\frac{1}{2n}\right)^{\alpha}$ may serve as a lower bound for the case where $a > 0$. We use this lower bound in the sequel.

Our FPTAS for the non-zero friction case is obtained by the following approach. We first show that any SumVar instance has a non-swapping optimal solution. Then, we show that we pay an approximation factor of $(1 + \varepsilon)$ for only considering a certain family of solutions. Finally, we design a dynamic programming algorithm that computes an optimal solution within this family.

Lemma 1. *Any* SumVar *instance has a non-swapping optimal solution.*

Proof. Let (x, t) be a SumVar instance, and let (y, r) be an optimal solution for (x, t) that minimizes the number of swaps. If there are no swaps, then we are done. Otherwise, we show that the number of swaps may be decreased. If there are swaps, then there must exist at least one swap due to a pair of adjacent sensors. Let i and j be such sensors. Consider a solution (y', r') swapping locations and radii of sensors i and j in (y, r), i.e., with $y'_i = y_j$ and $r'_i = r_j$, $y'_j = y_i$ and $r'_j = r_i$, and $y'_k = y_k$ and $r'_k = r_k$, for every $k \neq i, j$.

Clearly, the barrier $[0, 1]$ remains covered. We show that the energy sum does not increase, since the total distance traveled by the sensors does not increase. If both sensors move to the right in y, then we have that $x_i < x_j \leq y_j < y_i$.

In this case $(y_i' - x_i) + (y_j' - x_j) = (y_i - x_i) + (y_j - x_j)$, and we are done. The case where both sensors move to the left is symmetric. Suppose that i moves to the right while j moves to the left. If $x_i \leq y_j < y_i \leq x_j$ or $y_j < x_i < x_j < y_i$, then both i and j move less in y'. If $x_i \leq y_j \leq x_j < y_i$, then

$$(y_i - x_i) + (x_j - y_j) \geq y_i - x_i + y_j - x_j = (y_i' - x_i) + (y_j' - x_j) .$$

The case where $y_j < x_i \leq y_i \leq x_j$ is symmetric. It follows that (y', r') is an optimal solution with less swaps than (y, r). A contradiction.   □

Let m be a large integer to be determined later. We consider solutions in which the sensors must be located on certain points. More specifically, we define $\mathcal{G} = \{x_i : i \in \{1, \ldots, n\}\} \cup \{\frac{j}{m} : j \in \{0, \ldots, m\}\}$. The points in $\mathcal{G}$ are called *grid points*. Let $g_0, \ldots, g_{n+m}$ be an ordering of grid points such that $g_i \leq g_{i+1}$. Given a point $p \in [0, 1]$, let p^+ be the left-most grid point to the right of p, namely $p^+ = \min\{g \in \mathcal{G} : g \geq p\}$. Similarly, $p^- = \max\{g \in \mathcal{G} : g \leq p\}$ is the right-most grid point to the left of p. A solution (y, r) is called *discrete* if (i) $y_i \in \mathcal{G}$, for every sensor i, and (ii) for every $j \in \{1, \ldots, n + m\}$ there exists a sensor i such that $[g_{j-1}, g_j] \subseteq [y_i - r_i, y_i + r_i]$. That is, in a discrete solution sensors must be deployed at grid points, and a segment between grid points is contained in the covering interval of some sensor.

We show that we lose a factor of $(1 + \varepsilon)$ by focusing on discrete solutions.

Lemma 2. *Let $\varepsilon \in (0, 1)$, and let $m = 8 \lceil \alpha\mu/\varepsilon \rceil$, where $\mu = 2n/\varepsilon^{1/\alpha}$. Then, for any non-swapping solution (y, r) there exists a non-swapping discrete solution (y', r') such that $\sum_i E_i(y', r') \leq (1 + 2\varepsilon) \sum_i E_i(y, r)$.*

Proof. Given a SumVar instance (x, t) and a solution (y, r) we construct a discrete solution (y', r') as follows. First, each sensor i is taken back from y_i to the direction of x_i, until it hits a grid point: $y_i' = y_i^+$ if $y_i \leq x_i$ and $y_i' = y_i^-$ otherwise. Also, the radii are increased to compensate for the new deployment, and in order to obtain a discrete solution: $r_i' = \max\{y_i' - (y_i - r_i)^-, (y_i + r_i)^+ - y_i'\}$. The pair (y', r') is feasible, since $[y_i - r_i, y_i + r_i] \subseteq [y_i' - r_i', y_i' + r_i']$ by construction. Moreover, notice that if $(g_j, g_{j+1}) \cap [y_i - r_i, y_i + r_i] \neq \emptyset$, then $[g_j, g_{j+1}] \subseteq [y_i' - r_i', y_i' + r_i']$. Hence, (y', r') is discrete. We also note that (y', r') is non-swapping.

It remains to show that $\sum_i E_i(y', r') \leq (1 + 2\varepsilon) \sum_i E_i(y, r)$. Since y_i' can only be closer than y_i to x_i, we have that $|y_i' - x_i| \leq |y_i - x_i|$. In addition, the radius of sensor i may increase due to its movement from y_i to y_i' and due to covering up to grid points. Hence, $r_i' \leq r_i + \frac{2}{m}$.

If $r_i \geq \frac{1}{2\mu}$, then $r_i' \leq r_i + \frac{2}{m} \leq r_i + \frac{\varepsilon}{4\alpha\mu} \leq r_i (1 + \frac{\varepsilon}{2\alpha})$. Hence,

$$E_i(y', r') = a|y_i' - x_i| + t(r_i')^\alpha \leq a|y_i - x_i| + tr_i^\alpha \left(1 + \tfrac{\varepsilon}{2\alpha}\right)^\alpha \leq \left(1 + \tfrac{\varepsilon}{2\alpha}\right)^\alpha E_i(y, r)$$
$$\leq e^{\varepsilon/2} E_i(y, r) .$$

Otherwise, if $r_i < \frac{1}{2\mu}$, then $r_i' \leq r_i + \frac{2}{m} \leq \frac{1}{2\mu} + \frac{\varepsilon}{4\alpha\mu} \leq \frac{1}{\mu}$. Hence,

$$E_i(y', r') = a|y_i' - x_i| + t(r_i')^\alpha \leq a|y_i - x_i| + t\tfrac{1}{\mu^\alpha} = a|y_i - x_i| + t\tfrac{\varepsilon}{2^\alpha n^\alpha}$$

Putting it all together we get

$$\sum_i E_i(y', r') \le e^{\varepsilon/2} \sum_i E_i(y, r) + nt\frac{\varepsilon}{2^\alpha n^\alpha} \le (1+\varepsilon) \sum_i E_i(y, r) + \varepsilon \cdot \text{OPT} \quad (1)$$

where the second inequality follows from (i) $e^{\varepsilon/2} \le 1 + \varepsilon$, for any $\varepsilon \in (0,1)$, and (ii) $\text{OPT} \ge nt\frac{1}{(2n)^\alpha}$, as observed after Theorem 1. □

Lemma 2 implies that there is a discrete non-swapping solution which is $(1+\varepsilon)$-approximate. We now present a dynamic programming algorithm for finding the optimal discrete non-swapping solution.

Lemma 3. *There exists an $O(nm^4)$ time algorithm that finds the optimal discrete non-swapping solution.*

Proof. The dynamic programming table is denoted by Π, and it is constructed as follows. The entry $\Pi(i, \ell, k)$, where i is a sensor number, $\ell \in \{0, \ldots, n+m\}$, and $k \in \{0, \ldots, n+m\}$, stands for the minimum energy sum needed by a non-swapping discrete solution that uses the first i sensors, such that the ith sensor is located at $[0, g_\ell]$, to cover the interval $[0, g_k]$. Observe that the size of the table is $O(nm^2)$. Also, the optimum is given by $\Pi(n, n+m, n+m)$.

In the base case $\Pi(0, \ell, 0) = 0$, for all ℓ. Otherwise, we have

$$\Pi(i, \ell, k) = \min_{\ell' \le \ell} \left\{ a|g_{\ell'} - x_i| + \min\left\{ \Pi(i-1, \ell', k), \min_{k' < k}\{\Pi(i-1, \ell', k') + tr_i^\alpha\} \right\} \right\}$$
$$(2)$$

where $r_i = \max\{g_{\ell'} - g_{k'}, g_k - g_{\ell'}\}$. Notice that $g_{\ell'} - g_{k'}$ or $g_k - g_{\ell'}$ may be negative, but not both. The first term in (2) is the energy required by sensor i to arrive at $g_{\ell'}$. Then, we have two options, either i participates in the cover or it does not. In the first case, sensors 1 to $i - 1$ need to cover $[0, g_k]$, and $i - 1$ may stand anywhere in $[0, g_{\ell'}]$. Otherwise, r_i is determined such that i can cover $[g_{k'}, g_k]$ while standing at $g_{\ell'}$. The rest of the barrier, i.e. $[0, g_{k'}]$ is covered by sensors 1 to $i - 1$, and $i - 1$ may stand anywhere in $[0, g_{\ell'}]$.

Computing each entry takes $O(m^2)$ time. Hence, the total running time is $O(nm^4)$. We note that the above algorithm computes the minimum energy sum, but may also be used to compute the solution that achieves this value using standard techniques. □

Lemma 2 and the above algorithm lead to an FPTAS for SumVar.

Corollary 1. *There is an $O(n^5/\varepsilon^{4(1+1/\alpha)})$ time FPTAS for SumVar.*

In the case of static sensors (i.e., $a = \infty$) the dynamic programming can be simplified, since there is no reason to deal with the location of the sensors. In this case we have only $O(nm)$ entries, where $\Pi(i, k)$ stands for the minimum energy sum needed by a discrete solution that uses the first i sensor to cover the interval $[0, g_k]$. Also, (2) is changed to

$$\Pi(i, k) = \min\left\{ \Pi(i-1, k), \min_{k' < k}\{\Pi(i-1, k') + tr_i^\alpha\} \right\}, \quad (3)$$

where $r_i = \max\{x_i - g_{k'}, g_k - x_i\}$. An entry can be computed in $O(m)$, and the total running time is $O(nm^2)$. We get the following result.

Corollary 2. *There is an $O(n^3/\varepsilon^{2(1+1/\alpha)})$ time FPTAS for* SUMVAR$_\infty$.

3   Minimum Max Energy with Variable Radii

In this section we provide n FPTAS for MAXVAR with the non-zero finite friction that is based on the same approach that was used for SUMVAR. We first show that any MAXVAR instance has a non-swapping optimal solution. Then, we show that we pay an approximation factor of $(1 + \varepsilon)$ for considering non-swapping discrete solutions. Finally, we design a dynamic programming algorithm that computes an optimal non-swapping discrete solution.

We prove that there is no need to consider solutions which swap sensors, but as opposed to the proof of Lemma 1, the proof for MAXVAR is more involved and requires case analysis. Before proving that there is a non-swapping optimal solution for any MAXVAR instance we need the following definition. Given a solution (y, r) we define $d_i = |y_i - x_i|$. Also, given an energy level E and a position p, we define $\beta_i(p, E) = (E - a|p - x_i|)/t$ and $r_i(p, E) = \sqrt[\alpha]{\beta_i(p, E)}$. The radius $r_i(p, E)$ is the maximum possible radius that can be maintained for t time, assuming that i moves to p and that $E - a|p - x_i| > 0$.

We also use the following lemma that was proven in [7].

Lemma 4 ([7]). *Let $\beta_1, \beta_2, \gamma_1, \gamma_2 \geq 0$ such that (i) $\gamma_1 < \beta_1 \leq \beta_2$, and (ii) $\beta_1 + \beta_2 \geq \gamma_1 + \gamma_2$. Also let $\alpha \geq 1$. Then, $\sqrt[\alpha]{\beta_1} + \sqrt[\alpha]{\beta_2} \geq \sqrt[\alpha]{\gamma_1} + \sqrt[\alpha]{\gamma_2}$.*

Lemma 5. *Any MAXVAR instance has a non-swapping optimal solution.*

Proof. Let (x, t) be a MAXVAR instance, and let (y, r) be an optimal solution for (x, t) using maximum energy E, that minimizes the number of swaps. Throughout the proof we assume that the radius of sensor i is $r_i(y_i, E)$, for all i. If there are no swaps, then we are done. Otherwise, we show that the number of swaps can be decreased. Assume to the contrary that there are swaps, and consider a swap between a pair of adjacent sensors i and j. That is, $x_i < x_j$, $y_j < y_i$, and $y_k \notin (y_j, y_i)$ for every $k \neq i, j$. There are six possible configurations for such a pair of sensors as shown in Figure 2.

If the barrier can be covered without i, then i is moved to y_j. Sensor i has enough energy for moving to y_j, since either $|y_j - x_i| \leq |y_i - x_i|$ or $|y_j - x_i| \leq |y_j - x_j|$. Similarly, if the barrier is covered without j, then j is moved to y_i. Sensor j has enough energy to move to y_i, since either $|y_i - x_j| \leq |y_j - x_j|$ or $|y_i - x_j| \leq |y_i - x_i|$. In both cases we get a solution with less swaps than (y, r), hence we may assume in the following that both sensors are necessary for covering the barrier (i.e., the removal of either i or j breaks coverage). We define the coverage interval of i and j to be $[u, v] = [y_j - r_j, y_i + r_i]$.

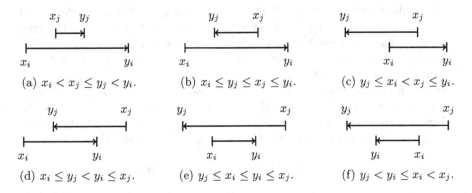

Fig. 2. Six configurations of a swapping pair

For each of the six cases (shown in Figure 2) we provide a solution (y', r') such that $y'_k = y_k$ and $r'_k = r_k$, for $k \neq i, j$, $y'_i \leq y'_j$, and the interval $[u, v]$ is covered by i and j. Moreover, $E_k(y', r') \leq E_k(y, r) = E$, for every k. Then, we eliminate any new swaps that may have been created by moving i and j. The resulting solution has less swaps than (y, r), and we get a contradiction.

We start with cases (c) and (d), since they are easier. Then, we move to deal with the other cases. Newly created swaps will be considered later on.

Case (c): $y_j \leq x_i < x_j \leq y_i$.
Swap the positions and radii of sensors i and j, namely set $y'_i = y_j$, $r'_i = r_j$, $y'_j = y_i$, and $r'_j = r_i$. Observe that $[u, v]$ is covered, and no new swaps are created. Also, $d'_i \leq d_j$ and $d'_j \leq d_i$, which means that $E_i(y', r') \leq E_j(y, r)$ and $E_j(y', r') \leq E_i(y, r)$.

Case (d): $x_i \leq y_j < y_i \leq x_j$.
First, notice that since both i and j participate in the cover, we have that $y_j \leq \frac{u+v}{2} \leq y_i$. Place sensor i at $y'_i = u + r_i$ with radius $r'_i = r_i$ and sensor j at $y'_j = v - r_j$ with radius $r'_j = r_j$. Observe that $[u, v]$ remains covered as $y'_j - y'_i = y_i - y_j$. Also, we have that $y'_i \leq \frac{u+v}{2} \leq y_i$ and $y'_j \geq \frac{u+v}{2} \geq y_j$. If $y'_i \geq x_i$, then $d'_i \leq d_i$. Otherwise, if $y'_i < x_i$, then

$$d'_i = x_i - y'_i \leq y_j - y'_i < y'_j - y'_i = y_i - y_j \leq y_i - x_i = d_i \ ,$$

which means that i moves less. Hence, $E_i(y', r') \leq E_i(y, r)$. A similar argument can be made for sensor j.

Cases (a): $x_i < x_j \leq y_j < y_i$.
First, place sensor i at the location y'_i such that $y'_i - r'_i(y'_i, E) = u$, namely to the point where the left endpoint of the covering interval of i is u while using energy E. Since $x_i \leq y_j$ and $r_i(x_i, E) > r_j(y_j, E)$, we have that $x_i - r_i(x_i, E) < y_j - r_j(y_j, E) = u$. Furthermore, $y_i - r_i(y_i, E) > u$. Since the function $g_i(z) = z - r_i(z, E)$ is continuous and also strictly increasing for $z \geq x_i$, there exists one location $y'_i \in [x_i, y_i]$, for which $y'_i - r_i(y'_i, E) = u$.

Next, place sensor j at the rightmost location y'_j such that $y'_j \leq y_i$ and $y'_j - r'_j(y'_j, E) \leq y'_i + r'_i$. We know that $y_j - r_j(y_j, E) = u < y'_i + r'_i$. Also, observe that j can reach $y_i > y'_i$, since i can. Since $g_j(z) = z - r_j(z, E)$ is continuous and strictly increasing for $z \geq x_j$, we have that there exists one location $y'_j > x_j$, for which $y'_j - r_j(y'_j, E) = y'_i + r'_i$.

If $y'_j = y_i$, we get that $y'_j + r'_j > v$. Otherwise, observe that $y'_i < y_i$, $y'_i < y_j$, and $y'_j < y_i$. It follows that $d'_i < d_i$, $d'_j < d_i$, and $d'_i + d'_j < d_i + d_j$. Hence, $\beta_i(y_i, E) \leq \beta_i(y'_i, E), \beta_j(y'_j, E)$, and $\beta_i(y_i, E) + \beta_j(y_j, E) \leq \beta_i(y'_i, E) + \beta_j(y'_j, E)$. By Lemma 4 we have that $r'_i + r'_j > r_i + r_j$, and thus

$$y'_j + r'_j = u + 2r'_i + 2r'_j > u + 2r_i + 2r_j \geq v .$$

Case (b): $x_i \leq y_j \leq x_j \leq y_i$.

In this case we have two options. First, if $d_j = x_j - y_j \geq y_j - x_i$, switch places and radii between i and j as done in case (c).

Otherwise, $d_j < y_j - x_i$. In this case place sensors i and j as done in case (a). Notice that it may be that $y'_j < x_j$. However, it is enough that $g_j(z)$ is continuous for our purposes. If $y'_j = y_i$, we get that $y'_j + r'_j > v$. Otherwise, observe that $y'_i < y_i$ and $y'_j \in (y_j, y_i)$, which means that $d'_i, d'_j < d_i$. Finally, if $y'_j \leq x_j$, then $d'_j < d_j$, and we have $d'_i + d'_j \leq d_i + d_j$. Otherwise, if $y'_j > x_j$, we have that

$$d'_i + d'_j = (y'_i - x_i) + (y'_j - x_j) < y_i - x_i = d_i ,$$

since $y'_i > x_i$. Again, apply Lemma 4 to show that $r'_i + r'_j > r_i + r_j$, and it follows that $y'_j + r'_j > v$.

Case (e): $y_j \leq x_i \leq y_i \leq x_j$. Symmetric to case (b).

Case (f): $y_j \leq y_i \leq x_i < x_j$. Symmetric to case (a).

It remains to deal with newly created swaps. If $y'_i < y_i$, there may be a sensor k such that $y_k \in (y'_i, y_i]$, and by moving i to y'_i a new swap is created, if $x_k < x_i$. Let $S_L = \{k : x_k < x_i \wedge y_k \in (y'_i, y_i]\}$, and let $S_R = \{k : x_k \geq x_i \wedge y_k \in (y'_i, y_i]\}$. By moving left to y'_i, i creates new swaps with sensors in S_L, but eliminates swaps with sensors in S_R. Let $\ell = \operatorname{argmin}_{k \in S_L}(y_k - r_k)$. If $y_\ell - r_\ell \geq u$, then the sensors in S_L are not needed for coverage and are moved left to y'_i. Consider a sensor $k \in S_L$. If $x_k \leq y'_i$, then y'_i is closer to x_i than y_k. Otherwise y'_i is closer to x_k than to x_i. Hence, in both cases k can reach y'_i. On the other hand, if $y_\ell - r_\ell < u$, it follows that $[y'_i - r'_i, y'_i + r'_i] \subset [y_\ell - r_\ell, y_\ell + r_\ell]$, which means that i is not needed for coverage, and can be moved to $\max_{k \in S_L} y_k$. In both cases all new swaps are eliminated. The case of $y'_i > y_i$ can be treated in a symmetric manner. Also, any new swaps created by j, can be eliminated in a similar manner.

Thus there is a solution with minimum maximum energy E with less swaps. A contradiction. □

Next, we show that we can focus on non-swapping discrete solutions.

Lemma 6. *Let $\varepsilon \in (0,1)$, and let $m = 8 \lceil \alpha\mu/\varepsilon \rceil$, where $\mu = 2n/\varepsilon^{1/\alpha}$. Then, for any non-swapping solution (y, r) there exists a non-swapping discrete solution (y', r') such that $\max_i E_i(y', r') \leq (1 + 2\varepsilon) \max_i E_i(y, r)$.*

Proof. The proof is almost the same as the proof of Lemma 2. The only difference is that Eqn (1) should be replaced by

$$\max_i E_i(y', r') \leq e^{\varepsilon/2} \max_i E_i(y, r) + t\frac{\varepsilon}{2^\alpha n^\alpha} \leq (1 + \varepsilon) \max_i E_i(y, r) + \varepsilon\mathrm{OPT} \ ,$$

where the second inequality is due to $e^{\varepsilon/2} \leq 1 + \varepsilon$, for any $\varepsilon \in (0,1)$, and $\mathrm{OPT} \geq t\frac{1}{(2n)^\alpha}$. $\qquad\square$

We use dynamic programming to find the best non-swapping discrete solution.

Lemma 7. *There exists an $O(nm^4)$ time algorithm that finds the optimal non-swapping discrete solution.*

Corollary 3. *There is an $O(n^5/\varepsilon^{4(1+1/\alpha)})$ time FPTAS for MAXVAR.*

4   Minimum Max Energy with Fixed Radii

In this section we study MAXFIX. Czyzowicz et al. [11] presented an algorithm for MAXFIX with uniform radii and $t = 0$. Chen et al. [10] improved upon the running time of the above problem and gave a polynomial time algorithm for MAXFIX with $t = 0$. We show that, for $a \in (0, \infty)$, MAXFIX is NP-hard even if $x = (\frac{1}{2})^n$, and that it is strongly NP-hard when radii are non-uniform. We note that our reductions are based on the fact that $t > 0$.

As mentioned earlier, Czyzowicz et al. [11] presented a polynomial time algorithm for MAXFIX with uniform radii and $t = 0$. Their result is based on showing that there exists a non-swapping optimal solution for the special case of uniform radii. We show that SUMFIX is NP-hard, even if $x = (\frac{1}{2})^n$, using a reduction from PARTITION. This implies that it is NP-hard to find an optimal ordering of a MAXFIX instance.

Theorem 2. *MAXFIX is NP-hard even if $x = (\frac{1}{2})^n$, for every $a \in (0, \infty)$ and $\alpha \geq 1$.*

Proof. Given a PARTITION instance $(s_1, \ldots, s_n)$, we construct a MAXFIX instance with $n + 1$ sensors as follows. $x_i = \frac{1}{2}$, for every i, $\rho_i = \frac{s_i}{4\sum_j s_j}$, for $i \leq n$, and $\rho_{n+1} = \frac{1}{4}$. Also, let $t = a4^\alpha$. The MAXFIX instance can be constructed in linear time. We show that $(s_1, \ldots, s_n) \in$ PARTITION if and only if there is a solution (y, r) such that $\max_i E_i(y, r) = a$.

Suppose that $(s_1, \ldots, s_n) \in$ PARTITION, and let $I \subseteq \{1, \ldots, n\}$ such that $\sum_{i \in I} s_i = \sum_{i \notin I} s_i$. Set $r_i = \rho_i$, for every i. Use sensor $n+1$ to cover the interval

$[\frac{1}{4}, \frac{3}{4}]$, the sensors that correspond to I to cover the interval $[0, \frac{1}{4}]$, and the rest of the sensors to cover the interval $[\frac{3}{4}, 1]$. This is possible, since $\sum_{i \in I} 2\rho_i = \sum_{i \in \{1,\ldots,n\} \setminus I} 2\rho_i = \frac{1}{4}$. A sensor i, where $i \leq n$, needs less than $\frac{a}{2}$ energy to move, and at most $a4^\alpha \cdot \frac{1}{8}^\alpha = \frac{a}{2^\alpha} \leq \frac{a}{2}$ for coverage, therefore it can stay alive for $a4^\alpha$ time. Sensor $n + 1$ stays put and requires $a4^\alpha \cdot \frac{1}{4^\alpha} = a$ energy. Hence, maintaining cover for $a4^\alpha$ time can be obtained with maximum energy a.

Now suppose that there exists a solution (y, r) such that $\max_i E_i(y, r) = a$. Notice that $\sum_i 2\rho_i = 1$, and thus it must be that $r_i = \rho_i$, for every i. Since sensors $n+1$ requires all its energy for covering, it must be that $y_{n+1} = x_{n+1} = \frac{1}{2}$. It follows that the interval $[0, \frac{1}{4}]$ is covered by a set of sensors I that satisfy $\sum_{i \in I} 2\rho_i = \frac{1}{4}$. Hence $\sum_{i \in I} s_i = \frac{1}{2} \sum_i s_i$, which means that $(s_1, \ldots, s_n) \in$ PARTITION.   □

We use a similar approach to describe a reduction from 3-PARTITION. This implies strong NP-hardness. The proof was omitted for lack of space.

Theorem 3. MAXFIX *is strongly NP-hard, for every* $a \in (0, \infty)$ *and* $\alpha \geq 1$.

5   Minimum Sum Energy with Fixed Radii

In this section we consider SUMFIX. Li et al. [15] solved SUMFIX$_\infty$ using an elegant reduction to the shortest path problem. In the full version of the paper we show that SUMFIX$_0$ is NP-hard, if $\alpha = 1$, but admits an FPTAS, for any α. We also prove that SUMFIX with uniform radii can be approximated to within an additive approximation $\varepsilon > 0$, for any constant $\varepsilon > 0$. This algorithm is based on the non-swapping property and on placing the sensors on grid points. However, as opposed to the variable case, we cannot change radii, only locations, which is problematic when there is very little excess coverage. We cope with this issue by considering two solution types, small excess and large excess.

Czyzowicz et al. [12] showed that it is NP-hard to approximate the special case of SUMFIX$_1$ where $t = 0$ to within any constant c. We extend their approach and obtain a stronger result, namely that it is NP-hard to approximate SUMFIX, for any $a \in (0, \infty)$, to within a factor of $O(n^c)$, for any constant c.

We note that the optimal solution and energy invested in movement may change dramatically with the increase of the required lifetime t. Assume $a = 1$ and consider an instance in which there are $n - 1$ sensors, where $x_i = \frac{i}{n-1}$ and $\rho_i = \frac{1}{2(n-1)}$, for $i \leq n - 1$, and $x_n = \frac{1}{2}$ and $\rho_n = \frac{1}{2}$. If $t = 0$, we can use sensor n to cover the barrier without moving any sensor. However, if t is large enough, it is better to deploy sensor i at $y_i = \frac{2i-1}{2(n-1)}$, for $i \leq n - 1$, and cover the barrier without the help of sensor n. In this case the optimal value is $\frac{1}{2} + \frac{t}{2^\alpha (n-1)^\alpha}$.

Next, we show that non-swapping does not hold in general for non-uniform instances. The proof was omitted for lack of space.

Lemma 8. *There are* SUMFIX *instances in which an optimal solution must be swapping, for any* $a > 0$ *and* $\alpha \geq 1$. *Moreover, the ratio between the value of best non-swapping solution and the optimum is* $\Omega(n)$.

Czyzowicz et al. [12] proved that the special case of SumFix in which $t = 0$ cannot be approximated within any constant. We show their approach can be used for a stronger result, namely that it is NP-hard to approximate SumFix, for any $a \in (0, \infty)$, to within a factor of $O(n^c)$, for any constant c. Our reduction is very similar to the reduction from [12].

Theorem 4. SumFix *cannot be approximated to within a factor of* $O(n^c)$, *for any constant* c, *for every* $a \in (0, \infty)$ *and* $\alpha \geq 1$, *unless P=NP.*

6 Open Problems

We briefly mention some research directions and open problems. An obvious open question is to come up with an approximation algorithm or a lower bound for MaxFix. Another research direction is to consider a model in which sensors are allowed to move and to change their covering radii at any given time. In another natural extension, sensors are located on a barrier and are required to cover a region (e.g., sensors on a coastline covering the sea). In the dual model, sensors could be located anywhere in the plane and are asked to cover a boundary (e.g., sensors in the sea covering the coastline). In an even more general model, a sensor network is required to cover a region in the plane and the initial locations of the sensors are anywhere in the plane.

References

1. Agnetis, A., Grande, E., Mirchandani, P.B., Pacifici, A.: Covering a line segment with variable radius discs. Computers & OR **36**(5), 1423–1436 (2009)
2. Anastasi, G., Conti, M., Di Francesco, M., Passarella, A.: Energy conservation in wireless sensor networks: A survey. Ad Hoc Networks **7**(3), 537–568 (2009)
3. Bar-Noy, A., Baumer, B.: Maximizing Network Lifetime on the Line with Adjustable Sensing Ranges. In: Erlebach, T., Nikoletseas, S., Orponen, P. (eds.) ALGOSENSORS 2011. LNCS, vol. 7111, pp. 28–41. Springer, Heidelberg (2012)
4. Bar-Noy, A., Baumer, B., Rawitz, D.: Changing of the Guards: Strip Cover with Duty Cycling. In: Even, G., Halldórsson, M.M. (eds.) SIROCCO 2012. LNCS, vol. 7355, pp. 36–47. Springer, Heidelberg (2012)
5. Bar-Noy, A., Baumer, B., Rawitz, D.: Set it and forget it: Approximating the set once strip cover problem. Tech. Rep. 1204.1082, CoRR (2012)
6. Bar-Noy, A., Brown, T., Johnson, M.P., Liu, O.: Cheap or Flexible Sensor Coverage. In: Krishnamachari, B., Suri, S., Heinzelman, W., Mitra, U. (eds.) DCOSS 2009. LNCS, vol. 5516, pp. 245–258. Springer, Heidelberg (2009)
7. Bar-Noy, A., Rawitz, D., Terlecky, P.: Maximizing Barrier Coverage Lifetime with Mobile Sensors. In: Bodlaender, H.L., Italiano, G.F. (eds.) ESA 2013. LNCS, vol. 8125, pp. 97–108. Springer, Heidelberg (2013)
8. Bhattacharya, B.K., Burmester, M., Hu, Y., Kranakis, E., Shi, Q., Wiese, A.: Optimal movement of mobile sensors for barrier coverage of a planar region. Theor. Comput. Sci. **410**(52), 5515–5528 (2009)

9. Buchsbaum, A.L., Efrat, A., Jain, S., Venkatasubramanian, S., Yi, K.: Restricted strip covering and the sensor cover problem. In: SODA, pp. 1056–1063 (2007)
10. Chen, D.Z., Gu, Y., Li, J., Wang, H.: Algorithms on Minimizing the Maximum Sensor Movement for Barrier Coverage of a Linear Domain. In: Fomin, F.V., Kaski, P. (eds.) SWAT 2012. LNCS, vol. 7357, pp. 177–188. Springer, Heidelberg (2012)
11. Czyzowicz, J., et al.: On Minimizing the Maximum Sensor Movement for Barrier Coverage of a Line Segment. In: Ruiz, P.M., Garcia-Luna-Aceves, J.J. (eds.) ADHOC-NOW 2009. LNCS, vol. 5793, pp. 194–212. Springer, Heidelberg (2009)
12. Czyzowicz, J., Kranakis, E., Krizanc, D., Lambadaris, I., Narayanan, L., Opatrny, J., Stacho, L., Urrutia, J., Yazdani, M.: On Minimizing the Sum of Sensor Movements for Barrier Coverage of a Line Segment. In: Nikolaidis, I., Wu, K. (eds.) ADHOC-NOW 2010. LNCS, vol. 6288, pp. 29–42. Springer, Heidelberg (2010)
13. Demaine, E.D., Hajiaghayi, M.T., Mahini, H., Sayedi-Roshkhar, A.S., Gharan, S.O., Zadimoghaddam, M.: Minimizing movement. ACM Transactions on Algorithms 5(3) (2009)
14. Dobrev, S., Durocher, S., Eftekhari, M., Georgiou, K., Kranakis, E., Krizanc, D., Narayanan, L., Opatrny, J., Shende, S., Urrutia, J.: Complexity of Barrier Coverage with Relocatable Sensors in the Plane. In: Spirakis, P.G., Serna, M. (eds.) CIAC 2013. LNCS, vol. 7878, pp. 170–182. Springer, Heidelberg (2013)
15. Fan, H., Li, M., Sun, X., Wan, P., Zhao, Y.: Barrier coverage by sensors with adjustable ranges. ACM Transactions on Sensor Networks 11(1) (2014)
16. Gibson, M., Varadarajan, K.: Decomposing coverings and the planar sensor cover problem. In: FOCS, pp. 159–168 (2009)
17. Lev-Tov, N., Peleg, D.: Polynomial time approximation schemes for base station coverage with minimum total radii. Computer Networks 47(4), 489–501 (2005)
18. Mehrandish, M., Narayanan, L., Opatrny, J.: Minimizing the number of sensors moved on line barriers. In: WCNC, pp. 653–658 (2011)
19. Phelan, B., Terlecky, P., Bar-Noy, A., Brown, T., Rawitz, D.: Should I stay or should I go? Maximizing lifetime with relays. In: 8th DCOSS, pp. 1–8 (2012)
20. Tan, X., Wu, G.: New Algorithms for Barrier Coverage with Mobile Sensors. In: Lee, D.-T., Chen, D.Z., Ying, S. (eds.) FAW 2010. LNCS, vol. 6213, pp. 327–338. Springer, Heidelberg (2010)

A Refined Complexity Analysis of Finding the Most Vital Edges for Undirected Shortest Paths

Cristina Bazgan[1,2], André Nichterlein[3]([✉]), and Rolf Niedermeier[3]

[1] PSL, Université Paris-Dauphine, LAMSADE UMR CNRS, 7243 Paris, France
bazgan@lamsade.dauphine.fr
[2] Institut Universitaire de France, Paris, France
[3] Institut für Softwaretechnik und Theoretische Informatik, TU Berlin, Germany
{andre.nichterlein,rolf.niedermeier}@tu-berlin.de

Abstract. We study the NP-hard SHORTEST PATH MOST VITAL EDGES problem arising in the context of analyzing network robustness. For an undirected graph with positive integer edge lengths and two designated vertices s and t, the goal is to delete as few edges as possible in order to increase the length of the (new) shortest st-path as much as possible. This scenario has been mostly studied from the viewpoint of approximation algorithms and heuristics, while we particularly introduce a parameterized and multivariate point of view. We derive refined tractability as well as hardness results, and identify numerous directions for future research. Among other things, we show that increasing the shortest path length by at least one is much easier than to increase it by at least two.

1 Introduction

SHORTEST PATHS, that is, given two distinguished vertices s and t in a graph with edge lengths with the task to find a shortest st-path, is arguably one of the most basic graph problems. We study the undirected case with positive integer edge lengths in the context of "most vital edges" or (equivalently) "interdiction" or "edge blocker" problems. That is, we are interested in the scenario where the goal is to delete (few) edges such that in the resulting graph the shortest st-path gets (much) longer. This is motivated by obvious applications in investigating robustness and critical infrastructure in the context of network design. Our results provide new insights with respect to classical, parameterized, and approximation complexity of this fundamental edge deletion problem which is known to be NP-hard in general [1,16].

The central decision problem we study is defined as follows.

SHORTEST PATH MOST VITAL EDGES (SP-MVE)
Input: An undirected graph $G = (V, E)$ with positive edge lengths $\tau \colon E \to \mathbb{N}$, two vertices $s, t \in V$, and integers $k, \ell \in \mathbb{N}$.
Question: Is there an edge subset $S \subseteq E$, $|S| \le k$, such that the length of a shortest st-path in $G - S$ is at least ℓ?

Work started during a visit (March 2014) of the second and third author at Université Paris-Dauphine.

V.Th. Paschos and P. Widmayer (Eds.): CIAC 2015, LNCS 9079, pp. 47–60, 2015.
DOI: 10.1007/978-3-319-18173-8_3

We set $b := \ell - \text{dist}_G(s,t)$ to be the number by how much the length of every shortest st-path needs to be increased. If all edges have length one, then we say that the graph has unit-length edges. Naturally, SP-MVE comes along with two optimization versions: either delete as few edges as possible in order to achieve a length increase of at least b (called MIN-COST SP-MVE) or getting maximum length increase under the constraint that k edges can be deleted (called MAX-LENGTH SP-MVE). For an instance of SP-MVE or MAX-LENGTH SP-MVE we assume that k is smaller than the size of any st-cut in the input graph. Otherwise, removing all edges of a minimum-size st-cut (which is polynomial-time computable) would lead to a solution disconnecting s and t.

Related Work. Due to the immediate practical relevance, there are numerous studies on "most vital edges (and vertices)" and related problems. We focus on shortest paths here, while there are also studies for problems such as MINIMUM SPANNING TREE [2,3,10,14,19] or MAXIMUM FLOW [14,24,26], to mention only two. With respect to shortest path computation, the following is known. First, we mention in passing that a (more general) result of Fulkerson and Harding [11] implies that allowing the subdivision of edges instead of edge deletions as modification operation makes the problem polynomial-time solvable. Notably, it also has been studied to find *one* most vital edge of a shortest path; this can be solved in almost linear time [21].

Bar-Noy et al. [1] showed that SP-MVE is NP-complete and also corrected some errors concerning algorithmic results from earlier work [20]. Khachiyan et al. [16] derived polynomial-time constant-factor inapproximability results for both optimization versions. For the case of directed graphs, Israeli and Wood [15] provided heuristic solutions based on mixed-integer programming together with experimental results. Pan and Schild [24] studied the restriction of the directed case to planar graphs and again obtained NP-hardness results.

Finally, we note that, while most algorithmic studies focussed on polynomial-time solvable special cases or polynomial-time approximability, there seem to be almost no studies concerning multivariate complexity aspects [8,23] of "most vital edges" ("edge interdiction") problems. We are only aware of the work of Guo and Shrestha [14] who performed a parameterized complexity analysis for minimum spanning tree, maximum matching, and maximum flow problems. They focus on standard parameterization by solution size, that is, the budget for the number of edge deletions, and derive several fixed-parameter tractability as well as parameterized hardness results.

Our Results. We perform an extensive study of multivariate complexity aspects of SP-MVE. More specifically, we perform a refined complexity analysis in terms of how certain problem-specific parameters influence the computational complexity of SP-MVE and its optimization variants. The parameters we study include aspects of graph structure as well as special restrictions on the problem parameters itself. Moreover, we also report a few findings on (parameterized) approximability. Let us feature two main conclusions from our work: First, harming the network significantly (that is, $b \geq 2$) is NP-hard while harming it only a little bit

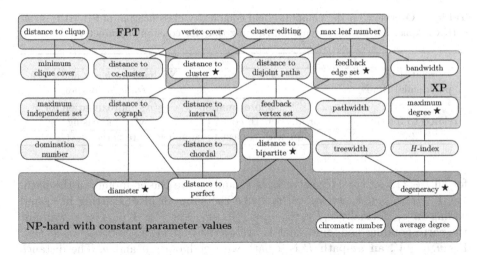

Fig. 1. The parameterized complexity of SP-MVE with unit-length edges with respect to different structural parameters. Herein, "distance to X" is the number of vertices that have to be deleted in order to transform the input graph into a graph from the graph class X. For two parameters that are connected by a line, the upper parameter is weaker (that is, larger) than the parameter below [17]. Refer to Sorge and Weller [25] for the formal definitions of the parameters. Results marked by a ★ are obtained in this work. XP means polynomial running time for constant parameter value.

(that is, $b = 1$) is doable in polynomial time. Second, while the parameterized complexity with respect to the standard parameter number k of edge deletions remains open, the cluster vertex deletion number, advocated by Doucha and Kratochvíl [6] as a parameterization between vertex cover number and clique-width, currently is our most interesting parameter that yields fixed-parameter tractability. Figure 1 surveys our current knowledge of the parameterized complexity of SP-MVE with respect to a number of well-known structural graph parameters, identifying numerous open questions. Moreover, towards the goal of spotting further fixed-parameter tractable special cases it also implicitly suggests to look for reasonable parameter combinations. To this end, a data-driven analysis of real-world parameter values would be valuable. In addition, Table 1 overviews our exact and approximate complexity results for SP-MVE.

Organization of the Paper. After introducing some preliminaries in Section 2, we prove in Section 3 some NP-hardness results and in Section 4 some polynomial-time solvable special cases. In Section 5 we provide fixed-parameter and approximation algorithms for SP-MVE. Conclusions and open questions are provided in Section 6. Due to the lack of space, several details are omitted.

2   Preliminaries

For an undirected graph $G = (V, E)$ we set $|V| := n$ and $|E| := m$. A path P of length $r - 1$ in G is a sequence of vertices $P = v_1\text{-}v_2\text{-}\ldots\text{-}v_r$ with $\{v_i, v_{i+1}\} \in E$

Table 1. Overview on the computational complexity classification of SP-MVE on n-vertex graphs. XP means polynomial running time for constant parameter value.

	k	ℓ
related to polynomial time	XP	NP-hard for $b = 2$ and $\ell = 9$ ℓ-approximation
related to fpt time	$n/2^{O(\sqrt{\log n})}$-approximation for unit-length edges	$r(n)$-approximation for every increasing r
2-3	fpt with respect to combined parameter (k, ℓ)	

for all $i \in \{1, \ldots, r-1\}$; the vertices v_1 and v_n are the endpoints of the path. For $1 \leq i < j \leq r$, we set $v_i P v_j$ to be the subpath of P starting in v_i and ending in v_j, formally $v_i P v_j := v_i\text{-}v_{i+1}\text{-}\ldots\text{-}v_j$. For $i = 1$ or $j = r$ we omit the corresponding endpoint, that is, we set $Pv_j := v_1 P v_j$ and $v_i P := v_1 P v_j$. For $u, v \in V$, an uv-path P is a path with endpoints u and v. The distance between u and v in G, denoted by $\text{dist}_G(u, v)$, is the length of a shortest uv-path. The diameter of G is the length of the longest shortest path in G.

For each vertex $v \in V$ we denote by $N_G(v)$ the set of neighbors of v and $N_G[v] = N_G(v) \cup \{v\}$ denotes v's the closed neighborhood. Two vertices $u, v \in V$ are called *true twins* if $N_G[u] = N_G[v]$ and *false twins* if $N_G(u) = N_G(v)$; they are called *twins* if they are either true or false twins. We denote by $G - S$ the graph obtained from G by removing the edge subset $S \subseteq E$. For $s, t \in V$, an edge subset S is called st-cut if $G - S$ contains no st-path. For $V' \subseteq V$ we denote by $G[V']$ the subgraph induced by V'. For $E' \subseteq V$ we denote by $G[E']$ the subgraph consisting of all endpoints of edges in E' and the edges in E'.

An edge set $F \subseteq E$ is a *feedback edge set* of G if $G - F$ is a tree or a forest. The feedback edge set number of G is the size of a minimum feedback edge set. Graph G is a *cluster graph* if G consists of disjoint cliques. A vertex set $X \subseteq V$ is called cluster vertex deletion set if $G[V \setminus X]$ is a cluster graph. The cluster vertex deletion number is the size of a minimum cluster vertex deletion set.

Parameterized Complexity. A parameterized problem is called *fixed-parameter tractable* (fpt) if there is an algorithm that decides any instance (I, k), consisting of the "classical" instance I and a parameter $k \in \mathbb{N}_0$, in $f(k) \cdot |I|^{O(1)}$ time, for some computable function f solely depending on k.

A core tool in the development of fixed-parameter algorithms is polynomial-time preprocessing by data reduction, called *kernelization* [13,18]. Here, the goal is to transform a given problem instance (I, k) in polynomial time into an equivalent instance (I', k') whose size is upper-bounded by a function of k. That is, (I, k) is a yes-instance if and only if (I', k'), $k' \leq g(k)$, and $|I'| \leq g(k)$ for some function g. Thus, such a transformation is a polynomial-time self-reduction with the constraint that the reduced instance is "small" (measured by $g(k)$). If such a transformation exists, then I' is called *kernel* of size $g(k)$. We refer to the monographs [7,9,22] for more details on parameterized complexity.

Approximation. Given an NP optimization problem and an instance I of this problem, we use $|I|$ to denote the size of I, opt(I) to denote the optimum value of I, and val(I, S) to denote the value of a feasible solution S of instance I. The *performance ratio* of S (or *approximation factor*) is $r(I, S) = \max \left\{ \frac{\text{val}(I,S)}{\text{opt}(I)}, \frac{\text{opt}(I)}{\text{val}(I,S)} \right\}$. For a function ρ, an algorithm $\mathcal{A}$ is an $\rho(|I|)$-*approximation*, if for every instance I of the problem, it returns a solution S such that $r(I, S) \leq \rho(|I|)$. If the problem comes with a parameter k and the algorithm $\mathcal{A}$ runs in $f(k) \cdot |I|^{O(1)}$ time, then $\mathcal{A}$ is called *parameterized $\rho(|I|)$-approximation*.

3   NP-Hard Cases

Adapting a reduction idea due to Khachiyan et al. [16] for the vertex deletion variant of SP-MVE, we prove that SP-MVE is NP-hard even for constant values of b, ℓ, and graph diameter.

Theorem 1. *SP-MVE is NP-hard, even for unit-length edges, $b = 2$, $\ell = 9$, and diameter 8.*

Proof. As Khachiyan et al. [16], we reduce from VERTEX COVER on three-partite graphs which remains NP-hard [12, GT1]. While the fundamental approach remains the same, the technical details when moving their vertex deletion scenario to our edge deletion scenario change to quite some extent. We refrain from a step-by-step comparison. Given a VERTEX COVER instance (G, h) with $G = (V_1 \cup V_2 \cup V_3, E)$ being a tripartite graph on n vertices, we construct an SP-MVE instance I' as follows. First, we set $k := h$ and $\ell := 9$. The graph $G' = (V', E')$ contains vertices $V' = V_1 \cup V_2 \cup V_3 \cup V_2' \cup \{s, t\}$, where s and t are two new vertices, and for each $v \in V_2$ we add a copy $v' \in V_2'$.

Before describing the edge set E', we introduce edge-gadgets. Here, by adding a length α *edge-gadget*, $\alpha \geq 2$, from the vertex u to vertex v, we mean to add n vertex-disjoint paths of length $\alpha - 2$ and to make u adjacent to the first vertex of each path and v adjacent to the last vertex of each path. If $\alpha = 2$, then each path is just a single vertex which is at the same time the first and last vertex. The idea behind this is that we will never delete edges in an edge-gadget.

We add the following edges and edge-gadgets to G' (see Figure 2 for a schematic representation of the constructed graph). For each vertex $v \in V_2$ we add the edge $\{v, v'\}$ of length one. We add edges of length one between s and every vertex $v \in V_1$ and between t and every vertex $v \in V_3$. We also add the following edge-gadgets: For each edge $\{u, v\} \in (V_1 \times V_2) \cap E$ we add the edge-gadget $e_{u,v}$ of length two, for each edge $\{u, v\} \in (V_2 \times V_3) \cap E$ we add the edge-gadget $e_{u',v}$ of length two, and for each edge $\{u, v\} \in (V_1 \times V_3) \cap E$ we add the edge-gadget $e_{u,v}$ of length five. Furthermore, we add edge-gadgets of length four between s and every vertex $v \in V_2$ and between t and every vertex $v' \in V_2'$. Observe that we have $\text{dist}_{G'}(s, t) = 7$ and thus $b = \ell - \text{dist}_G(s, t) = 2$.

We now show that G has a vertex cover of size at most h if and only if deleting $k = h$ edges in G' results in s and t having distance at least $\ell = 9$.

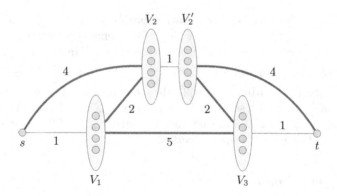

Fig. 2. A schematic representation of the graph G' constructed from the tripartite graph $G = (V_1 \cup V_2 \cup V_3, E)$. The vertices are grouped to the used sets. Each edge in G' is represented by a length-one edge in the picture. A bold edge indicates an edge-gadget and the corresponding number denotes its length.

"$\Rightarrow$:" Let $V'' \subseteq V$ be a vertex cover of size at most h in G. It is not hard to verify that for the set $E'' = \{\{s,v\} : v \in V_1 \cap V''\} \cup \{\{v,v'\} : v \in V_2 \cap V''\} \cup \{\{v,t\} : v \in V_3 \cap V''\}$ it holds that $\text{dist}_{G'-E''}(s,t) = 9$ and $|E''| = |V''| \leq h$.

"$\Leftarrow$:" Let $E'' \subseteq E'$ be a set of edges such that $\text{dist}_{G'-E''}(s,t) \geq 9$ and $|E''| \leq h$. If E'' contains edges from an edge-gadget $e_{u,v}$, then it must contain at least n edges from this gadget in order to have a chance to increase the solution value. Therefore, since $h < n$, we can assume that E'' does not contain any edge contained in an edge-gadget. Thus $E'' \subseteq (\{s\} \times V_1) \cup (V_2 \times V_2') \cup (V_3 \times \{t\})$. We construct a vertex cover V'' for G as follows: For each edge $\{s,v\} \in E''$ it follows that $v \in V_1$ and we add v to V''. Similarly, for each edge $\{v,t\} \in E''$ it follows that $v \in V_3$ and we add v to V''. Finally, for each edge $\{v,v'\} \in E'' \cap (V_2 \times V_2')$, we add v to V''.

Suppose towards a contradiction, that V'' is not a vertex cover in G, that is, there exists an edge $\{u,v\} \in E$ with $u,v \notin V''$. If $v \in V_1$ and $u \in V_2$, then the st-path s-v-u-u'-t of length $8 < \ell$ is contained in $G - E''$. If $v \in V_1$ and $u \in V_3$, then the st-path s-v-u-t of length $7 < \ell$ is contained in $G - E''$. Finally, if $v \in V_2$ and $u \in V_3$, then the st-path s-v-v'-u-t of length $8 < \ell$ is contained in $G - E''$. Each of the three cases contradicts the assumption that $\text{dist}_{G'-E''}(s,t) \geq 9$. □

When allowing length zero on edges, Khachiyan et al. [16] stated that it is NP-hard to approximate MAX-LENGTH SP-MVE within a factor smaller than two. We consider in this paper only positive edge lengths and, by adapting the construction given in the above proof by considering edge-gadgets of lengths polynomial in n (with high degree), we obtain the following.

Theorem 2. MAX-LENGTH SP-MVE *is not* $4/3 - 1/poly(n)$-*approximable in polynomial time, even for unit-length edges, unless $P = NP$.*

Concerning special graph classes, we can show that the problem remains NP-hard on restricted bipartite graphs. To this end, a graph G has *degeneracy* d if every subgraph of G contains a vertex of degree at most d. By subdividing every edge, we obtain the following.

Theorem 3. *SP-MVE is NP-hard, even for bipartite graphs with degeneracy two, unit-length edges, $b = 4$, $\ell = 18$, and diameter 8.*

Proof. We provide a self-reduction from SP-MVE with unit-length edges with $b = 2$, $\ell = 9$, and diameter 8. Let $I = (G = (V, E), k, \ell, s, t)$ be the given SP-MVE instance. We construct the instance $I' = (G', k, 2\ell, s, t)$ where G' is obtained from G by subdividing all edges, that is, each edge is replaced by a path of length two. The correctness of the reduction is easy to see as any minimal solution contains at most one edge of each induced path. Clearly, I' can be constructed in polynomial time. Furthermore, G' is bipartite and has degeneracy two.   □

4 Polynomial-Time Algorithms

In this section, we discuss two polynomial-time algorithms for special cases of SP-MVE. First, we consider bounded-degree graphs. Here, the basic observation is that the maximum degree Δ of the graph upper-bounds the solution size: a budget of Δ would allow to disconnect s from t by deleting all edges incident to s. Hence, the simple brute-force algorithm trying all possibilities to delete at most $\Delta - 1$ edges yields a polynomial-time algorithm in graphs with constant maximum degree.

Proposition 1. *SP-MVE can be solved in $O(m^{\Delta-1}(m + n \log n))$.*

The question whether one can replace $m^{\Delta-1}$ by $f(\Delta) \cdot m^{O(1)}$ for some function f, that is, whether SP-MVE is fixed-parameter tractable with respect to Δ, remains open.

The second, more interesting special case we consider is when we only want to increase the distance between s and t by one. In this case, we can exploit the connection between SP-MVE and minimum st-cuts. Observe that a minimum st-cut in the undirected graph can be larger than the edge set we are looking for, see left-hand side of Figure 3 for an example. Instead, the idea is to direct the edges from s to t (see right-hand side of Figure 3) and then search an st-cut; this only works if $b = 1$, as also witnessed by the NP-hardness for the case $b = 2$ (Theorem 1).

Theorem 4. *SP-MVE can be solved in $O(nm)$ time when $b = 1$.*

Proof. Let $I = (G = (V, E), k, \ell, s, t, \tau)$ be an instance of SP-MVE with $\ell = \text{dist}_G(s, t) + 1$, that is, $b = 1$. The task is to delete at least one edge in each shortest st-path. To achieve this goal, our algorithm works in three main steps:

1. Using an adaption of Dijkstra's algorithm, compute the subgraph $G' = (V', E')$ of the original graph containing *all* shortest path edges.
2. Direct all edges in these subgraphs from s to t (so that the tail of the each edge is closer to s than to t), obtaining a directed graph $D = (V', A)$.

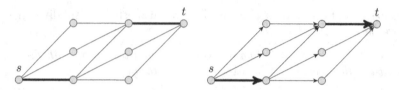

Fig. 3. The left-hand side shows a graph consisting of five st-paths of length three each. Removing the two bold edges increases the length of a shortest st-path to five. A minimum st-cut has size three, showing the difference between our vital edges scenario and minimum edge cuts. The right-hand side shows a digraph obtained from the graph on the left-hand side by directing all edges towards t. In this graph, the minimum st-cut has also size two. The proof of Theorem 4 shows that for $b = 1$ we can reduce SP-MVE to finding a minimum st-cut in a directed graph with unit arc-weights.

3. Compute a minimum st-cut in D and decide accordingly (if it contains more than k arcs, then there is no solution, otherwise the cut arcs form a solution).

We first discuss the correctness and then show that the algorithm runs indeed in $O(nm)$ time. To this end, we introduce the following notation. For an arc subset $A' \subseteq A$ we denote by $E(A')$ the underlying undirected edges of A'; thus $E(A') \subseteq E' \subseteq E$. For the correctness, we first show that for every st-cut $A' \subseteq A$ in D it holds that every shortest st-path in G contains an edge in $E(A')$: Suppose towards a contradiction that there is a shortest st-path P not containing an edge in $E(A')$. Recall that we require the edge lengths to be positive. Hence, for every edge $\{u, v\}$ on the path P it holds that $\text{dist}_G(s, u) \neq \text{dist}_G(s, v)$ since P is a shortest path. Assume without loss of generality that $\text{dist}_G(s, u) < \text{dist}_G(s, v)$. By construction of D, this implies that $(u, v) \in A$. Thus, P is also an st-path in D containing no arc from A'; a contradiction to the assumption that A' is an st-cut.

Conversely, let $S \subseteq E'$ be an edge set containing at least one edge of every shortest st-path in G. Then, the arc set $A' \subseteq A$ with $E(A') = S$ is an st-cut in D: Suppose towards a contradiction that A' is not an st-cut in D. Thus, there exists an st-path P in $D - A'$. Hence, P is also an st-path in $G' - S$. We next show that P is indeed a shortest st-path in G'. Suppose towards a contradiction that there exists a vertex $v \in V'$ that is the first vertex on P such that Pv (the path P until vertex v) is not a shortest sv-path. Clearly, $s \neq v$. Let $u \in V'$ be the predecessor of v on P, that is, $Pv = Pu\text{-}v$ (allowing $s = u$). By construction of G', each edge in E' is contained in at least one shortest st-path. Hence, also the edge $\{u, v\}$ is contained in some shortest st-path P'. This implies that $P'v$ is a shortest sv-path in G'. Since Pu is a shortest su-path, it follows that Pu and $P'u$ have same length. This implies that also $Pv = Pu\text{-}v$ and $P'v = P'u\text{-}v$ have same length. Hence, Pv is a shortest sv-path in G'; a contradiction. Thus, P is a shortest st-path in G'. Since P is also contained in $G' - S$, this contradicts the assumption that S contains an edge of each shortest st-path in G.

Summarizing, we showed that a set $A' \subseteq A$ is an st-cut in D if and only if $E(A')$ contains an edge of each shortest st-path in G. Hence, any minimum-size st-cut computed in Step 3 induces a minimum-size edge set S in G such that $\text{dist}_{G-S}(s,t)$ $> \text{dist}_G(s,t)$. This completes the correctness proof.

For the running time we show that Steps 1 and 2 can be performed in $O(n \log n + m)$ time: First, run in $O(n \log n + m)$ time two times Dijkstra's algorithm to compute for each vertex its distance to s and to t. Then, an edge $\{u,v\}$ is contained in a shortest st-path if either $\text{dist}_G(s,u) + \tau(\{u,v\}) + \text{dist}_G(v,t) = \text{dist}_G(s,t)$ (Case 1) or $\text{dist}_G(s,v) + \tau(\{u,v\}) + \text{dist}_G(u,t) = \text{dist}_G(s,t)$ (Case 2). Furthermore, in Case 1, the edge will be directed from u to v and in Case 2 from v to u. This can be done in $O(m)$ time by iterating over all edges. Using the Ford-Fulkerson-algorithm to compute the minimum cut, Step 3 can be performed in D in $O(nm)$ time. Overall, this gives a running time of $O(nm)$. □

5   Algorithms for NP-Hard Cases

In this section, we present fixed-parameter and approximation algorithms. As a warm-up, we show that SP-MVE is fixed-parameter tractable when combining the parameters number k of removed edges and minimum st-path length ℓ to be achieved.

Proposition 2. *SP-MVE can be solved in $O((\ell-1)^k \cdot (n \log n + m))$ time.*

Proof. We employ a simple depth-bounded search tree. The basic idea is to search for a shortest st-path and to "destroy" it by deleting one of the edges (trying all possibilities). This is repeated until every shortest st-path has length at least ℓ. For each such shortest path, we branch into at most $\ell-1$ possibilities to delete one of its edges, and the depth of the corresponding search tree is at most k (our "deletion budget") since otherwise we cannot find a solution with at most k edge deletions. The correctness is obvious. Hence, we arrive at a search tree of size at most $(\ell-1)^k$ where in each step we need to compute a shortest path. Using Dijkstra's algorithm, this can be done in $O(n \log n + m)$ time. The overall running time is thus $O((\ell-1)^k \cdot (n \log n + m))$. □

Using the search tree described in the proof of Proposition 2 to destroy all paths of length at most $2^{O(\sqrt{\log n})}$ yields the following.

Corollary 1. MAX-LENGTH SP-MVE *with unit-length edges is parameterized $n/2^{O(\sqrt{\log n})}$-approximable with respect to the parameter k.*

By deleting every edge on too short st-paths, we obtain an ℓ-approximation.

Proposition 3. MIN-COST SP-MVE *is polynomial-time ℓ-approximable.*

Combining the previous approximation algorithm with a tradeoff between running time and approximation factor [4, Lemma 2], we obtain the following.

Corollary 2. *For every increasing function r MIN-COST SP-MVE is parameterized $r(n)$-approximable with respect to the parameter ℓ.*

Parameter Feedback Edge set Number. We next provide a linear-size problem kernel for SP-MVE parameterized by the feedback edge set number. Recall that an edge set $F \subseteq E$ is called *feedback edge set* for a graph $G = (V, E)$ if $G - F$ is a tree or a forest. The feedback edge set number of G is the size of a minimum feedback edge set. Computing a spanning tree, one can determine a minimum feedback edge set in linear time. Hence, we assume in the following that we are given a feedback edge set F with $|F| = f$ for our input instance $(G = (V, E), k, \ell, s, t, \tau)$. We start with two simple data reduction rules dealing with degree-one and degree-two vertices.

Rule 1.   Let $(G = (V, E), k, \ell, s, t, \tau)$ be an SP-MVE instance and let $v \in V \setminus \{s, t\}$ be a vertex of degree one. Then, delete v.

The correctness of Rule 1 is obvious as no shortest path uses a degree-one vertex. We can deal with degree-two vertices as follows.

Rule 2.   Let $(G = (V, E), k, \ell, s, t, \tau)$ be an SP-MVE instance and let $v \in V \setminus \{s, t\}$ be a vertex of degree two with $N_G(v) = \{u, w\}$ and $\{u, w\} \notin E$. Then add the edge $\{u, w\}$ with the length $\tau(\{u, w\}) := \tau(\{u, v\}) + \tau(\{v, w\})$ and delete v.

The correctness of Rule 2 follows from the fact that on an induced path at most one edge will be deleted and it does not matter which one will get deleted. Applying both rules exhaustively can clearly be done in polynomial time and leads to the following problem kernel.

Theorem 5. *SP-MVE admits a problem kernel with $5f + 2$ vertices and $6f + 2$ edges.*

Corollary 3. *SP-MVE is fixed-parameter tractable with respect to the parameter feedback edge set number.*

Parameter Cluster Vertex Deletion Number. We now prove that SP-MVE restricted to unit-length edges is fixed-parameter tractable with respect to the parameter cluster vertex deletion number x. Recall that a graph G is a *cluster graph* if it is a union of disjoint cliques. A vertex set $X \subseteq V$ is called cluster vertex deletion set if $G[V \setminus X]$ is a cluster graph. The cluster vertex deletion number is the size of a minimum cluster vertex deletion set.

We assume in the following that for the input instance $(G = (V, E), k, \ell, s, t)$ we are given a cluster vertex deletion set X of size $|X| = x$. If X is not already given, then we can compute X in $O(1.92^x \cdot (n + m))$ time [5]. Our algorithm is based on the observation that twins can be handled equally in a solution. This follows from a more general statement provided in the following lemma. It shows that for any set $T \subseteq V \setminus \{s, t\}$ of vertices that have the same neighborhood in $V \setminus T$, we can assume that we do not delete edges in $G[T]$ and that the vertices in T behave the same, that is, one deletes either all or no edge between a vertex $v \in V \setminus T$ and the vertices in T.

Lemma 1. *Let $G = (V, E)$ be an undirected graph with unit-length edges, let $s, t \in V$ be two vertices, and let $T = \{v_1, \ldots, v_t\} \subseteq V \setminus \{s, t\}$ be a set of vertices such that $N_G(v_1) \setminus T = N_G(v_2) \setminus T = \ldots = N_G(v_t) \setminus T$. Then, for every edge subset $S \subseteq E$, there exists an edge subset $S' \subseteq E$ such that $\text{dist}_{G-S'}(s, t) \geq \text{dist}_{G-S}(s, t)$, $|S'| \leq |S|$, and $N_{G[S']}(v_1) = N_{G[S']}(v_2) = \ldots = N_{G[S']}(v_t)$.*

Theorem 6. *SP-MVE with unit-length edges is linear-time fixed-parameter tractable with respect to the parameter cluster vertex deletion number.*

Proof. Let $(G = (V, E), k, \ell, s, t)$ be the input instance of SP-MVE and let $X \subseteq V$ be a cluster vertex deletion set of size $|X| = x$. Hence, $G - X$ is a cluster graph and the vertex sets $C_1, \ldots, C_r$ form the cliques (clusters) for some $r \in \mathbb{N}$. We set $\mathcal{C} := \{C_1, \ldots, C_r\}$. Assume that there is an SP-MVE solution $S \subseteq E$ of size at most k; otherwise the algorithm will output no as it finds no solution. We describe an algorithm that finds S.

Our algorithm is based on the following observation. Let P be an arbitrary shortest st-path that goes through a clique $C \in \mathcal{C}$ in $G - S$. Then, P contains at most 2^x vertices from C: By Lemma 1, we can assume that the twins in G are still twins in $G - S$. Since P is a shortest path, P does not contain two vertices that are twins. As the vertices in C form a clique, they only differ in how they are connected to vertices in X. Thus, C contains at most 2^x "different" vertices, that is, vertices with pairwise different neighborhoods.

Now, consider two non-adjacent vertices $u, v \in X$. From the above considerations it follows that in $G - S$ an uv-path avoiding the vertices in X has length between one and $2^x + 1$ as it can pass through at most one clique. Our algorithm tries for each vertex pair from X all possibilities for the distance it has in $G - S$ and then tries to realize the current possibility. After the current possibility is realized, the cliques in $\mathcal{C}$ are obsolete and thus the instance size can be bounded in a function of x. More precisely, our algorithm works as follows:

1. Branch into all possibilities to delete edges contained in $G[X]$. Decrease the budget k accordingly.
2. Branch into all possibilities to add for each pair u, v of non-adjacent vertices in X an edge with a length of $\{2, 3, \ldots, 2^x, 2^x + 1, \infty\}$ indicating the length of a shortest path between u and v that does not contain any vertex in X.
3. Delete for each clique containing neither s nor t the *minimum number* of edges to ensure that a shortest path between each pair of vertices in X is completely contained in $G[X]$. Decrease the budget k accordingly.
4. Remove all cliques except the ones that contain s or t. Do *not* change the budget k.
5. Solve the problem on the remaining graph with the remaining budget (that was not spent in Steps 1 and 3).

Note that Step 2 is performed for each possibility in Step 1. Hence, in Steps 1 and 2 at most $2^{x^2} \cdot (2^x + 1)^{x^2}$ possibilities are considered and for each of these possibilities Step 3 is invoked.

In Step 3, the algorithm tries to realize the prediction made in Step 2. To this end, let $C \in \mathcal{C}$ be a clique containing neither s nor t. The algorithm branches into all possibilities to delete edges in $G[C]$ or edges with one endpoint in C and the other endpoint in X. Since $G[C]$ contains at most 2^x different vertices, it follows from Lemma 1 that at most $2^{(2^x)^2 + 2^x \cdot x} = 2^{(4^x) + 2^x \cdot x}$ possibilities need to be considered to delete edges. For each possibility, the algorithm checks in $x^{O(1)}$ time whether all shortest paths between a pair of vertices of X go through C. If yes, then the algorithm discards the currently considered branch; if no, then the current branch is called valid. From all valid branches for C, the algorithm picks the one that deletes the minimum amount of edges and proceeds with the next clique. Observe that since X is a vertex separator for all cliques in $\mathcal{C}$, the algorithm can solve Step 3 for each clique independently of the outcome in the other cliques. Hence, the overall running time for Step 3 is $2^{2^{O(x)}} \cdot n$ as $|\mathcal{C}| \le n$.

As discussed above, the cliques in $\mathcal{C}$ containing neither s nor t are now obsolete as there is always a shortest path avoiding these cliques. Hence, the algorithm removes these cliques (Step 4). This can be done in linear time. The remaining instance consists of the vertices in X and the at most two cliques containing s and t. As the algorithm deleted the edges within $G[X]$ in Step 1, it remains to consider deleting edges within the two cliques or between the two cliques and the vertices in X. Again, by Lemma 1, the algorithm only needs to branch into $2^{2 \cdot (4^x + x \cdot 2^x)}$ possibilities to delete edges and check for each branch whether s and t have distance at least ℓ and the overall budget k is not exceeded. If one branch succeeds, then the algorithm found a solution and returns it. If no branch succeeds, then there exists no solution of size k since the algorithm performed an exhaustive search. Overall, the running time is $2^{2^{O(x)}} \cdot (n + m)$. $\qquad\square$

6 Conclusion

SHORTEST PATH MOST VITAL EDGES (SP-MVE) is a natural edge deletion problem that clearly deserves further study from a parameterized complexity perspective. While we showed that SP-MVE remains NP-hard for even constant values of the parameters b and ℓ relating to the length increase, we left open whether SP-MVE is fixed-parameter tractable with respect to the "standard parameter" k (number of edge deletions). Even fixed-parameter tractability with respect to the combined parameter (b, k) remains open. Figure 1 in the introductory section depicts a wide range of structural graph parameters for which the parameterized complexity status of SP-MVE is unknown. Also concerning the approximation point of view not much is known. There is a huge gap between the known lower and upper bounds of the approximation factor achievable in polynomial time. Further, from a practical point of view it would make sense to extend our studies by restricting the input to planar graphs [24]. Finally, also in terms of parameterized approximability SP-MVE offers a number of interesting challenges, altogether making it an excellent candidate problem for a full-fledged multivariate complexity analysis.

References

1. Bar-Noy, A., Khuller, S., Schieber, B.: The complexity of finding most vital arcs and nodes. Technical report, College Park, MD, USA (1995)
2. Bazgan, C., Toubaline, S., Vanderpooten, D.: Efficient determination of the k most vital edges for the minimum spanning tree problem. Computers and Operations Research **39**(11), 2888–2898 (2012)
3. Bazgan, C., Toubaline, S., Vanderpooten, D.: Critical edges/nodes for the minimum spanning tree problem: complexity and approximation. Journal of Combinatorial Optimization **26**(1), 178–189 (2013)
4. Bazgan, C., Chopin, M., Nichterlein, A., Sikora, F.: Parameterized approximability of maximizing the spread of influence in networks. Journal of Discrete Algorithms **27**, 54–65 (2014)
5. Boral, A., Cygan, M., Kociumaka, T., Pilipczuk, M.: A Fast Branching Algorithm for Cluster Vertex Deletion. In: Hirsch, E.A., Kuznetsov, S.O., Pin, J.É., Vereshchagin, N.K. (eds.) CSR 2014. LNCS, vol. 8476, pp. 111–124. Springer, Heidelberg (2014)
6. Doucha, M., Kratochvíl, J.: Cluster Vertex Deletion: A Parameterization between Vertex Cover and Clique-Width. In: Rovan, B., Sassone, V., Widmayer, P. (eds.) MFCS 2012. LNCS, vol. 7464, pp. 348–359. Springer, Heidelberg (2012)
7. Downey, R.G., Fellows, M.R.: Fundamentals of Parameterized Complexity. Springer (2013)
8. Fellows, M.R., Jansen, B.M.P., Rosamond, F.A.: Towards fully multivariate algorithmics: Parameter ecology and the deconstruction of computational complexity. European Journal of Combinatorics **34**(3), 541–566 (2013)
9. Flum, J., Grohe, M.: Parameterized Complexity Theory. Springer (2006)
10. Frederickson, G.N., Solis-Oba, R.: Increasing the weight of minimum spanning trees. In: Proc. 7th SODA, pp. 539–546 (1996)
11. Fulkerson, D., Harding, G.C.: Maximizing the minimum source-sink path subject to a budget constraint. Mathematical Programming **13**, 116–118 (1977)
12. Garey, M.R., Johnson, D.S.: Computers and Intractability: A Guide to the Theory of NP-Completeness. Freeman (1979)
13. Guo, J., Niedermeier, R.: Invitation to data reduction and problem kernelization. SIGACT News **38**(1), 31–45 (2007)
14. Guo, J., Shrestha, Y.R.: Parameterized Complexity of Edge Interdiction Problems. In: Cai, Z., Zelikovsky, A., Bourgeois, A. (eds.) COCOON 2014. LNCS, vol. 8591, pp. 166–178. Springer, Heidelberg (2014)
15. Israeli, E., Wood, R.K.: Shortest-path network interdiction. Networks **40**(2), 97–111 (2002)
16. Khachiyan, L., Boros, E., Borys, K., Elbassioni, K.M., Gurvich, V., Rudolf, G., Zhao, J.: On short paths interdiction problems: Total and node-wise limited interdiction. Theory of Computing Systems **43**(2), 204–233 (2008)
17. Komusiewicz, C., Niedermeier, R.: New Races in Parameterized Algorithmics. In: Rovan, B., Sassone, V., Widmayer, P. (eds.) MFCS 2012. LNCS, vol. 7464, pp. 19–30. Springer, Heidelberg (2012)
18. Kratsch, S.: Recent developments in kernelization: A survey. Bulletin of the EATCS **113**, 58–97 (2014)
19. Liang, W.: Finding the k most vital edges with respect to minimum spanning trees for fixed k. Discrete Applied Mathematics **113**(2–3), 319–327 (2001)

20. Malik, K., Mittal, A., Gupta, S.K.: The k most vital arcs in the shortest path problem. Operations Research Letters **8**, 223–227 (1989)
21. Nardelli, E., Proietti, G., Widmayer, P.: A faster computation of the most vital edge of a shortest path. Information Processing Letters **79**(2), 81–85 (2001)
22. Niedermeier, R.: Invitation to Fixed-Parameter Algorithms. Oxford University Press (2006)
23. Niedermeier, R.: Reflections on multivariate algorithmics and problem parameterization. In: Proc. 27th STACS, vol. 5. LIPIcs, pp. 17–32. Schloss Dagstuhl-Leibniz-Zentrum für Informatik (2010)
24. Pan, F., Schild, A.: Interdiction Problems on Planar Graphs. In: Raghavendra, P., Raskhodnikova, S., Jansen, K., Rolim, J.D.P. (eds.) RANDOM 2013 and APPROX 2013. LNCS, vol. 8096, pp. 317–331. Springer, Heidelberg (2013)
25. Sorge, M., Weller, M.: The graph parameter hierarchy. Manuscript (2013). http://fpt.akt.tu-berlin.de/msorge/parameter-hierarchy.pdf
26. Wood, R.K.: Deterministic network interdiction. Mathematical and Computer Modeling **17**(2), 1–18 (1993)

Orthogonal Graph Drawing with Inflexible Edges

Thomas Bläsius$^{(\boxtimes)}$, Sebastian Lehmann, and Ignaz Rutter

Faculty of Informatics, Karlsruhe Institute of Technology (KIT), Karlsruhe, Germany
{thomas.blasius,sebastian.lehmann,ignaz.rutter}@kit.edu

Abstract. We consider the problem of creating plane orthogonal drawings of *4-planar graphs* (planar graphs with maximum degree 4) with constraints on the number of bends per edge. More precisely, we have a *flexibility function* assigning to each edge e a natural number flex(e), its *flexibility*. The problem FLEXDRAW asks whether there exists an orthogonal drawing such that each edge e has at most flex(e) bends. It is known that FLEXDRAW is NP-hard if flex(e) = 0 for every edge e [7]. On the other hand, FLEXDRAW can be solved efficiently if flex(e) ≥ 1 [2] and is trivial if flex(e) ≥ 2 [1] for every edge e.

To close the gap between the NP-hardness for flex(e) = 0 and the efficient algorithm for flex(e) ≥ 1, we investigate the computational complexity of FLEXDRAW in case only few edges are *inflexible* (i.e., have flexibility 0). We show that for any $\varepsilon > 0$ FLEXDRAW is NP-complete for instances with $O(n^{\varepsilon})$ inflexible edges with pairwise distance $\Omega(n^{1-\varepsilon})$ (including the case where they induce a matching). On the other hand, we give an FPT-algorithm with running time $O(2^k \cdot n \cdot T_{\text{flow}}(n))$, where $T_{\text{flow}}(n)$ is the time necessary to compute a maximum flow in a planar flow network with multiple sources and sinks, and k is the number of inflexible edges having at least one endpoint of degree 4.

1 Introduction

Bend minimization in orthogonal drawings is a classical problem in the field of graph drawing. We consider the following problem called OPTIMALFLEXDRAW. The input is a 4-planar graph G (from now on all graphs are 4-planar) together with a cost function $\text{cost}_e \colon \mathbb{N} \to \mathbb{R} \cup \{\infty\}$ assigned to each edge. We want to find an orthogonal drawing Γ of G such that $\sum \text{cost}_e(\beta_e)$ is minimal, where β_e is the number of bends of e in Γ. The basic underlying decision problem FLEXDRAW restricts the cost function of every edge e to $\text{cost}_e(\beta) = 0$ for $\beta \in [0, \text{flex}(e)]$ and $\text{cost}_e(\beta) = \infty$ otherwise, and asks whether there exists a *valid* drawing (i.e., a drawing with finite cost). The value flex(e) is called the *flexibility* of e. Edges with flexibility 0 are called *inflexible*. Note that FLEXDRAW represents the important base case of testing for the existence of a drawing with cost 0 that is included in solving OPTIMALFLEXDRAW.

Garg and Tamassia [7] show that FLEXDRAW is NP-hard in this generality, by showing that it is NP-hard if every edge is inflexible. For special cases, namely

Partially supported by grant WA 654/21-1 of the German Research Foundation (DFG).

V.Th. Paschos and P. Widmayer (Eds.): CIAC 2015, LNCS 9079, pp. 61–73, 2015.
DOI: 10.1007/978-3-319-18173-8_4

planar graphs with maximum degree 3 and series-parallel graphs, Di Battista et al. [5] give an algorithm minimizing the total number of bends, which solves OPTIMALFLEXDRAW with $\mathrm{cost}_e(\beta) = \beta$ for each edge e. Their approach can be used to solve FLEXDRAW, as edges with higher flexibility can be modeled by a path of inflexible edges. Biedl and Kant [1] show that every 4-planar graph (except for the octahedron) admits an orthogonal drawing with at most two bends per edge. Thus, FLEXDRAW is trivial if the flexibility of every edge is at least 2. Bläsius et al. [2,3] tackle the NP-hard problems FLEXDRAW and OPTIMALFLEXDRAW by not counting the first bend on every edge. They give a polynomial time algorithm solving FLEXDRAW if the flexibility of every edge is at least 1 [2]. Moreover, they show how to efficiently solve OPTIMALFLEXDRAW if the cost function of every edge is convex and allows the first bend for free [3].

When restricting the allowed drawings to those with a specific planar embedding, the problem OPTIMALFLEXDRAW becomes significantly easier. Tamassia [9] shows how to find a drawing with as few bends as possible by computing a flow in a planar flow network. This flow network directly extends to a solution of OPTIMALFLEXDRAW with fixed planar embedding, if all cost functions are convex. Cornelsen and Karrenbauer [4] recently showed, that this kind of flow network can be solved in $O(n^{3/2})$ time.

Contribution and Outline. In this work we consider OPTIMALFLEXDRAW for instances that may contain inflexible edges, closing the gap between the general NP-hardness result [7] and the polynomial-time algorithms in the absence of inflexible edges [2,3]. After presenting some preliminaries in Section 2, we show in Section 3 that FLEXDRAW remains NP-hard even for instances with only $O(n^\varepsilon)$ (for any $\varepsilon > 0$) inflexible edges that are distributed evenly over the graph, i.e., they have pairwise distance $\Omega(n^{1-\varepsilon})$. This includes the cases where the inflexible edges are restricted to form very simple structures such as a matching.

On the positive side, we describe a general algorithm that can be used to solve OPTIMALFLEXDRAW by solving smaller subproblems (Section 4). This provides a framework for the unified description of bend minimization algorithms which covers both, previous work and results presented in this paper. We use this framework in Section 5 to solve OPTIMALFLEXDRAW for series-parallel graphs with monotone cost functions. This extends the algorithm by Di Battista et al. [5] to non-biconnected series-parallel graphs and thus solves one of their open problems. Moreover, we allow a significantly larger set of cost functions (in particular, the cost functions may be non-convex).

In Section 6, we present our main result, which is an FPT-algorithm with running time $O(2^k \cdot n \cdot T_{\mathrm{flow}}(n))$, where k is the number of inflexible edges incident to degree-4 vertices, and $T_{\mathrm{flow}}(n)$ is the time necessary to compute a maximum flow in a planar flow network of size n with multiple sources and sinks. Note that we can allow an arbitrary number of edges whose endpoints both have degree at most 3 to be inflexible without increasing the running time. Thus, our algorithm can also test the existence of a 0-bend drawing (all edges are inflexible) in FPT-time with respect to the number of degree-4 nodes. This partially solves another open problem of Di Battista et al. [5]. We conclude with open questions in Section 7.

Due to space constraints, we omit or only sketch several proofs. A full version with detailed proofs is available [3].

2   Preliminaries

Connectivity and the Composition of Graphs. A graph G is *connected* if there exists a path between every pair of vertices. A *separating k-set S* is a subset of vertices of G such that $G - S$ is not connected. Separating 1-sets are called *cutvertices* and separating 2-sets *separation pairs*. A connected graph without cutvertices is *biconnected* and a biconnected graph without separation pairs is *triconnected*. The *blocks* of a connected graph are its maximal (with respect to inclusion) biconnected subgraphs.

An *st-graph G* is a graph with two designated vertices s and t, its *poles*, such that $G + st$ is biconnected and planar. Let G_1 and G_2 be two *st*-graphs with poles s_1, t_1 and s_2, t_2, respectively. The *series composition G* of G_1 and G_2 is the union of G_1 and G_2 where t_1 is identified with s_2. Clearly, G is again an *st*-graph with the poles s_1 and t_2. In the *parallel composition G* of G_1 and G_2 the vertices s_1 and s_2 and the vertices t_1 and t_2 are identified with each other and form the poles of G. An *st*-graph is *series-parallel*, if it is a single edge or the series or parallel composition of two series-parallel graphs.

To be able to compose all *st*-graphs, we need a third composition. Let $G_1, \ldots, G_\ell$ be a set of *st*-graphs with poles s_i and t_i associated with G_i. Let H be an *st*-graph with poles s and t such that $H + st$ is triconnected and let $e_1, \ldots, e_\ell$ be the edges of H. The *rigid composition G* with respect to the so-called *skeleton H* is obtained by replacing each edge e_i of H by the graph G_i, identifying the endpoints of e_i with the poles of G_i. It follows from the theory of SPQR-trees that every *st*-graph is either a single edge or the series, parallel or rigid composition of *st*-graphs [6].

Orthogonal Representation. To handle orthogonal drawings of a graph G, we use the abstract concept of orthogonal representations neglecting distances in a drawing. Orthogonal representations were introduced by Tamassia [9], however, we use a slight modification that makes it easier to work with, as bends of edges and bends at vertices are handled the same. Let Γ be a *normalized* orthogonal drawing of G, i.e., every edge has only bends in one direction. If additional bends cannot improve the drawing (i.e., costs are monotonically increasing), a normalized optimal drawing exists [9]. All orthogonal drawings we consider are normalized. We assume that G is biconnected. This simplifies the description, as each edge and vertex has at most one incidence to a face. All definitions extend to connected graphs.

Let e be an edge in G that has β bends in Γ and let f be a face incident to e. We define the *rotation* of e in f as $\mathrm{rot}(e_f) = \beta$ and $\mathrm{rot}(e_f) = -\beta$ if the bends of e form $90°$ and $270°$ angles in f, respectively. For a vertex v forming the angle

α in the face f, we define $\operatorname{rot}(v_f) = 2 - \alpha/90°$. Note that, when traversing a face of G in clockwise (counter-clockwise for the outer face) direction, the right and left bends correspond to rotations of 1 and -1, respectively (we may have two left bends at once at vertices of degree 1). The values for the rotations we obtain from a drawing Γ satisfy the following properties; see Fig. 1a.

(1) The sum over all rotations in a face is 4 (-4 for the outer face).
(2) For every edge e with incident faces f_ℓ and f_r we have $\operatorname{rot}(e_{f_\ell}) + \operatorname{rot}(e_{f_r}) = 0$.
(3) The sum of rotations around a vertex v is $2 \cdot \deg(v) - 4$.
(4) The rotations at vertices lie in the range $[-2, 1]$.

Let $\mathcal{R}$ be a structure consisting of an embedding of G plus a set of values fixing the rotation for every vertex-face and edge-face incidence. We call $\mathcal{R}$ an *orthogonal representation* of G if the rotation values satisfy the above properties (1)–(4). Given an orthogonal representation $\mathcal{R}$, a drawing inducing the specified rotation values exists and can be computed efficiently [9].

Orthogonal Representations and Bends of st-Graphs. We extend the notion of rotation to paths; conceptually this is very similar to spirality [5]. Let π be a path from vertex u to vertex v. We define the rotation of π (denoted by $\operatorname{rot}(\pi)$) to be the number of bends to the right minus the number of bends to the left when traversing π from u to v.

There are two special paths in an st-graph G. Let s and t be the poles of G and let $\mathcal{R}$ be an orthogonal representation with s and t on the outer face. Then $\pi(s,t)$ denotes the path from s to t when traversing the outer face of G in counter-clockwise direction. Similarly, $\pi(t,s)$ is the path from t to s. We define the *number of bends* of $\mathcal{R}$ to be $\max\{|\operatorname{rot}(\pi(s,t))|, |\operatorname{rot}(\pi(t,s))|\}$. Note that a single edge $e = st$ is also an st-graph. Note further that the notions of the number of bends of the edge e and the number of bends of the st-graph e coincide. Thus, the above definition is consistent.

When considering orthogonal representations of st-graphs, we always require the poles s and t to be on the outer face. We say that the vertex s has σ *occupied incidences* if $\operatorname{rot}(s_f) = \sigma - 3$ where f is the outer face. We also say that s has $4 - \sigma$ *free incidences* in the outer face. If the poles s and t have σ and τ occupied incidences in $\mathcal{R}$, respectively, we say that $\mathcal{R}$ is a (σ, τ)-*orthogonal representation*; see Fig. 1b.

Note that $\operatorname{rot}(\pi(s,t))$ and $\operatorname{rot}(\pi(t,s))$ together with the number of occupied incidences σ and τ basically describe the outer shape of G and thus how it has to be treated if it is a subgraph of some larger graph. Using the bends of $\mathcal{R}$ instead of the rotations of $\pi(s,t)$ and $\pi(t,s)$ implicitly allows to mirror the orthogonal representation (and thus exchanging $\pi(s,t)$ and $\pi(t,s)$).

Thick Edges. In the basic formulation of an orthogonal representation, every edge *occupies* exactly one incidence at each of its endpoints. We introduce *thick edges* that may occupy more than one incidence at each endpoint to represent larger subgraphs. Let $e = st$ be an edge in G. We say that e is a (σ, τ)-*edge* if e is defined to occupy σ and τ incidences at s and t, respectively. Note that the total

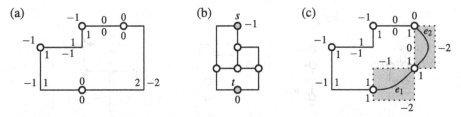

Fig. 1. (a) An orthogonal drawing together with its orthogonal representation given by the rotation values. (b) A $(2,3)$-orthogonal representation (s and t have 2 and 1 free incidences, respectively). (c) An orthogonal representation with thick edges e_1 and e_2. The gray boxes indicate how many attachments the thick edges occupy, i.e., e_1 is a $(2,3)$-edge and e_2 is a $(2,2)$-edge. Both thick edges have two bends.

amount of occupied incidences of a vertex in G must not exceed 4. With this extended notion of edges, we define a structure $\mathcal{R}$ consisting of an embedding of G plus a set of values for all rotations to be an *orthogonal representation* if it satisfies the following (slightly extended) properties; see Fig. 1c.

(1) The sum over all rotations in a face is 4 (-4 for the outer face).
(2) For every (σ, τ)-edge e with incident faces f_ℓ and f_r we have $\mathrm{rot}(e_{f_\ell}) + \mathrm{rot}(e_{f_r}) = 2 - (\sigma + \tau)$.
(3) The sum of rotations around a vertex v with incident edges $e_1, \ldots, e_\ell$ occupying $\sigma_1, \ldots, \sigma_\ell$ incidences of v is $\sum(\sigma_i + 1) - 4$
(4) The rotations at vertices lie in the range $[-2, 1]$.

Note that requiring every edge to be a $(1, 1)$-edge in this definition of an orthogonal representation exactly yields the previous definition without thick edges. The *number of bends* of a (thick) edge e incident to the faces f_ℓ and f_r is $\max\{|\mathrm{rot}(e_{f_\ell})|, |\mathrm{rot}(e_{f_r})|\}$. Note that this is again consistent with the definition of number of bends of st-graphs and normal edges. Unsurprisingly, replacing a (σ, τ)-edge with β bends in an orthogonal representation by a (σ, τ)-orthogonal representation with β bends of an arbitrary st-graph yields a valid orthogonal representation [2, Lemma 5].

3   A Matching of Inflexible Edges

In this section, we show that FLEXDRAW is NP-complete even if the inflexible edges form a matching. In fact, we show the stronger result of NP-hardness of instances with $O(n^\varepsilon)$ inflexible edges (for $\varepsilon > 0$) even if these edges are distributed evenly over the graph, i.e., they have pairwise distance $\Omega(n^{1-\varepsilon})$.

We adapt the proof of NP-hardness by Garg and Tamassia [7] for the case that all edges of an instance of FLEXDRAW are inflexible. For a given instance of NAE-3SAT (Not All Equal 3SAT) they show how to construct a graph G that admits an orthogonal representation without bends if and only if the instance of NAE-3SAT is satisfiable. The graph G is obtained by first constructing a graph F that has a unique planar embedding [7, Lemma 5.1] and replacing the edges of F by larger st-graphs. These graphs have degree-1 poles and their embedding is fixed up to a flip, which implies the following lemma.

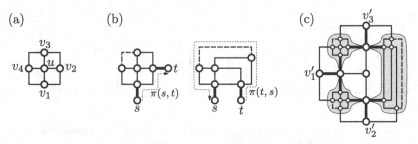

Fig. 2. The bold edges are inflexible; dashed edges have flexibility 2; all other edges have flexibility 1. (a) The wheel W_4. (b) The bend gadget $B_{1,2}$. (c) The gadget W_3' for replacing degree-3 vertices. The marked subgraphs are bend gadgets.

Lemma 1 (Garg and Tamassia [7]). FLEXDRAW *is NP-hard, even if the order of edges around each vertex is fixed up to reversal.*

We assume that our instances do not contain degree-2 vertices; their incident edges can be replaced by a single edge with higher flexibility. In the following, we first show how to replace vertices of degree 3 by graphs of constant size such that each inflexible edge is incident to two vertices of degree 4. Afterwards, we can replace degree-4 vertices by smaller subgraphs with positive flexibility, which increases the distance between the inflexible edges.

The shape of the outer face of the wheel W_4 (with flexibility 1 on each edge) is the same in every valid orthogonal representation [2]; see Fig. 2a. It follows directly that the st-graph in Fig. 2b, which we call *bend gadget* and denote by $B_{1,2}$, has either one or two bends in each orthogonal representation. This ensures that replacing a degree-3 vertex by the construction W_3' (which is basically an enhanced wheel of size 3) shown in Fig. 2c results in an equivalent instance of FLEXDRAW. Note that, after replacing each degree-3 vertex by a copy of W_3', all endpoints of inflexible edges have degree 4. We obtain the following lemma.

Lemma 2. FLEXDRAW *is NP-hard, even if the endpoints of each inflexible edge have degree 4 and if the order of edges around each vertex is fixed up to reversal.*

Similarly, replacing the vertices incident to inflexible edges by copies of W_4 (whose edges have flexibility 1), increases the distance between any pair of inflexible edges by at least 1. Note that this does not increase the number of inflexible edges. Iterative replacement yields the following theorem.

Theorem 1. FLEXDRAW *is NP-complete even for instances of size n with $O(n^\varepsilon)$ inflexible edges with pairwise distance $\Omega(n^{1-\varepsilon})$.*

The instances described above may contain edges with flexibility larger than 1. By iteratively replacing an edge e with flexibility $\mathrm{flex}(e) > 1$ by a chain consisting of an edge with flexibility 1, the wheel W_4, and an edge with flexibility $\mathrm{flex}(e) - 1$, we obtain an equivalent instance where all edges have flexibility 1 or 0.

4   The General Algorithm

In this section we describe a general algorithm that can be used to solve OPTIMAL-FLEXDRAW by solving smaller subproblems for the different types of graph compositions. To this end, we start with the definition of cost functions for subgraphs. The *cost function* cost($\cdot$) of an *st*-graph G is defined such that cost(β) is the minimum cost of all orthogonal representations of G with β bends. The (σ, τ)-*cost function* cost$_\tau^\sigma(\cdot)$ of G is defined analogously by setting cost$_\tau^\sigma(\beta)$ to the minimum cost of all (σ, τ)-orthogonal representations of G with β bends. Clearly, $\sigma, \tau \in \{1, \ldots 4\}$, though, for a fixed graph G, not all values may be possible. If for example deg(s) = 1, then σ is 1 for every orthogonal representation of G. Note that there is a lower bound on the number of bends depending on σ and τ. For example, a $(2, 2)$-orthogonal representation has at least one bend and thus cost$_2^2(0)$ is undefined. We formally set undefined values to ∞.

With *the cost functions* of G we refer to the collection of (σ, τ)-cost functions of G for all possible combinations of σ and τ. Let G be the composition of two or more (for a rigid composition) graphs $G_1, \ldots, G_\ell$. Computing the cost functions of G assuming that the cost functions of $G_1, \ldots, G_\ell$ are known is called *computing cost functions of a composition*. The following theorem states that the ability to compute cost functions of compositions suffices to solve OPTIMALFLEXDRAW. The terms T_S, T_P and $T_R(\ell)$ denote the time for computing the cost functions of a series, a parallel, and a rigid composition with skeleton of size ℓ, respectively.

Theorem 2. *Let G be an st-graph containing the edge st. An optimal (σ, τ)-orthogonal representation of G with st on the outer face can be computed in $O(nT_S + nT_P + T_R(n))$ time.*

Applying Theorem 2 for each pair of adjacent nodes as poles in a given instance of OPTIMALFLEXDRAW yields the following corollary.

Corollary 1. OPTIMALFLEXDRAW *can be solved in $O(n \cdot (nT_S + nT_P + T_R(n)))$ time for biconnected graphs.*

In the following, we extend this result to the case where G may contain cutvertices. The extension is straightforward, however, there is one pitfall. Given two blocks B_1 and B_2 sharing a cutvertex v such that v has degree 2 in B_1 and B_2, we have to ensure for both blocks that v does not form an angle of 180°. Thus, for a given graph G, we get for each block a list of vertices and we restrict the set of all orthogonal representations of G to those where these vertices form 90° angles. We call these orthogonal representations *restricted orthogonal representations*. Moreover, we call the resulting cost functions *restricted cost functions*. We use the terms T_S^r, T_P^r and $T_R^r(\ell)$ to denote the time necessary to compute the *restricted* cost functions of a series, a parallel, and a rigid composition, respectively. We get the following theorem.

Theorem 3. OPTIMALFLEXDRAW *is $O(n \cdot (nT_S^r + nT_P^r + T_R^r(n)))$-time solvable.*

Note that Theorem 3 provides a framework for uniform treatment of bend minimization over all planar embeddings in orthogonal drawings. In particular, the polynomial-time algorithm for FLEXDRAW with positive flexibility [2] can be expressed in this way. There, all resulting cost functions of st-graphs are 0 on a non-empty interval containing 0 (with one minor exception) and ∞, otherwise. Thus, the cost functions of the compositions can be computed using Tamassia's flow network. The results on OPTIMALFLEXDRAW [3] can be expressed similarly. When restricting the number of bends of each st-graph occurring in the composition to 3, all resulting cost functions are convex (with one minor exception). Thus, Tamassia's flow network can again be used to compute the cost functions of the compositions. The overall optimality follows from the fact that there exists an optimal solution that can be composed in such a way. In the following sections we see two further applications of this framework.

5   Series-Parallel Graphs

In this section we show that the cost functions of a series composition (Lemma 3) and a parallel composition (Lemma 4) can be computed efficiently. Using our framework, this leads to a polynomial-time algorithm for OPTIMALFLEXDRAW for series-parallel graphs with monotone cost functions (Theorem 4). We note that this is only a slight extension to the results by Di Battista et al. [5]. However, it shows the easy applicability of the above framework before diving into the more complicated FPT-algorithm in the following section.

To get the running time claimed in Theorem 4, it is necessary to bound the maximum number ℓ of bends of an st-graph. Using Tamassia's flow network it can be seen that $\ell \in O(n)$.

Lemma 3. *If the (restricted) cost functions of two st-graphs are ∞ for bend numbers larger than ℓ, the (restricted) cost functions of their series composition can be computed in $O(\ell^2)$ time.*

Proof. We first consider the case of non-restricted cost functions. Let G_1 and G_2 be the two st-graphs with poles s_1, t_1 and s_2, t_2, respectively, and let G be their series composition with poles $s = s_1$ and $t = t_2$. For each of the constantly many valid combinations of σ and τ, we compute the (σ, τ)-cost function separately. Assume for the following, that σ and τ are fixed. Since G_1 and G_2 both have at most ℓ bends, G can only have $O(\ell)$ possible values for the number of bends β. We fix the value β and show how to compute $\mathrm{cost}_\tau^\sigma(\beta)$ in $O(\ell)$ time.

Let $\mathcal{R}$ be a (σ, τ)-orthogonal representation with β bends and let $\mathcal{R}_1$ and $\mathcal{R}_2$ be the (σ_1, τ_1)- and (σ_2, τ_2)-orthogonal representations induced for G_1 and G_2, respectively. Obviously, $\sigma_1 = \sigma$ and $\tau_2 = \tau$ holds. However, there are the following other parameters that may vary (although they may restrict each other). The parameters τ_1 and σ_2; the number of bends β_1 and β_2 of $\mathcal{R}_1$ and $\mathcal{R}_2$, respectively; the possibility that for $i \in \{1, 2\}$ the number of bends of $\mathcal{R}_i$ are determined by $\pi(s_i, t_i)$ or by $\pi(t_i, s_i)$, i.e., $\beta_i = -\mathrm{rot}(\pi(s_i, t_i))$ or $\beta_i = -\mathrm{rot}(\pi(t_i, s_i))$; and finally, the rotations at the vertex v in the outer face, where v is the vertex of G belonging to both, G_1 and G_2.

Assume we fixed the parameters τ_1 and σ_2, the choice by which paths β_1 and β_2 are determined, the rotations at the vertex v, and the number of bends β_1 of $\mathcal{R}_1$. Then there is no choice left for the number of bends β_2 of $\mathcal{R}_2$, as choosing a different value for β_2 also changes the number of bends β of G, which was fixed. As each of the parameters can have only a constant number of values except for β_1, which can have $O(\ell)$ different values, there are only $O(\ell)$ possible choices in total. For each of these choices, we get a (σ, τ)-orthogonal representation of G with β bends and cost $\mathrm{cost}_{\tau_1}^{\sigma_1}(\beta_1) + \mathrm{cost}_{\tau_2}^{\sigma_2}(\beta_2)$. By taking the minimum cost over all these choices we get the desired value $\mathrm{cost}_{\tau}^{\sigma}(\beta)$ in $O(\ell)$ time.

For restricted cost functions, we may have to restrict the angle at v. Obviously, this constraint can be easily added to the described algorithm.   □

Lemma 4. *If the (restricted) cost functions of two st-graphs are ∞ for bend numbers larger than ℓ, the (restricted) cost functions of their parallel composition can be computed in $O(\ell)$ time.*

Theorem 4. *For series-parallel graphs with monotone cost functions* OPTIMALFLEXDRAW *can be solved in $O(n^4)$ time.*

6   An FPT-Algorithm for General Graphs

Let G be an instance of FLEXDRAW. We call an edge in G *critical* if it is inflexible and at least one of its endpoints has degree 4. We call G *k-critical*, if it contains exactly k critical edges. An inflexible edge that is not critical is *semi-critical*. The poles s and t of an *st*-graph G are considered to have additional neighbors (which comes from the fact that we usually consider *st*-graphs to be subgraphs of larger graphs). Inflexible edges incident to the pole s (or t) are already *critical* if $\deg(s) \geq 2$ (or $\deg(t) \geq 2$). In the following, we study cost functions of k-critical *st*-graphs and give an FPT-algorithm for k-critical instances.

6.1   The Cost Functions of k-Critical Instances

Let G be an st-graph and let $\mathcal{R}$ be a valid orthogonal representation of G. We define an operation that transforms $\mathcal{R}$ into another valid orthogonal representation of G. Let $G^\star$ be the *double directed* dual graph of G, i.e., each edge e of G with incident faces g and f corresponds to the two dual edges (g, f) and (f, g). We call a dual edge $e^\star = (g, f)$ of e *valid* if one of the following conditions holds.

  (I) $\mathrm{rot}(e_f) < \mathrm{flex}(e)$ (which is equivalent to $-\mathrm{rot}(e_g) < \mathrm{flex}(e)$).

  (II) $\mathrm{rot}(v_f) < 1$ where v is an endpoint of e but not a pole.

A simple directed cycle $C^\star$ in $G^\star$ consisting of valid edges is called *valid cycle*. Then *bending along $C^\star$* changes the orthogonal representation $\mathcal{R}$ as follows; see Fig. 3a. Let $e^\star = (g, f)$ be an edge in $C^\star$ with primal edge e. If $e^\star$ is valid due to Condition (I), we reduce $\mathrm{rot}(e_g)$ by 1 and increase $\mathrm{rot}(e_f)$ by 1. Otherwise, if Condition (II) holds, we reduce $\mathrm{rot}(v_g)$ by 1 and increase $\mathrm{rot}(v_f)$ by 1, where v is the vertex incident to e with $\mathrm{rot}(v_f) < 1$.

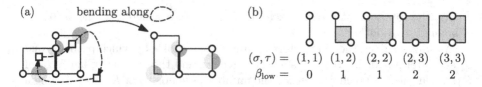

Fig. 3. (a) An orthogonal representation (the bold edge is inflexible, other edges have flexibility 1) and a valid cycle C^* (dashed). Bending along C^* increases the green and decreases the red angles. (b) Illustration of Fact 1 for some values of σ and τ.

Lemma 5. *Let G be an st-graph with a valid (σ, τ)-orthogonal representation $\mathcal{R}$. Bending along a valid cycle C^* yields a valid (σ, τ)-orthogonal representation.*

As mentioned in Section 4, depending on σ and τ, there is a lower bound β_{low} on the number of bends of (σ, τ)-orthogonal representations; see Fig. 3b.

Fact 1. *A (σ, τ)-orthogonal representation has at least $\beta_{\text{low}} = \left\lceil \dfrac{\sigma + \tau}{2} \right\rceil - 1$ bends.*

For a valid orthogonal representation with a large number of bends, the following lemma states that we can reduce its bends by bending along a valid cycle. This can later be used to show that the cost function of an st-graph is 0 on a significantly large interval. Or in other words, arbitrary alterations of cost 0 and cost ∞ that are hard to handle only occur on a small interval (depending on k). The lemma and its proof are a generalization of Lemma 1 from [2] that incorporates inflexible edges. For $\sigma = \tau = 3$ a slightly weaker result holds.

Lemma 6. *Let G be a k-critical st-graph and let $\mathcal{R}$ be a valid (σ, τ)-orthogonal representation with $\sigma + \tau \leq 5$. If $- \text{rot}(\pi(t, s)) \geq \beta_{\text{low}} + k + 1$ holds, then there exists a valid cycle C^* such that bending $\mathcal{R}$ along C^* reduces $- \text{rot}(\pi(t, s))$ by 1.*

Lemma 7. *Let G be a k-critical st-graph and let $\mathcal{R}$ be a valid $(3, 3)$-orthogonal representation. If $- \text{rot}(\pi(t, s)) \geq \beta_{\text{low}} + k + 2$ holds, then there exists a valid cycle C^* such that bending $\mathcal{R}$ along C^* reduces $- \text{rot}(\pi(t, s))$ by 1.*

The previous lemmas basically show that the existence of a valid orthogonal representation with a lot of bends implies the existence of valid orthogonal representations for a "large" interval of bend numbers. This is made more precise in the following.

Let $\mathcal{B}_\tau^\sigma$ be the set containing an integer β if and only if G admits a valid (σ, τ)-orthogonal representation with β bends. Assume G admits a valid (σ, τ)-orthogonal representation, i.e., $\mathcal{B}_\tau^\sigma$ is not empty. We define the *maximum bend value* $\beta_{\max}$ to be the maximum in $\mathcal{B}_\tau^\sigma$. Moreover, let $\beta \in \mathcal{B}_\tau^\sigma$ be the smallest value, such that every integer between β and $\beta_{\max}$ is contained in $\mathcal{B}_\tau^\sigma$. Then we call the interval $[\beta_{\text{low}}, \beta - 1]$ the (σ, τ)-*gap* of G. The value $\beta - \beta_{\text{low}}$ is also called the (σ, τ)-*gap* of G; see Fig. 4.

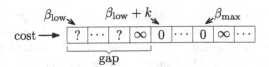

Fig. 4. A cost function with gap k

Lemma 8. *The (σ, τ)-gap of a k-critical st-graph G is at most k if $\sigma + \tau \leq 5$. The $(3,3)$ gap of G is at most $k + 1$.*

The following lemma basically expresses the gap of an st-graph in terms of the rotation along $\pi(s,t)$ instead of the number of bends.

Lemma 9. *Let G be an st-graph with (σ, τ)-gap k. The set $\{\rho \mid G$ admits a valid (σ, τ)-orthogonal representation with $\mathrm{rot}(\pi(s,t)) = \rho\}$ is the union of at most $k + 1$ intervals.*

6.2 Computing the Cost Functions of Compositions

Let G be a graph with fixed planar embedding. We describe a flow network, similar to the one by Tamassia [9] that can be used to compute orthogonal representations of graphs with thick edges. In general, we consider a flow network to be a directed graph with a lower and an upper bound assigned to every edge and a demand assigned to every vertex. The bounds and demands can be negative. An assignment of flow-values to the edges is a feasible flow if it satisfies the following properties. The flow-value of each edge is at least its lower and at most its upper bound. For every vertex the flow on incoming edges minus the flow on outgoing edges must equal its demand.

We define the flow network N as follows. The network N contains a node for each vertex of G, the *vertex nodes*, each face of G, the *face nodes*, and each edge of G, the *edge nodes*. Moreover, N contains arcs from each vertex to all incident faces, the *vertex-face arcs*, and similarly from each edge to both incident faces, the *edge-face arcs*. We interpret an orthogonal representation $\mathcal{R}$ of G as a flow in N. A rotation $\mathrm{rot}(e_f)$ of an edge e in the face f corresponds to the same amount of flow on the edge-face arc from e to f. Similarly, for a vertex v incident to f the rotation $\mathrm{rot}(v_f)$ corresponds to the flow from v to f.

Obviously, the properties (1)–(4) of an orthogonal representation are satisfied if and only if the following conditions hold for the flow (note that we allow G to have thick edges).

(1) The total amount of flow on arcs incident to a face node is 4 (-4 for the outer face).
(2) The flow on the two arcs incident to an edge node stemming from a (σ, τ)-edge sums up to $2 - (\sigma + \tau)$.
(3) The total amount of flow on arcs incident to a vertex node, corresponding to the vertex v with incident edges $e_1, \ldots, e_\ell$ occupying $\sigma_1, \ldots, \sigma_\ell$ incidences of v is $\sum(\sigma_i + 1) - 4$.
(4) The flow on vertex-face arcs lies in the range $[-2, 1]$.

Properties (1)–(3) are equivalent to the flow conservation requirement when setting appropriate demands. Property (4) is equivalent to the capacity constraints in a flow network when setting the lower and upper bounds of vertex-face arcs to -2 and 1, respectively. In the following, we use this flow network to compute the cost function of a rigid composition of graphs. The term $T_{\text{flow}}(\ell)$ denotes the time necessary to compute a maximal flow in a planar flow network of size ℓ.

Lemma 10. *The (restricted) cost functions of a rigid composition of ℓ graphs can be computed in $O(2^k \cdot T_{\text{flow}}(\ell))$ time if the resulting graph is k-critical.*

Proof (sketch). Let H be the skeleton of the rigid composition of the graphs $G_1, \ldots, G_\ell$ and let G be the resulting graph with poles s and t. In the following, we show that the (restricted) cost functions can be determined by computing flows in $O(2^k)$ flow networks. We first show that, similar to the proof of Lemma 3, the following parameters lead to only constantly many combinations: the number of occupied incidences of the composed graphs and the graph itself; the decision whether $\pi(s,t)$ or $\pi(t,s)$ determines the number of bends of G. The remaining degrees of freedom are the number of bends β of G and for each graph G_i an interval for its rotation along $\pi(s_i, t_i)$, where s_i and t_i are the poles of G_i. It is not hard to see that all these constraints can be expressed by capacity constraints in the flow network. It remains to count the number of combinations.

Let k_i be the number of critical edges in G_i. Due to Lemma 9, we get $k_i + 1$ intervals for each of these graphs G_i leading to $\prod(k_i + 1)$ combinations. As critical edges in G_i remain critical in G (and the G_i are pairwise edge-disjoint), we have $\sum k_i \le k$. One can show that with this restriction $\prod(k_i + 1) \in O(2^k)$.

Constructing all these flow networks for every possible number of bends β would lead to $O(\beta_{\max} \cdot 2^k)$ flow networks. However, once the rotation intervals for the G_i are fixed, one can compute a maximum and minimum rotation for G with these intervals. It follows from basic flow theory that all intermediate rotation values are also possible. Thus with two flow computations, we obtain all possible rotation values for the chosen intervals of the G_i. This leads to $O(2^k)$ intervals whose union describes the cost function of G. From that, an explicit representation of the cost function of G can be computed in $O(2^k)$ time, which yields a total running time of $O(2^k \cdot T_{\text{flow}}(\ell))$. □

Lemma 11. *The (restricted) cost functions of a series and a parallel composition can be computed in $O(k^2 + 1)$ time if the resulting graph is k-critical.*

Theorem 5. FLEXDRAW *for k-critical graphs is $O(2^k \cdot n \cdot T_{\text{flow}}(n))$-time solvable.*

Proof. By Theorem 3, we get an algorithm with the running time $O(n \cdot (n \cdot T_S + n \cdot T_P + T_R(n)))$, where $T_S, T_P \in O(k^2 + 1)$ (Lemma 11) and $T_R(\ell) = 2^k \cdot T_{\text{flow}}(\ell)$ (Lemma 10) holds. This obviously yields the running time $O((k^2 + 1) \cdot n^2 + 2^k \cdot n \cdot T_{\text{flow}}(n)) = O(2^k \cdot n \cdot T_{\text{flow}}(n))$. □

7   Conclusion

We want to conclude with the open question whether there exists an FPT-algorithm for OPTIMALFLEXDRAW for the case where all cost functions are convex and where the first bend causes cost only for k edges. One might think that this works similar as for FLEXDRAW by showing that the cost functions of st-graphs are only non-convex if they contain inflexible edges. Then, when encountering a rigid composition, one could separate these non-convex cost functions into convex parts and consider all combinations of these convex parts. Unfortunately, the cost functions of st-graphs may already be non-convex, even though they do not contain inflexible edges. The reason why OPTIMALFLEXDRAW can still be solved efficiently if there are no inflexible edges [3] is that, in this case, the cost functions need to be considered only up to three bends (and for this restricted intervals, the cost functions are convex). However, a single subgraph with inflexible edges in a rigid composition may force arbitrary other subgraphs in this composition to have more than three bends, potentially resulting in linearly many non-convex cost functions that have to be considered. Thus, although the algorithms for FLEXDRAW and OPTIMALFLEXDRAW are very similar, the latter does not seem to allow even a small number of inflexible edges.

Acknowledgments. We thank Marcus Krug for discussions on FLEXDRAW.

References

1. Biedl, T., Kant, G.: A better heuristic for orthogonal graph drawings. Comput. Geom.: Theory Appl. **9**(3), 159–180 (1998)
2. Bläsius, T., Krug, M., Rutter, I., Wagner, D.: Orthogonal graph drawing with flexibility constraints. Algorithmica 68(4) (2014)
3. Bläsius, T., Lehmann, S., Rutter, I.: Orthogonal graph drawing with inflexible edges. CoRR abs/1404.2943, 1–23 (2014). http://arxiv.org/abs/1404.2943
4. Bläsius, T., Rutter, I., Wagner, D.: Optimal orthogonal graph drawing with convex bend costs. In: Fomin, F.V., Freivalds, R., Kwiatkowska, M., Peleg, D. (eds.) ICALP 2013, Part I. LNCS, vol. 7965, pp. 184–195. Springer, Heidelberg (2013)
5. Cornelsen, S., Karrenbauer, A.: Accelerated bend minimization. J. Graph Alg. Appl. **16**(3), 635–650 (2012)
6. Di Battista, G., Liotta, G., Vargiu, F.: Spirality and optimal orthogonal drawings. SIAM J. Comput. **27**(6), 1764–1811 (1998)
7. Di Battista, G., Tamassia, R.: On-line maintenance of triconnected components with SPQR-trees. Algorithmica **15**(4), 302–318 (1996)
9. Garg, A., Tamassia, R.: On the computational complexity of upward and rectilinear planarity testing. SIAM J. Comput. **31**(2), 601–625 (2001)
9. Tamassia, R.: On embedding a graph in the grid with the minimum number of bends. SIAM J. Comput. **16**(3), 421–444 (1987)

Linear Time Constructions of Some d-Restriction Problems

Nader H. Bshouty[✉]

Technion, Haifa, Israel
bshouty@ca.technion.ac.il

Abstract. We give new linear time globally explicit constructions for perfect hash families, cover-free families and separating hash functions.

Keywords: Derandomization · d-Restriction problems · Perfect hash · Cover-free families · Separating hash functions

1 Introduction

A *d-restriction problem* [7,13,58] is a problem of the following form:
Given an alphabet Σ of size $|\Sigma| = q$, an integer n and a class $\mathcal{M}$ of nonzero functions $f : \Sigma^d \to \{0,1\}$.
Find a small set $A \subseteq \Sigma^n$ such that: For every $1 \le i_1 < i_2 < \cdots < i_d \le n$ and $f \in \mathcal{M}$ there is $a \in A$ such that $f(a_{i_1}, \ldots, a_{i_d}) \ne 0$.

A $(1 - \epsilon)$-*dense d-restriction problem* is a problem of the following form:
Given an alphabet Σ of size $|\Sigma| = q$, an integer n and a class $\mathcal{M}$ of nonzero functions $f : \Sigma^d \to \{0,1\}$.
Find a small set $A \subseteq \Sigma^n$ such that: For every $1 \le i_1 < i_2 < \cdots < i_d \le n$ and $f \in \mathcal{M}$

$$\mathbf{Pr}_{a \in A}[f(a_{i_1}, \ldots, a_{i_d}) \ne 0] > 1 - \epsilon$$

where the probability is over the choice of a from the uniform distribution on A.

We give new constructions for the following three $((1-\epsilon)$-dense) d-restriction problems: Perfect hash family, cover-free family and separating hash family.

Perfect hash families were introduced by Mehlhorn [50] in 1984 and used as database management. They were used in compiler design to prove lower bounds on the size of a program that constructs a hash function suitable for fast retrieval of fixed data such as library function names [27]. Perfect hash families have been also applied to circuit complexity problems [59], derandomize some probabilistic algorithms [6], broadcast encryption [39] and threshold cryptography [11,12].

Cover-free families were first introduced in 1964 by Kautz and Singleton [47] to investigate superimposed binary codes. Cover-free families have been used to solve some problems in cryptography and communications, including blacklisting, broadcast encryption, broadcast anti-jamming, source authentication in a network setting, group key predistribution and pooling designs over complexes. See [25,29,33,41,42,46,52,62–64,68,69].

© Springer International Publishing Switzerland 2015
V.Th. Paschos and P. Widmayer (Eds.): CIAC 2015, LNCS 9079, pp. 74–88, 2015.
DOI: 10.1007/978-3-319-18173-8_5

A construction is *global explicit* if it runs in deterministic polynomial time in the size of the construction. A *local explicit construction* is a construction where one can find any bit in the construction in time poly-log in the size of the construction. The constructions in this paper are linear time global explicit constructions.

To the best of our knowledge, our constructions have sizes that are less than the ones known from the literature.

1.1 Learning Hypergraphs

In this section we give one application in computational learning theory.

A hypergraph is $H = (V, E)$ where V is the set of vertices and $E \subseteq 2^V$ is the set of edges. The dimension of the hypergraph H is the cardinality of the largest set in E. For a set $S \subseteq V$, the *edge-detecting queries* $Q_H(S)$ is answered "Yes" or "No", indicating whether S contains all the vertices of at least one edge of H. Learning a hidden hypergraph of constant dimension r with s edges using edge-detecting queries is equivalent to another important problem in learning theory [4]: Learning the class s-term r-MDNF (Monotone DNF with s terms of size r) with *membership queries* (the learner can ask about the value of the function in some point).

This problem has many applications in chemical reactions and genome sequencing. In chemical reactions, we are given a set of chemicals, some of which react and some which do not. When multiple chemicals are combined in one test tube, a reaction is detectable if and only if at least one set of the chemicals in the tube reacts. The goal is to identify which sets react using as few experiments as possible. See [2–5,16,18,24,26,28,34,35,40,53,61] for more details on the problem and many other applications.

This problem is also called "sets of positive subsets" [70] "complex group testing" [53] and "group testing in hypergraph" [35].

It is known that a cover-free families can be used as queries to solve the above problem. Several algorithms with non-optimal query complexity are known from the literature. See [28] and references within. Our construction is the first linear time construction that construct an optimal query set for learning hypergraphs.

2 Old and New Results

2.1 Perfect Hash Family

Let H be a family of functions $h : [n] \rightarrow [q]$. For $d \leq q$ we say that H is an (n, q, d)-*perfect hash family* $((n, q, d)$-PHF$)$ [7] if for every subset $S \subseteq [n]$ of size $|S| = d$ there is a *hash function* $h \in H$ such that $h|_S$ is injective (one-to-one) on S, i.e., $|h(S)| = d$.

Blackburn and Wild [23] gave an optimal explicit construction when $q \geq exp(\sqrt{d \log d \log n})$. Stinson et al., [65], gave an explicit construction of (n, q, d)-PHF of size $d^{\log^* n} \log n$ for $q \geq d^2 \log n / \log q$. It follows from the technique used in [1] with Reed-Solomon codes that an explicit (n, q, d)-PHF of size $d^2 \log n / \log q$ exist for $q \geq d^2 \log n / \log q$. In [7,9,58] it was shown that there are $(n, \Omega(d^2), d)$-PHF of size $O(d^6 \log n)$ that can be constructed in $poly(n)$ time. Wang and Xing

[71] used algebraic function fields and gave an (n, d^4, d)-PHF of size $O((d^2/\log d)$ $\log n)$ for infinite sequence of integers n. Their construction is not linear time construction. The above constructions are either for large q or are not linear time constructions.

Bshouty in [13] shows that for a constant $c > 1$, the following (third column in the table) (n, q, d)-PHF can be locally explicitly constructed in almost linear time (within $poly(\log)$)

n	q	Linear time. Size $= O()$	Upper Bound	Lower Bound
I.S.	$q \geq \frac{c}{4}d^4$	$d^2 \frac{\log n}{\log q}$	$d\frac{\log n}{\log q}$	$d\frac{\log n}{\log q}$
all	$q \geq \frac{c}{4}d^4$	$d^4 \frac{\log n}{\log q}$	$d\frac{\log n}{\log q}$	$d\frac{\log n}{\log q}$
I.S.	$q \geq \frac{c}{2}d^2$	$d^4 \frac{\log n}{\log d}$	$d\frac{\log n}{\log(2q/(d(d-1)))}$	$d\frac{\log n}{\log q}$
all	$q \geq \frac{c}{2}d^2$	$d^6 \frac{\log n}{\log d}$	$d\frac{\log n}{\log(2q/(d(d-1)))}$	$d\frac{\log n}{\log q}$
I.S.	$q = \frac{d(d-1)}{2} + 1 + o(d^2)$	$d^6 \frac{\log n}{\log d}$	$d\log n$	$d\frac{\log n}{\log q}$
all	$q = \frac{d(d-1)}{2} + 1 + o(d^2)$	$d^8 \frac{\log n}{\log d}$	$d\log n$	$d\frac{\log n}{\log q}$

The upper bound in the table follows from union bound [13]. The lower bound is from [10,51] (see also [17,22,23,37,44,45,55]). We note here that all the lower bounds in this paper are true even for non-explicit constructions. I.S. stands for "true for infinite sequence of integers n". Here we prove

Theorem 1. *Let q be a power of prime. If $q > 4(d(d-1)/2+1)$ then there is a (n, q, d)-PHF of size*

$$O\left(\frac{d^2\log n}{\log(q/e(d(d-1)/2+1))}\right)$$

that can be constructed in linear time.

If $d(d-1)/2+2 \leq q \leq 4(d(d-1)/2+1)$ then there is a (n, q, d)-PHF of size

$$O\left(\frac{q^2 d^2 \log n}{(q - d(d-1)/2 - 1)^2}\right)$$

that can be constructed in linear time.

In particular, for any constants $c > 1$, $\delta > 0$ and $0 \leq \eta < 1$, the following (n, q, d)-PHF can be constructed in linear time (the third column in the following table)

n	q	Linear time. Size $= O()$	Upper Bound	Lower Bound
all	$q \geq d^{2+\delta}$	$d^2 \frac{\log n}{\log q}$	$d\frac{\log n}{\log q}$	$d\frac{\log n}{\log q}$
all	$q \geq \frac{c}{2}d^2$	$d^2\log n$	$d\log n$	$d\frac{\log n}{\log q}$
all	$q = \frac{d(d-1)}{2} + 1 + d^{2\eta}$	$d^{6-4\eta}\log n$	$d\log n$	$d\frac{\log n}{\log q}$
all	$q = \frac{d(d-1)}{2} + 2$	$d^6 \frac{\log n}{\log d}$	$d\log n$	$d\frac{\log n}{\log q}$

Notice that for $q > cd^2/2$, $c > 1$ the sizes in the above theorem is within a factor of d of the lower bound. Constructing almost optimal (within $poly(d)$) (n, q, d)-PHF for $q = o(d^2)$ is still a challenging open problem. Some nearly optimal constructions of (n, q, d)-PHF for $q = o(d^2)$ are given in [49, 58].

The (n, q, d)-perfect hash families for $d \leq 6$ are studied in [8, 10, 20, 21, 23, 49, 54, 65]. In this paper we prove

Theorem 2. *If q is prime power and $d \leq \log n/(8 \log \log n)$ then there is a linear time construction of (n, q, d)-PHF of size $O\left(d^3 \log n/g(q, d)\right)$ where $g(q, d) = (1 - 1/q)(1 - 2/q) \cdots (1 - (d-1)/q)$.*

Using the lower bound in [37] we show that the size in the above theorem is within a factor of d^4 of the lower bound when $q = d + O(1)$ and within a factor of d^3 for $q > cd$ for some $c > 1$.

2.2 Dense Perfect Hash Family

We say that H is an $(1 - \epsilon)$-*dense* (n, q, d)-PHF if for every subset $S \subseteq [n]$ of size $|S| = d$ there are at least $(1 - \epsilon)|H|$ hash functions $h \in H$ such that $h|_S$ is injective on S.

The following improves the results that can be obtained from [13, 14]

Theorem 3. *Let q be a power of prime. If $\epsilon > 4(d(d - 1)/2 + 1)/q$ then there is a $(1 - \epsilon)$-dense (n, q, d)-PHF of size*

$$O\left(\frac{d^2 \log n}{\epsilon \log(\epsilon q/e(d(d - 1)/2 + 1))}\right)$$

that can be constructed in linear time.

If $(d(d-1)/2+1)/(q-1) \leq \epsilon \leq 4(d(d-1)/2+1)/q$ then there is a $(1-\epsilon)$-dense (n, q, d)-PHF of size

$$O\left(\frac{q^2 d^2 \log n}{\epsilon(q - (d(d - 1)/2 + 1)/\epsilon)^2}\right)$$

that can be constructed in linear time.

We also prove (what we believe) two folklore results that show that the bounds on the size and ϵ in the above theorem are almost tight. First, we show that the size of any $(1 - \epsilon)$-dense (n, q, d)-PHF is $\Omega(d \log n/(\epsilon \log q))$. Second, we show that no $(1 - \epsilon)$-dense (n, q, d)-PHF exists when $\epsilon < d(d - 1)/(2q) + O((d^2/q)^2)$. Notice that for $q \geq (d/\epsilon)^{1+c}$, where $c > 1$ is any constant, the size of the construction in Theorem 3, $O\left(d^2 \log n/(\epsilon \log q)\right)$, is within a factor d of the lower bound. Also the bound on ϵ is asymptotically tight.

For the rest of this section we will only state the results for the non-dense d-restriction problems. Results similar to Theorem 3 can be easily obtained using the same technique.

2.3 Cover-Free Families

Let X be a set with N elements and let $\mathcal{B}$ be a set of subsets (blocks) of X. We say that $(X, \mathcal{B})$ is (w, r)-*cover-free family* $((w, r)$-CFF), [47], if for any w blocks $B_1, \ldots, B_w \in \mathcal{B}$ and any other r blocks $A_1, \ldots, A_r \in \mathcal{B}$, we have

$$\bigcap_{i=1}^{w} B_i \not\subseteq \bigcup_{j=1}^{r} A_j.$$

Let $N((w, r), n)$ denotes the minimum number of points in any (w, r)-CFF having n blocks. Here we will study CFF when $w = o(r)$ (or $r = o(w)$). We will write $(n, (w, r))$-CFF when we want to emphasize the number of blocks.

When $w = 1$, the problem is called *group testing*. The problem of group testing which was first presented during World War II was presented as follows [30,56]: Among n soldiers, at most r carry a fatal virus. We would like to blood test the soldiers to detect the infected ones. Testing each one separately will give n tests. To minimize the number of tests we can mix the blood of several soldiers and test the mixture. If the test comes negative then none of the tested soldiers are infected. If the test comes out positive, we know that at least one of them is infected. The problem is to come up with a small number of tests.

This problem is equivalent to $(n, (1, r))$-CFF and is equivalent to finding a small set $\mathcal{F} \subseteq \{0, 1\}^n$ such that for every $1 \le i_1 < i_2 < \cdots < i_d \le n$ and every $1 \le j \le d$ there is $a \in \mathcal{F}$ such that $a_{i_k} = 0$ for all $k \neq j$ and $a_{i_j} = 1$.

Group testing has the following lower bound [31,32,36]

$$N((1, r), n) \ge \Omega\left(\frac{r^2}{\log r} \log n\right). \tag{1}$$

It is known that a group testing of size $O(r^2 \log n)$ can be constructed in linear time [30,43,60].

An $(n, (w, r))$-CFF can be regarded as a set $\mathcal{F} \subseteq \{0, 1\}^n$ such that for every $1 \le i_1 < i_2 < \cdots < i_d \le n$ where $d = w + r$ and every $J \subset [d]$ of size $|J| = w$ there is $a \in \mathcal{F}$ such that $a_{i_k} = 0$ for all $k \notin J$ and $a_{i_j} = 1$ for all $j \in J$. Then $N((w, r), n)$ is the minimum size of such $\mathcal{F}$.

It is known that, [67],

$$N((w, r), n) \ge \Omega\left(\frac{d\binom{d}{w}}{\log \binom{d}{w}} \log n\right).$$

Using union bound it is easy to show

Lemma 1. *For* $d = w + r = o(n)$ *we have* $N((w, r), n) \le O\left(\sqrt{wrd} \cdot \binom{d}{w} \log n\right).$

It follows from [65], that for infinite sequence of integers n, an $(n, (w, r))$-CFF of size $M = O\left((wr)^{\log^* n} \log n\right)$ can be constructed in polynomial time. For constant d, the (n, d)-universal set over $\Sigma = \{0, 1\}$ constructed in [57] of size $M = O(2^{3d} \log n)$ (and in [58] of size $M = 2^{d+O(\log^2 d)} \log n$) is $(n, (w, r))$-CFF for any w and r of size $O(\log n)$. See also [48]. In [13], Bshouty gave the following locally explicit constructions of $(n, (w, r))$-CFF that can be constructed in (almost) linear time in their sizes (the third column in the table).

n	w	Linear time Size=	Upper Bound	Lower Bound
I.S	$O(1)$	$\frac{r^{w+2}}{\log r} \log n$	$r^{w+1} \log n$	$\frac{r^{w+1}}{\log r} \log n$
all	$O(1)$	$\frac{r^{w+3}}{\log r} \log n$	$r^{w+1} \log n$	$\frac{r^{w+1}}{\log r} \log n$
I.S.	$o(r)$	$\frac{w^2(ce)^w r^{w+2}}{\log r} \log n$	$\frac{r^{w+1}}{(w/e)^{w-1/2}} \log n$	$\frac{r^{w+1}}{(w/e)^{w+1} \log r} \log n$
all	$o(r)$	$\frac{w^3(ce)^w r^{w+3}}{\log r} \log n$	$\frac{r^{w+1}}{(w/e)^{w-1/2}} \log n$	$\frac{r^{w+1}}{(w/e)^{w+1} \log r} \log n$

In the table, $c > 1$ is any constant. We also added to the table the non-constructive upper bound in the forth column and the lower bound in the fifth column.

In this paper we prove

Theorem 4. *For any constant $c > 1$, the following $(n, (w, r))$-CFF can be constructed in linear time in their sizes*

n	w	Linear time. Size=$O(\ )$	Upper Bound	Lower Bound
all	$O(1)$	$r^{w+1} \log n$	$r^{w+1} \log n$	$\frac{r^{w+1}}{\log r} \log n$
all	$o(r)$	$(ce)^w r^{w+1} \log n$	$\frac{r^{w+1}}{(w/e)^{w-1/2}} \log n$	$\frac{r^{w+1}}{(w/e)^{w+1} \log r} \log n$

Notice that when $w = O(1)$ the size of the construction matches the upper bound obtained with union bound and is within a factor of $\log r$ of the lower bound.

See the results for Separating Hash Family in the full paper [15].

3   Preliminary Constructions

A *linear code* over the field $\mathcal{F}_q$ is a linear subspace $C \subset \mathcal{F}_q^m$. Elements in the code are called *words*. A linear code C is called $[m, k, d]_q$ *linear code* if $C \subset \mathcal{F}_q^m$ is a linear code, $|C| = q^k$ and for every two words v and u in the code dist$(v, u) := |\{i \mid v_i \neq u_i\}| \geq d$. The q-ary entropy function is

$$H_q(p) = p \log_q \frac{q-1}{p} + (1-p) \log_q \frac{1}{1-p}.$$

The following is from [60] (Theorem 2)

Lemma 2. *Let q be a prime power, m and k positive integers and $0 \leq \delta \leq 1$. If $k \leq (1 - H_q(\delta))m$, then an $[m, k, \delta m]_q$ linear code can be globally explicit constructed in time $O(mq^k)$.*

All the results in this paper uses Lemma 2 and therefore they are globally explicit constructions. In the full paper, [15], we prove the following

Lemma 3. *Let q be a prime power, $1 < h < q/4$ and*

$$m = \left\lceil \frac{h \ln(q(n+1))}{\ln q - \ln h - 1} \right\rceil.$$

A

$$\left[m, \left\lceil \frac{\log(n+1)}{\log q} \right\rceil, \left(1 - \frac{1}{h}\right) m \right]_q$$

linear code can be constructed in time $O(hqn \log(qn))$.

When $h = \Theta(q)$ we show

Lemma 4. *Let q be a prime power, $2 \le q/4 \le h \le q - 1$ and*

$$m = \left\lceil \frac{4(q-1)^2 h \ln(q(n+1))}{(q-h)^2} \right\rceil.$$

A

$$\left[m, \left\lceil \frac{\log(n+1)}{\log q} \right\rceil, \left(1 - \frac{1}{h}\right) m \right]_q$$

linear code can be constructed in time $O(h(q^2/(q-h)^2)n \log(qn))$.

4   Main Results

In this section we give two main results that will be used throughout the paper
   Let $I \subseteq [n]^2$. Define the following homogeneous polynomial $H_I = \prod_{(i_1, i_2) \in I} (x_{i_1} - x_{i_2})$. We denote by $\mathcal{H}_d \subseteq \mathcal{F}_q[x_1, \ldots, x_n]$ the class of all such polynomials of degree at most d. A *hitting set* for $\mathcal{H}_d$ over $\mathcal{F}_q$ is a set of assignment $A \subseteq \mathcal{F}_q^n$ such that for every $H \in \mathcal{H}_d, H \not\equiv 0$, there is $a \in A$ where $H(a) \ne 0$. A $(1 - \epsilon)$-*dense hitting set* for $\mathcal{H}_d$ over $\mathcal{F}_q$ is a set of assignment $A \subseteq \mathcal{F}_q^n$ such that for every $H \in \mathcal{H}_d, H \not\equiv 0$,

$$\mathbf{Pr}_{a \in A}[H(a) \ne 0] > 1 - \epsilon$$

where the probability is over the choice of a from the uniform distribution on A. When $H(a) \ne 0$ then we say that the assignment a *hits* H and H is *not zero on* a.
   We prove

Lemma 5. *Let $n > q, d$. If $q > 4(d+1)$ is prime power then there is a hitting set for $\mathcal{H}_d$ of size*

$$m = \left\lceil \frac{(d+1) \log(q(n+1))}{\log(q/e(d+1))} \right\rceil = O\left(\frac{d \log n}{\log(q/e(d+1))} \right)$$

that can be constructed in time $O(mn) = O(dqn \log(qn))$.
   If $d + 2 \le q \le 4(d+1)$ is prime power then there is a hitting set for $\mathcal{H}_d$ of size

$$m = \left\lceil \frac{4(q-1)^2(d+1) \ln(q(n+1))}{(q-d-1)^2} \right\rceil = O\left(\frac{dq^2 \log n}{(q-d-1)^2} \right)$$

that can be constructed in time $O(mn) = O(d(q^2/(q-d-1)^2)n \log(qn))$.

Proof. Consider the code C

$$\left[m, \left\lceil \frac{\log(n+1)}{\log q} \right\rceil, \left(1 - \frac{1}{d+1}\right)m\right]_q$$

constructed in Lemma 3 and Lemma 4. The number of non-zero words in the code is at least n. Take any n distinct non-zero words $c^{(1)}, \cdots, c^{(n)}$ in C and define the assignments $a^{(i)} \in \mathcal{F}_q^n$, $i = 1, \ldots, m$ where $a_j^{(i)} = c_i^{(j)}$. Let $H_I \in \mathcal{H}_d, H_I \not\equiv 0$. Then $H_I = \prod_{(i_1,i_2)\in I}(x_{i_1} - x_{i_2}) \not\equiv 0$ where $|I| \leq d$. For each $t := x_{i_1} - x_{i_2}$ we have $(t(a^{(1)}), \ldots, t(a^{(m)}))^T = c^{(i_1)} - c^{(i_2)} \in C$ is a non-zero word in C and therefore t is zero on at most $m/(d+1)$ assignments. Therefore H_I is zero on at most $dm/(d+1) < m$ assignment. This implies that there is an assignment in A that hits H_I. □

Notice that the size of the hitting set is mn and therefore the time complexity in the above lemma is linear in the size of the hitting set.

In the same way one can prove

Lemma 6. *Let q be a prime power. If $q > 4(d+1)/\epsilon$ be a prime power. Let $n > q, d$. There is a $(1 - \epsilon)$-dense hitting set for $\mathcal{H}_d$ of size*

$$m = \left\lceil \frac{(d+1)\log(q(n+1))}{\epsilon \log(\epsilon q/e(d+1))} \right\rceil = O\left(\frac{d \log n}{\epsilon \log(\epsilon q/e(d+1))}\right)$$

that can be constructed in time $O(dqn \log(qn)/\epsilon)$.

If $(d+1)/\epsilon + 1 \leq q \leq 4(d+1)/\epsilon$ be a prime power. Let $n > q, d$. There is a $(1 - \epsilon)$-dense hitting set for $\mathcal{H}_d$ of size

$$m = \left\lceil \frac{4(q-1)^2(d+1)\ln(q(n+1))}{(q-(d+1)/\epsilon)^2\epsilon} \right\rceil = O\left(\frac{dq^2 \log n}{(q-(d+1)/\epsilon)^2\epsilon}\right)$$

that can be constructed in time $O(d(q^2/(q-d-1)^2)n \log(qn)/\epsilon)$.

We note here that such result cannot be achieved when $q < d/\epsilon$ [13].

5 Proof of the Theorems

5.1 Perfect Hash Family

Here we prove Theorem 1

Proof. Consider the set of functions

$$\mathcal{F} = \{\Delta_{\{i_1,\ldots,i_d\}}(x_1,\ldots,x_n) \mid 1 \leq i_1 < \cdots < i_d \leq n\}$$

in $\mathcal{F}_q[x_1, x_2, \ldots, x_n]$ where

$$\Delta_{\{i_1,\ldots,i_d\}}(x_1,\ldots,x_n) = \prod_{1\leq k<j\leq d} (x_{i_k} - x_{i_j}).$$

It is clear that a hitting set for $\mathcal{F}$, when each assignment is regarded as functions $f : [n] \to \mathcal{F}_q$, is (n, q, d)-PHF. Now since $\mathcal{F} \subseteq \mathcal{H}_{d(d-1)/2+1}$ the result follows from Lemma 5. □

When $q > d(d-1)/2$ is not a power of prime number then we can take the nearest prime $q' < q$ and construct an (n, q', d)-PHF that is also (n, q, d)-PHF. It is known that the nearest prime $q' \geq q - \Theta(q^{.525})$, [19], and therefore the result in the above table is also true for any integer $q \geq d(d+1)/2 + O(d^{1.05})$.

5.2   Perfect Hash Family for Small d

We now prove Theorem 2

Proof. If $q > d^2$ then the construction in Theorem 1 has the required size. Let $q \leq d^2$. We first use Theorem 1 to construct an (n, d^3, d)-PHF H_1 of size $O(d^2 \log n / \log d)$ in linear time. Then a (d^3, q, d)-PHF H_2 of size $O(d \log d / g(q, d))$ can be constructed in time, [7,58],

$$\binom{d^3}{d} q^{1+\lceil \log d^3 / \log q \rceil (d-1)} \leq d^{3d} q^d d^{3d} \leq d^{8d} < n.$$

Then $H = \{h_2(h_1) \mid h_2 \in H_2, h_1 \in H_1\}$ is (n, q, d)-PHF of the required size.   □

It follows from [37] that this bound is within a factor of d^4 of the lower bound when $q = d + O(1)$ and within a factor of $d^3 \log d$ of the lower bound when $q > cd$ for some constant $c > 1$. See details in the following

Lemma 7. *[37] Let $n > d^{2+\epsilon}$ for some constant $\epsilon > 0$. Any (n, q, d)-PHF is of size at least*

$$\Omega \left(\frac{(q-d+1)}{q \log(q-d+2)} \frac{\log n}{g(q, d)} \right).$$

In particular, for $q = d + O(1)$ the bound is

$$\Omega \left(\frac{\log n}{dg(q, d)} \right)$$

and for $q > cd$ for some constant $c > 1$ the bound is

$$\Omega \left(\frac{\log n}{(\log d) g(q, d)} \right).$$

5.3   Dense Perfect Hash

Using Lemma 6 with the same proof as in Theorem 1 we get Theorem 3.

In the appendix we show that the size in the above Theorem and the constraint on ϵ are tight.

For the rest of the paper we will only state the results for the non-dense d-restriction problems. The results for the dense d-restriction problems follows immediately from applying Lemma 6.

5.4   Cover-Free Families

We now prove the following

Theorem 5. *Let $q \geq wr + 2$ be a prime power. Let $S \subseteq \mathcal{F}_q^n$ be a hitting set for $\mathcal{H}_{wr}$. Given a $(q, (w, r))$-CFF of size M that can be constructed in linear time one can construct an $(n, (w, r))$-CFF of size $M \cdot |S|$ that can be constructed in linear time.*

In particular, there is an (w, r)-CFF of size

$$\binom{q}{w} \cdot |S|$$

that can be constructed in linear time in its size.

In particular, for any constant $c > 1$, the following (w, r)-CFF can be constructed in linear time in their sizes

n	w	Linear time. Size=$O(\ )$	Upper Bound	Lower Bound
all	$O(1)$	$r^{w+1} \log n$	$r^{w+1} \log n$	$\frac{r^{w+1}}{\log r} \log n$
all	$o(r)$	$(ce)^w r^{w+1} \log n$	$\frac{r^{w+1}}{(w/e)^{w-1/2}} \log n$	$\frac{r^{w+1}}{(w/e)^{w+1} \log r} \log n$

Proof. Let $d = r + w$. Consider the set of non-zero functions

$$\mathcal{M} = \{\Delta_{\mathbf{i}} \mid \mathbf{i} \in [n]^d, \ i_1, i_2, \ldots, i_d \text{ are distinct}\}$$

where

$$\Delta_{\mathbf{i}}(x_1, \ldots, x_n) = \prod_{1 \leq k \leq w \text{ and } w < j \leq d} (x_{i_k} - x_{i_j}).$$

Then S is a hitting set for $\mathcal{M}$.

Let $\mathcal{F} \subseteq \{0, 1\}^q$ be a $(q, (w, r))$-CFF of size M. Regard each $f \in \mathcal{F}$ as a function $f : \mathcal{F}_q \to \{0, 1\}$. It is easy to see that

$$\{(f(b_1), f(b_2), \ldots, f(b_n)) \mid b \in S, f \in \mathcal{F}\} \subseteq \{0, 1\}^n$$

is (w, r)-CFF of size $|\mathcal{F}| \cdot |S| = M \cdot |S|$.

Now for every subset $R \subseteq \mathcal{F}_q$ define the function $\chi_R : \mathcal{F}_q \to \{0, 1\}$ where for $\beta \in \mathcal{F}_q$ we have $\chi_R(\beta) = 1$ if $\beta \in R$ and $\chi_R(\beta) = 0$ otherwise. Then $\{\chi_R \mid R \subseteq \mathcal{F}_q, |R| = w\} \subseteq \{0, 1\}^{\mathcal{F}_q}$ is a $(q, (w, r))$-CFF of size $\binom{q}{w}$. Therefore

$$C = \{(\chi_R(b_1), \chi_R(b_2), \ldots, \chi_R(b_n)) \mid b \in S, R \subseteq \mathcal{F}_q, |R| = w\}$$

is (w, r)-CFF of size

$$|C| \leq \binom{q}{w} |S|.$$

Now for the results in the table consider a constant $c > c' > 1$ and let q be a power of prime such that $q = c'wr + o(wr)$. This is possible by [19]. By Lemma 5 there is a hitting set S for $\mathcal{H}_{wr}$ of size $O(wr \log n)$. This gives a (w, r)-CFF of size

$$O\left(\binom{q}{w} \cdot wr \log n\right) = O\left(\left(\frac{qe}{w}\right)^w wr \log n\right) = O\left((ce)^w r^{w+1} \log n\right)$$

that can be constructed in linear time in its size.   □

5.5   Open Problems

Here we give some open problems

1. Find a polynomial time almost optimal (within $poly(d)$) construction of (n, q, d)-PHF for $q = o(d^2)$. Using the techniques in [58] it is easy to give an almost optimal construction for (n, q, d)-PHF when $q = d^2/c$ for any constant $c > 1$. Unfortunately the size of the construction is within a factor of $d^{O(c)}$ of the lower bound.

2. In this paper we gave a construction of $(n, (w, r))$-CFF of size

$$\min((2e)^w r^{w+1}, (2e)^r w^{r+1}) \log n$$
$$= \binom{w+r}{r} 2^{\min(w \log w, r \log r)(1+o(1))} \log n \qquad (2)$$

that can be constructed in linear time. Fomin et. al. in [38] gave a construction of size

$$\binom{w+r}{r} 2^{O\left(\frac{r+w}{\log \log(r+w)}\right)} \log n \qquad (3)$$

that can be constructed in linear time. The former bound, (2), is better than the latter when $w \geq r \log r \log \log r$ or $r \geq w \log w \log \log w$. We also note that the former bound, (2), is almost optimal, i.e.,

$$\binom{w+r}{r}^{1+o(1)} \log n = N^{1+o(1)} \log n,$$

where $N \log n$ is the optimal size, when $r = w^{\omega(1)}$ or $r = w^{o(1)}$ and the latter bound, (3), is almost optimal when

$$o(w \log \log w \log \log \log w) = r = \omega\left(\frac{w}{\log \log w \log \log \log w}\right).$$

Find a polynomial time almost optimal (within $N^{o(1)}$) construction for (w, r)-CFF when $w = \omega(1)$.

3. A construction is global explicit if it runs in deterministic polynomial time in the size of the construction. A local explicit construction is a construction where one can find any bit in the construction in time poly-log in the size of the construction. The constructions in this paper are linear time global explicit constructions. Some almost linear time almost optimal local explicit constructions follows from my recently published paper [14]. It is interesting to find other explicit constructions that are more optimal.

References

1. Alon, N.: Explicit construction of exponential sized families of k-independent sets. Discrete Math. **58**, 191–193 (1986)
2. Alon, N., Asodi, V.: Learning a hidden subgraph. In: Díaz, J., Karhumäki, J., Lepistö, A., Sannella, D. (eds.) ICALP 2004. LNCS, vol. 3142, pp. 110–121. Springer, Heidelberg (2004)
3. Alon, N., Beigel, R., Kasif, S., Rudich, S., Sudakov, B.: Learning a Hidden Matching. SIAM J. Comput. **33**(2), 487–501 (2004)
4. Angluin, D., Chen, J.: Learning a Hidden Hypergraph. Journal of Machine Learning Research **7**, 2215–2236 (2006)
5. Angluin, D., Chen, J.: Learning a hidden graph using $O(\log n)$ queries per edge. J. Comput. Syst. Sci. **74**(4), 546–556 (2008)
6. Alon, N., Naor, M.: Rerandomization, witnesses for Boolean matrix multiplication and construction of perfect hash functions. Algorithmica **16**, 434–449 (1996)
7. Alon, N., Moshkovitz, D., Safra, S.: Algorithmic construction of sets for k-restrictions. ACM Transactions on Algorithms **2**(2), 153–177 (2006)
8. Atici, M., Magliveras, S.S., Stinson, D.R., Wei, W.-D.: Some Recursive Constructions for Perfect Hash Families. Journal of Combinatorial Designs **4**(5), 353–363 (1996)
9. Alon, N., Bruck, J., Naor, J., Naor, M., Roth, R.M.: Construction of asymptotically good low-rate error-correcting codes through pseudo-random graphs. IEEE Transactions on Information Theory **38**(2), 509–516 (1992)
10. Blackburn, S.R.: Perfect Hash Families: Probabilistic Methods and Explicit Constructions. Journal of Combinatorial Theory, Series A **92**, 54–60 (2000)
11. Blackburn, S.R.: Combinatorics and threshold cryptography. In: Holroyd, F.C., Quinn, K.A.S., Rowley, C., Webb, B.S. (eds.) Combinatorial Designs and Their Applications. Research Notes in Mathematics, vol. 403, pp. 44–70. CRC Press, London (1999)
12. Blackburn, S.R., Burmester, M., Desmedt, Y.G., Wild, P.R.: Efficient multiplicative sharing schemes. In: Maurer, U.M. (ed.) EUROCRYPT 1996. LNCS, vol. 1070, pp. 107–118. Springer, Heidelberg (1996)
13. Bshouty, N.H.: Testers and their applications. In: ITCS 2014, pp. 327–352 (2014). Full version: Electronic Colloquium on Computational Complexity (ECCC) 19, 11 (2012)
14. Bshouty, N.H.: Dense Testers: Almost Linear Time and Locally Explicit Constructions. Electronic Colloquium on Computational Complexity (ECCC) 22: 6 (2015)
15. Bshouty, N.H.: Linear time Constructions of some d-Restriction Problems. CoRR abs/1406.2108. (2014)
16. Beigel, R., Alon, N., Kasif, S., Apaydin, M.S., Fortnow, L.: An optimal procedure for gap closing in whole genome shotgun sequencing. RECOMB **2001**, 22–30 (2001)
17. Blackburn, S.R., Etzion, T., Stinson, D.R., Zaverucha, G.M.: A bound on the size of separating hash families. Journal of Combinatorial Theory, Series A **115**(7), 1246–1256 (2008)
18. Bouvel, M., Grebinski, V., Kucherov, G.: Combinatorial Search on Graphs Motivated by Bioinformatics Applications: A Brief Survey. WG **2005**, 16–27 (2005)
19. Baker, R.C., Harman, G., Pintz, J.: The difference between consecutive primes. II. Proceedings of the London Mathematical Society 83(3), 532–562 (2001)
20. Barwick, S.G., Jackson, W.-A.: Geometric constructions of optimal linear perfect hash families. Finite Fields and Their Applications **14**(1), 1–13 (2008)

21. Barwick, S.G., Jackson, W.-A., Quinn, C.T.: Optimal Linear Perfect Hash Families with Small Parameters. Journal of Combinatorial Designs **12**(5), 311–324 (2004)

22. Bazrafshan, M., van Trung, T.: Bounds for separating hash families. Journal of Combinatorial Theory, Series A **118**(3), 1129–1135 (2011)

23. Blackburn, S.R., Wild, P.R.: Optimal linear perfect hash families. Journal of Combinatorial Theory, Series A **83**(2), 233–250 (1998)

24. Chang, H., Chen, H.-B., Fu, H.-L., Shi, C.-H.: Reconstruction of hidden graphs and threshold group testing. J. Comb. Optim. **22**(2), 270–281 (2011)

25. Canetti, R., Garay, J., Itkis, G., Micciancio, D., Naor, M., Pinkas, B.: Multicast security: a taxonomy and some efficient constructions. In: Proceedings of INFOCOM 1999, vol. 2, pp. 708–716 (1999)

26. Chen, H.-B., Hwang, F.K.: A survey on nonadaptive group testing algorithms through the angle of decoding. J. Comb. Optim. **15**(1), 49–59 (2008)

27. Czech, Z.J., Havas, G., Majewski, B.S.: Perfect hashing. Theoret. Comput. Sci. **182**, 1–143 (1997)

28. Chin, F.Y.L., Leung, H.C.M., Yiu, S.-M.: Non-adaptive complex group testing with multiple positive sets. Theor. Comput. Sci. **505**, 11–18 (2013)

29. Dyer, M., Fenner, T., Frieze, A., Thomason, A.: On key storage in secure networks. J. Cryptol. **8**, 189–200 (1995)

30. Du, D.Z., Hwang, F.K.: Combinatorial group testing and its applications. Series on Applied Mathematics. 2nd edn., vol. 12. World Scientific, New York (2000)

31. Dýachkov, A.G., Rykov, V.V.: Bounds on the length of disjunctive codes. Problemy Peredachi Inf. **18**(3), 7–13 (1982)

32. Dýachkov, A.G., Rykov, V.V., Rashad, A.M.: Superimposed distance codes. Problems Control Inform. Theory/Problemy Upravlen. Teor. Inform. **18**(4), 237–250 (1989)

33. Desmedt, Y., Safavi-Naini, R., Wang, H., Batten, L., Charnes, C., Pieprzyk, J.: Broadcast anti-jamming systems. Comput. Networks 35, 223–236 (2001)

34. D'yachkov, A., Vilenkin, P., Macula, A., Torney, D.: Families of finite sets in which no intersection of ℓ sets is covered by the union of s others. J. Comb. Theory Ser. A. 99, 195–218 (2002)

35. Gao, H., Hwang, F.K., Thai, M.T., Wu, W., Znati, T.: Construction of d(H)-disjunct matrix for group testing in hypergraphs. J. Comb. Optim. **12**(3), 297–301 (2006)

36. Füredi, Z.: On r-cover-free families. Journal of Combinatorial Theory, Series A **73**(1), 172–173 (1996)

37. Fredman, M.L., Komlós, J.: On the size of seperating systems and families of perfect hash function. SIAM J. Algebraic and Discrete Methods **5**(1), 61–68 (1984)

38. Fomin, F.V., Lokshtanov, D., Saurabh, S.: Efficient Computation of Representative Sets with Applications in Parameterized and Exact Algorithms. SODA **2014**, 142–151 (2014)

39. Fiat, A., Naor, M.: Broadcast encryption. In: Stinson, D.R. (ed.) CRYPTO 1993. LNCS, vol. 773, pp. 480–491. Springer, Heidelberg (1994)

40. Grebinski, V., Kucherov, G.: Reconstructing a Hamiltonian Cycle by Querying the Graph: Application to DNA Physical Mapping. Discrete Applied Mathematics **88**(1–3), 147–165 (1998)

41. Garay, J.A., Staddon, J., Wool, A.: Long-Lived Broadcast Encryption. In: Bellare, M. (ed.) CRYPTO 2000. LNCS, vol. 1880, pp. 333–352. Springer, Heidelberg (2000)

42. Huang, T., Wang, K., Weng, C.-W.: A class of error-correcting pooling designs over complexes. J. Comb. Optim. **19**(4), 486–491 (2010)

43. Indyk, P., Ngo, H.Q., Rudra, A.: Efficiently decodable non-adaptive group testing. In: The 21st Annual ACM-SIAM Symposium on Discrete Algorithms (SODA 2010), pp. 1126–1142 (2010)
44. Körner, J.: Fredman-Komlós bounds and information theory. SIAM J. Algebraic and Discrete Methods **7**(4), 560–570 (1986)
45. Körner, J., Marton, K.: New bounds for perfect hashing via information theory. Europ. J. of Combinatorics **9**(6), 523–530 (1988)
46. Kumar, R., Rajagopalan, S., Sahai, A.: Coding constructions for blacklisting problems without computational assumptions. In: Wiener, M. (ed.) CRYPTO 1999. LNCS, vol. 1666, pp. 609–623. Springer, Heidelberg (1999)
47. Kautz, W.H., Singleton, R.C.: Nonrandom binary superimposed codes. IEEE Trans. Inform. Theory **10**(4), 363–377 (1964)
48. Liu, L., Shen, H.: Explicit constructions of separating hash families from algebraic curves over finite fields. Designs, Codes and Cryptography **41**(2), 221–233 (2006)
49. Martirosyan, S.: Perfect Hash Families, Identifiable Parent Property Codes and Covering Arrays. Dissertation zur Erlangung des Grades eines Doktors der Naturwissenschaften (2003)
50. Mehlborn, K.: Data Structures and Algorithms. 1. Sorting and Searching. Springer, Berlin (1984)
51. Mehlhorn, K.: On the program size of perfect and universal hash functions. In: Proceedings of the 23rd IEEE Symposium on Foundations of Computer Science (FOCS 1982), pp. 170–175 (1982)
52. Mitchell, C.J., Piper, F.C.: Key storage in secure networks. Discrete Appl. Math. **21**, 215–228 (1988)
53. Macula, A.J., Popyack, L.J.: A group testing method for finding patterns in data. Discret Appl Math. **144**, 149–157 (2004)
54. Martirosyan, S., van Trung, T.: Explicit constructions for perfect hash families. Designs, Codes and Cryptography **46**(1), 97–112 (2008)
55. Nilli, A.: Perfect hashing and probability. Combinatorics, Probability and Computing **3**(3), 407–409 (1994)
56. Ngo, H.Q., Du, D.Z.: A survey on combinatorial group testing algorithms with applications to DNA library screening. Theoretical Computer Science **55**, 171–182 (2000)
57. Naor, J., Naor, M.: Small-bias probability spaces: efficient constructions and applications. SIAM J. Comput. **22**(4), 838–856 (1993)
58. Naor, M., Schulman, L.J., Srinivasan, A.: Splitters and Near-optimal Derandomization. FOCS **95**, 182–191 (1995)
59. Newman, I., Wigderson, A.: Lower bounds on formula size of Boolean functions using hypergraph entropy. SIAM J. Discrete Math. **8**, 536–542 (1995)
60. Porat, E., Rothschild, A.: Explicit Nonadaptive Combinatorial Group Testing Schemes. IEEE Transactions on Information Theory **57**(12), 7982–7989 (2011)
61. Reyzin, L., Srivastava, N.: Learning and verifying graphs using queries with a focus on edge counting. In: Hutter, M., Servedio, R.A., Takimoto, E. (eds.) ALT 2007. LNCS (LNAI), vol. 4754, pp. 285–297. Springer, Heidelberg (2007)
62. Stinson, D.R.: On some methods for unconditionally secure key distribution and broadcast encryption. Des. Codes Cryptogr. **12**, 215–243 (1997)
63. Stinson, D.R., Wei, R.: Combinatorial properties and constructions of traceability schemes and frameproof codes. SIAM J. Discrete Math. **11**, 41–53 (1998)
64. Safavi-Naini, R., Wang, H.: Multireceiver authentication codes: models, bounds, constructions, and extensions Inform. Comput. **151**, 148–172 (1999)

65. Stinson, D.R., Wei, R., Zhu, L.: New constructions for perfect hash families and related structures using combintorial designs and codes. J. Combin. Designs. **8**(3), 189–200 (2000)
66. Stinson, D.R., Wei, R., Chen, K.: On generalised separating hash families. Journal of Combinatorial Theory, Series A **115**(1), 105–120 (2008)
67. Stinson, D.R., Wei, R., Zhu, L.: Some new bounds for cover-free families. Journal of Combinatorial Theory, Series A **90**(1), 224–234 (2000)
68. Stinson, D.R., van Trung, T.: Some new results on key distribution patterns and broadcast encryption. Des. Codes Cryptogr. **14**, 261–279 (1998)
69. Stinson, D.R., van Trung, T., Wei, R.: Secure frameproof codes, key distribution patterns, group testing algorithms and related structures. J. Stat. Planning and Inference **86**(2), 595–617 (2000)
70. Torney, D.C.: Sets pooling designs. Ann. Comb. **3**, 95–101 (1999)
71. Wang, H., Xing, C.P.: Explicit Constructions of perfect hash families from algebraic curves over finite fields. J. of Combinatorial Theory, Series A 93(1), 112–124 (2001)

Efficiently Testing T-Interval Connectivity in Dynamic Graphs

Arnaud Casteigts[1]([✉]), Ralf Klasing[1], Yessin M. Neggaz[1],
and Joseph G. Peters[2]

[1] LaBRI, CNRS, University of Bordeaux, Talence, France
`arnaud.casteigts@labri.fr`
[2] School of Computing Science, Simon Fraser University, Burnaby, BC, Canada

Abstract. Many types of dynamic networks are made up of durable entities whose links evolve over time. When considered from a *global* and *discrete* standpoint, these networks are often modelled as evolving graphs, i.e. a sequence of static graphs $\mathcal{G} = \{G_1, G_2, ..., G_\delta\}$ such that $G_i = (V, E_i)$ represents the network topology at time step i. Such a sequence is said to be T-interval connected if for any $t \in [1, \delta - T + 1]$ all graphs in $\{G_t, G_{t+1}, ..., G_{t+T-1}\}$ share a common connected spanning subgraph. In this paper, we consider the problem of deciding whether a given sequence $\mathcal{G}$ is T-interval connected for a given T. We also consider the related problem of finding the largest T for which a given $\mathcal{G}$ is T-interval connected. We assume that the changes between two consecutive graphs are arbitrary, and that two operations, *binary intersection* and *connectivity testing*, are available to solve the problems. We show that $\Omega(\delta)$ such operations are required to solve both problems, and we present optimal $O(\delta)$ online algorithms for both problems.

Keywords: T-interval connectivity · Dynamic graphs · Time-varying graphs

1 Introduction

Dynamic networks consist of entities making contact over time with one another. The types of dynamics resulting from these interactions are varied in scale and nature. For instance, some of these networks remain connected at all times [9]; others are always disconnected [6] but still offer some kind of connectivity over time and space (*temporal* connectivity); others are recurrently connected, periodic, etc. All of these contexts can be represented as dynamic graph classes. A dozen such classes were identified in [4] and organized into a hierarchy.

Part of this work was done while Joseph G. Peters was visiting the LaBRI as a guest professor of the University of Bordeaux. This work was partially funded by the ANR projects DISPLEXITY (ANR-11-BS02-014) and DAISIE (ANR-13-ASMA-0004). This study has been carried out in the frame of "the Investments for the future" Programme IdEx Bordeaux CPU (ANR-10-IDEX-03-02).
A long version is available as ArXiv e-print No 1502.00089

V.Th. Paschos and P. Widmayer (Eds.): CIAC 2015, LNCS 9079, pp. 89–100, 2015.
DOI: 10.1007/978-3-319-18173-8_6

Given a dynamic graph, a natural question to ask is to which of the classes this graph belongs. This question is interesting because most of the known classes of dynamic graphs correspond to necessary or sufficient conditions for given distributed problems or algorithms (broadcast, election, spanning trees, token forwarding, etc.). Thus, being able to classify a graph in the hierarchy is useful for determining which problems (or algorithms) can be successfully solved (executed) on that graph. Furthermore, classification tools, such as testing algorithms for given classes, can be useful for choosing a good algorithm in settings where the evolution of a network is not known in advance. An algorithm designer can record topological traces from the real world and then test whether the corresponding dynamic graphs are included in classes that correspond to the topological conditions for the problem at hand [3]. Alternatively, online algorithms that process dynamic graphs as they evolve could accomplish the same goal without the need to collect traces.

Dynamic graphs can be modelled in a number of ways. It is often convenient, when looking at the topology from a global standpoint (e.g. a recorded trace), to represent a dynamic graph as a sequence of static graphs $\mathcal{G} = \{G_1, G_2, ..., G_\delta\}$, each of which corresponds to a given time in the discrete domain (also known as *untimed* evolving graphs [2]). Solutions for testing the inclusion of such a dynamic graph in a handful of basic classes were provided in [3]; these classes are those in which a *journey* (temporal path) or *strict journey* (a journey that traverses at most one edge per G_i) exists between any pair of nodes. In this particular case, the problem reduces to testing whether the transitive closure of (strict) journeys is a complete graph. The transitive closure itself can be computed efficiently in a number of ways [1, 2, 10].

Recently, the class of *T-interval connected* graphs was identified in [8] as playing an important role in several distributed problems, such as determining the size of a network or computing a function of the initial inputs of the nodes. Informally, T-interval connectivity requires that, for every T consecutive graphs in the sequence $\mathcal{G}$, there exists a common connected spanning subgraph. This class generalizes the class of dynamic graphs that are connected at all time instants [9]. Indeed, the latter corresponds to the case that $T = 1$. From a set-theoretic viewpoint, however, every $T > 1$ induces a class of graphs that is a strict subset of that of [9] since a graph that is T-interval connected is obviously 1-interval connected. Hence, T-interval connectivity is more specialized in that sense.

In this paper, we look at the problem of deciding whether a given sequence $\mathcal{G}$ is T-interval connected for a given T. We also consider the related problem of finding the largest T for which the given $\mathcal{G}$ is T-interval connected. We assume that the changes between two consecutive graphs are arbitrary and we do not make any assumptions about the data structures that are used to represent the sequence of graphs. As such, we focus on high-level strategies that work directly at the graph level. Precisely, we consider two graph-level operations as building blocks, which are *binary intersection* (given two graphs, compute their intersection) and *connectivity testing* (given a graph, decide whether it is connected). Put together, these operations have a strong and natural connection with the problems at stake. We first show that both problems require $\Omega(\delta)$ such operations using the basic argument that every graph of the sequence must be considered at least once. More surprisingly, we show that both problems can be solved using only $O(\delta)$ such operations and we develop optimal online algorithms that achieve these matching bounds. Hence, the cost of the operations – both linear in the number of edges – is counterbalanced by efficient high-level logic that could, for instance, benefit from dedicated circuits (or optimized code) for both operations.

The paper is organized as follows. Section 2 presents the main definitions and makes some basic observations, including the fact that both problems can be solved using $O(\delta^2)$ operations (intersections or connectivity tests) by a naive strategy that examines $O(\delta^2)$ intermediate graphs. Section 3 presents a second strategy, yielding upper bounds of $O(\delta \log \delta)$ operations for both problems. Its main interest is in the fact that it can be parallelized, and this allows us to classify both problems as being in **NC** (i.e. Nick's class). Finally, in Section 4 we present an optimal strategy which we use to solve both problems online in $O(\delta)$ operations. This strategy exploits structural properties of the problems to construct carefully selected subsequences of the intermediate graphs. In particular, only $O(\delta)$ of the $O(\delta^2)$ intermediate graphs are selected for evaluation by the algorithms.

2   Definitions and Basic Observations

Graph Model. In this work, we consider dynamic graphs that are given as untimed evolving graphs, that is, a sequence $\mathcal{G} = \{G_1, G_2, ..., G_\delta\}$ of static graphs such that $G_i = (V, E_i)$ accounts for the network topology at (discrete) time i. Observe that V is non-varying; only the set of edges varies. We consider *undirected* edges throughout the paper, which is the setting in which T-interval connectivity was originally introduced. The parameter δ is called the *length* of the sequence $\mathcal{G}$. It corresponds to the number of time steps this graph covers.

Definition 1 (Intersection graph). *Given a (finite) set S of graphs $\{G' = (V, E'), G'' = (V, E''), \dots\}$, we call the graph $(V, \cap\{E', E'', \dots\})$ the intersection graph of S and denote it by $\cap\{G', G'', \dots\}$. When the set consists of only two graphs, we talk about binary intersection and use the infix notation $G' \cap G''$. If the intersection involves a consecutive subsequence $\{G_i, G_{i+1}, \dots, G_j\}$ of a dynamic graph $\mathcal{G}$, then we denote the intersection graph $\cap\{G_i, G_{i+1}, \dots, G_j\}$ simply as $G_{(i,j)}$.*

Definition 2 (T-interval connectivity). *A dynamic graph $\mathcal{G}$ is said to be T-interval connected if the intersection graph $G_{(t,t+T-1)}$ is connected for every $t \in [1, \delta - T + 1]$. In other words, all graphs in $\{G_t, G_{t+1}, ..., G_{t+T-1}\}$ share a common connected spanning subgraph.*

Testing T-Interval Connectivity. We will use T-INTERVAL-CONNECTIVITY to refer to the problem of deciding whether a dynamic graph $\mathcal{G}$ is T-interval connected for a given T, and INTERVAL-CONNECTIVITY to refer to the problem of finding $\max\{T : \mathcal{G} \text{ is } T\text{-interval connected}\}$ for a given $\mathcal{G}$.

Let $\mathcal{G}^T = \{G_{(1,T)}, G_{(2,T+1)}, ..., G_{(\delta-T+1,\delta)}\}$. We call $\mathcal{G}^T$ the Tth *row* in $\mathcal{G}$'s intersection hierarchy, as depicted in Fig. 1. A particular case is $\mathcal{G}^1 = \mathcal{G}$. For any $1 \leq i \leq \delta - T + 1$, we define $\mathcal{G}^T[i] = G_{(i,i+T-1)}$. We call $\mathcal{G}^T[i]$ the ith *element of row* $\mathcal{G}^T$ and i is called the *index of* $\mathcal{G}^T[i]$ *in row* $\mathcal{G}^T$.

Observation 1   *A sequence of graphs $\mathcal{G}$ is T-interval connected if and only if all graphs in $\mathcal{G}^T$ are connected.*

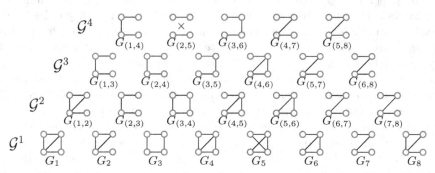

Fig. 1. Example of an intersection hierarchy for a given dynamic graph $\mathcal{G}$ of length $\delta = 8$. Here, $\mathcal{G}$ is 3-interval connected, but not 4-interval connected; $\mathcal{G}^4$ contains a disconnected graph $G_{(2,5)}$ which implies that G_2, G_3, G_4, G_5 share no connected spanning subgraph.

Computational Model. As shown in Observation 1, the concept of T-interval connectivity can be reformulated quite naturally in terms of the connectivity of some intersection graphs. For this reason, we consider two building block operations: *binary intersection* (given two static graphs, compute their intersection) and *connectivity testing* (given a static graph, decide whether it is connected). This approach is suitable for a high-level study of these problems when the details of changes between successive graphs in a sequence are arbitrary. If more structural information about the evolution of the dynamic graphs is known, for example, if it is known that the number of changes between each pair of consecutive graphs is bounded by a constant, then algorithms could benefit from the use of sophisticated data structures and a lower-level approach might be more appropriate.

Observation 2 (Cost of the operations). *Using an adjacency list data structure for the graphs, a binary intersection can be performed in linear time in the number of edges. Checking connectivity of an undirected graph can also be done in linear time in the number of edges by building a DFS tree from an arbitrary node and testing whether all nodes were reached through it. Hence, both operations have similar costs. In what follows, we will refer to them as* elementary operations. *One advantage of using them is that the high-level logic of the algorithm becomes elegant and simple. Also, their cost can be counterbalanced by the* *fact that they are highly generic and thus could benefit from dedicated circuits (e.g. FPGA) or optimized code.*

Naive Upper Bound. One can easily see that both problems are solvable using $O(\delta^2)$ elementary operations based on a naive strategy. It suffices to compute the rows of $\mathcal{G}'s$ intersection hierarchy incrementally, using the fact that each graph $G_{(i,j)}$ can be obtained as $G_{(i,j-1)} \cap G_{(i+1,j)}$. For instance, $G_{(3,6)} = G_{(3,5)} \cap G_{(4,6)}$ in Fig. 1. Hence, each row k can be computed from row $k-1$ using $O(\delta)$ binary intersections. In the case of T-INTERVAL-CONNECTIVITY, one simply has to repeat the operation until the Tth row, then answer true iff all graphs in this row are connected. The total cost is $O(\delta T) = O(\delta^2)$ binary intersections, plus $\delta - T + 1 = O(\delta)$ connectivity tests for the Tth row. Solving INTERVAL-CONNECTIVITY is similar except that one needs to test the connectivity of all new graphs during the process. Whenever a disconnected graph is found in row k, the answer is $k-1$. If all graphs are connected up to row δ, then δ is the answer. Since there are $O(\delta^2)$ graphs in the intersection hierarchy, the total number of connectivity tests and binary intersections is $O(\delta^2)$.

Lower Bound. The following lower bound is valid for any algorithm that uses only the two elementary operations *binary intersection* and *connectivity test*.

Lemma 1. $\Omega(\delta)$ *elementary operations are necessary to solve* T-INTERVAL-CONNECTIVITY.

Proof (by contradiction). Let $\mathcal{A}$ be an algorithm that uses only elementary operations and that decides whether any sequence of graphs is T-interval connected in $o(\delta)$ operations. Then, for any sequence $\mathcal{G}$, at least one graph in $\mathcal{G}$ is never accessed by $\mathcal{A}$. Let $\mathcal{G}_1$ be a sequence that is T-interval connected and suppose that $\mathcal{A}$ decides that $\mathcal{G}$ is T-interval connected without accessing graph G_k. Now, consider a sequence $\mathcal{G}_2$ that is identical to $\mathcal{G}_1$ except G_k is replaced by a disconnected graph G'_k. Since G'_k is never accessed, the executions of $\mathcal{A}$ on $\mathcal{G}_1$ and $\mathcal{G}_2$ are identical and $\mathcal{A}$ incorrectly decides that $\mathcal{G}_2$ is T-interval connected. □

A similar argument can be applied for INTERVAL-CONNECTIVITY, by making the answer T dependent on the graph G_k that is never accessed.

3 Row-Based Strategy

In this section, we present a basic strategy that improves over the previous naive strategy, yielding upper bounds of $O(\delta \log \delta)$ operations for both problems. Its main interest is in the fact that it can be parallelized, and this allows us to show that both problems are in **NC** (i.e. parallelizable on a PRAM with a polylogarithmic running time). We first describe the algorithms for a sequential machine (RAM). The general strategy is to compute only some of the rows of $\mathcal{G}$'s intersection hierarchy based on the following lemma.

Lemma 2. *If some row $\mathcal{G}^k$ is already computed, then any row $\mathcal{G}^\ell$ for $k + 1 \leq l \leq 2k$ can be computed with $O(\delta)$ elementary operations.*

Proof. Assume that row $\mathcal{G}^k$ is already computed and that one wants to compute row $\mathcal{G}^\ell$ for some $k + 1 \leq \ell \leq 2k$. Note that row $\mathcal{G}^\ell$ consists of the entries $\mathcal{G}^\ell[1], \ldots, \mathcal{G}^\ell[\delta - \ell + 1]$. Now, observe that for any $1 \leq i \leq \delta - \ell + 1$, $\mathcal{G}^\ell[i] = \mathcal{G}^k[i] \cap \mathcal{G}^k[i + \ell - k]$. Hence, $\delta - \ell + 1 = O(\delta)$ intersections are sufficient to compute all of the entries of row $\mathcal{G}^\ell$.   □

T-**Interval-Connectivity.** Using Lemma 2, we can incrementally compute rows $\mathcal{G}^{2^i}$ ("power rows") for all i from 1 to $\lceil \log_2 T \rceil - 1$ without computing the intermediate rows. Then, we compute row $\mathcal{G}^T$ directly from row $\mathcal{G}^{2^{\lceil \log_2 T \rceil - 1}}$ (again using Lemma 2). This way, we compute $\lceil \log_2 T \rceil = O(\log \delta)$ rows using $O(\delta \log \delta)$ intersections, after which we perform $O(\delta)$ connectivity tests.

Interval-Connectivity. Here, we incrementally compute rows $\mathcal{G}^{2^i}$ until we find a row that contains a disconnected graph (thus, a connectivity test is performed after each intersection). By Lemma 2, each of these rows can be computed using $O(\delta)$ intersections. Suppose that row $\mathcal{G}^{2^{j+1}}$ is the first power row that contains a disconnected graph, and that $\mathcal{G}^{2^j}$ is the row computed before $\mathcal{G}^{2^{j+1}}$. Next, we do a binary search among the rows between $\mathcal{G}^{2^j}$ and $\mathcal{G}^{2^{j+1}}$ to find the row $\mathcal{G}^T$ with the highest row number T such that all graphs on this row are connected (see Fig. 2 for an illustration of the algorithm). The computation of each of these rows is based on row $\mathcal{G}^{2^j}$ and takes $O(\delta)$ intersections by Lemma 2. Overall, we compute at most $2\lceil \log_2 T \rceil = O(\log \delta)$ rows using $O(\delta \log \delta)$ intersections and the same number of connectivity tests.

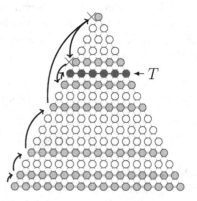

Fig. 2. Example of interval connectivity testing based on the computation of power rows. Here $\delta = 16$ and $T = 11$. The computation of power rows stops upon reaching $\mathcal{G}^{16}$ which contains a disconnected graph ($\times$). A binary search between rows $\mathcal{G}^8$ and $\mathcal{G}^{16}$ is then used to find $\mathcal{G}^{11}$, the highest row where all graphs are connected.

Now we establish that these problems are in **NC** by showing that our algorithms are efficiently parallelizable.

Lemma 3. *If some row $\mathcal{G}^k$ is already computed, then any row between $\mathcal{G}^{k+1}$ and $\mathcal{G}^{2k}$ can be computed in $O(1)$ time on an EREW PRAM with $O(\delta)$ processors.*

Proof. Assume that row $\mathcal{G}^k$ is already computed, and that one wants to compute row $\mathcal{G}^\ell$, consisting of the entries $\mathcal{G}^\ell[1], \ldots, \mathcal{G}^\ell[\delta - \ell + 1]$, for some $k+1 \leq \ell \leq 2k$. Since $\mathcal{G}^\ell[i] = \mathcal{G}^k[i] \cap \mathcal{G}^k[i + \ell - k]$, $1 \leq i \leq \delta - \ell + 1$, the computation of row $\mathcal{G}^\ell$ can be implemented on an EREW PRAM with $\delta - \ell + 1$ processors in two rounds as follows. Let P_i, $1 \leq i \leq \delta - \ell + 1$, be the processor dedicated to computing $\mathcal{G}^\ell[i]$. In the first round P_i reads $\mathcal{G}^k[i]$, and in the second round P_i reads $\mathcal{G}^k[i + \ell - k]$. This guarantees that each P_i has exclusive access to the entries of row $\mathcal{G}^k$ that it needs for its computation. Hence, row $\mathcal{G}^\ell$ can be computed in $O(1)$ time on an EREW PRAM using $O(\delta)$ processors. □

T-Interval-Connectivity on an EREW PRAM. We compute the same rows in the same order as the sequential algorithm. We can use $O(\delta)$ processors in parallel to compute row $\mathcal{G}^{2^{i+1}}$ from row $\mathcal{G}^{2^i}$ in $O(1)$ time on an EREW PRAM by Lemma 3. The (potentially missing) row $\mathcal{G}^T$ can also be computed in $O(1)$ time using $O(\delta)$ processors. Overall, we compute $O(\log \delta)$ rows, and all necessary rows (and hence all necessary intersections) can be computed in $O(\log \delta)$ time with $O(\delta)$ processors. The $O(\delta)$ connectivity tests for row $\mathcal{G}^T$ can be done in $O(1)$ time with $O(\delta)$ processors and then the processors can establish whether or not all graphs in row $\mathcal{G}^T$ are connected in time $O(\log \delta)$. The total time is $O(\log \delta)$ on an EREW PRAM with $O(\delta)$ processors.

Interval-Connectivity on an EREW PRAM. The sequential algorithm for this problem computes $O(\log \delta)$ rows. By Lemma 3, each of these rows can be computed in $O(1)$ time on an EREW PRAM with $O(\delta)$ processors. Therefore, all of the rows (and hence all necessary intersections) can be computed in $O(\log \delta)$ time with $O(\delta)$ processors. A connectivity test is done for each of the computed graphs (rather than just those of the last row) and it has to be determined for each computed row whether or not all of the graphs are connected. This takes $O(\log \delta)$ time for each of the $O(\log \delta)$ computed rows. The total time is $O(\log^2 \delta)$ on an EREW PRAM with $O(\delta)$ processors.

CRCW PRAM. Using standard techniques and results for computing the logical AND on a CRCW PRAM (see [5,7]), our results for an EREW PRAM can be improved to $O(1)$ time on a CRCW PRAM using $O(\delta^3)$ processors.

4 Optimal Solution

We now present our strategy for solving both T-INTERVAL-CONNECTIVITY and INTERVAL-CONNECTIVITY using a linear number of elementary operations (in the length δ of $\mathcal{G}$), matching the $\Omega(\delta)$ lower bound presented in Section 2. The strategy relies on the concept of *ladder*. Informally, a ladder is a sequence of graphs that "climbs" the intersection hierarchy bottom-up.

Definition 3. *The* right ladder *of length l at index i, denoted by $\mathcal{R}^l[i]$, is the sequence of intersection graphs $\{\mathcal{G}^k[i], k = 1, 2, \ldots, l\}$. The* left ladder *of length l at index i, denoted by $\mathcal{L}^l[i]$, is the sequence $\{\mathcal{G}^k[i - k + 1], k = 1, 2, \ldots, l\}$.*

A right (resp. left) ladder of length $l - 1$ at index i is said to be incremented *when graph $\mathcal{G}^l[i]$ (resp. $\mathcal{G}^l[i - l + 1]$) is added to it, and the resulting sequence of intersection graphs is called the* increment *of that ladder.*

Lemma 4. *A ladder of length l can be computed using $l - 1$ binary intersections.*

Proof. Consider a right ladder (a symmetrical argument holds for left ladders). For any k in $[2, l]$ it holds that $\mathcal{G}^k[i] = \mathcal{G}^{k-1}[i] \cap G_{i+k-1}$. Indeed, by definition, $\mathcal{G}^{k-1}[i] = \cap\{G_i, G_{i+1}, ..., G_{i+k-2}\}$. The ladder can thus be built bottom-up using a single new intersection at each level. □

Lemma 5. *Given a left ladder of length l_ℓ at index i_ℓ and a right ladder of length l_r at index $i_r = i_\ell + 1$. For any pair (i, k) such that $i_r - l_\ell \leq i < i_r$ and $i_r - i < k \leq i_r - i + l_r$, $\mathcal{G}^k[i]$ can be computed by a single binary intersection, namely $\mathcal{G}^k[i] = \mathcal{G}^{i_r-i}[i] \cap \mathcal{G}^{k-i_r+i}[i_r]$ (see example below).*

Proof. By definition, $\mathcal{G}^k[i] = \cap\{G_i, G_{i+1}, ..., G_{i+k-1}\}$ and $\mathcal{G}^{i_r-i}[i] = \cap\{G_i, G_{i+1}, ..., G_{i_r-1}\}$ and $\mathcal{G}^{k-i_r+i}[i_r] = \cap\{G_{i_r}, G_{i_r+1}, ..., G_{i+k-1}\}$. It follows that $\mathcal{G}^k[i] = \mathcal{G}^{i_r-i}[i] \cap \mathcal{G}^{k-i_r+i}[i_r]$. □

Informally, the constraints $i_r - l_\ell \leq i < i_r$ and $i_r - i < k \leq i_r - i + l_r$ in Lemma 5 define a rectangle delimited by two ladders and two lines that are parallel to the two ladders as shown in the figure to the right. The pairs (i, k) defined by the constraints, shown in light grey in the figure, include all pairs that are strictly inside the rectangle, and all pairs on the parallel lines, but pairs on the two ladders are excluded.

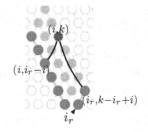

T-Interval-Connectivity. We describe our optimal algorithm for this problem with reference to Fig. 3 below, which shows two examples of execution of the algorithm. The algorithm traverses the Tth row in the intersection hierarchy from left to right, starting at $\mathcal{G}^T[1]$. If a disconnected graph is found, the algorithm returns `false` and terminates. If the algorithm reaches the last graph in the row, i.e. $\mathcal{G}^T[\delta - T + 1]$, and no disconnected graph was found, then it returns `true`. The graphs $\mathcal{G}^T[1], \mathcal{G}^T[2], ..., \mathcal{G}^T[\delta - T + 1]$ are computed based on the set of ladders $\mathcal{S} = \{\mathcal{L}^T[T], \mathcal{R}^{T-1}[T + 1], \mathcal{L}^T[2T], \mathcal{R}^{T-1}[2T + 1], ...\}$, which are constructed as follows. Each left ladder is built entirely (from bottom to top) when the traversal arrives at its top location in row T (i.e. where the last increment is to take place). For instance, $\mathcal{L}^T[T]$ is built when the walk is at index 1 in row T, $\mathcal{L}^T[2T]$ is built at index $T + 1$, and so on. If a disconnected graph is found in the process, the execution terminates returning `false`.

Differently from left ladders, right ladders are constructed gradually as the traversal proceeds. Each time that the traversal moves right to a new index in the Tth row, the current right ladder is incremented and the new top element of this right ladder is used immediately to compute the graph at the current index in the Tth row (using Lemma 5). This continues until the right ladder reaches row $T - 1$ after which a new left ladder will be built.

The set $\mathcal{S}$ of ladders constructed by this process includes at most δ/T left ladders and δ/T right ladders, each of length at most T. By Lemma 4, the set of ladders $\mathcal{S}$ can be computed using less than 2δ binary intersections. Based on Lemma 5, each of the $\delta - T + 1$ graphs $\mathcal{G}^T[i]$ in row T can be computed at the cost of a single intersection of two graphs in $\mathcal{S}$. At most $\delta - T + 1$ connectivity tests are performed for row T. This establishes the following result which matches the lower bound of Lemma 1.

Theorem 1. T-INTERVAL-CONNECTIVITY *can be solved with $O(\delta)$ elementary operations, which is optimal (up to a constant factor).*

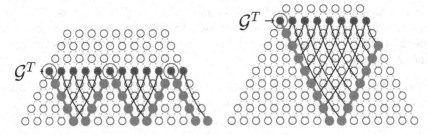

Fig. 3. Examples of the execution of the optimal algorithm for T-INTERVAL-CONNECTIVITY with $T < \delta/2$ (left) and $T \geq \delta/2$ (right). $\mathcal{G}$ is T-interval connected in both examples.

Interval-Connectivity. The strategy of our optimal algorithm for this problem is in the same spirit as the one for T-INTERVAL-CONNECTIVITY. However, it is more complex and corresponds to a walk in the two dimensions of the intersection hierarchy. It is best understood with reference to Fig. 4 which shows an example of the execution of the algorithm.

The walk starts at the bottom left graph $\mathcal{G}^1[1]$ and builds a right ladder incrementally until it encounters a disconnected graph. If $\mathcal{G}^\delta[1]$ is reached and is connected, then $\mathcal{G}$ is δ-interval connected and execution terminates returning δ. Otherwise, suppose that a disconnected graph is first found in row $k+1$. Then k is an upper bound on the connectivity of $\mathcal{G}$ and the walk drops down a level to $\mathcal{G}^k[2]$ which is the next graph in row k that needs to be checked. This requires the construction of a left ladder $\mathcal{L}^k[k+1]$ of length k ending at $\mathcal{G}^k[2]$. The walk proceeds rightward on row k using a similar traversal strategy to the algorithm for T-INTERVAL-CONNECTIVITY. Here, however, every time that a disconnected graph is found, the walk drops down by one row. The dropping down operation, say, from some $\mathcal{G}^k[i]$, is made in two steps (bent line on Fig. 4). First it goes to $\mathcal{G}^{k-1}[i]$, which is necessarily connected, then it moves one unit right to $\mathcal{G}^{k-1}[i+1]$. If the walk eventually reaches the rightmost graph of some row and this graph is connected, then the algorithm terminates returning the corresponding row number as T. Otherwise the walk will terminate at a disconnected graph in row 1 and $\mathcal{G}$ is not T-interval connected for any T. In this case, the algorithm returns $T = 0$.

Similar to the algorithm for T-INTERVAL-CONNECTIVITY, the computations of the graphs in a walk by this algorithm for INTERVAL-CONNECTIVITY using binary intersections are based on Lemmas 4 and 5. If the algorithm returns that $\mathcal{G}$ is T-interval connected, then each graph $\mathcal{G}^T[1], \mathcal{G}^T[2], \ldots, \mathcal{G}^T[\delta - T + 1]$ must be connected. The graphs that are on the walk are checked directly by the algorithm. For each graph $\mathcal{G}^T[i]$ on row T that is below the walk, there is a graph $\mathcal{G}^j[i]$ with $j > T$ that is on the walk and is connected and this implies that $\mathcal{G}^T[i]$ is connected.

The ranges of the indices covered by the left ladders that are constructed by this process are disjoint, so their total length is O(δ). The first right ladder has length at most δ and each subsequent right ladder has length less than the left ladder that precedes it so the total length of the right ladders is also O(δ). Therefore, this algorithm performs at most O(δ) binary intersections and O(δ) connectivity tests. This establishes the following result which matches the lower bound of Lemma 1.

Theorem 2. INTERVAL-CONNECTIVITY *can be solved with O(δ) elementary operations, which is optimal (up to a constant factor).*

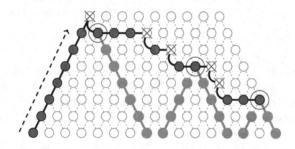

Fig. 4. Example of the execution of the optimal algorithm for INTERVAL-CONNECTIVITY. *(It is a coincidence that the rightmost ladder matches the outer face.)*

Online Algorithms. The optimal algorithms for T-INTERVAL-CONNECTIVITY and INTERVAL-CONNECTIVITY can be adapted to an online setting in which the sequence of static graphs $G_1, G_2, G_3, \ldots$ of a dynamic graph $\mathcal{G}$ is processed in the order that the graphs are received. In the case of T-INTERVAL-CONNECTIVITY, the algorithm cannot provide an answer until at least T static graphs have been received. When the Tth graph is received, the algorithm builds the first left ladder using $T - 1$ binary intersections. It can then perform a connectivity test and answer whether or not the sequence is T-interval connected so far. After this initial period, a T-connectivity test can be performed for the T most recently received static graphs (corresponding to a graph in row T) after the receipt of each new static graph. At no time does the number of intersections performed to build left ladders exceed the number of static graphs received and the same is true for right ladders. Furthermore, each new static graph after the first $T-1$ corresponds to a graph in row T which can be computed with one intersection by Lemma 5. In summary, the amortized cost is $O(1)$ elementary operations for each static graph received and for each T-connectivity test after the initial period. The analysis for INTERVAL-CONNECTIVITY is similar except the algorithm can report the connectedness of the sequence so far starting with the first graph received.

Theorem 3. T-INTERVAL-CONNECTIVITY *and* INTERVAL-CONNECTIVITY *can be solved online with* $O(1)$ *elementary operations per static graph received.*

5   Conclusions

In this paper, we studied the problem of testing whether a given dynamic graph $\mathcal{G} = \{G_1, G_2, ..., G_\delta\}$ is T-interval connected. We also considered the related problem of finding the largest T for which a given $\mathcal{G}$ is T-interval connected. We assumed that the dynamic graph $\mathcal{G}$ is a sequence of *independent* static graphs and we investigated algorithmic solutions that use two elementary operations, *binary intersection* and *connectivity testing*, to solve the problems. We developed efficient algorithms that use only $O(\delta)$ elementary operations, asymptotically matching the lower bound of $\Omega(\delta)$. We also presented PRAM algorithms that show that both problems can be solved efficiently in parallel and online algorithms that use $O(1)$ elementary operations per static graph received.

In our study, we focused on algorithms using only the two elementary operations *binary intersection* and *connectivity testing*. This approach is suitable for a high-level study of these problems when the details of changes between successive graphs in a sequence are arbitrary. If the evolution of the dynamic is constrained in some ways (e.g. bounding the number of changes in-between graphs), then one could benefit from the use of more sophisticated data structures to lower the complexity of the problem.

A natural extension of our investigation of T-interval connectivity would be a similar study for other classes of dynamic graphs, as identified in [4]. Distributed algorithms for all of these problems, in which a vertex in the graph only sees its local neighbourhood, would also be of high interest.

Acknowledgments. We would like to thank the referees for constructive suggestions that have improved the presentation of this paper.

References

1. Barjon, M., Casteigts, A., Chaumette, S., Johnen, C., Neggaz, Y.M.: Testing temporal connectivity in sparse dynamic graphs. CoRR abs/1404.7634, 8p (2014). a French version appeared in Proc. of ALGOTEL (2014)
2. Bui-Xuan, B., Ferreira, A., Jarry, A.: Computing shortest, fastest, and foremost journeys in dynamic networks. Int. J. of Foundations of Computer Science **14**(2), 267–285 (2003)

3. Casteigts, A., Chaumette, S., Ferreira, A.: Characterizing topological assumptions of distributed algorithms in dynamic networks. In: Kutten, S., Žerovnik, J. (eds.) SIROCCO 2009. LNCS, vol. 5869, pp. 126–140. Springer, Heidelberg (2010)
4. Casteigts, A., Flocchini, P., Quattrociocchi, W., Santoro, N.: Time-varying graphs and dynamic networks. Int. J. of Parallel, Emergent and Distributed Systems 27(5), 387–408 (2012)
5. Gibbons, A., Rytter, W.: Efficient parallel algorithms. Cambridge University Press (1988)
6. Jain, S., Fall, K., Patra, R.: Routing in a delay tolerant network. In: Proc. of SIGCOMM, pp. 145–158 (2004)
7. JáJá, J.: An Introduction to Parallel Algorithms. Addison-Wesley (1992)
8. Kuhn, F., Lynch, N., Oshman, R.: Distributed computation in dynamic networks. In: Proc. of STOC, pp. 513–522. ACM, Cambridge (2010)
9. O'Dell, R., Wattenhofer, R.: Information dissemination in highly dynamic graphs. In: Proc. of DIALM-POMC, pp. 104–110. ACM, Cologne (2005)
10. Whitbeck, J., Dias de Amorim, M., Conan, V., Guillaume, J.L.: Temporal reachability graphs. In: Proc. of MOBICOM, pp. 377–388. ACM (2012)

Competitive Strategies for Online Clique Clustering

Marek Chrobak[1], Christoph Dürr[2,3], and Bengt J. Nilsson[4(✉)]

[1] University of California at Riverside, Riverside, USA
[2] Sorbonne Universités, UPMC Univ Paris 06, UMR 7606, LIP6, Paris, France
[3] CNRS, UMR 7606, LIP6, Paris, France
[4] Department of Computer Science, Malmö University, Malmö, Sweden
`bengt.nilsson.ts@mah.se`

Abstract. A *clique clustering* of a graph is a partitioning of its vertices into disjoint cliques. The quality of a clique clustering is measured by the total number of edges in its cliques. We consider the online variant of the clique clustering problem, where the vertices of the input graph arrive one at a time. At each step, the newly arrived vertex forms a singleton clique, and the algorithm can merge any existing cliques in its partitioning into larger cliques, but splitting cliques is not allowed. We give an online strategy with competitive ratio 15.645 and we prove a lower bound of 6 on the competitive ratio, improving the previous respective bounds of 31 and 2.

1 Introduction

A *clique clustering* of a graph $G = (V, E)$ is a partitioning of the vertex set V into disjoint cliques $C_1, C_2, ..., C_k$. The *profit* of this clustering is defined to be the total number of edges in these cliques, that is $\sum_{i=1}^{k} \binom{|C_i|}{2} = \frac{1}{2} \sum_{i=1}^{k} |C_i|(|C_i|-1)$. In the *clique clustering problem* the objective is to compute a clique clustering of the given graph that maximizes this profit value. For a graph G, by $\mathsf{O}(G)$ we denote the optimal profit for G.

We consider the online variant of the clique clustering problem, where the input graph G is not known in advance. (See [3], for more background on online problems). The vertices of G arrive one at a time. Let v_t denote the vertex that arrives at time t, for $t = 1, 2, ...$. When v_t arrives, its edges to all preceding vertices $v_1, ..., v_{t-1}$ are revealed as well. In other words, after step t, the subgraph of G induced by $v_1, v_2, ..., v_t$ is known, but no other information about G is available.

Our objective is to construct a procedure that incrementally constructs and outputs a clustering based on the information acquired so far. Specifically, when v_t arrives at step t, the procedure first creates a singleton clique $\{v_t\}$. Then it is allowed to merge any number of cliques (possibly none) in its current partitioning into larger cliques. No other modifications of the clustering are allowed.

M. Chrobak—Research supported by NSF grants CCF-0729071 and CCF-1217314.

V.Th. Paschos and P. Widmayer (Eds.): CIAC 2015, LNCS 9079, pp. 101–113, 2015.
DOI: 10.1007/978-3-319-18173-8_7

We avoid using the word algorithm for our procedure, since it evokes connotations with computational limits in terms of complexity and computability. In fact, we place no limits on the computational power of our procedure and to emphasize this, we use the word *strategy* rather than algorithm. This approach allows us to focus specifically on the limits posed by the lack of complete information about the input. Similar considerations played a role in some earlier work on online computation, for example for online medians [6,7,12], minimum-latency tours [5], and several other online optimization problems (see [8]).

Throughout the paper we will implicitly assume that any graph G has its vertices ordered $v_1, v_2, ..., v_n$, according to the ordering in which they arrive on input. For an online strategy S let $\mathsf{profit}_S(G)$ be the profit of S when the input graph is G. We say that an online strategy S is R-*competitive* if for any input graph G we have

$$R \cdot \mathsf{profit}_S(G) + \beta \geq \mathsf{O}(G), \tag{1}$$

for some constant β independent of G. The competitive ratio of S is the smallest R for which S is R-competitive[1]. This concept is sometimes referred to as the *asymptotic competitive ratio* in the literature, but we will omit the term "asymptotic" in the paper. If $\beta = 0$, then R is called the *absolute competitive ratio*.

The online model for clique clustering was studied by Fabijan *et al.* [10], who designed an online strategy with competitive ratio 31 and proved that no online strategy can have competitive ratio better than 2. They also showed that the greedy strategy's competitive ratio is linear with respect to the graph size, and they studied an alternative model where the objective is to minimize the number of edges that are not in the clusters.

The clique clustering problem arises in applications to gene expression profiling and DNA clone classification [2,11,14]. The offline variant is known to be NP-hard, and in fact not even approximable within factor $n^{1-o(1)}$ under some reasonable complexity-theoretic assumptions [9].

Our Results. We provide two new bounds on the competitive ratio of online clique clustering, considerably improving the results in [10]. First, we present an online strategy with competitive ratio 15.645. The idea of the strategy is based on the "doubling" technique. Roughly (but not exactly), we divide the computation into phases, where the optimal profit of the set of vertices from phase j grows exponentially with j. After each phase j the cliques computed from this optimal clustering are added to the strategy's clustering of the current graph. We give an example showing that the competitive ratio of our strategy is no better than 10.92. We then also show that there is no deterministic online strategy for clique clustering with competitive ratio smaller than 6.

Related Work. Clustering is a dynamic and important field of research with multiple applications in almost all areas of sciences, humanities and engineering.

---

[1] Earlier papers on online clustering define the competitive ratio as the maximum value of $\mathsf{profit}_S(G)/\mathsf{O}(G)$, which is the inverse of the value we use.

There are many clustering models in the literature, with varying criteria for data similarity (which determines whether two data items can be clustered together), quality measures for clustering, and requirements for the number of clusters.

Approximation algorithms for incremental clustering, where the only operations allowed are to create singleton clusters and merge existing clusters, were first studied by Charikar et al. [4], although for a different clustering model than ours. Mathieu et al. [13] applied this incremental approach in the model of online correlation clustering, initially introduced in [1,2]. In correlation clustering, as in our model, the similarity relation is represented by an undirected graph, but the objective function is equal to the sum of the number of edges in the clusters plus the number of non-edges outside clusters. The results in [13] include a lower bound of 1.245 and an upper bound slightly below 2 on the competitive ratio (the ratio 2 can be achieved with a greedy strategy).

2   A Competitive Strategy

In this section we give our competitive online strategy OCC. Roughly, the strategy works in phases. In each phase we consider the "batch" of nodes that have not yet been clustered with other nodes, compute an optimal clustering for this batch, and add these new clusters to the strategy's clustering. The phases are defined so that the profit for consecutive phases increases exponentially.

The overall idea can be thought of as an application of the "doubling" strategy (see [8], for example), but in our case a subtle modification is required. Unlike in other doubling approaches, in our strategy the phases are not completely independent: the clustering computed in each phase, in addition to the new nodes, needs to include the singleton nodes from earlier phases as well. This is needed, because in our objective function singleton clusters do not bring any profit.

We remark that one could alternatively consider using profit value $\frac{1}{2}p^2$ for a clique of size p, which is a very close approximation to our function if p is large. This would lead to a simpler strategy and much simpler analysis. However, this function is a bad approximation when the clustering involves many small cliques, which is also in fact the most challenging scenario in the analysis of our algorithm, and instances with this property are also used in the lower bound proof.

The Strategy OCC. Formally, our method works as follows. Fix some constant parameter $\gamma > 1$ of the strategy which we will later optimize. The strategy works in phases, starting with phase $j = 0$. At any moment the clustering maintained by the strategy contains a set U of *singleton* cliques. Each arriving vertex is added into U. As soon as there is a clustering of U of profit at least γ^j, the strategy creates these clusters, adds them to its current clustering, and moves to phase $j + 1$.

Note that phase 0 ends as soon as one edge is released, since then it is possible for OCC to create a clustering with $\gamma^0 = 1$ edge. The last phase may not be complete; as a result all nodes released in this phase will be clustered as singletons. Note also that the strategy never merges non-singleton cliques.

Asymptotic Analysis of OCC. It is convenient to think of the computation as lasting forever. We then want to show that at each step of the computation, the optimal profit is at most R times the profit of OCC, plus some absolute additive constant, where $R \approx 15.645$ is the claimed competitive ratio.

For every phase $j = 0, 1, \ldots$, denote by Δ_j the optimal profit of the vertices that arrived in phase j. Let $\mathsf{S}_j = \Delta_0 + \ldots + \Delta_j$ be the total profit of the strategy and O_j the total profit of the adversary at the end of phase j. By the definition of OCC, for all phases j we have $\Delta_j \geq \gamma^j$ and $\mathsf{S}_j \geq (\gamma^{j+1} - 1)/(\gamma - 1)$.

We fix some instance and start with some observations. First, at the end of phase 0 the strategy is optimal. Also, in each step, except for the last step of a phase, the strategy's profit does not change while the optimum profit can only increase. Therefore it suffices to compare the optimal profit O_j at the end of a phase $j \geq 1$, with the strategy's profit right before the end of the phase, which is equal to S_{j-1}.

After any phase j, the optimal clustering of U may include some singletons. If this is so, the adversary can release those vertices during the next phase instead, and the behavior of OCC will remain unchanged. We can thus assume without loss of generality that the optimal clustering of U does not contain any singletons. As a result, after each phase j, all clusters of OCC have at least two vertices.

With the above assumption, we can divide the vertices into disjoint *batches*, where batch B_j contains the vertices released in phase j. During phase j, the clustering of OCC is then the union of clusterings of all its batches $B_0, B_1, \ldots, B_{j-1}$, plus the singletons released in phase j.

Let $\bar{B}_j = B_0 \cup B_1 \cup \ldots \cup B_j$ be the set of vertices released in phases $0, 1, \ldots, j$. Consider the optimal clustering of $\bar{B}_j$. In this clustering, every cluster C has some number a of nodes in $\bar{B}_{j-1}$ and some number b of nodes in B_j. Let $k_{a,b}$ be the number of clusters of this form in the optimal clustering. Then we have the following bounds, where the sums range over all integers $a, b \geq 0$.

$$\mathsf{O}_j = \sum \binom{a+b}{2} k_{a,b} \quad (2) \qquad\qquad \Delta_j \geq \sum \binom{b}{2} k_{a,b} \quad (4)$$

$$\mathsf{O}_{j-1} \geq \sum \binom{a}{2} k_{a,b} \quad (3) \qquad\qquad \mathsf{S}_{j-1} \geq \tfrac{1}{2} \sum a k_{a,b} \quad (5)$$

Equality (2) is the definition of O_j. Inequality (3) holds because the right hand side represents the profit of the optimal clustering of $\bar{B}_j$ restricted to $\bar{B}_{j-1}$, so it cannot exceed the optimal profit O_{j-1} for $\bar{B}_{j-1}$. Similarly, inequality (4) holds because the right hand side is the profit of the optimal clustering of $\bar{B}_j$ restricted to B_j, while Δ_j the optimal profit of B_j. The last bound (5) follows from the fact that the strategy does not have any singleton clusters in $\bar{B}_{j-1}$. This means that in the strategy's clustering of $\bar{B}_{j-1}$ (which has $\sum a k_{a,b}$ vertices) each vertex has an edge included in some cluster, so the number of these edges must be at least $\tfrac{1}{2} \sum_{a \geq 0} a k_{a,b}$.

We can also bound Δ_j, the strategy's profit increase, from above. We have $\Delta_0 = 1$ and for each phase $j \geq 1$

$$\Delta_j < \gamma^j + \sqrt{2}\gamma^{j/2} + 2 - \sqrt{2}. \tag{6}$$

To show (6), suppose that phase j ends at step t (that is, right after v_t is revealed). Consider the optimal partitioning $\mathcal{P}$ of B_j, and let the cluster C of v_t in $\mathcal{P}$ have size $p+1$. If we remove v_t from this partitioning, we obtain a partitioning of the batch after step $t-1$, whose profit must be strictly smaller than γ^j. So the profit of $\mathcal{P}$ is smaller than $\gamma^j + p$. In this new partitioning, cluster $C - \{v_t\}$ has size p. We thus obtain that $\binom{p}{2} < \gamma^j$, which gives us $p < \sqrt{2}\gamma^{j/2} + 2 - \sqrt{2}$, thus proving (6).

From (6), by adding up all profits from phases $0, \ldots, j$, we obtain an upper bound on the total profit of the strategy:

$$S_j < \frac{\gamma^{j+1} - 1}{\gamma - 1} + \sqrt{2} \cdot \frac{\gamma^{(j+1)/2} - \gamma^{1/2}}{\gamma^{1/2} - 1} + (2 - \sqrt{2})j. \tag{7}$$

When phase 0 ends we have $O_0 = S_0 = 1$. As explained earlier, for $j \geq 1$ the worst case ratio occurs right before phase j ends. At this point, OCC has accrued a profit of S_{j-1}, since all vertices released during phase j are put into singleton clusters. The optimal solution, on the other hand, is bounded by O_j. The ratio $R_j = O_j/S_{j-1}$ is therefore also an upper bound on the competitive ratio throughout phase j. Our goal now is to upper bound R_j, for all j. We will use the following technical lemma.

Lemma 1. *For any pair of non-negative integers a and b, the inequality*

$$\binom{a+b}{2} \leq (x+1)\binom{a}{2} + \frac{x+1}{x}\binom{b}{2} + a$$

holds for any $0 < x \leq 1$.

Proof. Define the function

$$F(a, b, x) = 2x(x+1)\binom{a}{2} + 2(x+1)\binom{b}{2} + 2ax - 2x\binom{a+b}{2}$$

$$= a^2x^2 - ax^2 + 2ax + b^2 - b - 2abx = (b - ax)^2 + ax(2 - x) - b,$$

i.e., twice x times the difference between the right hand side and the left hand side of the inequality above. It is sufficient to show that $F(a, b, x)$ is non-negative for integers $a, b \geq 0$ and $0 < x \leq 1$.

Consider first the cases when $a \in \{0, 1\}$ or $b \in \{0, 1\}$. $F(0, b, x) = b(b-1) \geq 0$, for any non-negative integer b and any x. $F(a, 0, x) = ax(ax - x + 2) \geq ax(ax + 1) > 0$, for any positive integer a and $0 < x \leq 1$. $F(a, 1, x) = x^2a(a-1) \geq 0$, for any positive integer a and any x. $F(1, 2, x) = 2 - 2x \geq 0$, for $0 < x \leq 1$, and $F(1, b, x) = b^2 - b + 2x - 2bx \geq b^2 - 3b \geq 0$, for any integer $b \geq 3$ and $0 < x \leq 1$.

Thus, it only remains to show that $F(a, b, x)$ is non-negative when both $a \geq 2$ and $b \geq 2$. The function $F(a, b, x)$ is quadratic and hence has one local minimum at $x_0 = \frac{b-1}{a-1}$, as can be easily verified by differentiating F in x. Therefore, in the case when $a \leq b$, $F(a, b, x) \geq F(a, b, 1) = (b-a)^2 - (b-a) \geq (b-a) - (b-a) = 0$, for $0 < x \leq 1$. In the case when $a > b$, we have that $F(a, b, x) \geq F(a, b, \frac{b-1}{a-1}) = \frac{(a-b)(b-1)}{a-1} > 0$, which completes the proof. □

Now, to find an upper bound on all R_j's, we will establish a recurrence relation for the sequence $R_1, R_2, \ldots$. The value of R_1 is some constant (its exact value is not important since we are interested in the asymptotic ratio). Suppose that $j \geq 2$ and fix some parameter x, $0 < x < 1$, whose value we will determine later. Using Lemma 1 and the bounds (2)-(5) we obtain

$$R_j S_{j-1} = O_j = \sum \binom{a+b}{2} k_{a,b}$$

$$\leq (x+1) \sum \binom{a}{2} k_{a,b} + \frac{x+1}{x} \sum \binom{b}{2} k_{a,b} + \sum a k_{a,b}$$

$$\leq (x+1) O_{j-1} + \frac{x+1}{x} \Delta_j + 2 S_{j-1} \tag{8}$$

$$= (x+1) R_{j-1} S_{j-2} + \frac{x+1}{x} \Delta_j + 2 S_{j-1}.$$

Thus R_j satisfies the recurrence

$$R_j \leq \frac{x+1}{x S_{j-1}} \left[x S_{j-2} R_{j-1} + \Delta_j \right] + 2. \tag{9}$$

From inequalities (6) and (7), we have $\Delta_i = \gamma^i (1 + o(1))$ and $S_i = \frac{\gamma^{i+1}(1+o(1))}{\gamma-1}$ for all i. We use the notation $o(1)$ to denote any function that tends to 0 as the number of phases goes to infinity. Substituting into the above recurrence, we get

$$R_j \leq \frac{(x+1)(1+o(1))}{\gamma} R_{j-1} + \frac{(x+1)(\gamma-1)}{x} + 2 + o(1). \tag{10}$$

Assuming that $x + 1 < \gamma$, (10) implies that the sequence R_j converges and, denoting its limit by $R = \lim_{j \to \infty} R_j$, we then get

$$R \leq \frac{\gamma(\gamma x + x + \gamma - 1)}{x(\gamma - x - 1)}. \tag{11}$$

This expression is minimized for parameters $x = (5 - \sqrt{13})/2 \approx 0.697$ and $\gamma = (3 + \sqrt{13})/2 \approx 3.303$, yielding the asymptotic competitive ratio

$$R \leq \tfrac{1}{6}(47 + 13\sqrt{13}) \approx 15.645.$$

Summarizing this analysis, we obtain the following theorem.

Theorem 1. *The asymptotic competitive ratio of* OCC *is at most* 15.645.

Table 1. Some initial upper bound values for the absolute competitive ratio

Phase	1	2	3	4	5	6	7	8
Bound	10.000	17.493	23.157	24.854	24.521	22.539	20.474	18.793

Absolute Competitive Ratio. In fact, for parameters $x = (5 - \sqrt{13})/2$ and $\gamma = (3 + \sqrt{13})/2$, Strategy OCC has a low absolute competitive ratio as well. We show that this ratio is at most 24.854.

When phase 0 ends, the competitive ratio is 1. For $j \geq 1$, let O'_j be the optimal profit right before phase j ends. (Earlier we used O_j to estimate this value, but O_j also includes the profit for the last step of phase j.) It remains to show that for phases $j \geq 1$ we have $R'_j \leq 24.854$, where $R'_j = O'_j/S_{j-1}$.

By analyzing the behavior of Strategy OCC in phase 1 and exhaustively enumerating the possible configurations, given that $\gamma \approx 3.303$, we can establish that $R'_1 = 10$.

For phases $j \geq 2$, we can tabulate upper bounds for R'_j by explicitly computing the ratios O'_j/S_{j-1} using a modification of recurrence (9), where we take advantage of the fact that some quantities in inequalities (6) and (7) are integral, so their estimates can be rounded down. We show the first few estimates in Table 1.

To bound the sequence $\{R'_j\}_{j>0}$ we use (9), (6) and (7), to obtain the recurrence

$$R'_j \leq (x+1)\alpha_j R'_{j-1} + \beta_j,$$

where $\alpha_j \leq \dfrac{\gamma^{j-1} + \sqrt{5\gamma^j} + 3j/2}{\gamma^j - 1}$ and $\beta_j \leq \dfrac{(x+1)(\gamma-1)}{x} \cdot \dfrac{\gamma^j + \sqrt{2\gamma^j} + 1}{\gamma^j - 1} + 2$.

For $j \geq 6$ it is not hard to show that $\beta_j \leq 8$. Consider the denominator $\gamma^j - 1$ of α_j. We have that $\gamma^j - 1 > \frac{9}{10}\gamma^j$ for $j \geq 2$. Hence, $R'_j \leq \hat{R}_j$, where $\hat{R}_j$ is given by the recurrence

$$\hat{R}_j \leq \frac{10(x+1)(\gamma^{j-1} + \sqrt{5}\gamma^{j/2} + 3j/2)}{9\gamma^j}\hat{R}_{j-1} + 8 \leq \frac{3}{5}\hat{R}_{j-1} + 8 = 20 - 19\left(\frac{3}{5}\right)^j$$

for $j \geq 8$. The sequence $\{\hat{R}_j\}_{j \geq 0}$, with $\hat{R}_0 = 1$, grows monotonically to the limit $\lim_{j \to \infty} \hat{R}_j = 20$ and hence $\hat{R}_j \leq 20$ for every $j \geq 8$. Combining this with the earlier bounds, we see that the largest bound on R'_j is 24.854, given in Table 1 for $j = 4$. We can thus conclude that the absolute competitive ratio is at most 24.854.

3   A Lower Bound of 6

We now prove that any deterministic online strategy $\mathcal{S}$ for the clique clustering problem has competitive ratio at least 6. We present the proof for the absolute competitive ratio; later we explain how to extend it to the asymptotic ratio. The lower bound is established by showing, for any constant $R < 6$, an adversary

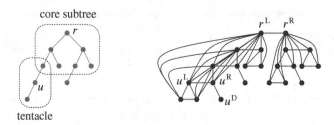

Fig. 1. On the left, an example of a skeleton tree T. The core subtree of T has depth 2 and two tentacles, one of length 2 and one of length 1. On the right, the corresponding graph $\mathcal{G}_T$.

strategy for constructing an input graph on which the optimal profit is at least R times the profit of S.

Fix some $R < 6$ and let D be a non-negative integer (that depends on R) whose value will be specified later. It is convenient to describe the graph constructed by the adversary in terms of its underlying *skeleton tree* T, which is a rooted binary tree. The root of T will be denoted by r. For a node $v \in T$, define the *depth* or *level* of v to be the number of edges on the simple path from v to r. The adversary will only use skeleton trees of the following special form: each non-leaf node at depths $0, 1, \ldots, D-1$ has two children, and each non-leaf node at levels at least D has one child. Such a tree can be thought of as consisting of its *core subtree*, which is a complete binary tree of depth D, with paths attached to its leaves at level D. The nodes of T at depth D are the leaves of the core subtree. If v is a node of the core subtree of T then the path extending from v down to a leaf of T is called a *tentacle* – see Figure 1. (Thus v belongs both to the core subtree and to a tentacle attached to v.) The length of a tentacle is the number of its edges. The nodes in the tentacles are all considered left children of their parents.

The graph represented by a skeleton tree T will be denoted by $\mathcal{G}_T$. We differentiate between the *nodes* of T and the *vertices* of $\mathcal{G}_T$. The relation between T and $\mathcal{G}_T$ is illustrated in Figure 1. $\mathcal{G}_T$ is obtained from T as follows:

- For each node $u \in T$ we create two vertices u^L and u^R in $\mathcal{G}_T$, with an edge between them. This edge (u^L, u^R) is called the *cross edge* corresponding to u.
- Suppose that $u, v \in T$. If u is in the left subtree of v then (u^L, v^L) and (u^R, v^L) are edges of $\mathcal{G}_T$. If u is in the right subtree of v then (u^L, v^R) and (u^R, v^R) are edges of $\mathcal{G}_T$. These edges are called *upward edges*.
- If $u \in T$ is a node in a tentacle of T and is not a leaf, then $\mathcal{G}_T$ has a vertex u^D with edge (u^D, u^R). This edge is called a *whisker*.

The adversary constructs $\mathcal{G}_T$ gradually, in response to S's choices. Initially, T is a single node r, and thus $\mathcal{G}_T$ is a single edge (r^L, r^R). At this time, $\mathsf{profit}_S(T) = 0$ and $\mathsf{O}(T) = 1$, so S is forced to collect this edge (that is, it creates a 2-clique $\{r^L, r^R\}$).

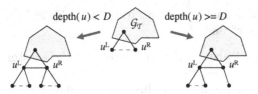

Fig. 2. Adversary moves. Upward edges from new vertices are not shown, to avoid clutter. Dashed lines represent cross edges that are not collected by $\mathcal{S}$, while thick lines represent those that are already collected by $\mathcal{S}$.

In general, the strategy will be able to collect only cross edges. Suppose that, at some step, $\mathcal{S}$ collects a cross edge $(u^{\mathrm{L}}, u^{\mathrm{R}})$, corresponding to node u of $\mathcal{T}$. If u is at depth less than D, the adversary extends $\mathcal{T}$ by adding two children of u. If u is at depth at least D, the adversary only adds the left child of u, thus extending the tentacle ending at u. In terms of $\mathcal{G}_{\mathcal{T}}$, the first move adds two triangles to u^{L} and u^{R}, with all corresponding upward edges. The second move adds a triangle to u^{L} and a whisker to u^{R} (see Figure 2).

Thus the adversary will be building the core binary skeleton tree down to level D, and from then on, it will extend the tentacles. Our objective is to prove that after each step the ratio between the adversary profit and the strategy's profit is at least $6 - O(1/D)$. This is enough to prove the lower bound. The reason is this: If the strategy stops collecting edges at some point, the ratio is $6 - O(1/D)$, and we are done. Otherwise, suppose that the game lasts for a very long time, and since D is fixed, then at least one tentacle will grow without bound. But the optimal cost is at least quadratic with respect to the maximum tentacle length s, while $\mathcal{S}$'s profit is only linear in s. Thus eventually the adversary can simply stop playing, and even if the strategy collects the remaining cross edges (and there will be at most $2^D \cdot s$ of those), the ratio will be larger than 6.

Denote by $\mathcal{T}_v$ the subtree of $\mathcal{T}$ rooted at v. To simplify the computation of the adversary (or optimal) profit, we will assume that the adversary computes his clustering recursively, as follows:

(opt1) If x is a leaf of $\mathcal{T}$, then x^{L} and x^{R} are in the same cluster.

(opt2) Suppose that x is an internal node of $\mathcal{T}$ and let y be the left child of x. Assume that the clustering of $\mathcal{T}_y$ is already computed. If x has a right child, let z be this child and assume that the clustering of $\mathcal{T}_z$ is already computed. Then

    (opt2.a) x^{L} is added either to the cluster of $\mathcal{T}_y$ containing y^{L} or to the cluster containing y^{R}. (When we estimate the adversary profit, we will specify which choice we use.) This is correct, since all neighbors of y^{L} and y^{R} that correspond to nodes in $\mathcal{T}_y$ are also neighbors of x^{L}. Note that in the special case when y is a leaf, the clusters of y^{L} and y^{R} are the same.

    (opt2.b) If x has the right child z, then the rule for adding x^{R} to the clustering of $\mathcal{T}_z$ is symmetric to (opt2.a). If x does not have the right child (so x is in a tentacle), then we create the "whisker" cluster consisting of two vertices x^R and x^D.

Observe that, in particular, all clusters, except for the whisker clusters, have at least three vertices.

We stress that the profit of the clustering computed as above (even for the way we specify the adversary choices in (opt2.a) and (opt2.b)) may not be actually maximized, but this does not matter, since for the purpose of our proof we only need a lower bound on the adversary profit.

We now claim that before the core tree reaches its target height D the ratio is at least 6. Indeed, consider one step, when $\mathcal{S}$ collects an edge (u^L, u^R). (See Figure 2.) The strategy's profit increases by 1. As for the adversary, he can increase his profit as follows:

(i) Create a new clique that is a triangle consisting of u^R and two new vertices, increasing the profit by 3.

(ii) In the current clique that contained u^L and u^R, replace u^R by the two new vertices connected to u^L. This current clique had size at least 3 (the adversary will maintain the invariant that in his clustering each cross edge is in a clique of size at least 3) and its size increases by 1, so its profit increases by at least 3.

Overall, the adversary's profit increases by at least 6, proving the claim.

Thus from now on it is sufficient to analyze skeleton trees of height strictly larger than D, namely trees that have at least one tentacle already started. Let $\mathcal{T}$ be such a skeleton tree. We will focus on analyzing the profits of the adversary and the strategy on such trees $\mathcal{T}_v$, where v is a node in the core subtree of $\mathcal{T}$. If $\mathcal{T}_v$ ends at depth $D+1$ or more, we call it a *bottom subtree*. The *core depth* of a bottom subtree $\mathcal{T}_v$ is defined as the depth of the part of $\mathcal{T}_v$ within the core subtree of $\mathcal{T}$. If h and s are, respectively, the core depth of $\mathcal{T}_v$ and its maximum tentacle length, then $0 \leq h \leq D$ and $s \geq 1$.

For a subtree $X = \mathcal{T}_v$, let $O(X)$ be the optimal profit in X, computed according to the description above, and $S(X)$ be $\mathcal{S}$'s profit (the number of cross edges). The lemma below is key in our argument.

Lemma 2. *Let X be a bottom subtree of height $h \geq 0$ and maximum tentacle length $s \geq 1$. Then*

$$O(X) + 2(h+s) \geq 6 \cdot S(X).$$

Before proving the lemma, let us argue first that this lemma is sufficient to establish our lower bound. Indeed, since we are now considering the case when $\mathcal{T}$ is a bottom subtree itself, the lemma implies that $O(\mathcal{T}) + 2(D+s) \geq 6 \cdot S(\mathcal{T})$, where s is the maximum tentacle length of $\mathcal{T}$. But $O(\mathcal{T})$ is at least quadratic in $D+s$. So for large D the ratio $O(\mathcal{T})/S(\mathcal{T})$ approaches 6.

So now we prove Lemma 2. The proof is by induction on h, the core height of X. Consider first the base case, for $h = 0$ (when X is just a tentacle). The adversary has one clique of $s+2$ vertices, namely all x^L vertices in the tentacle

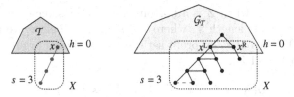

Fig. 3. Illustration of the inductive proof, the base case. Subtree X on the left, the corresponding subgraph on the right.

(there are $s+1$ of these), plus one z^R vertex for the leaf z. He also has s whiskers, so his profit for X is $\binom{s+2}{2} + s = \frac{1}{2}(s^2 + 5s + 2)$. The strategy collects only s cross edges, namely all cross edges in X except last. (See Figure 3.) Solving the quadratic inequality and using the integrality of s, we get $O(X) + 2s \geq 6s = 6 \cdot S(X)$. Note that this inequality is in fact tight for $s = 1, 2$.

In the inductive step, consider a bottom subtree $X = \mathcal{T}_u$. Let Y and Z be its left and right subtrees, respectively. Without loss of generality, we can assume that Y is a bottom tree with height $h-1$ and the same maximum tentacle length s as X, while Z is either not a bottom tree (that is, it has no tentacles), or it is a bottom tree with maximum tentacle length at most s.

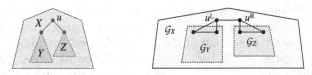

Fig. 4. Illustration of the inductive proof, the inductive step. Subtrees X, Y, Z on the left, the corresponding subgraphs on the right.

By the inductive assumption, we have $O(Y)+2(h-1+s) \geq 6 \cdot S(Y)$. Regarding Z, if Z is not a bottom tree then $O(Z) \geq 6 \cdot S(Z)$, and if Z is a bottom tree (necessarily of height $h-1$) then $O(Z)+2(h-1+s') \geq 6 \cdot S(Z)$, where s' is Z's maximum tentacle length, such that $1 \leq s' \leq s$.

Consider first the case when Z is not a bottom tree. Note that

$$S(X) = S(Y) + S(Z) + 1 \quad \text{and} \quad O(X) \geq O(Y) + O(Z) + h + s + 4$$

The first equation is trivial, because for X the strategy gets all cross edges in Y and Z, plus one more cross edge (u^L, u^R). The second inequality holds because u^L can be added to Y's largest cluster which has $(h-1)+s+2 = h+s+1$ vertices, and u^R can be added to Z's largest cluster that has at least 3 vertices. Then we get (since $h, s \geq 1$):

$$\begin{aligned}
O(X) + 2(h+s) &\geq [O(Y) + O(Z) + h + s + 4] + 2(h+s) \\
&= [O(Y) + 2(h-1+s)] + O(Z) + 6 \\
&\geq 6 \cdot S(Y) + 6 \cdot S(Z) + 6 = 6 \cdot S(X).
\end{aligned}$$

The second case is when Z is a bottom tree (of the same core height $h - 1$) and maximum tentacle length s', where $1 \leq s' \leq s$. As before, we have $S(X) = S(Y) + S(Z) + 1$. The optimum profit satisfies (by a similar argument as before, applied to both Y and Z):

$$O(X) \geq O(Y) + O(Z) + 2h + s + s' + 2.$$

Then we get (using $s \geq s'$):

$$
\begin{aligned}
O(X) + 2(h + s) &\geq [O(Y) + O(Z) + 2h + s + s' + 2] + 2(h + s) \\
&\geq [O(Y) + 2(h - 1 + s)] + [O(Z) + 2(h - 1 + s')] + 6 \\
&\geq 6 \cdot S(Y) + 6 \cdot S(Z) + 6 = 6 \cdot S(X).
\end{aligned}
$$

This completes the proof of Lemma 2, for the case of the absolute competitive ratio.

We still need to explain how to extend our proof so that it also applies to the asymptotic competitive ratio. This is quite simple: Choose some large constant M. The adversary will create M instances of the above game, playing each one independently. Our construction above used the fact that at each step the strategy was forced to collect one of the pending cross edges, for otherwise its competitive ratio would exceed ratio R (where R was arbitrarily close to 6). Now, for M sufficiently large, the strategy will be forced to collect cross edges in all except for some finite number of copies of the game, where this number depends on the additive constant in the competitiveness bound.

Note: Our construction is very tight, in the following sense. Suppose that the strategy maintains T as balanced as possible. Then the ratio is exactly 6 when the depth of T is 1 or 2. Further, suppose that D is very large and the strategy constructs T to have depth D or more. Then the ratio is $6 - o(1)$ for $s = 1$ and $s = 2$. The intuition is that when the adversary plays optimally, he will only allow the online strategy to collect isolated edges (cliques of size 2). For this reason, we conjecture that 6 is the optimal competitive ratio.

4   Conclusions

We have shown an improved strategy with competitive ratio 15.645 for the problem of clique clustering where the objective is to maximize the number of edges in the cliques. Our strategy uses doubling to guarantee that the optimal measure does not become significantly larger than the strategy's measure. In fact, it is possible to prove (this result is omitted from this paper because of space constraints) that any strategy that uses doubling in this manner cannot achieve a competitive ratio better than 10.927.

We also prove that no strategy whatsoever can achieve a competitive ratio better than 6. Evidently, tightening these bounds would be of significant interest.

References

1. Bansal, N., Blum, A., Chawla, S.: Correlation clustering. Machine Learning **56** (1–3), 89–113 (2004)
2. Ben-Dor, A., Shamir, R., Yakhini, Z.: Clustering gene expression patterns. Journal of Computational Biology **6**(3/4), 281–297 (1999)
3. Borodin, A., El-Yaniv, R.: Online computation and competitive analysis. Cambridge University Press (1998)
4. Charikar, M., Chekuri, C., Feder, T., Motwani, R.: Incremental clustering and dynamic information retrieval. SIAM J. Comput. **33**(6), 1417–1440 (2004)
5. Chaudhuri, K., Godfrey, B., Rao, S., Talwar, K.: Paths, trees, and minimum latency tours. In: 44th Symposium on Foundations of Computer Science (FOCS 2003), Proceedings, Cambridge, MA, USA, October 11–14, pp. 36–45 (2003)
6. Chrobak, M., Hurand, M.: Better bounds for incremental medians. Theor. Comput. Sci. **412**(7), 594–601 (2011)
7. Chrobak, M., Kenyon, C., Noga, J., Young, N.E.: Incremental medians via online bidding. Algorithmica **50**(4), 455–478 (2008)
8. Chrobak, M., Kenyon-Mathieu, C.: SIGACT news online algorithms column 10: competitiveness via doubling. SIGACT News **37**(4), 115–126 (2006)
9. Dessmark, A., Jansson, J., Lingas, A., Lundell, E.-M., Persson, M.: On the approximability of maximum and minimum edge clique partition problems. Int. J. Found. Comput. Sci. **18**(2), 217–226 (2007)
10. Fabijan, A., Nilsson, B.J., Persson, M.: Competitive online clique clustering. In: Proc. 8th International Conference on Algorithms and Complexity (CIAC 2013), pp. 221–233 (2013)
11. Figueroa, A., Borneman, J., Jiang, T.: Clustering binary fingerprint vectors with missing values for DNA array data analysis. Journal of Computational Biology **11**(5), 887–901 (2004)
12. Lin, G., Nagarajan, C., Rajaraman, R., Williamson, D.P.: A general approach for incremental approximation and hierarchical clustering. SIAM J. Comput. **39**(8), 3633–3669 (2010)
13. Mathieu, C., Sankur, O., Schudy, W.: Online correlation clustering. In: 27th International Symposium on Theoretical Aspects of Computer Science (STACS 2010), pp. 573–584 (2010)
14. Valinsky, L., Vedova, G.D., Scupham, R.J., Alvey, S., Figueroa, A., Yin, B., Jack Hartin, R., Chrobak, M., Crowley, D.E., Jiang, T., Borneman, J.: Analysis of bacterial community composition by oligonucleotide fingerprinting of rRNA genes. Applied and Environmental Microbiology 68, 2002 (2002)

Scheduling with Gaps: New Models and Algorithms

Marek Chrobak[1]($\boxtimes$), Mordecai Golin[2], Tak-Wah Lam[3], and Dorian Nogneng[4]

[1] Department of Computer Science, University of California at Riverside,
Riverside, USA
marek@cs.ucr.edu
[2] Department of Computer Science and Engineering,
Hong Kong University of Science and Technology, Hong Kong, China
[3] Department of Computer Science, University of Hong Kong, Hong Kong, China
[4] LIX, École Polytechnique, Palaiseau, France

Abstract. We initiate the study of scheduling problems where the number or size of the gaps in the schedule is taken into consideration. We focus on the model with unit jobs. First we examine scheduling problems with release times and deadlines, where we consider variants of minimum-gap scheduling, including maximizing throughput with a budget for gaps or minimizing the number of gaps with a throughput requirement. We then turn to other objective functions. For example, in some scenarios, gaps in a schedule may be actually desirable, leading to the problem of maximizing the number of gaps. The second part of the paper examines the model without deadlines, where we focus on the tradeoff between the number of gaps and flow time.

For all these problems we provide polynomial algorithms. The solutions involve a spectrum of algorithmic techniques, including different dynamic programming formulations, speed-up techniques based on searching Monge arrays, searching $X + Y$ matrices, or implicit binary search. Throughout the paper, we also draw a connection between our scheduling problems and their continuous analogues, namely hitting set problems for intervals of real numbers.

1 Introduction

We consider scheduling of unit jobs, with given release times, where the number or size of the gaps in the schedule is taken into consideration, either in the objective function or as a constraint. The problem of minimizing the number of gaps arises naturally in minimum-energy scheduling with deadlines, because in some scenarios the energy consumption associated with a schedule is a linear function of the number of gaps. The problem was proposed by Irani and Pruhs [7]. The first polynomial time algorithm, with running time $O(n^7)$, was given by Baptiste [2]. The time complexity was subsequently reduced to $O(n^4)$ in [3].

Research supported by NSF grants CCF-0729071 and CCF-1217314 and by grant FSGRF14EG28 (Hong Kong).

V.Th. Paschos and P. Widmayer (Eds.): CIAC 2015, LNCS 9079, pp. 114–126, 2015.
DOI: 10.1007/978-3-319-18173-8_8

A greedy algorithm was analyzed in [4] and shown to have approximation ratio 2. Other variants of this problem have been studied, for example the multiprocessor case [6] or the case when the jobs have agreeable deadlines [1].

To our knowledge, the above power-down scheduling framework is the only scheduling model in the literature that considers gaps in the schedule as a performance measure. As we show in this paper, however, one can formulate a number of other natural, but not yet studied variants of gap scheduling problems. Some of these problems can be solved using dynamic-programming techniques resembling those used for minimizing the number of gaps. Other require new techniques, giving rise to new and interesting algorithmic problems.

Throughout the paper, we focus on the model with unit-length jobs. The first type of problems we study involve unit jobs with release times and deadlines. In this category, we study the following problems:

• In Section 3, we study maximizing throughput (the number or total weight of executed jobs) with a budget γ for the number of gaps. We give an $O(\gamma n^6)$-time algorithm for this problem.

• In Section 4 we study the variant where we need to minimize the number of gaps under a throughput requirement. We show that this problem can be solved in time $O(n^7)$.

• In certain applications gaps in a schedule may be actually desirable, which motivates the gap scheduling model where we wish to *maximize* the number of gaps. We study this problem in Section 5, and we provide an algorithm that computes an optimal schedule in time $O(n^5)$.

• Instead of the total number of gaps, the *size* of gaps may be a useful attribute of a schedule. In Section 6 we study the problem of minimizing the maximum gap size, for which we give an $O(n^2 \log n)$-time algorithm.

In the full version of this paper (see [5]), we also provide an $O(n^4)$-time algorithm for minimizing the number of gaps, simplifying the one in [3].

Next, we consider scheduling problems where jobs have no deadlines and all jobs must be scheduled, focusing on the tradeoffs between the number of gaps and the flow time measure, where the flow of a job is the time elapsed between its release and completion times. We address four problems in this category:

• Minimizing total flow time with a budget γ for the number of gaps (Section 7). As we show, this problem can be solved in time $O(\gamma n + n \log n)$.

• Minimizing the number of gaps with a budget for total flow (Section 8). We show that this problem can be solved in time $O(g^* n \log n)$, where g^* is the optimum solution.

• Minimizing the number of gaps with a bound on the maximum flow time (Section 9). We show that this problem can be solved in time $O(n \log n)$.

• Minimizing maximum flow time with a budget γ for the number of gaps (Section 10). For this problem we give an algorithm with running time $O(n \log n)$.

The solutions involve a spectrum of algorithmic techniques, including different dynamic programming formulations, as well as speed-up techniques based on searching Monge arrays, searching $X + Y$ matrices, and implicit binary search.

Throughout the paper, we draw a connection between our scheduling problems and their continuous analogues, which are variants of the hitting set problems for intervals. For some problems, the continuous variants provide insights leading to efficient algorithms for the corresponding discrete versions, while in other problems new techniques are needed in the discrete version.

Due to space limitations, full descriptions of some algorithms, as well as most proofs, are omitted. These can be found in the full version of this paper [5].

2   Preliminaries

The time is discrete, divided into unit time intervals $[t, t+1)$, for $t = 1, 2, ...$, that we call *slots*. We refer to $[t, t + 1)$ simply as *time slot* t, or as *time* t. By $\mathcal{J}$ we denote the instance, consisting of a set of unit-length jobs numbered $1, 2, ..., n$, each job j with an integer release time r_j (the first slot where j can be executed).

A *schedule* S of $\mathcal{J}$ is defined by an assignment of jobs to time slots such that (i) if a job j is assigned to a slot t then $t \geq r_j$, and (ii) no two jobs are assigned to the same slot. In most scheduling problems we assume that all jobs are scheduled. In problems that involve throughput, we will consider partial schedules, where only some of the jobs may be scheduled. For a schedule S, time slots where jobs are scheduled are called *busy*, while all other slots are called *idle*. A maximal finite time interval of idle slots in S is called a *gap* of S. A maximal time interval of busy slots is called a *block* of S.

Jobs with Deadlines and Feasibility. In some of the scheduling problems, the jobs in $\mathcal{J}$ will also have specified deadlines. The *deadline* of a job j, denoted d_j, is the last slot where j can be executed.

For instances with deadlines, we can restrict our attention to schedules S that satisfy the *earliest-deadline first property (EDF)*: at any time t, either S is idle at t or it schedules a pending job with the earliest deadline. Using the standard exchange argument, any schedule can be converted into one that satisfies the EDF property and has the same set of busy slots. Without loss of generality, we can make the following assumptions about $\mathcal{J}$: (i) $r_j \leq d_j$ for each j, and (ii) all jobs are ordered according to deadlines, that is $d_1 \leq ... \leq d_n$. Further, we can assume that (iii) all release times are distinct and all deadlines are distinct. Any schedule can be converted in time $O(n \log n)$ into one that satisfies these conditions, see [5].

Instances without Deadlines. For schedules involving objective functions other than throughput, we can assume that the jobs are ordered according to non-decreasing release times. For the total-flow objective function we can assume that all release times are different. This is because, although modifying the release times as above may change the total flow value, this change will be uniform for all schedules, so the optimality will not be affected. The maximum flow values though may change non-uniformly, so we will not be using this assumption in the Sections 9 and 10, where maximum flow of jobs is considered.

Shifting Blocks. To improve the running time, some of our algorithms use assumptions about possible locations of the blocks in an optimal schedule. The general idea is that each block can be shifted, without affecting the objective function, to a location where it will contain either a deadline or a release time. The following lemma is useful for this purpose. We formulate the lemma for leftward shifts; an analogous lemma can be formulated for rightward shifts and for deadlines instead of release times (see [3,5] for the proof).

Lemma 1. *Assume that all jobs on input have different release times. Let $B = [u,v]$ be a block in a schedule such that the job scheduled at v has release time strictly before v. Then B can be shifted leftward by one slot, in the sense that the jobs in B can be scheduled in the interval $[u-1, v-1]$.*

Interval Hitting. We also examine the "continuous" analogues of our scheduling problems, obtained by assuming that all release times and deadlines are spread very far apart; thus in the limit we can think of jobs as having length 0. Each r_j and d_j is a point in time, and to "schedule" j we choose a point in the interval $[r_j, d_j]$. Two jobs scheduled right next to each other end up being on the same point. This problem is then equivalent to computing hitting sets for a given collection of intervals on the real line, with some conditions involving gaps between the points in the hitting set.

Formally, in the hitting-set problem we are given a collection of intervals $I_j = [r_j, d_j]$ on the real line, where r_j, d_j are real numbers. Our objective is to compute a set H of points such that $H \cap I_j \neq \emptyset$ for all j. This set H is called a *hitting set* of the intervals $I_1, I_2, ..., I_n$. (This formulation corresponds to scheduling problems with deadlines and where all jobs need to be scheduled. Other scheduling problems that we study have similar continuous analogues.)

If H is a hitting set of $I_1, I_2, ..., I_n$, then for each j we can pick a *representative* $h_j \in H \cap I_j$. Sorting these from left to right, $h_{i_1} \leq h_{i_2} \leq ... \leq h_{i_n}$, the non-zero differences $h_{i_{b+1}} - h_{i_b}$ between consecutive representatives are called *gaps* of H.

For each gap scheduling problem we can then consider the corresponding hitting-set problem. These interval-hitting problems are conceptually easier to deal with than their discrete counterparts. As we show, some algorithms for interval-hitting problems extend to the corresponding gap scheduling problems, while for other the discrete cases require different techniques.

3   Maximizing Throughput with Budget for Gaps

In this section we consider a variant of gap scheduling where we want to maximize throughput, given a budget γ for the number of gaps. We first show that the continuous version of this problem can be solved in time $O(\gamma n^2)$. For the discrete case we give an algorithm with running time $O(\gamma n^6)$.

Continuous Case. Formally, the problem is defined as follows. We are given a collection of intervals $I_j = [r_j, d_j]$, $j = 1, 2, ..., n$ and a positive integer γ. The objective is to compute a set H of at most γ points that hits the maximum number of intervals. As explained earlier, without loss of generality we only need to consider sets $H \subseteq \{d_1, d_2, ..., d_n\}$.

Using dynamic programming, we can solve it as follows. Order the intervals according to the deadlines, that is $d_1 \leq d_2 \leq ... \leq d_n$. For $g = 1, 2, ..., \gamma$ and

$b = 1, 2, ..., n$, let $\mathsf{T}_{b,g}$ be the maximum number of intervals that are hit with at most g points from $\{d_1, d_2, ..., d_b\}$, assuming that one of these points is d_b. For all b, we first initialize $\mathsf{T}_{b,1}$ to be the number of intervals that contain d_b. Then, for $g \geq 2$ and all b, we can compute $\mathsf{T}_{b,g}$ using the recurrence: $\mathsf{T}_{b,g} = \max_{a<b}\{\mathsf{T}_{a,g-1} + \Delta_{a,b}\}$, where $\Delta_{a,b}$ is the number of intervals I_i such that $d_a < r_i \leq d_b \leq d_i$. The output value is $\max_b \mathsf{T}_{b,\gamma}$.

All values $\Delta_{a,b}$ can be computed in time $O(n^2)$. To accomplish this, first sort all release times and deadlines. For each a, consider only intervals I_i with $r_i > d_a$. Then start with $x = r_a$, and keep incrementing x to the next release time or deadline, whichever is earlier, at each step updating the number of intervals hit by x, and recording these values for each $x = d_b$. This sweep costs time $O(n)$.

This gives us an algorithm with running time $O(\gamma n^2)$, because we have $O(\gamma n)$ values $\mathsf{T}_{b,g}$ to compute, each computation taking time $O(n)$.

Discrete Case. For the discrete case, a more intricate dynamic programming approach (similar to minimizing the number of gaps) is needed.

As before, denote by $\mathcal{J}$ the set of jobs on input. For each job k and times $u \leq v$, let $\mathcal{K}_{k,u,v}$ denote the sub-instance of $\mathcal{J}$ that consists of all jobs $j \in \{1, 2, ..., k\}$ that satisfy $u \leq r_j \leq v$. Define $\mathsf{T}_{k,u,v,g}$ to be the maximum number of jobs from $\mathcal{K}_{k,u,v}$ that can be scheduled in the interval $[u, v]$ with the number of gaps not exceeding g. Here, the initial and final gap (between u and the first job, and between the last job and v) are also counted.

The idea to compute $\mathsf{T}_{k,u,v,g}$ is this: Consider an optimal schedule for $\mathcal{K}_{k,u,v}$. If k is not scheduled, then $\mathsf{T}_{k,u,v,g} = \mathsf{T}_{k-1,u,v,g}$. So assume that k is scheduled, say at time t, where $u \leq t \leq v$. Letting h be the number of gaps before t, we have $\mathsf{T}_{k,u,v,g} = \mathsf{T}_{k-1,u,t-1,h} + \mathsf{T}_{k-1,t+1,v,g-h} + 1$. If k is scheduled last, then either $d_k \geq v$ and k is scheduled at v, in which case $\mathsf{T}_{k,u,v,g} = \mathsf{T}_{k-1,u,v-1,g} + 1$, or k is scheduled at some time $t' < v$ and $\mathsf{T}_{k,u,v,g} = \mathsf{T}_{k-1,u,t'-1,g-1} + 1$.

Algorithm MaxThrpt. For all $k = 0, 1, ..., n$ and time slots u, v, where $v \geq u$, we process all instances $\mathcal{K}_{k,u,v}$ in order of increasing k, and for each k in order of increasing interval length, $v - u$. (We will explain shortly how to choose u and v to make the algorithm more efficient.) For each instance $\mathcal{K}_{k,u,v}$ and each gap budget g we compute the corresponding value $\mathsf{T}_{k,u,v,g}$.

First, if $\mathcal{K}_{k,u,v} = \emptyset$, we let $\mathsf{T}_{k,u,v,g} = 0$. Assume that $\mathcal{K}_{k,u,v} \neq \emptyset$. If $k \notin \mathcal{K}_{k,u,v}$ (which means that $r_k \notin [u, v]$) then $\mathsf{T}_{k,u,v,g} = \mathsf{T}_{k-1,u,v,g}$. Otherwise, let

$$\mathsf{T}_{k,u,v,g} = \max \left\{ \begin{array}{l} \mathsf{T}_{k-1,u,v,g} \\ \max_{\substack{r_k \leq t \leq \min(d_k,v) \\ 0 \leq h \leq g}} \{\mathsf{T}_{k-1,u,t-1,h} + \mathsf{T}_{k-1,t+1,v,g-h}\} + 1 \\ \max_{r_k \leq t' \leq \min(d_k,v-1)} \{\mathsf{T}_{k-1,u,t'-1,g-1}\} + 1 \\ \mathsf{T}_{k-1,u,v-1,g} + 1 \qquad \text{if } d_k \geq v \end{array} \right\}$$

To achieve polynomial running time we need to show that only a small set of time parameters u, v and t needs to be considered. As shown in [5], by appropriately restricting the ranges of these parameters, one can achieve running time $O(\gamma n^6)$.

Weighted Throughput. Both algorithms, for the continuous and discrete cases, easily extend to the model where jobs have non-negative weights and the objective is to maximize the weighted throughput.

4 Minimizing Number of Gaps with Throughput Requirement

Suppose that now we want to minimize the number of gaps, under a throughput requirement, that is we want to find a schedule that schedules at least m jobs and minimizes the number of gaps. We assume that the maximum throughput is at least m; this condition can be verified in time $O(n \log n)$ by the earliest-deadline-first algorithm. We can solve this problem by using the algorithms from the previous section. We explain the solution for the continuous variant; the solution of the discrete case can be obtained in a similar manner.

Recall that $T_{b,g}$ was defined to be the maximum number of intervals that are hit with at most g points from $\{d_1, d_2, ..., d_b\}$, assuming that one of them is d_b. We can use these values to compute T_g, which is the maximum number of intervals hit with at most g points. Then, given our requirement m on the throughput, we compute the smallest g for which $T_g \geq m$. This g is the output of the algorithm. The total running time will be $O(n^3)$.

An analogous approach gives an algorithm for the discrete case with running time $O(n^7)$. In both cases, the algorithms extend to weighted throughput.

5 Maximizing Number of Gaps

Earlier we were looking for schedules with as few gaps as possible. However, in some applications, gaps in the schedule may actually be desirable. This arises, for example, for input streams consisting of high-priority jobs that are executed in pre-computed slots and low-priority jobs that use remaining slots, if available. One such scenario appears in coordinating access to a channel through point coordination function (PCF) [9] where gaps (in our terminology) in the high-priority traffic are forced to allow low-priority traffic to access the channel.

Thus in this section we will examine the problem where the objective is to create as many gaps as possible in the schedule. The continuous version of this problem is trivial: for intervals $I_j = [r_j, d_j]$ with $r_j = d_j$, we must of course choose $h_j = r_j$; those with $r_j < d_j$ can be assigned unique points $h_j \in I_j$.

We now focus only on the discrete model. Specifically, we are again given an instance with n unit jobs with release times and deadlines, and we assume that the instance is feasible, that is all jobs can be scheduled. The objective is to find a schedule that maximizes the number of gaps.

As before, we will assume that all jobs have different deadlines and different release times, and that they are ordered according to increasing deadlines. We can also assume that jobs 1 and n satisfy $r_1 = d_1 = \min_{j>1} r_j - 2$ and $d_n = r_n = \max_{j<n} d_j + 2$, that is, they are tight jobs executed at the beginning and end of the schedule, separated by gaps from other jobs.

For any job $k = 1, 2..., n$ and time steps $u \leq v$ define $\mathcal{K}_{k,u,v}$ to be the sub-instance of $\mathcal{J}$ that consists of all jobs $j \in \{1, 2, ..., k\}$ that satisfy $u \leq r_j \leq v$. Define $D_{k,u,v}$ to be the maximum number of gaps in a schedule of $\mathcal{K}_{k,u,v}$ in the interval $[u, v]$. In $D_{k,u,v}$ we include the extremal gaps in the schedule, namely the gap between u and the first job and the gap between the last job and v.

Lemma 2. *For any sub-instance $\mathcal{K}_{k,u,v}$ there is a schedule S with $D_{k,u,v}$ gaps in the interval $[u, v]$ that has the EDF property and satisfies the following two conditions: (i) For any job $j \in \mathcal{K}_{k,u,v}$, if j is scheduled at time S_j then all gaps in the interval $[r_j, S_j]$ have length at most 2 (including the gap between r_j and the first job). (ii) For each block B of S, either all jobs in B are scheduled at their release times or, assuming that B does not start at u, the gap immediately to the left of B has length 1.*

The proof of this lemma appears in the full version of this paper [5]. The fundamental idea of our algorithm is similar to the gap minimization algorithm from [2,5] and is based on dynamic programming, using Lemma 2 to achieve polynomial running time. We compute all values $D_{k,u,v}$. We can assume that $k \in \mathcal{K}_{k,u,v}$, for otherwise $D_{k,u,v} = D_{k-1,u,v}$. Suppose that, in an optimal schedule, k is scheduled at some time $t \in [u, v]$. Then, by the EDF property, t cannot be a release time of any job $i \in \mathcal{K}_{k,u,v}$ other than k. Further, also using the EDF property, all jobs scheduled in $[u, t-1]$ belong to $\mathcal{K}_{k-1,u,t-1}$ and all jobs scheduled in $[t + 1, v]$ belong to $\mathcal{K}_{k-1,t+1,v}$, thus giving us a partition of $\mathcal{K}_{k,u,v}$ into $\{k\}$ and two disjoint sub-instances $\mathcal{K}_{k-1,u,t-1}$ and $\mathcal{K}_{k-1,t+1,v}$, whose schedules must maximize the number of gaps in intervals $[u, t - 1]$ and $[t + 1, v]$, respectively. We thus conclude that $D_{k,u,v} = D_{k-1,u,t-1} + D_{k-1,t+1,v}$. Since we do not know t a priori, we can maximize the expression on the right hand side over all choices of t.

A complete description of the algorithm, with its implementation in time $O(n^5)$, appears in the full version of the paper [5].

6   Minimizing Maximum Gap

In earlier sections we focussed on the number of gaps in the schedule. For certain applications, the *size* of the gaps is also of interest. In this section we will study the problem where the objective is to minimize the maximum gap in the schedule. Such schedules will tend to produce many short gaps, which may be useful in applications discussed in Section 5, where a good schedule should leave some gaps between high-priority jobs, to allow other jobs to access the processor.

The general setting is as before. We have an instance $\mathcal{J}$ consisting of n unit jobs, where job j has release time r_j and deadline $d_j \geq r_j$. The objective is to compute a schedule of all jobs that minimizes the maximum gap size, or to report that no feasible schedule exists.

Interestingly, this problem is structurally different from these in the previous sections, because now, intuitively, a good schedule should spread the jobs more-or-less evenly in time. For example, if we have n jobs released at 0 and with deadline $D \gg n$, with a tight job at the beginning and a tight job at the end, we would then schedule them at times $i\frac{D}{n-1}$. In contrast, the algorithms in Section 4 attempted to group the jobs into a small number of batches.

In this section we give an $O(n^2 \log n)$-time algorithm for computing schedules that minimize the maximum gap. We first give an algorithm for the continuous model, and then extend it to the discrete model.

Continuous Case. The continuous analogue of our scheduling problem can be formulated as follows. The input consists of n intervals $I_1, I_2, ..., I_n$. As before, $I_j = [r_j, d_j]$, for each j. The objective is to compute a hitting set H for which the maximum gap between its points is minimized. Another way to think about this problem is as computing a representative $h_j \in H \cap I_j$ for each interval I_j. Except for degenerate situations (two equal intervals of length 0), we can assume that all representatives are different, although we will not be using this property in our algorithm, and we treat H as a multiset.

Without losing generality, we can assume that $d_1 \leq d_2 \leq ... \leq d_n$. Further, we only need to be concerned with sets H that contain d_1, because if H contains any points before d_1 then we can replace them all by d_1 without increasing the maximum gap in H. Also, if $\max_i r_i \leq d_1$ then there is a singleton hitting set, $H = \{d_1\}$, in which case the maximum gap is vacuously equal 0. Thus we can also assume that $\max_i r_i > d_1$, so that we need at least two points in H.

Consider first the decision version: *"Given $\lambda > 0$, is there a hitting set H for $I_1, I_2, ..., I_n$ in which all gaps are at most λ?"* If λ has this property, we will call it *viable*. We first give an algorithm for this decision version and later we will show how to use it to obtain an efficient minimization algorithm.

Algorithm VIABLE(λ). We initialize $h_1 = d_1$, $H = \{h_1\}$, and $U = \{2, 3, ..., n\}$. U is the set containing the indices of the intervals that still do not have representatives selected. We move from left to right, at each step assigning a representative to one interval in U, and we remove this interval from U.

Specifically, at each step, we proceed as follows. Let $z = \max(H) + \lambda$. If all $j \in U$ satisfy $r_j > z$, declare failure by returning false. Otherwise, choose $j \in U$ with $r_j \leq z$ that minimizes d_j and remove j from U. We now have two cases. If $d_j \leq z$, let $h_j = d_j$, and otherwise (that is, when $r_j \leq z < d_j$) let $h_j = z$. Then add h_j to H, and continue. If the process completes with $U = \emptyset$, return true. (The solution is $H = \{h_1, h_2, ..., h_n\}$.)

The (simple) correctness proof can be found in the full version of this paper [5]. We claim that the algorithm can be implemented in time $O(n \log n)$. Instead of U, the algorithm can maintain a set U' consisting only of those intervals j with $r_j \leq \max(H) + \lambda$ that do not have yet a representative. Then choosing the j in the algorithm and removing it from U' takes time $O(\log n)$. When $\max(H)$ is incremented (after adding h_j), new intervals are inserted into U' in order of release times, each insertion taking time $O(\log n)$.

The idea is to use the above procedure as an oracle in the binary search on λ's. To this end, we need a small set of candidate values for the optimal λ. Let

$$\Lambda = \{ (r_i - d_j)/k : k \in \{1, 2, ..., n-1\}, i, j \in \{1, 2, ..., n\}, r_i > d_j \}.$$

It is not hard to show (see [5]) that Λ contains an optimal gap length. This yields an $O(n^3 \log n)$-time algorithm. This algorithm first computes the set Λ, sorts it, and then finds the optimal λ through binary search in Λ. In the full version of this paper [5], we show that this running time can be improved to $O(n^2 \log n)$, by a more careful search in Λ that avoids constructing Λ explicitly.

7   Minimizing Total Flow Time with Budget for Gaps

Unlike in earlier sections, we now consider jobs without deadlines and focus on the tradeoff between the number of gaps and delay. Formally, an instance $\mathcal{J}$ is given by a collection of n unit length jobs. For each job $j = 1, 2, ..., n$ we are given its release time r_j. If, in some schedule, job j is executed at time S_j then $F_j = S_j - r_j$ is called the *flow time* of j. We are also given a budget value γ. The objective is to compute a schedule for $\mathcal{J}$ in which the number of gaps is at most γ and the total flow time $\sum_j F_j$ is minimized.

Continuous Case. The continuous variant of this problem is equivalent to the k-medians problem on a directed line: given points $r_1, r_2, ..., r_n$, find a set H of k points that minimizes the sum $\sum_{i=1}^{n} \min_{h \in H \wedge h \geq r_i} (h - r_i)$, where the ith term of the sum represents the distance between r_i and the first point in H after r_i. (Here, the value of k corresponds to $\gamma - 1$, the number of blocks.) This problem can be solved in time $O(kn)$ if the points are sorted [10].

Discrete Case. The discrete case differs from its continuous analogue because the jobs executed in the same block do not occupy a single point. We show, however, that the techniques for computing k-medians can be adapted to our problem, resulting in an algorithm with running time $O(n \log n + \gamma n)$.

We can assume that all release times are different and ordered in increasing order: $r_1 < r_2 < ... < r_n$ (see Section 2). This is the only part of the algorithm that requires time $O(n \log n)$; the remaining part will run in time $O(\gamma n)$.

We first give a simple dynamic programming formulation with running time $O(\gamma n^2)$, and then show how to improve it to $O(\gamma n)$. To reduce the running time, we need to show that the blocks of the schedule can only be located at a small number of possible places. For this, we will need the following lemma, that can be easily derived from Lemma 1.

Lemma 3. *There is an optimal schedule with the following properties: (i) all jobs are scheduled in order of their release times, and (ii) the last job of each block is scheduled at its release time.*

Based on this lemma, each block consists of consecutive jobs, say $i, i+1, ..., j$, with the last job j scheduled at time r_j. The hth job of the block is scheduled at time $r_j - j + h$. So the contribution of this block to the total flow is $W_{i,j} = \sum_{h=i}^{j-1} F_h = \sum_{h=i}^{j-1} (r_j - j + h - r_h) = (j - i) r_j - \binom{j-i+1}{2} - R_{j-1} + R_{i-1}$, where $R_b = \sum_{a=1}^{b} r_a$, for each job b.

Algorithm MINTOTFLOW. Let $\mathsf{F}_{j,g}$ denote the minimum flow of a schedule for the sub-instance consisting of jobs $1, 2, ..., j$ with the number of gaps at most g.

We initialize $F_{0,g} = 0$ and $F_{j,0} = W_{1,j}$ for $j > 0$. Then, for $j = 1, 2, ..., n$ and $g = 1, ..., \gamma$, we compute $F_{j,g} = \min_{1 \le i \le j} \{F_{i-1,g-1} + W_{i,j}\}$.

To justify correctness, we need to explain why the above recurrence holds. Consider a schedule that realizes $F_{j,g}$. From Lemma 3, since we are minimizing the total flow, we can assume that job j is scheduled at r_j. Let i be the first job of the last block. As we calculated earlier, the contribution of this block to the total flow is $W_{i,j}$. The minimum flow of the remaining jobs $1, 2, ..., i-1$, is independent of how jobs $i, ..., j$ are scheduled, so (inductively) it is equal $F_{i-1,g-1}$.

We now consider the running time. All values $W_{i,j}$ can be precomputed in time $O(n^2)$. We have at most γ choices for g and at most n choices for j, so there are $O(\gamma n)$ values $F_{j,g}$ to compute. Computing each value takes time $O(n)$, for the total running time $O(\gamma n^2)$.

To improve the running time to $O(\gamma n)$, in the full version of this paper [5] we show that the values $W_{i,j}$ satisfy the quadrangle inequality. This implies that $F_{j,g}$ has the Monge property, which can be used to speed up the dynamic programming algorithm to $O(\gamma n)$, see [10]. Adding the time $O(n \log n)$ needed for preprocessing, the overall running time of Algorithm MINTOTFLOW is $O(n \log n + \gamma n)$.

8   Minimizing Number of Gaps with Bound on Total Flow

An alternative way to formulate the tradeoff in the previous section would be to minimize the number of gaps, given a budget f for the total flow. This can be reduced to the previous problem by finding the smallest g for which there is a schedule with at most g gaps and flow at most f. The solution is the same for both the continuous and discrete versions, so we focus on the discrete variant. Specifically, let $\hat{F}(g)$ be the minimum flow of a schedule that has at most g gaps. Then $\hat{F}(g)$ is monotonely decreasing as g increases. Using binary search, where at each step we use Algorithm MINTOTFLOW as an oracle, we can then find the smallest g for which $\hat{F}(g) \le f$. The resulting algorithm will have running time $O(n^2 \log n)$. This can be improved to $O(g^* n \log n)$, where g^* is the minimum number of gaps: By using the oracle for consecutive powers of 2, find ℓ such that $\hat{F}(2^\ell) \le f < \hat{F}(2^{\ell+1})$, and then do the binary search in the interval $[2^\ell, 2^{\ell+1})$.

9   Minimizing Number of Gaps with Bound on Max Flow

Now, instead of minimizing the *total* flow time, we consider the objective function equal to the *maximum* flow time, $F_{\max} = \max_j (S_j - r_j)$. At the same time, we would also like to minimize the number of gaps. This leads to two optimization problems, by placing a bound on one value and minimizing the other. In this section we give $O(n \log n)$-time algorithm for minimizing the number of gaps when an upper bound on the flow of each job is given.

Formally, we are given an instance consisting of n unit jobs with release times, and a threshold value f. The objective is to compute a schedule of these jobs in which each job's flow time is at most f and the number of gaps is minimized. As before, we can assume that the jobs are sorted according to their release times.

Continuous Case. We start by giving an $O(n \log n)$-time algorithm for the continuous case. Here we are given a collection of n real numbers $r_1, r_2, ..., r_n$, and a number f, and we want to compute a set H of minimum cardinality such that $\min \{h \in H : h \geq r_i\} \leq r_i + f$, for all $i = 1, 2, ..., n$.

We show that this can be solved in time $O(n)$, assuming the input is sorted, using a simple greedy algorithm: Initialize $H = \{r_1 + f\}$, and then in each step, choose i to be smallest index for which $r_i > \max(H)$ and add $r_i + f$ to H. A routine inductive argument shows that the computed set H has indeed minimum cardinality. If $r_1 \leq r_2 \leq ... \leq r_n$, then the algorithm is essentially a linear scan through this sequence, so its running time is $O(n)$. With sorting, the time will be $O(n \log n)$.

Discrete Case. Next, we want to show that we can achieve the same running time for the discrete variant, where we schedule unit jobs. Our approach is to reduce the problem to the gap minimization problem for the version with deadlines. To this end, we define the deadline d_j for each job j as $d_j = r_j + f$. We now need to solve the gap minimization problem for jobs with deadlines, for which an $O(n^4)$-time algorithm is known [3,5]. However, we can do better than this. The instances we created satisfy the "agreeable deadline" property, which means that the ordering of the deadlines is the same as the ordering of release times. For such instances a minimum-gap schedule can be computed in time $O(n \log n)$ (unpublished folklore result, see [1]). This will thus give us an $O(n \log n)$-time algorithm for gap minimization with a bound on maximum flow.

In the full version of this paper [5] we present an alternative $O(n \log n)$-time algorithm for this problem, which has the property that its running time is actually $O(n)$ if the jobs are already sorted in non-decreasing order of release times. This algorithm will be useful in the next section.

10 Minimizing Maximum Flow with Budget for Gaps

We now consider an alternative variant of the tradeoff between maximum flow and the number of gaps, where, for a given collection of n unit jobs with release times $r_1, r_2, ..., r_n$, and a budget γ, we want to compute a schedule that minimizes the maximum flow time $F_{\max} = \max_j(S_j - r_j)$ and has at most γ gaps.

Continuous Case. In the continuous case, $r_1, r_2, ..., r_n$ are points on the real line, ordered from left to right, and we want to compute a set H of at most γ points that minimizes $F_{\max}(H) = \max_i \min_{x \in H, x \geq r_i} |x - r_i|$. (This is a special case of the k-center problem on the line.) Then the idea of the algorithm is to do binary search for the optimal value f^* of $F_{\max}(H)$, at each step of the binary search using the algorithm from the previous section as an oracle.

For binary search, we need a small set of candidate values for f^*. If H is an optimal set of points, then, without loss of generality, we can assume that H contains only release times, since any other point in H can be shifted left until it reaches a release time. Thus we only need to consider the set Φ of all values of the form $r_j - r_i$ for $j > i$. Since $|\Phi| = O(n^2)$ and we need to sort Φ before doing binary search, we would obtain an $O(n^2 \log n)$-time algorithm.

Fortunately, we do not need to construct Φ explicitly. Observe that the elements of Φ can be thought of as forming an $X + Y$ matrix with sorted rows and columns, where X is the vector of release times and $Y = -X$. We can thus use the $O(n)$-time selection algorithm for $X + Y$ matrices [8] to speed up computation. Specifically, at each step we will have two indices p, q, with $1 \leq p \leq q \leq n(n-1)/2$, and we will know that the optimal value of f^* is between the pth and qth smallest values in Φ, inclusive. We let $l = \lfloor (p + q)/2 \rfloor$ and we use the algorithm from [8] to find the lth smallest element in Φ, say f. We now determine whether $f^* \leq f$ by applying the $O(n)$ algorithm from the previous section to answer the query "is there a set H with at most γ gaps and with $F_{\max}(H) \leq f$?" If the answer is "yes", we let $q = l$, otherwise we let $p = l + 1$. This will give us an algorithm with running time $O(n \log n)$.

Discrete Case. The scheduling variant can be solved in time $O(n \log n)$ as well. Although our set Φ of candidate flow values is different now, we can show that Φ can be still expressed as an $X + Y$ set. Another change is that, to answer decision queries in the binary search, we now use Algorithm MINGAPMAXFLOW. The details of the algorithm are in the full version [5].

11   Final Comments

Many open problems remain. The most fundamental and intriguing question is whether the running time for minimizing the number of gaps for unit jobs (see [3,5]) can be improved to below $O(n^4)$. Speeding up the algorithms in Sections 3, 4, 5, and 6 would also be of considerable interest. There is a number of other variants of gap scheduling, even for unit jobs, that we have not addressed in our paper, including the problem of maximizing the minimum gap, the tradeoff between throughput and gap size, or between flow time (total or maximum) and gap size. Another direction of research would be to study variants of gap scheduling for jobs of arbitrary length, for models with preemptive or non-preemptive jobs.

References

1. Angel, E., Bampis, E., Chau, V.: Low complexity scheduling algorithm minimizing the energy for tasks with agreeable deadlines. In: Fernández-Baca, D. (ed.) LATIN 2012. LNCS, vol. 7256, pp. 13–24. Springer, Heidelberg (2012)
2. Baptiste, P.: Scheduling unit tasks to minimize the number of idle periods: a polynomial time algorithm for offline dynamic power management. In: Proceedings of the 17th Annual ACM-SIAM Symposium on Discrete Algorithms (SODA), pp. 364–367 (2006)
3. Baptiste, P., Chrobak, M., Dürr, C.: Polynomial time algorithms for minimum energy scheduling. In: Arge, L., Hoffmann, M., Welzl, E. (eds.) ESA 2007. LNCS, vol. 4698, pp. 136–150. Springer, Heidelberg (2007)
4. Chrobak, M., Feige, U., Taghi Hajiaghayi, M., Khanna, S., Li, F., Naor, S.: A greedy approximation algorithm for minimum-gap scheduling. In: Spirakis, P.G., Serna, M. (eds.) CIAC 2013. LNCS, vol. 7878, pp. 97–109. Springer, Heidelberg (2013)

5. Chrobak, M., Golin, M.J., Lam, T.-W., Nogneng, D.: Scheduling with gaps: New models and algorithms. CoRR, abs/1410.7092 (2014)
6. Demaine, E., Ghodsi, M., Hajiaghayi, M., Sayedi-Roshkhar, A., Zadimoghaddam, M.: Scheduling to minimize gaps and power consumption. In: Proceedings of the ACM Symposium on Parallelism in Algorithms and Architectures (SPAA), pp. 46–54 (2007)
7. Irani, S., Pruhs, K.R.: Algorithmic problems in power management. SIGACT News **36**(2), 63–76 (2005)
8. Mirzaian, A., Arjomandi, E.: Selection in X+Y and matrices with sorted rows and columns. Inf. Process. Lett. **20**(1), 13–17 (1985)
9. Wikipedia. Point coordination function. http://en.wikipedia.org/wiki/Point_coordination_function
10. Woeginger, G.: Monge strikes again: optimal placement of web proxies in the internet. Oper. Res. Lett. **27**(3), 93–96 (2000)

MinMax-Distance Gathering on Given Meeting Points

Serafino Cicerone[1], Gabriele Di Stefano[1], and Alfredo Navarra[2(✉)]

[1] Dipartimento di Ingegneria e Scienze dell'Informazione e Matematica,
Università degli Studi dell'Aquila, L'Aquila, Italy
{serafino.cicerone,gabriele.distefano}@univaq.it
[2] Dipartimento di Matematica e Informatica, Università degli Studi di Perugia,
Perugia, Italy
alfredo.navarra@unipg.it

Abstract. We consider a set of oblivious robots moving in the plane that have to gather at one point among a predetermined set of so called *meeting points*. Robots operate in asynchronous Look-Compute-Move cycles. In one cycle, a robot perceives the current configuration in terms of relative positions of robots and meeting points (Look), decides whether to move toward some direction (Compute), then makes the computed move, eventually (Move). Robots are anonymous and execute the same distributed algorithm that must guarantee to gather all robots at a meeting point by minimizing the longest distance traveled by a single robot. This is a new metric for evaluating the quality of the gathering, and we start characterizing configurations where optimal gathering can be achieved. We then provide a distributed algorithm to optimally solve most of such configurations.

1 Introduction

Gathering a team of robots/agents at some place is a basic task for distributed systems. The problem has been extensively investigated under different assumptions (e.g., see [2,5,6,8,10,11]).

In this paper, we are interested in robots placed in the plane. Initially, no robots occupy the same location. Robots operate in *Look-Compute-Move* cycles (see, e.g. [8]). The model assumes that in each cycle a robot takes a snapshot of the current global configuration in terms of occupied points in the plane (Look), then, based on the perceived configuration, takes a decision to stay idle or to move toward a specific direction (Compute), and in the latter case it moves, eventually (Move). Cycles are performed asynchronously, i.e., the time between Look, Compute, and Move operations is finite but unbounded, and it is decided by an adversary for each robot. Moreover, during the Look phase, a robot does

Work partially supported by the following Research Grants: 2010N5K7EB "PRIN 2010" ARS TechnoMedia (Algoritmica per le Reti Sociali Tecno-mediate) and 2012C4E3KT "PRIN 2012" Amanda (Algorithmics for MAssive and Networked DAta), both from the Italian Ministry of University and Research.

not perceive whether other robots are moving or not. Hence, robots may move based on outdated perceptions.

Robots are assumed to be oblivious (without memory of the past), uniform (running the same deterministic algorithm), autonomous (without a common coordinate system, identities or chirality), asynchronous (without central coordination), without the capability to communicate.

Within this model, the gathering problem has been shown to be unsolvable if no further assumptions are made. For instance, the scheduler determining the Look-Compute-Move cycles timing must be assumed to be fair, that is, each robot performs its cycle within finite time and infinitely often. Contrary, if there are robots never moving, it is impossible to gather them. Moreover, during the Move phase, an adversary can decide to prevent a robot to reach its destination. However, there must exist an (arbitrarily small) constant $\delta > 0$ such that if the destination point is closer than δ, the robot will reach it, otherwise the robot will be closer to the destination point of at least δ. If no such constant δ exists, gathering is unsolvable.

Still the problem is unsolvable if no further assumptions are considered [12]. In [4], the authors fully characterize the problem when robots are empowered by the so called *global weak multiplicity detection*. That is, during the Look phase, robots perceive whether a same location is occupied by more than one robot without acquiring the exact number. The *global strong* version would provide the robots with the exact number of robots within the same location, while the *local weak/global* versions provide similar information but limited to the robots composing the multiplicity.

In [3], a slightly different task has been introduced. In fact, during the Look phase robots are assumed to perceive another finite set of points from now on called *meeting points*. These represent the only locations where gathering can be finalized. As for robots, the meeting points are perceived as relative positions with respect to the robot performing a Look phase. Moreover, a metric to evaluate the quality of the computed gathering is introduced, that is, it was requested that the overall distance traveled by all robots is minimized to accomplish the gathering. The concept of *exact* and *optimal* gathering has been introduced. The former refers to the gathering performed at a meeting point m with minimum eccentricity from all robots, by making move all robots straightly toward m. The latter refers to those configurations that cannot be gathered on meeting points with minimum eccentricity, even though there exists a gathering algorithm that minimizes the total traveled distance.

The interest in such a model is twofold. From the one hand, it is theoretically challenging as it is a hybrid scenario in between the classical environment where robots freely move in the plane (see, e.g., [1,4]), and the more structured one where robots move on the vertices of a graphs (see, e.g., [7,9]) implemented here by the set of meeting points. On the other hand, meeting points might be a practical choice when robots move in particular environments where not all places can be candidate to serve as gathering points.

In this paper, we are still interested in studying the gathering at meeting points, but we change the metric to evaluate the quality of the computed solutions. In particular, we ask for minimizing the maximum traveled distance by each single robot. To this respect we can define again similar concepts of exact and optimal gathering. As we are going to show, the new metric completely changes the approach required. First of all, we do not assume any multiplicity detection capability. Then, we start characterizing when exact gathering with respect to the new metric can be accomplished. To this respect, we inherited some negative results from [3] about ungatherable configurations. Whereas, for most of the configurations not proved to be ungatherable we provide a new distributed algorithm that ensures exact gathering.

Outline. The next section introduces the required notation and gives some definitions, along with basic impossibility results. Section 3 introduces some concepts and properties that must be taken into account when designing exact gathering algorithms. Section 4 provides our new gathering algorithm, designed for most of the configurations not proved to be ungatherable. It is presented in terms of few different strategies according to different types of configurations. Due to the lack of space, we describe in detail only one type of the tackled configurations by means of formal description, pseudo-code, and correctness proof. For the other types of configurations we only sketch on the general ideas. A dissertation on the configurations not addressed by our algorithm is also provided. Finally, Section 5 concludes the paper.

2   Definitions

In this section we provide notation and basic concepts used in the paper.

Notation. The system is composed of a set of n mobile *robots*. At any time, the multiset $R = \{r_1, r_2, \ldots, r_n\}$, with $r_i \in \mathbb{R}^2$, contains the *positions* of all robots. The set $U(R) = \{x \mid x \in R\}$ contains the *unique* robots' positions. M is a finite set of fixed *meeting points* in the plane representing the only locations in which robots can be gathered. The pair $\mathcal{C} = (R, M)$ represents a system *configuration*. A configuration $\mathcal{C}$ is *initial* at time t if at that time all robots have distinct positions (i.e., $|U(R)| = n$). A configuration $\mathcal{C}$ is *final* at time t if (i) at that time each robot computes or performs a null movement and (ii) there exists a point $m \in M$ such that $r_i = m$ for each $r_i \in R$; in this case we say that the robots have gathered on point m at time t.

We study the GATHERING OVER MEETING POINTS problem (shortly, GMP), that is, the problem of transforming an initial configuration into a final one. A gathering algorithm for the GMP problem is a deterministic distributed algorithm that brings the robots in the system to a final configuration in a finite number of cycles from any given initial configuration. We assume that during the *Look* phase, robots are able to distinguish points hosting robots from points in M, whereas no multiplicity detection capabilities are assumed.

Efficiency of gathering algorithms. Here we measure the efficiency of a gathering algorithm according to the travel distance of each robot. In particular, given a gathering algorithm A for an initial configuration $C = (R, M)$, we denote by $td(A, C, r)$ the *total travel distance* performed by the robot $r \in R$ during the execution of A starting from C until finalizing the gathering task, and by $td(A, C) = \max_{r \in R} td(A, C, r)$. We use $td(A, C)$ as a measure for the efficiency of the algorithm A on a configuration C. Of course, the challenge is to define a gathering algorithm A with $td(A, C)$ as small as possible for each possible initial configuration C.

We now introduce a new but natural concept to model the efficiency of a gathering algorithm. Given a configuration $C = (R, M)$, the *minmax-distance of* C is defined as the value

$$\Delta(C) = \min_{m \in M} \max_{r \in R} d(r, m)$$

where symbol $d(\cdot, \cdot)$ denotes the Euclidean distance between two points.

Definition 1. *A gathering algorithm A for an initial configuration $C = (R, M)$ is* optimal *if it guarantees the minimum possible value $td(A, C)$. Since $\Delta(C)$ is a lower bound for some robots, then we say that an algorithm is* exact *if it achieves the gathering with at most $\Delta(C)$ as travel distance for each robot.*

Let $C = (R, M)$ be a configuration. A point $m \in M$ is a *minmax-point of C* if $\max_{r \in R} d(r, m) = \Delta(C)$, and $\mathrm{MM}(C)$ is the set containing all the minmax-points of C. According to Definition 1, it follows that any exact gathering algorithm for GMP has to select a minmax-point as point where to gather all robots and make move all robots toward the selected point along paths of length not greater than $\Delta(C)$.[1]

Configuration view and symmetries. In [3] it has been introduced the notion of *configuration view*, a data structure, computable by all robots, containing all the information about any initial configuration. It makes use of $cg(M)$, the *center of gravity* of all points in M. Informally, to each point $p \in R \cup M$ it is first associated the cyclic sequence $V^+(p) = (p_0, p_1, \ldots, p_{k-1})$, with $p_0 = cg(M)$,[2] that represents the order in which p views all the points in $C \setminus \{p\}$ starting from $cg(M)$ and turning clockwise. Then, from $V^+(p)$ we directly get the string $\mathcal{V}^+(p)$, that is the *clockwise view* of p, as follows: replace p_i in $V^+(p)$ by the triple α_i, d_i, x_i in $\mathcal{V}^+(p)$, where $\alpha_i = \sphericalangle(p_0, p, p_i)$,[3] $d_i = d(p, p_i)$, and $x \in \{\text{'}r\text{'}, \text{'}m\text{'}, \text{'}c\text{'}\}$ according whether p_i is a robot position, a meeting point or the center of $cg(M)$,

---

[1] This means only robots at distance $\Delta(C)$ from the gathering point are required to reach it along shortest paths, all other robots could potentially make deviations from the shortest paths.

[2] We are assuming that p and $cg(M)$ are not coincident; due to lack of space we omit to recall how $V^+(p)$ is defined in such a case (details can be found in [3]).

[3] The half-line starting at point u (but excluding the point u) and passing through v is denoted by $hline(u, v)$. Given two lines $line(c, u)$ and $line(c, v)$, $\sphericalangle(u, c, v)$ represents the convex angle centered in c and with sides $hline(c, u)$ and $hline(c, v)$.

respectively. Similarly, the *counter-clockwise view* of p, denoted by $\mathcal{V}^-(p)$, is defined.

The *view of $\mathcal{C}$* is denoted as $\mathcal{V}(\mathcal{C})$ and consists of all the views of points in $R \cup M$. It is possible to define a total order (e.g. lexicographic) on the set of all distinct views. Accordingly, the *view of p* is defined as the minimum between $\mathcal{V}^+(p)$ and $\mathcal{V}^-(p)$. The next theorem provides a relationship between symmetries and the configuration view.

Theorem 1. [3] *An initial configuration $\mathcal{C} = (R, M)$ is symmetric if and only if there exist two distinct points p and q, belonging both to R or to M, such that either $\mathcal{V}^+(p) = \mathcal{V}^-(q)$ or $\mathcal{V}^+(p) = \mathcal{V}^+(q)$.*

It is now useful to formalize the concept of configuration symmetry by means of isometries in the plane. In general, an *automorphism* of a configuration $\mathcal{C} = (R, M)$ is an isometry from $\mathbb{R}^2$ to itself, that maps robots to robots (i.e., points of R into R) and meeting points to meeting points (i.e., points of M into M). The set of all automorphisms of $\mathcal{C}$ is denoted by $\text{Aut}(\mathcal{C})$. The isometries in $\text{Aut}(\mathcal{C})$ are the identity, rotations, reflections and their compositions. An isometry φ is a rotation if there exists a unique point x such that $\varphi(x) = x$ (and x is called *center of rotation*); it is a reflection if there exists a line ℓ such that $\varphi(x) = x$ for each point $x \in \ell$ (and ℓ is called *axis of symmetry*). If $|\text{Aut}(\mathcal{C})| = 1$, i.e. $\mathcal{C}$ admits only the identity automorphism, then $\mathcal{C}$ is said *asymmetric*, otherwise it is said *symmetric* (i.e., $\mathcal{C}$ admits rotations or reflections).

Theorem 2. [3] *Let $\varphi \in \text{Aut}(\mathcal{C})$ be an isometry of a configuration $\mathcal{C} = (R, M)$. $\mathcal{C}$ is ungatherable if (1) φ is a rotation and the center $c \notin R \cup M$, or (2) φ is a reflection with axis ℓ and $\ell \cap (R \cup M) = \emptyset$.*

Weber points. Let $\mathcal{C} = (R, M)$ be a configuration. The *Weber distance* between R and a point $m \in M$ is denoted by $wd(R, m)$ and it is defined as $wd(R, m) = \sum_{r \in R} d(r, m)$, while the Weber distance of $\mathcal{C}$ is is denoted by $wd(\mathcal{C})$ and it is defined as $wd(\mathcal{C}) = \min_{m \in M} wd(R, m)$. A point $m \in M$ is a *Weber point* of $\mathcal{C}$ if $m = \text{argmin}_{m' \in M} wd(R, m')$, and the set containing all the Weber points is denoted by $wp(\mathcal{C})$.

A k-ellipse is the plane curve consisting of all points p whose sum of distances from k given points $p_1, p_2, \ldots, p_k$ is a fixed number. It follows that $\sum_{r \in R} d(p, r) = \lambda$ is a $|R|$-ellipse consisting of all points p whose sum of distances from all robots is a fixed number λ. If we set $\lambda = wd(\mathcal{C})$, then the previous equation represents the $|R|$-ellipse containing all the Weber points in $wp(\mathcal{C})$: such an ellipse is denoted by $\mathcal{E}(\mathcal{C})$. In [13], it is shown that $\mathcal{E}(\mathcal{C})$ is a strictly-convex curve, provided the foci p_i are not collinear. If the foci are collinear, $\mathcal{E}(\mathcal{C})$ degenerates into a segment corresponding to the *median of the foci*, denoted by $med(R)$[4]. The following result characterizes the set of all Weber points after a robot moved toward one of such points.

---

[4] $med(R)$ is the segment $[r_1, r_2]$, where r_1 and r_2 are the median points of R when $|R|$ is even, and degenerates to a single point when $|R|$ is odd ($r_1 = r_2$).

Lemma 1. [3] *Let $\mathcal{C} = (R, M)$ be a configuration. Assume that a robot $r \in R$ moves toward a point $m \in wp(\mathcal{C})$ and this move creates a configuration $\mathcal{C}' = (R', M)$. Then:*

- *if $\mathcal{E}(\mathcal{C})$ is a strictly-convex curve with non-empty interior, then $wp(\mathcal{C}')$ contains one or two Weber points only: one is m and the other (if any) lies on $hline(r, m)$;*
- *if $\mathcal{E}(\mathcal{C})$ is a single point (i.e, $wp(\mathcal{C}) = \{m\}$), then $wp(\mathcal{C}')$ contains only m;*
- *if $\mathcal{E}(\mathcal{C})$ is $med(R)$, then $wp(\mathcal{C}') = med(R') \cap M$.*

3   Basic Properties

Given a point $m \in M$, we denote by $C(m, \rho)$ the circle centered in m of radius $\rho \geq 0$. By definition, all points in R are inside or on the boundary of the circle $C(m, \Delta(\mathcal{C}))$ if and only if m is a minmax-point. We use the following short-hands: if m is a minmax-point, then $C(m, \Delta(\mathcal{C}))$ is simply denoted as $bC(m)$ and called *black-circle* centered on m; if m is not a minmax-point, then $C(m, \Delta(\mathcal{C}))$ is simply denoted as $gC(m)$ and called *gray-circle* centered on m. It follows that there is at least one robot lying on the boundary of a black-circle. Then, given a black-circle $bC(m)$:

- any robot in the interior of $bC(m)$ is called *internal-robot* of m;
- any robot on the boundary of $bC(m)$ is called *border-robot* of m;
- $B(m)$ is the set containing all the border-robots of m.

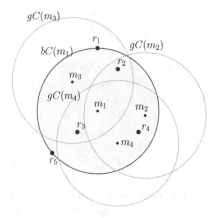

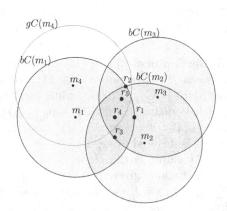

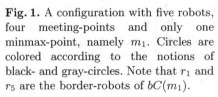

Fig. 1. A configuration with five robots, four meeting-points and only one minmax-point, namely m_1. Circles are colored according to the notions of black- and gray-circles. Note that r_1 and r_5 are the border-robots of $bC(m_1)$.

Fig. 2. A configuration with multiple minmax-points, namely m_1, m_2, and m_3. Note that all robots are in the intersection of all the black-circles.

See Figures 1 and 2 for a visualization of the defined concepts.

In the remainder, we use the simple sentence *robot r moves toward a meeting point m* to mean that "*r* performs a straight move toward *m* and the final position of *r* lies on the interval $(r, m]$". We remind that a robot either reaches the destination or moves of at least δ. In Section 4, we devise an exact gathering algorithm for the GMP problem that basically (1) according to some property $\mathcal{P}$, selects a minmax point *m* where to gather and then (2) makes move each robot toward *m*. It follows that, during the execution of the algorithm, *m* must remain a minmax-point and the property $\mathcal{P}$ must always select *m* as gathering point. This is not a simple task since, given an initial configuration $\mathcal{C}$:

– if a robot moves toward a minmax-point *m*, it is possible that after the movement the new configuration $\mathcal{C}'$ admits new minmax-points (see Fig. 3.a). These moves should be avoided so that, likely, the property $\mathcal{P}$ may still select *m*;

– if a robot moves toward a minmax-point *m*, it is possible that *m* is not a minmax-point in the new configuration (see Fig. 3.b when *r* reaches *p*). These moves have to be avoided otherwise the total travel distance of a robot moving toward a different minmax-point in a sequence of configurations could be larger of $\Delta(\mathcal{C})$, where $\mathcal{C}$ is the initial configuration, and then exact gathering cannot be guaranteed.

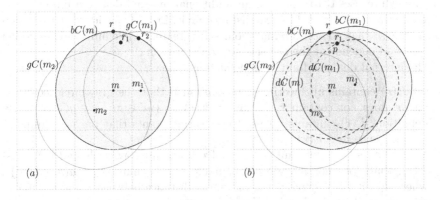

Fig. 3. Two configurations ($\mathcal{C}_1$ and $\mathcal{C}_2$, from left to right) to be analyzed according to the movement of robots toward a minmax-point *m* (*a*) – Here MM($\mathcal{C}_1$) = {*m*}. If *r* moves toward *m* and reaches the boundary of $gC(m_1)$, then a configuration $\mathcal{C}'_1$, with MM($\mathcal{C}'_1$) = {m, m_1}, is created. Moving *r* toward *m* without reaching the boundaries of both $gC(m_1)$ and $gC(m_2)$ does not create new minmax-points. (*b*) – Here MM($\mathcal{C}_2$) = {m, m_1}. If *r* moves toward *m* and reaches point *p* within one Look-Compute-Move cycle, since $d(m_1, r_1) < d(m, r_1)$, then a configuration $\mathcal{C}'_2$ with MM($\mathcal{C}'_2$) = {m_1} is created. Instead, if *r* is prevented to overtake the boundary of the dashed circle centered on *m*, then MM($\mathcal{C}'_2$) = {*m*}.

According to the above arguments, when a robot *r* moves toward a minmax-point *m* there are some *distance limits* that the move should take into consideration.

Procedure: MOVE

Input: Configuration $\mathcal{C} = (R, M)$, minmax-point $m \in$ MM($\mathcal{C}$), distance limit ℓ_2;

```
/* Computing data to set distance limit ℓ₁                          */
```
1 Compute $\Delta(\mathcal{C})$;
2 Compute $E(r) = \{m' \in M \mid r \notin gC(m')\}$;
3 Compute $\delta(r) = \min_{m' \in E(r)} d(r, gC(m'))$;
4 Set limits $\ell_1 = \delta(r)/2$;
5 Set $\ell = \min\{\ell_1, \ell_2\}$;
```
/* Use limit ℓ to determine a destination point p                  */
```
6 **if** $d(r, m) \leq \ell$ **then**
7 $\quad\mid\quad$ set p as m ;
8 **else**
9 $\quad\mid\quad$ set p as the point belonging to $(r, m]$ such that $d(r, p) = \ell$;
10 Move r toward p ;

Fig. 4. Procedure MOVE for moving a robot r toward a minmax-point m

---

Hence, we define a basic algorithm (Procedure MOVE, see Fig. 4) that takes as input a configuration $\mathcal{C}$, a minmax-point m, and a distance limit ℓ_2, and allows a robot r to perform a basic move toward m by taking ℓ_2 into consideration. The procedure uses $\mathcal{C}$ to compute an additional internal distance limit ℓ_1 and indeed uses $\ell = \min\{\ell_1, \ell_2\}$ as real distance limit. Limit ℓ_1 avoids that r enters a gray-circle, so that no new minmax-points are created, while limit ℓ_2 is provided as input and depends of the the specific configuration $\mathcal{C}$.

The internal distance limit ℓ_1 is defined as follows:

- set $E(r)$ contains all meeting-points m' such that the gray-circle centered on m' does not contain r. Formally, $E(r) = \{m' \in M \mid r \notin gC(m')\}$. By definition, $E(r)$ contains a subset of meeting-points that are not minmax-points, hence $m \notin E(r)$.
- Value $\delta(r)$ is the minimum distance of r from any gray-circle $gC(m')$, $m' \in E(r)$, that does not include r. Formally, $\delta(r) = \min_{m' \in E(r)} d(r, gC(m'))$, where $d(r, gC(m'))$ denotes the minimum distance between r and a point of the gray-circle $gC(m')$, $m' \in E(r)$.
- The distance limit ℓ_1 is defined as $\ell_1 = \delta(r)/2$.

Basically, the algorithm makes move r toward a point p on the interval $(r, m]$ at distance $\ell = \min\{\ell_1, \ell_2\}$ from r (of course, if m is closer than ℓ to r, then r moves toward m). Notice that, according to the definition of the distance limit ℓ_1, it is easy to see that the execution of Procedure MOVE never creates new minmax-points.

4   Algorithm

In this section, we present an exact gathering algorithm solving GMP in most of the possible initial configurations. The initial configurations processed by the algorithm, along with configurations created during the execution, are partitioned into classes. For defining such classes, given a configuration $\mathcal{C} = (R, M)$, we often select a subset $\text{MM}_V(\mathcal{C})$ of $\text{MM}(\mathcal{C})$ as follows:

- $\text{MM}_B(\mathcal{C}) \subseteq \text{MM}(\mathcal{C})$ is the set of minmax-points with minimum number of border-robots;
- $\text{MM}_W(\mathcal{C}) \subseteq \text{MM}_B(\mathcal{C})$ is the set of minmax-points in $\text{MM}_B(\mathcal{C})$ with minimal Weber distance;
- $\text{MM}_V(\mathcal{C}) \in \text{MM}_W(\mathcal{C})$ is the minmax-point in $\text{MM}_W(\mathcal{C})$ with minimal view.

The classes are defined as follows:

$\mathcal{S}_1$: any configuration $\mathcal{C}$ such that $|\text{MM}(\mathcal{C})| = 1$;
$\mathcal{S}_2$: any configuration $\mathcal{C} \notin \mathcal{S}_1$ such that $|\text{MM}_V(\mathcal{C})| = 1$;
$\mathcal{S}_3$: initial configurations $\mathcal{C} \notin \bigcup_{1 \leq i \leq 2} \mathcal{S}_i$ such that $\mathcal{C}$ admits a reflection with robots on the axis;
$\mathcal{S}_4$: initial configurations $\mathcal{C} \notin \bigcup_{1 \leq i \leq 3} \mathcal{S}_i$ such that $\mathcal{C}$ admits a rotation with a robot as center.

Note that all asymmetric configurations are in $\mathcal{S}_1 \cup \mathcal{S}_2$ as an asymmetric configuration either has a single minmax-point or the minimal view identifies a single minmax-point, that is $\text{MM}_V(\mathcal{C}) = \{m\}$ for some $m \in M$. Moreover, rotational configuration with a meeting point in the center are included in $\mathcal{S}_1$. In fact, if $p \in M$ is the center of the rotation, then p results to be a minmax-point (it simply follows by observing that $C(p, \Delta(\mathcal{C}))$ is the smallest circle that includes all robots). It is easy to see that p is the unique minmax-point in the configuration and then $\mathcal{C}$ belongs to class $\mathcal{S}_1$.

According to the above classes, the strategy of the algorithm is:

- if $\text{MM}(\mathcal{C}) = \{m\}$ (i.e., $\mathcal{C} \in \mathcal{S}_1$), then move concurrently all robots toward m;
- if $\text{MM}_V(\mathcal{C}) = \{m\}$ for some $m \in M$ but $|\text{MM}(\mathcal{C})| > 1$ (i.e., $\mathcal{C} \in \mathcal{S}_2$), then move concurrently all the robots in $B(m)$ toward m so that a configuration $\mathcal{C}'$ is obtained such that $\text{MM}(\mathcal{C}') = \{m\}$;
- if $\mathcal{C}$ is symmetric (in particular, if $\mathcal{C} \in \mathcal{S}_3 \cup \mathcal{S}_4$), then move a single robot (on the axis/center of the symmetry) so that a configuration $\mathcal{C}'$ is obtained such that $|\text{MM}_V(\mathcal{C}')| = 1$.

The algorithm (see Procedure COMPUTE in Fig. 5) is divided into various subprocedures, each of that designed to process configurations belonging to a given class. Since such procedures determine the Compute phase of a robot, after each instruction determining the move of the executing robot, as well as at the end of each procedure, instruction *exit()* must be implicitly considered. Moreover, all sub-procedures are invoked after having computed the class $\mathcal{S}_i$ which the configuration $\mathcal{C}$ belongs to. For this task robots can exploit the computation of $\text{MM}(\mathcal{C})$ (for class $\mathcal{S}_1$), of $\text{MM}(\mathcal{C})$, $wp(\mathcal{C})$, and $\text{MM}_V(\mathcal{C})$ (for class $\mathcal{S}_2$), of $\mathcal{V}(\mathcal{C})$ and Theorem 1 (for classes $\mathcal{S}_3$ and $\mathcal{S}_4$).

Due to lack of space, in the remainder we provide full details of Procedure SINGLE-MINMAX only, and then we give its correctness.

Procedure: COMPUTE
Input: Configuration $\mathcal{C}$

1 Compute $GC(\mathcal{C})$, $\mathcal{V}(\mathcal{C})$, $wp(\mathcal{C})$, $\text{MM}(\mathcal{C})$, $\text{MM}_V(\mathcal{C})$;
2 **if** $C \in \mathcal{S}_1$ **then** SINGLE-MINMAX$(\mathcal{C})$;
3 **if** $C \in \mathcal{S}_2$ **then** SINGLE-MINVIEW$(\mathcal{C})$;
4 **if** $C \in \mathcal{S}_3$ **then** SYMM$(\mathcal{C})$;
5 **if** $C \in \mathcal{S}_4$ **then** ROTATE$(\mathcal{C})$;

Fig. 5. Procedure COMPUTE executed by any robot r during the Compute phase

---

Procedure: SINGLE-MINMAX
Input: Configuration $\mathcal{C} = (R, M)$ with $|\text{MM}(\mathcal{C})| = 1$;

1 Compute $\text{MM}(\mathcal{C})$;
2 Let m be the only element of $\text{MM}(\mathcal{C})$;
3 Execute MOVE$(\mathcal{C}, m, \infty)$; /* each robot in $\mathcal{C}$ executes this basic move */

Fig. 6. Procedure SINGLE-MINMAX executed by a generic robot r belonging to a configurations $\mathcal{C} \in \mathcal{S}_1$

4.1   Class $\mathcal{S}_1$: $|\text{MM}(\mathcal{C})| = 1$

Class $\mathcal{S}_1$ consists of all configurations that have exactly one minmax-point. Notice that its definition also contains some symmetric configurations. As the next lemma states, there exists a procedure (SINGLE-MINMAX, see Fig. 6) that provides exact gathering for all configurations in class $\mathcal{S}_1$. In particular, for a configuration $\mathcal{C}$ having only one minmax-point m, Procedure MOVE can be used with just limit ℓ_1 to maintain m as unique minmax-point, even if the procedure is executed concurrently by all robots in $\mathcal{C}$.

Lemma 2. *Let $C = (R, M)$ be a configuration such that $\text{MM}(\mathcal{C}) = \{m\}$, and let $\mathcal{C}'$ be the configuration obtained from $\mathcal{C}$ by letting each robot to execute* MOVE$(\mathcal{C}, m, \infty)$ *concurrently, an arbitrary number of times. Then, $\text{MM}(\mathcal{C}') = \{m\}$.*

Proof. Since $\text{MM}(\mathcal{C}) = \{m\}$, for each meeting-point $m' \neq m$ there exists a robot r such that $m' \in E(r)$. Let us assume that, at a given time, r executes MOVE$(\mathcal{C}, m, \infty)$ and m' is still in $E(r)$. Suppose that r computes $d = \Delta(\overline{C})$, where $\overline{C}$ is the configuration detected by r during the last Look phase. Since $\ell_2 = \infty$, r sets $\ell = \ell_1 = \delta(r)/2$, and then performs the move. According to definition of $\delta(r)$, r does not reach the gray-circle $gC(m')$ (which has radius $\Delta(\overline{C})$) and hence m' is still in $E(r)$. It is worth to remark that this property is not affected by possible concurrent moves of other robots, since $\Delta(\overline{C})$ can only decrease (note that this occurs if, during the current Look-Compute-Move cycle of r, all the border-robots of m at distance $\Delta(\overline{C})$ moved toward m). If $m' \in E(r)$ after the execution of Procedure MOVE, then m' is not a minmax-point. According to the generality of m', this implies $\text{MM}(\mathcal{C}') = \{m\}$.   □

Lemma 3. *Let $\mathcal{C} = (R, M)$ be a configuration with $\mathrm{MM}(\mathcal{C}) = \{m\}$. There exists an algorithm (see Procedure SINGLE-MINMAX) that guarantees exact gathering.*

Proof. Let t_1 be the time when the first robot starts the Look phase, and t_2 the minimum time after which all robots ended at least one entire Look-Compute-Move cycle. We call the interval $[t_1, t_2]$ *first turn*. In general, $[t_i, t_{i+1}]$ is the i-th turn, where t_{i+1} is the minimum time, starting from t_i, after which all robots ended at least one entire Look-Compute-Move cycle. The robot model assures that each turn is finite. Let $\mathcal{C}^i$, $i = 1, 2, \ldots$, be the configuration at the beginning of the i-th turn, with $\mathcal{C}^1 = \mathcal{C}$. Moreover, given a generic robot $r \in R$, we denote by r^i the position of r in $\mathcal{C}^i$. Let $E^i(r)$ be $E(r)$ in $\mathcal{C}^i$. Note that $E^i(r)$ may differ from $E^{i+1}(r)$. In particular, $E^{i+1}(r) \supseteq E^i(r)$. This may happen when border robots move, hence reducing the radii of the black and grey circles. It follows that a robot might be overcome by the border of a grey circle. Remembering that a robot never enters a grey circle, the set $E(r)$ can only grow during the run of Procedure SINGLE-MINMAX until all the robots are out of all grey circles of nil radii.

Let us assume first the case in which all robots are out of all grey circles. Let $\bar{\delta} = \min\{\delta, \min_{r \in R}\{\delta(r^1)/2\}\}$. We say that a robot during the i-th turn has *target position* either the point on $hline(r, m)$ at distance $i \cdot \bar{\delta}$ from the initial position of r (i.e., from r^1), or m if m is closer. According to Lemma 2, the move toward the target position is such that no new minmax-points can be created. We now prove by induction on the number of turns that exact gathering is accomplished within $\lceil \Delta(\mathcal{C})/\bar{\delta} \rceil$ turns. In particular, we prove that at $\mathcal{C}^i$, $i = 1, 2, \ldots, \lceil \Delta(\mathcal{C})/\bar{\delta} \rceil$, all robots have reached the $(i-1)$-th target position, and hence $\Delta(\mathcal{C}^i) \leq \Delta(\mathcal{C}) - (i-1)\bar{\delta}$.

Base. For $i = 1$, in the initial configuration $\mathcal{C}^1 = \mathcal{C}$ all robots are at their 0-th target positions that is the initial one, and hence $\Delta(\mathcal{C}^1) \leq \Delta(\mathcal{C}) - 0\bar{\delta} = \Delta(\mathcal{C})$.

Inductive hypothesis. Let us assume the claim true in $\mathcal{C}^k$. This means that in $\mathcal{C}^k$ all robots have moved toward m for a distance of at least $(k-1)\bar{\delta}$, or they have reached m. As a consequence, $\Delta(\mathcal{C})$ is decreased of at least the same quantity.

Let us consider a generic robot $r \in R$. If r^k is at a distance of at least $k\bar{\delta}$ from r^1 or it coincides with m, then we are done. So we consider r^k is in between the $(k-1)$-th and the k-th target positions of the corresponding robot, that is $(k-1)\bar{\delta} \leq d(r^k, r^1) < k\bar{\delta}$. Moreover, the distance d between r^1 and the closest gray-circle whose center is in $E(r^k)$ is $\delta(r^1) + (k-1)\bar{\delta}$; since $\delta(r^1) \geq 2\bar{\delta}$, then $d \geq (k+1)\bar{\delta}$.

Our algorithm, by means of Procedure MOVE makes move r toward the target position represented by the point on $hline(r, m)$ at distance $\frac{d - d(r^k, r^1)}{2} \geq \frac{(k+1)\bar{\delta} - d(r^k, r^1)}{2}$ from its actual position r^k. It follows that the k-th target position from r is at distance at least

$$d(r^k, r^1) + \frac{(k+1)\bar{\delta} - d(r^k, r^1)}{2} = \frac{(k+1)\bar{\delta} + d(r^k, r^1)}{2} \geq \frac{(k+1)\bar{\delta} + (k-1)\bar{\delta}}{2} = k\bar{\delta}.$$

Hence, during the k-th turn, all robots will reach their target positions as the adversary can not stop a robot before it travels for a distance at least $\bar{\delta}$. Hence at the beginning of the $(k+1)$-th turn $\Delta(\mathcal{C}^k) \leq \Delta(\mathcal{C}) - k\bar{\delta}$.

Now, let us assume there is at least one robot r inside a grey circle. During a generic turn, robot r might be overcome by the grey circle or not. If not, the same analysis as above holds. Otherwise, r may perceive a new minimal distance from a grey circle. Half of such a distance will be guaranteed during the current turn until a new overcome occurs or the gathering is reached.

Since there are $|M|$ meeting points and $|R|$ robots, there might be at most $|R|(|M|-1)$ overcomes. After that, all robots are out of all the grey circles, and the above arguments hold.   □

4.2   Correctness

The next theorem shows that our algorithm (Procedure COMPUTE) provides exact gathering for configurations in $\bigcup_{1 \leq i \leq 4} \mathcal{S}_i$.

Theorem 3 (correctness). *There exists an exact gathering algorithm for each configuration* $\mathcal{C} \in \bigcup_{1 \leq i \leq 4} \mathcal{S}_i$.

Notice that, according to Theorem 2, there are initial configurations that cannot be gathered. These correspond to configurations $\mathcal{C}$ such that either (1) $\mathcal{C}$ admits a rotation with center c, and there are neither robots nor meeting points on c, or (2) $\mathcal{C}$ admits a reflection on axis ℓ, and there are neither robots nor meeting points on ℓ. We denote such set of configurations by $\mathcal{NG}$.

Define now an additional class $\mathcal{S}_5$ of initial configurations: $\mathcal{C} \in \mathcal{S}_5$ if and only if $\mathcal{C} \notin \bigcup_{1 \leq i \leq 4} \mathcal{S}_i$ and $\mathcal{C}$ admits a reflection with meeting points on the axis. We easily get that $\{\mathcal{NG}, \mathcal{S}_1, \mathcal{S}_2, \mathcal{S}_3, \mathcal{S}_4, \mathcal{S}_5\}$ is a partition of all initial configurations, as its elements are pairwise disjoint by definition and they cover all the initial configurations.

The next result tells that some configurations in $\mathcal{S}_5$, even if they are potentially gatherable, are not gatherable in the exact way.

Theorem 4. *There exist configurations in $\mathcal{S}_5$ that do not admit exact gathering, even if there are minmax-points on the axis of symmetry.*

5   Conclusion

We have studied a new variant of the gathering problem under the Look-Compute-Move cycle model. Robots are required to gather at some predetermined points while minimizing on the maximum distance traveled by each single robot. This further constraint makes the problem challenging, and very different from the previous work. We have proposed a distributed algorithm working for most of the initial configurations that provides exact gathering without any multiplicity detection capability. Other configurations have been proven to be ungatherable, while few cases remain open.

References

1. Bouzid, Z., Das, S., Tixeuil, S.: Gathering of mobile robots tolerating multiple crash faults. In: IEEE 33rd Int. Conf. on Distributed Computing Systems (ICDCS), pp. 337–346 (2013)
2. Chalopin, J., Das, S., Widmayer, P.: Rendezvous of mobile agents in directed graphs. In: Lynch, N.A., Shvartsman, A.A. (eds.) DISC 2010. LNCS, vol. 6343, pp. 282–296. Springer, Heidelberg (2010)
3. Cicerone, S., Stefano, G.D., Navarra, A.: Minimum-traveled-distance gathering of oblivious robots over given meeting points. In: Gao, J., Efrat, A., Fekete, S.P., Zhang, Y. (eds.) ALGOSENSORS 2014, LNCS 8847. LNCS, vol. 8847, pp. 57–72. Springer, Heidelberg (2015)
4. Cieliebak, M., Flocchini, P., Prencipe, G., Santoro, N.: Distributed computing by mobile robots: Gathering. SIAM Journal on Computing **41**(4), 829–879 (2012)
5. Czyzowicz, J., Gasieniec, L., Pelc, A.: Gathering few fat mobile robots in the plane. Theoretical Computer Science **410**(6–7), 481–499 (2009)
6. D'Angelo, G., Di Stefano, G., Navarra, A.: Gathering asynchronous and oblivious robots on basic graph topologies under the look-compute-move model. In: Search Theory: A Game Theoretic Perspective, pp. 197–222. Springer (2013)
7. D'Angelo, G., Di Stefano, G., Navarra, A.: Gathering on rings under the look-compute-move model. Distributed Computing **27**(4), 255–285 (2014)
8. Flocchini, P., Prencipe, G., Santoro, N.: Distributed Computing by Oblivious Mobile Robots. Synthesis Lectures on Distributed Computing Theory. Morgan & Claypool Publishers (2012)
9. Klasing, R., Markou, E., Pelc, A.: Gathering asynchronous oblivious mobile robots in a ring. Theoretical Computer Science **390**, 27–39 (2008)
10. Kranakis, E., Krizanc, D., Markou, E.: The Mobile Agent Rendezvous Problem in the Ring. Morgan & Claypool (2010)
11. Pelc, A.: Deterministic rendezvous in networks: A comprehensive survey. Networks **59**(3), 331–347 (2012)
12. Prencipe, G.: Impossibility of gathering by a set of autonomous mobile robots. Theoretical Computer Science **384**, 222–231 (2007)
13. Sekino, J.: n-ellipses and the minimum distance sum problem. Amer. Math. Monthly **106**(3), 193–202 (1999)

Evacuating Robots from a Disk
Using Face-to-Face Communication
(Extended Abstract)

J. Czyzowicz[1], K. Georgiou[2(✉)], E. Kranakis[3], L. Narayanan[4],
J. Opatrny[4], and B. Vogtenhuber[5]

[1] Dépt. d'Informatique, Université du Québec en Outaouais, Gatineau, QC, Canada
Jurek.Czyzowicz@uqo.ca
[2] Department of Combinatorics Optimization,
University of Waterloo, Waterloo, ON, Canada
k2georgiou@uwaterloo.ca
[3] School of Computer Science, Carleton University, Ottawa, ON, Canada
kranakis@scs.carleton.ca
[4] Department of Computer Science and Software Engineering,
Concordia University, Montreal, QC, Canada
{lata,opatrny}@cs.concordia.ca
[5] Institute for Softwaretechnology, Graz University of Technology, Graz, Austria
bvogt@ist.tugraz.at

Abstract. Assume that two robots are located at the centre of a unit disk. Their goal is to *evacuate* from the disk through an *exit* at an unknown location on the boundary of the disk. At any time the robots can move anywhere they choose on the disk, independently of each other, with maximum speed 1. The robots can cooperate by exchanging information whenever they meet. We study algorithms for the two robots to minimize the *evacuation time*: the time when *both* robots reach the exit. In [9] the authors gave an algorithm defining trajectories for the two robots yielding evacuation time at most 5.740 and also proved that any algorithm has evacuation time at least $3 + \frac{\pi}{4} + \sqrt{2} \approx 5.199$. We improve both the upper and lower bounds on the evacuation time of a unit disk. Namely, we present a new non-trivial algorithm whose evacuation time is at most 5.628 and show that any algorithm has evacuation time at least $3 + \frac{\pi}{6} + \sqrt{3} \approx 5.255$. To achieve the upper bound, we designed an algorithm which non-intuitively proposes a forced meeting between the two robots, even if the exit has not been found by either of them.

1 Introduction

The goal of traditional search problems is to find an object which is located in a specific domain. This subject of research has a long history and there is a

This work was partially supported by NSERC grants.
This work was initiated during the 13th *Workshop on Routing* held in July 2014 in Querétaro, México.

V.Th. Paschos and P. Widmayer (Eds.): CIAC 2015, LNCS 9079, pp. 140–152, 2015.
DOI: 10.1007/978-3-319-18173-8_10

plethora of models investigated in the mathematical and theoretical computer science literature with emphasis on probabilistic search in [17], game theoretic applications in [3], cops and robbers in [8], classical pursuit and evasion in [16], search problems and group testing in [1], and many more.

In this paper, we investigate the problem of searching for a stationary point target called an *exit* at an unknown location using two robots. This type of collaborative search is advantageous in that it reduces the required search time by distributing the search effort between the two robots. In previous work on collaborative search, the goal has generally been to minimize the time taken by the *first* robot to find the object of the search. In contrast, in this work, we are interested in minimizing the time when the *last robot* finds the exit. In particular, suppose two robots are in the interior of a region with a single exit. The robots need to evacuate the region but the location of the exit is unknown to them. The robots can cooperate to search for the exit, but it is not enough for one robot to find the exit, we require *both* robots to reach the exit as soon as possible.

We study the problem of two robots that start at the same time at the centre of a unit disk and attempt to reach an exit placed at an unknown location on the boundary of the disk. At any time the robots can move anywhere they choose within the disk. Indeed, they can take short-cuts by moving in the interior of the disk if desired. We assume their maximum speed is 1. The robots can communicate with each other only if they are at the same point at the same time: we call this communication model *face-to-face communication*. Our goal is to schedule the trajectories of the robots so as to minimize the *evacuation time*, which is the time it takes both robots to reach the exit (for the worst case location of the exit).

1.1 Related Work

The most related work to ours is [9], where the evacuation problem for a set of robots all starting from the centre of a unit disk was introduced and studied. Two communication models are introduced in [9]. In the *wireless* model, the two robots can communicate at any time regardless of their locations. In particular, a robot that finds the exit can immediately communicate its location to the other robot. The other model is called the *non-wireless or local* model in [9], and is the same as our face-to-face model: two robots can only communicate when they are face to face, that is, they are at the same point location at the same time. In [9], for the case of 2 robots, an algorithm with evacuation time $1 + \frac{2\pi}{3} + \sqrt{3} \approx 4.826$ is given for the wireless model; this is shown to be optimal. For the face-to-face model, they prove an upper bound of 5.740 and a lower bound of 5.199 on the evacuation time.

Baeza-Yates *et al* posed the question of minimizing the worst-case trajectory of a single robot searching for a target point at an unknown location in the plane [4]. This was generalized to multiple robots in [15], and more recently has been studied in [12,14]. However, in these papers, the robots cannot communicate, and moreover, the objective is for the first robot to find the target. Two seminal and influential papers (that appeared almost at the same time) on probabilistic search are [5], and [6] and concern minimizing the *expected time* for the robot to find the target. Useful surveys on search theory can also be found in [7] and [11]. In addition, the latter citation has an interesting classification of search problems by search objectives, distribution of effort, point target (stationary, large, moving), two-sided search, etc. The evacuation problem considered in our paper is related to searching on a line, in that we are searching on the boundary of a disk but with the additional ability to make short-cuts in order to enable the robots to meet sooner and thus evacuate faster.

Our problem is also related to the *rendezvous problem* and the problem of *gathering* [2,13]. Indeed our problem can be seen as a version of a rendezvous problem for three robots, where one of them remains stationary.

1.2   Preliminaries and Notation

We assume that two robots R_1 and R_2 are initially at the center of a disk with radius 1, and that there is an exit at some location X on the boundary of the disk. The robots do not know X, but do know each other's algorithms. The robots move at a speed subject to a maximum speed, say 1. They cannot communicate except if they are at the same location at the same time. Finally, both robots are equipped with deterministic processors that can numerically solve trigonometric equations, and as such they are assumed to have the required memory. The *evacuation problem* is to define trajectories for the two robots that minimize the *evacuation time*.

For two points A and B on the unit circle, the length of an arc AB is denoted by $\overset{\frown}{AB}$, while the length of the corresponding chord (line segment) will be denoted by $\overline{AB}$ (arcs on the circle are always read clockwise, i.e., arc AB together with arc BA cover the whole circle).

1.3   Outline and Results of the Paper

In [9] an algorithm is given defining a trajectory for two robots in the face-to-face communication model with evacuation time 5.740 and it is also proved that any such algorithm has evacuation time at least $3 + \frac{\pi}{4} + \sqrt{2} > 5.199$.

Our main contribution in this paper is to improve both the upper and lower bounds on the evacuation time. Namely, we give a new algorithm whose evacuation time is at most 5.628 (see Section 2) and also prove that any algorithm has evacuation time at least $3 + \frac{\pi}{6} + \sqrt{3} > 5.255$ (see Section 3). To prove our lower bound on the disk, we first give tight bounds for the problem of evacuating a regular hexagon where the exit is placed at an unknown vertex. We observe that, surprisingly, in our optimal evacuation algorithm for the hexagon, the two robots are forced to meet after visiting a subset of vertices, even if an exit has not been found at that time. We use the idea of such a forced meeting in the design of our disk evacuation algorithm.

Omitted proofs can be found in the full version of the paper [10].

2   Evacuation Algorithms

In this section we give two new evacuation algorithms for two robots in the face-to-face model that take evacuation time approximately 5.644 and 5.628 respectively. We begin by presenting Algorithm $\mathcal{A}$ proposed by [9] which has been shown to have evacuation time 5.740. Our goal is to understand the worst possible configuration for this algorithm, and subsequently to modify it accordingly so as to improve its performance.

All the algorithms we present follow the same general structure: The two robots R_1 and R_2 start by moving together to an arbitrary point A on the boundary of the disk. Subsequently R_1 explores the arc $A'A$, where A' is the antipodal point of A, by moving along some trajectory defined by the algorithm. At the same time, R_2 explores the arc AA', following a trajectory that is the reflection of R_1's trajectory. If either of the robots finds the exit, it immediately uses the *Meeting Protocol* defined below to meet the other robot (note that the other robot has not yet found the exit and hence keeps exploring). After meeting, the two robots travel together on the shortest path to the exit, thereby completing the evacuation. At all times, the two robots travel at unit speed. Without loss of generality, we assume that R_1 finds the exit and then *catches* R_2 for our analysis.

Meeting Protocol for R_1 : *If at any time t_0 R_1 finds the exit at point X, it computes the shortest additional time t such that R_2, after traveling distance $t_0 + t$, is located at point M satisfying $\overline{XM} = t$. Robot R_1 moves along the segment XM. At time $t_0 + t$ the two robots meet at M and traverse directly back to the exit at X incurring total time cost $t_0 + 2t$.*

2.1   Evacuation Algorithm $\mathcal{A}$ of [9]

We proceed by describing the trajectories of the two robots in Algorithm $\mathcal{A}$. As mentioned above, both robots start from the centre O of the disk and move together to an arbitrary position A on the boundary of the disk. R_2 then moves clockwise along the boundary of the disk up to distance π, see left-hand side of Figure 1, and robot R_1 moves counter clockwise on the trajectory which is a reflection of R_2's trajectory with respect to the line passing through O and A. When R_1 finds the exit, it invokes the *meeting protocol* in order to meet R_2, after which the evacuation is completed.

The meeting-protocol trajectory of R_1 in Algorithm $\mathcal{A}$ is depicted in the right-hand side of Figure 1. Clearly, for the two robots to meet, we must have $\overset{\frown}{AB} = \overset{\frown}{EA} + \overline{EB}$. Next we want to analyze the performance of the algorithm, with respect to $x := \overset{\frown}{AB}$, i.e. the length x of the arc that R_2 travels, before it is met by R_1. We also set $f(x) := \overline{EB}$.

It follows that $\overset{\frown}{EA} = x - f(x)$, and since $2\sin(\overset{\frown}{EB}/2) = \overline{EB}$ we conclude that

$$f(x) \;=\; z \text{ where } z \text{ is a solution of the equation } z = 2\sin\left(x - \frac{z}{2}\right). \qquad (1)$$

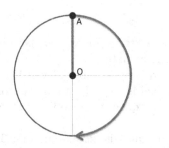 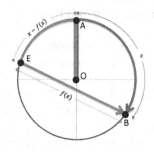

Fig. 1. Evacuation Algorithm $\mathcal{A}$ with exit position E. The trajectory of robot R_2 is depicted on the left. The movement paths of robots R_1, R_2 are shown on the right, till the moment they meet at point B on the circle.

In other words, $f(x)$ is the length of interval EB that R_1 needs to travel in the interior of the disk after locating the exit at E, to meet R_2 at point B.

Then, the cost of Algorithm $\mathcal{A}$, given that the two robots meet at time x after they together reached the boundary of the disk at A, is $1 + x + f(x)$. Given that distance $x - f(x)$ traveled by R_1 until finding the exit is between 0 and π, it directly follows that x can take any value between 0 and π as well. Hence, the worst case performance of Algorithm $\mathcal{A}$ is determined by $\sup_{x \in [0,\pi]} \{x + f(x)\}$ The next lemma, along with its proof, follows from [9].

Lemma 1. *Expression $F(x) := x + f(x)$ attains its supremum at $x_0 \approx 2.85344$ (which is $\approx 0.908279\pi$). In particular, $F(x)$ is strictly increasing when $x \in [0, x_0]$ and strictly decreasing when $x \in [x_0, \pi]$*

By Lemma 1, the evacuation time of Algorithm $\mathcal{A}$ is $1 + x_0 + f(x_0) < 5.740$. The worst case is attained for $x_0 - f(x_0) \approx 0.308\pi$.

2.2   New Evacuation Algorithm $\mathcal{B}(\chi, \phi)$

We now show how to improve the previously described algorithm and obtain evacuation time at most 5.644. The main idea for improving Algorithm $\mathcal{A}$ is to change the trajectory of the robots when the distance traveled on the boundary of the disk approaches the critical value x_0 of Lemma 1. Informally, robot R_2 could meet R_1 earlier if it makes a linear detour inside the interior of the disk towards R_1 a little before traversing distance x_0.

We describe a generic family of algorithms that realizes this idea. The specific trajectory of each algorithm is determined by two parameters χ and ϕ where $\chi \in [\pi/2, x_0]$ and $\phi \in [0, f(\chi)/2]$, whose optimal values will be determined later. For ease of exposition, we assume R_1 finds the exit. The trajectory of R_2 (assuming it has not yet met R_1) is partitioned into four phases that we call the *deployment*, *pre-detour*, *detour* and *post-detour* phases. The description of the phases rely on the left-hand side of Figure 2.

Algorithm $\mathcal{B}(\chi, \phi)$(with a linear detour). R_2's trajectory until it meets R_1 is described below:

⋆ *Deployment phase:* Robot R_2 starts from the centre O of the disk and moves to an arbitrary position A on the boundary of the disk.

⋆ *Pre-detour phase:* R_2 moves clockwise along the boundary of the disk until having explored an arc of length χ.

⋆ *Detour phase:* Let D be the reflection of B with respect to AA' (where A' is the antipodal point of A). Then, R_2 moves on a straight line towards the interior of the disk and towards the side where O lies, forming an angle of ϕ with line BD, until R_2 touches line AA' at point C. From C it follows a straight line segment to B. Note that C is indeed in the interior of the line segment AA' by the restrictions on ϕ.

⋆ *Post-detour phase:* Robot R_2 continues moving clockwise on the arc BA'.

At the same time R_1 follows a trajectory that is the reflection of R_2's trajectory along the line AA'. When at time t_0 it finds the exit, it follows the Meeting Protocol defined above.

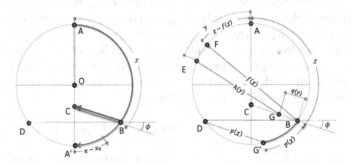

Fig. 2. Illustrations for Algorithm $\mathcal{B}(\chi, \phi)$

Notably, the two robots may meet at point C without having located the exit. Next we consider three cases as to where R_2 can be caught by R_1 while moving on its trajectory (after R_1 has located the exit). For all three cases, the reader can consult the right-hand side of Figure 2. As the time needed for the deployment phase is independent of where the exit is located, we ignore this extra cost of 1 during the case distinction.

Case 1: R_2 is caught during its pre-detour phase: The meeting point is anywhere on the arc AB. Recall that $\chi \le x_0$, so by Lemma 1 the location F of the exit on the arc FA that maximizes the cost of $\mathcal{B}(\chi, \phi)$ is the one at at distance $\chi - f(\chi)$ from A (see right-hand side of Figure 2). The cost then is $\widehat{AB} + \overline{BF} = \widehat{FA} + 2\overline{BF} = \chi + f(\chi)$.

Case 2: R_2 is caught during its detour phase: Let G be the point on BC where the robots meet. Further, let E be the position of the exit on the arc $A'A$, and let $y := \widehat{EA}$. In the following, $h(y) := \overline{EG}$ denotes the length of the trajectory of R_1 in its attempt to catch R_2 after it finds the exit. Also, $q(y) := \overline{BG}$ denotes the distance that R_2 travels on BC till it is caught by R_1. Note that the functions h and q also depend on χ and ϕ; however, while those are fixed, y varies with the position of the exit. Lemma 2 below states that $h(y)$ and $q(y)$ are well defined.

Lemma 2. $y \in [\chi - f(\chi), \chi]$ *if and only if the meeting point G of the robots is on the line segment BC. Moreover, robot R_2 can be caught by R_1 only while moving from B to C.*

We conclude that if the exit is located at point E, then the cost of the algorithm is $y + 2h(y)$. Hence, in case 2, the cost of the algorithm is at most $\sup_{y \in [\chi - f(\chi), \chi]}\{y + 2h(y)\}$. We emphasize again that $h(y)$ and $q(y)$ also depend on the fixed parameters χ and ϕ.

Case 3: R_2 is caught during its post-detour phase: Clearly, in this case the exit lies in the interior of the arc $A'D$ or coincides with A'. At time $t_d = \chi + 2q(\chi)$, robots R_1 and R_2 are located at points D and B, respectively. Then they move towards each other on the arc BD till R_1 finds the exit. Note that, since $\overline{DB}/2 = \sin(\chi)$, we have $q(\chi) = \sin(\chi) / \cos(\phi)$. Clearly, the closer the exit is to D is, the higher is the cost of the evacuation algorithm. In the limit (as the position of the exit approaches D), the cost of case 3 approaches t_d plus the time it takes R_1 to catch R_2 if the exit was located at D, and if they started moving from points D and B respectively. Let G' be the meeting point on the arc BD in this case, i.e. $\overline{DG'} = \overset{\frown}{BG'}$. We define $p(x)$ to be the distance that R_1 needs to travel in the interior of the disk to catch R_2, if the exit is located at distance x from A. Clearly[1]

$$p(x) := \text{ unique } z \text{ satisfying } z = 2\sin\left(x + \frac{z}{2}\right). \qquad (2)$$

Note also that $\overline{DG'} = p(\chi)$ so that the total cost in this case is at most $t_d + 2p(\chi) = \chi + 2\sin(\chi) / \cos(\phi) + 2p(\chi)$.

The following two lemmata summarize the above analysis and express $h(y)$ in explicit form (in dependence of χ and ϕ), respectively.

Lemma 3. *The evacuation time of Algorithm $\mathcal{B}(\chi, \phi)$ is*

$1 + \max\left\{\chi + f(\chi), \sup_{y \in [\chi - f(\chi), \chi]}\{y + 2h(y)\}, \quad \chi + 2\sin(\chi) / \cos(\phi) + 2p(\chi)\right\},$

where $h(y)$ (that also depends on the choice of χ, ϕ) denotes the time that a robot needs from the moment it finds the exit till it meets the other robot when following the meeting protocol.

Lemma 4. *For every $\chi > 0$ and for every $\chi - f(\chi) \le y \le \chi$, the distance $h(y)$ that R_1 travels from A until finding R_2 when following the meeting protocol in Algorithm $\mathcal{B}(\chi, \phi)$ is*

$h(y) = \frac{2 + (\chi - y)^2 - 2\cos(\chi + y) + 2(\chi - y)(\sin(\phi + y) - \sin(\phi - \chi))}{2(\chi - y - \sin(\phi - \chi) + \sin(\phi + y))}$. *In particular, $h(y)$ is strictly decreasing for $0 \le \phi \le f(\chi)/2$.*

---

[1] Uniqueness of the root of the equation defining $p(x)$ is an easy exercise.

The first natural attempt in order to beat Algorithm $\mathcal{A}$ would be to consider $\mathcal{B}(\chi, 0)$, i.e. make BC perpendicular to AA' in Figure 2. In light of Lemma 4 and using Lemma 3, we state the following claim to build some intuition for our next, improved, algorithm.

Claim 1. *The performance of algorithm $B(\chi, 0)$ is optimized when $\chi = \chi_0 \approx 2.62359$, and its cost is $1 + 4.644 = 5.644$. The location of the exit inducing the worst case for $B(\chi, 0)$ is when $y = \widehat{EA} \approx 0.837\pi$. The meeting point of the two robots takes place at point G (see Figure 2, and set $\phi = 0$), where $q(y) = \overline{BG} \approx 0.117 \approx 0.236\overline{BC}$. In particular, the cost of the algorithm, if the meeting point of the robots is during R_2's pre-detour, detour and post-detour phase, is (approximately) 5.621, 5.644 and 5.644 respectively.*

Note that χ_0 of Claim 1 is strictly smaller than x_0 of Lemma 1. In other words, the previous claim is in coordination with our intuition that if the robots moved towards the interior of the disk a little before the critical position of the meeting point x_0 of Algorithm $\mathcal{A}$, then the cost of the algorithm could be improved.

2.3   New Evacuation Algorithm $\mathcal{C}(\chi, \phi, \lambda)$

Claim 1 is instructive for the following reason. Note that the worst meeting point G for Algorithm $\mathcal{B}(\chi_0, 0)$ satisfies $\overline{BG} \approx 0.236\overline{BC}$. This suggests that if we consider algorithm $\mathcal{B}(\chi_0, \phi)$ instead, where $\phi > 0$, then we would be able to improve the cost if the meeting point happened during the detour phase of R_2. On one hand, this further suggests that we can decrease the detour position χ_0 (note that the increasing in χ cost $\chi + f(\chi)$ is always a lower bound to the performance of our algorithms when $\chi < x_0$). On the other hand, that would have a greater impact on the cost when the meeting point is in the post-detour phase of R_2, as in this case the cost of moving from B to C and back to B would be $2\sin(\chi)/\cos(\phi)$ instead of just $2\sin(\chi)$. A compromise to this would be to follow the linear detour trajectory of R_2 in $\mathcal{B}(\chi_0, \phi)$ only up to a certain threshold-distance λ, after which the robot should reach the diameter segment AA' along a linear segment perpendicular to segment AA' then return to the detour point B along a linear segment. Thus the detour forms a triangle. This in fact completes the high level description of Algorithm $\mathcal{C}(\chi, \phi, \lambda)$ that we formally describe below.

In that direction, we fix χ, ϕ, λ, with $\chi \in [\pi/2, x_0], \phi \in [0, f(\chi)/2]$ and $\lambda \in [0, \sin(\chi)/\cos(\phi)]$. As before, we only describe the trajectory of robot R_2. The meeting protocol that R_1 follows once it finds the exit is the same as for Algorithms $\mathcal{A}$ and $\mathcal{B}(\chi, \phi)$.

The trajectory of robot R_2 (that has neither found the exit nor met R_1 yet) can be partitioned into roughly the same four phases as for Algorithm $\mathcal{B}(\chi, \phi)$; so we again call them *deployment, pre-detour, detour* and *post-detour* phases. The description of the phases refers to the left-hand side of Figure 3, which is a partial modification of Figure 2.

Algorithm $\mathcal{C}(\chi,\phi,\lambda)$ (with a triangular detour). The phases of robot R_2's trajectory are:

⋆ *Deployment phase:* Same as in Algorithm $\mathcal{B}(\chi,\phi)$. At time 1, R_2 is at point A.

⋆ *Pre-detour phase:* Same as in Algorithm $\mathcal{B}(\chi,\phi)$. In additional time χ, R_2 is in point B.

⋆ *Detour phase:* This phase is further split into three subphases.

  ⋄ *Subphase-1:* Up to additional time λ, R_2 moves along a line segment exactly as in the detour phase of Algorithm $\mathcal{B}(\chi,\phi)$. Let G be the position of the robot at the end of this phase.

  ⋄ *Subphase-2:* Let C be the projection of G onto AA'. R_2 follows line segment GC till it reaches point C.

  ⋄ *Subphase-3 (Recovering phase):* Robot follows line segment CB back to point B.

⋆ *Post-detour phase:* Same as in Algorithm $\mathcal{B}(\chi,\phi)$. After additional time $\overset{\frown}{EA'}$, R_2 reaches point A'.

At the same time R_1 follows a trajectory that is the reflection of R_2's trajectory along the line AA'. If a robot finds the exit, it follows the meeting protocol defined earlier.

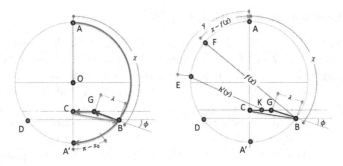

Fig. 3. Illustrations for Algorithm $\mathcal{C}(\chi,\phi,\lambda)$

Obviously, Algorithm $\mathcal{C}\left(\chi,\phi,\frac{\sin(\chi)}{\cos(\phi)}\right)$ is identical to Algorithm $\mathcal{B}(\chi,\phi)$. Moreover, as before robots may meet at point C without having located the exit.

Notice that an immediate consequence of the definition of $\mathcal{C}(\chi,\phi,\lambda)$ is that if robot R_1 finds the exit and meets R_2 during its detour subphase-2 in some point K (as in the right-hand side of Figure 3), then $\overset{\frown}{EA} + \overline{EK} = \overset{\frown}{AB} + \overline{BG} + \overline{GK}$.

When R_1 finds the exit somewhere on the arc $A'A$, it catches R_2 on its trajectory so that they return together to the exit. Note that since robots meet at point C, if the exit is not in the arc DB, it is impossible for a robot to be caught by the other robot in subphase-3 of its detour phase. Hence, there are four cases as to where R_2 can be caught by R_1 that found the exit. As before, we omit the extra cost 1 which is the time needed for the deployment phase during the case distinction.

Case 1: R_2 is caught in its pre-detour phase: The cost of Algorithm $\mathcal{C}(\chi,\phi,\lambda)$ is at most $\chi + f(\chi)$, exactly as in the analogous case of Algorithm $\mathcal{B}(\chi,\phi)$.

Case 2: R_2 is caught in its detour subphase-1: As in case 2 of the analysis of Algorithm $\mathcal{B}(\chi, \phi)$, if E is the position of the exit then, $y = \widehat{EA}$ satisfies $y \geq \chi - f(\chi)$. As longs as $q(y)$ remains less than λ, the cost of the algorithm remains $y + 2h(y)$. In order to find the maximum y for which this formula remains valid, we need to solve the equation $\lambda = q(y)$. This is possible by recalling that $h(y) = \chi + q(y) - y$, and by invoking the formula of $h(y)$ as it appears in Lemma 4. By the monotonicity of $h(y)$, we have that there exists unique ψ satisfying $h(\psi) = \chi + \lambda - \psi$. It follows that the cost of the algorithm in this case is at most $\sup_{\chi-f(\chi) \leq y \leq \psi}\{y + 2h(y)\}$.

Case 3: R_2 is caught in its detour subphase-2: In this case, the relevant figure is the right-hand side of Figure 3. Let the exit be at point E, and let K denote the meeting point of the robots on the line segment GC. We set $h'(y) := \overline{EK}$, which is calculated next in Lemma 5 (a). We conclude that in this case the cost of the algorithm is at most $\sup_{\psi \leq y \leq \chi}\{y + 2h'(y)\}$.

Case 4: R_2 is caught in its post-detour phase: Let t_d again be the total time a robot needs till it enters its post-detour phase. As in case 3 of Algorithm $\mathcal{B}(\chi, \phi)$, the cost of Algorithm $\mathcal{C}(\chi\phi, \lambda)$ for this case is at most $t_d + 2p(\chi)$. It thus remains to show how to calculate $t_d = \widehat{AB} + \overline{BG} + \overline{GC} + \overline{CB}$, which is done in Lemma 5 (b).

Lemma 5. *The following statements hold for Algorithm $\mathcal{C}(\chi, \phi, \lambda)$:*
(a) Suppose that R_1 finds the exit and meets R_2 in its detour subphase-2. Then the time $h'(y)$ that R_1 needs from finding the exit until meeting R_2 is $h'(y) = N(\chi, y, \lambda, \phi)/D(\chi, y, \lambda, \phi)$, where

$$N(\chi, y, \lambda, \phi) := \quad 2 + \lambda^2 + (\lambda + \chi - y)^2 + 2\lambda(\sin(\phi - \chi) - \sin(\phi + y))$$
$$+ 2(\lambda + \chi - y)(\sin(\chi) + \sin(y) - \lambda\cos(\phi)) - 2\cos(\chi + y),$$
$$D(\chi, y, \lambda, \phi) := 2(\lambda + \chi - y + \sin(\chi) + \sin(y) - \lambda\cos(\phi)).$$

(b) Suppose that R_1 finds the exit and meets R_2 in its post-detour phase. Then, the total time that R_2 spends in its detour phase is $\lambda + \sin(\chi) - \lambda\cos(\phi) + \sqrt{\sin^2(\chi) + \lambda^2\sin^2(\phi)}$.

Before stating our main theorem, we summarize the total time required by Algorithm $\mathcal{C}(\chi, \phi, \lambda)$ in the following lemma.

Lemma 6. *The cost of Algorithm $\mathcal{C}(\chi, \phi, \lambda)$ can be expressed as*

$$1 + \max \begin{cases} \chi + f(\chi) & \text{(pre-detour phase)} \\ \sup_{\chi-f(\chi) \leq y \leq \psi}\{y + 2h(y)\} & \text{(detour subphase-1)} \\ \sup_{\psi \leq y \leq \chi}\{y + 2h'(y)\} & \text{(detour subphase-2)} \\ \chi + \lambda + \sin(\chi) - \lambda\cos(\phi) + \sqrt{\sin^2(\chi) + \lambda^2\sin^2(\phi)} + 2p(\chi) \\ & \text{(post-detour phase)} \end{cases},$$

where the functions f and p are as in (1) and (2), respectively; functions $h(y)$ and $h'(y)$ are expressed explicitly in Lemmas 4 and 5 (a), respectively; and ψ is the unique solution to the equation $h(\psi) = \chi + \lambda - \psi$.

Using the statement of Lemma 6 and numerical optimization, we obtain the following improved upper bound.

Theorem 1. *For $\chi_0 = 2.631865, \phi_0 = 0.44916$ and $\lambda_0 = 0.05762$, the evacuation algorithm $\mathcal{C}(\chi_0, \phi_0, \lambda_0)$ has cost no more than 5.628.*

3 Lower Bound

In this section we show that any evacuation algorithm for two robots in the face-to-face model takes time at least $3 + \frac{\pi}{6} + \sqrt{3} \approx 5.255$. We first prove a result of independent interest about an evacuation problem on a hexagon.

Theorem 2. *Consider a hexagon of radius 1 with an exit placed at an unknown vertex. The worst case evacuation time for two robots starting at any two arbitrary vertices of the hexagon is at least $2 + \sqrt{3}$.*

Proof sketch. Assume an arbitrary deterministic algorithm $\mathcal{D}$ for the problem. $\mathcal{D}$ solves the problem for any input, i.e., any placement of the exit. The basic idea for proving this theorem is to construct inputs for $\mathcal{D}$ and show that for at least one of them, the required evacuation time is at least $2 + \sqrt{3}$. First, we let $\mathcal{D}$ run without placing an exit at any vertex, so as to find out in which order the robots are exploring all the vertices of the hexagon. We label the vertices of the hexagon according to this order (if two vertices are explored simultaneously then we just order them arbitrarily) and assume w.l.o.g. that v_5 was visited by R_1. The two inputs for proving Theorem 2 are then placing the exit on (1) the vertex of the hexagon v_6 (that was visited last), and (2) the vertex that was visited last by R_2 before (or while) the fifth vertex v_5 was visited by R_1.   ∎

It is worth noting that the lower bound from Theorem 2 matches the upper bound of evacuating a regular hexagon, when the initial starting vertices may be chosen by the algorithm. Consider a hexagon $ABCDEF$ and suppose that the trajectory of one robot, as long as no exit was found, is $ABDC$. Similarly, the other robot follows the symmetric trajectory $FECD$; cf. left-hand side of Fig. 4. By symmetry it is sufficient to consider exits at vertices A, B or C. An exit at C is reached by each robot independently, while both robots proceed to an exit at A or B after meeting at point M, the intersection of segments BD and EC. Altogether, they need a total time of at most $\max\{1 + 4/\sqrt{(3)}, 1 + (2 + \sqrt{7})/\sqrt{3}, 1 + \sqrt{3} + 1\}$ to evacuate from the hexagon. It is easy then to verify that, in each case, the evacuation time of this algorithm is always upper bounded by $2 + \sqrt{3}$.

In the above algorithm, the robots meet at M, regardless of whether the exit has been already found or not. The idea of our algorithm for disk evacuation presented in the previous section was influenced by this non-intuitive presence of a forced meeting.

Combining Theorem 2 with some reasoning from measure theory, we obtain the following lower bound for our evacuation problem.

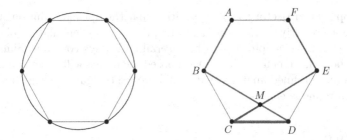

Fig. 4. The trajectories for R_1 (red) and R_2 (blue) for the hexagon evacuation algorithm having evacuation time $2 + \sqrt{3}$, while the exit has not been found, are depicted on the left. Right-hand side: At time $1 + \frac{\pi}{6} - \varepsilon$, there is regular hexagon all of whose vertices are unexplored and lie on the boundary of the disk.

Theorem 3. *Assume you have a unit disk with an exit placed somewhere on the boundary. The worst case evacuation time for two robots starting at the centre in the face-to-face model is at least $3 + \frac{\pi}{6} + \sqrt{3} \approx 5.255$.*

Proof. It takes 1 time unit for the robots to reach the boundary of the hexagon. By time $t = 1 + \frac{\pi}{6}$, any algorithm could have explored at most $\frac{2\pi}{6}$ of the boundary of the disk. Hence for any ε with $0 < \varepsilon < t$, there exists a regular hexagon with all vertices on the boundary of the disk and all of whose vertices are unexplored at time $t - \varepsilon$; see the right-hand side of Figure 4. Now, invoking Theorem 2 gives the bound of at least $1 + \frac{\pi}{6} + 2 + \sqrt{3}$ to evacuate both robots. ∎

4   Conclusion

In this paper we studied evacuating two robots from a disk, where the robots can collaborate using face-to-face communication. Unlike evacuation for two robots in the wireless communication model, for which the tight bound $1 + \frac{2\pi}{3} + \sqrt{3}$ is proved in [9], the evacuation problem for two robots in the face-to-face model is much harder to solve. We gave a new non-trivial algorithm for the face-to-face communication model which improved the upper bound in [9]. We used a novel, non-intuitive idea of a forced meeting between the robots, regardless of whether the exit was found before the meeting. We also provided a different analysis that improved the lower bound in [9].

We believe that none of our bounds are close to be tight. More specifically, we do know that our upper bound is not optimal, since by disallowing robots to meet without having found the exit (by slightly truncating their trajectory), we can provably improve the performance of our algorithm. Unfortunately, the improvement we obtain this way is negligible (affecting the third significant decimal digit) while the additional required technicalities would be overwhelming, without offering new insights for the problem. This also suggests that the choice of the parameters we choose for our algorithm are not optimal. We are also certain that the proposed algorithm, i.e. family of trajectories we consider, cannot give the optimal trajectory, as it is intuitive that the optimal solution should be related to a properly defined differential equation ensuring that if robots meet during the deployment phase then the overall cost stays constant. Similarly for the lower bound, we believe that our proposed technique will serve as a guideline towards a more refined analysis that would reduce the gap. To conclude, a tight bound still remains elusive.

References

1. Ahlswede, R., Wegener, I.: Search problems. Wiley-Interscience (1987)
2. Alpern, S., Beck, A.: Asymmetric rendezvous on the line is a double linear search problem. Mathematics of Operations Research **24**(3), 604–618 (1999)
3. Alpern, S., Gal, S.: The theory of search games and rendezvous, vol. 55. Springer, New York (2003)
4. Yates, R.B., Culberson, J., Rawlins, G.: Searching in the plane. Information and Computation **106**(2), 234–252 (1993)
5. Beck, A.: On the linear search problem. Israel Journal of Mathematics **2**(4), 221–228 (1964)
6. Bellman, R.: An optimal search. Siam Review **5**(3), 274–274 (1963)
7. Benkoski, S., Monticino, M., Weisinger, J.: A survey of the search theory literature. Naval Research Logistics (NRL) **38**(4), 469–494 (1991)
8. Bonato, A., Nowakowski, R.: The game of cops and robbers on graphs. American Mathematical Soc. (2011)
9. Czyzowicz, J., Gasieniec, L., Gorry, T., Kranakis, E., Martin, R., Pajak, D.: Evacuating robots via unknown exit in a disk. In: Kuhn, F. (ed.) DISC 2014. LNCS, vol. 8784, pp. 122–136. Springer, Heidelberg (2014)
10. Czyzowicz, J., Georgiou, K., Kranakis, E., Narayanan, L., Opatrny, J., Vogtenhuber, B.: Evacuating using face-to-face communication. CoRR, abs/1501.04985 (2015)
11. Dobbie, J.: A survey of search theory. Operations Research **16**(3), 525–537 (1968)
12. Emek, Y., Langner, T., Uitto, J., Wattenhofer, R.: Solving the ANTS problem with asynchronous finite state machines. In: Esparza, J., Fraigniaud, P., Husfeldt, T., Koutsoupias, E. (eds.) ICALP 2014, Part II. LNCS, vol. 8573, pp. 471–482. Springer, Heidelberg (2014)
13. Flocchini, P., Prencipe, G., Santoro, N., Widmayer, P.: Gathering of asynchronous robots with limited visibility. Theoretical Computer Science **337**(1), 147–168 (2005)
14. Lenzen, C., Lynch, N., Newport, C., Radeva, T.: Trade-offs between selection complexity and performance when searching the plane without communication. In: Proceedings of PODC, pp. 252–261 (2014)
15. López-Ortiz, A., Sweet, G.: Parallel searching on a lattice. In: Proceedings of CCCG, pp. 125–128 (2001)
16. Nahin, P.: Chases and Escapes: The Mathematics of Pursuit and Evasion. Princeton University Press (2012)
17. Stone, L.: Theory of optimal search. Academic Press, New York (1975)

Planarity of Streamed Graphs

Giordano Da Lozzo[1]([✉]) and Ignaz Rutter[2]

[1] Department of Engineering, Roma Tre University, Rome, Italy
dalozzo@dia.uniroma3.it
[2] Karlsruhe Institute of Technology (KIT), Karlsruhe, Germany
rutter@kit.edu

Abstract. In this paper we introduce a notion of planarity for graphs that are presented in a streaming fashion. A *streamed graph* is a stream of edges $e_1, e_2, \ldots, e_m$ on a vertex set V. A streamed graph is ω-*stream planar* with respect to a positive integer window size ω if there exists a sequence of planar topological drawings Γ_i of the graphs $G_i = (V, \{e_j \mid i \leq j < i + \omega\})$ such that the common graph $G_\cap^i = G_i \cap G_{i+1}$ is drawn the same in Γ_i and in Γ_{i+1}, for $1 \leq i < m - \omega$. The STREAM PLANARITY Problem with window size ω asks whether a given streamed graph is ω-stream planar. We also consider a generalization, where there is an additional *backbone graph* whose edges have to be present during each time step. These problems are related to several well-studied planarity problems.

We show that the STREAM PLANARITY Problem is $\mathcal{NP}$-complete even when the window size is a constant and that the variant with a backbone graph is $\mathcal{NP}$-complete for all $\omega \geq 2$. On the positive side, we provide $O(n + \omega m)$-time algorithms for (i) the case $\omega = 1$ and (ii) all values of ω provided the backbone graph consists of one 2-connected component plus isolated vertices and no stream edge connects two isolated vertices. Our results improve on the Hanani-Tutte-style $O((nm)^3)$-time algorithm proposed by Schaefer [GD'14] for $\omega = 1$.

1 Introduction

In this work we consider the following problem concerning the drawing of evolving networks. We are given a stream of edges $e_1, e_2 \ldots, e_m$ with their endpoints in a vertex set V and an integer *window size* $\omega > 0$. Intuitively, edges of the stream are assigned a fixed "lifetime" of ω time intervals. Namely, for $1 \leq i < |V| - \omega$, edge e_i will *appear* at the i-th time instant and *disappear* at the $(i + \omega)$-th time instant. We aim at finding a sequence of drawings Γ_i of the graphs $G_i = (V, \{e_j \mid i \leq j < i + \omega\})$, for $1 \leq i < |V| - \omega$, showing the vertex set and the subset of the edges of the stream that are "alive" at each time instant i, with

Giordano Da Lozzo was supported by the MIUR project AMANDA "Algorithmics for MAssive and Networked DAta", prot. 2012C4E3KT_001. Ignaz Rutter was supported by a fellowship within the Postdoc-Program of the German Academic Exchange Service (DAAD). This research was done at the Department of Applied Mathematics at Charles University in Prague.

V.Th. Paschos and P. Widmayer (Eds.): CIAC 2015, LNCS 9079, pp. 153–166, 2015.
DOI: 10.1007/978-3-319-18173-8_11

the following two properties: (i) each drawing Γ_i is planar and (ii) the drawing of the common graphs $G_\cap^i = G_i \cap G_{i+1}$ is the same in Γ_i and in Γ_{i+1}. We call such a sequence of drawings an ω-*streamed drawing* (ω-SD).

The introduced problem, which we call STREAM PLANARITY (SP, for short), captures the practical need of displaying evolving relationships on the same set of entities. As large changes in consecutive drawings might negatively affect the ability of the user to effectively cope with the evolution of the dataset to maintain his/her mental map, in this model only one edge is allowed to enter the visualization and only one edge is allowed to exit the visualization at each time instant, visible edges are represented by the same curve during their lifetime, and each vertex is represented by the same distinct point. Thus, the amount of relational information displayed at any time stays constant. However, the magnitude of information to be simultaneously presented to the user may significantly depend on the specific application as well as on the nature of the input data. Hence, an interactive visualization system would benefit from the possibility of selecting different time windows. On the other hand, it seems generally reasonable to consider time windows whose size is fixed during the whole animation.

To widen the application scenarios, we consider the possibility of specifying portions of a streamed graph that are alive during the whole animation. These could be, e.g., context-related substructures of the input graph, like the backbone network of the Internet (where edges not in the backbone disappear due to faults or congestion and are later replaced by new ones), or sets of edges directly specified by the user. We call this variant of the problem STREAM PLANARITY WITH BACKBONE (SPB, for short) and the sought sequence of drawings an ω-streamed drawing with backbone (ω-SDB).

Related Work. The problem is similar to on-line planarity testing [9], where one is presented a stream of edge insertions and deletions and has to answer queries whether the current graph is planar. Brandes *et al.* [6] study the closely related problem of computing planar straight-line grid drawings of trees whose edges have a fixed lifetime under the assumption that the edges are presented one at a time and according to an Eulerian tour of the tree. The main difference, besides using topological rather than straight-line drawings, is that in our model the sequence of edges determining the streamed graph is known in advance and no assumption is made on the nature of the stream.

It is worth noting that the SP Problem can be conveniently interpreted as a variant of the much studied SIMULTANEOUS EMBEDDING WITH FIXED EDGES (SEFE) Problem (see [4] for a recent survey). In short, an instance of SEFE consists of a sequence of graphs $G_1, \ldots, G_k$, sharing some vertices and edges, and the task is to find a sequence of planar drawings Γ_i of G_i such that Γ_i and Γ_j coincide on $G_i \cap G_j$. It is not hard to see that deciding whether a streamed graph is ω-stream planar is equivalent to deciding whether the graphs induced by the edges of the stream that are simultaneously present at each time instant admit a SEFE. Unfortunately, positive results on SEFE mostly concentrate on the variant with $k = 2$, whose complexity is still open, and the problem is NP-hard for $k \geq 3$ [10]. However, while the SEFE problem allows the edge sets of the input graphs to significantly differ from each other, in our model only small changes in the subsets of the edges of the stream displayed at consecutive time instants are permitted. In this sense, the problems we study can be seen as an attempt to overcome the hardness of SEFE for $k \geq 3$ to enable visualization of graph sequences consisting of several steps, when any two consecutive graphs exhibit a strong similarity.

We note that the ω-stream planarity of the stream $e_1, \ldots, e_m$ on vertex set V and backbone edges S is equivalent to the existence of a drawing of the (multi)graph $G_\cup = (V, \{e_1, \ldots, e_m\} \cup S)$ such that (i) two edges cross only if neither of them is in S and (ii) if e_i and e_j cross, then $|i - j| \geq \omega$. As such the problem is easily seen to be a special case of the WEAK REALIZABILITY Problem, which given a graph $G = (V, E)$ and a symmetric relation $R \subseteq E \times E$ asks whether there exists a topological drawing of G such that no pair of edges in R crosses. It follows that SP and SPB are contained in $\mathcal{NP}$ [12]. For $\omega = 1$, the problem amounts to finding a drawing of $G_\cup$, where a subset of the edges, namely the edges of S, are not crossed. This problem has recently been studied under the name PARTIAL PLANARITY [1,11]. Angelini et al. [1] mostly focus on straight-line drawings, but they also note that the topological variant can be solved efficiently if the non-crossing edges form a 2-connected graph. Recently Schaefer [11] gave an $O((nm)^3)$-time testing algorithm for the general case of PARTIAL PLANARITY via a Hanani-Tutte-style approach. He further suggests to view the relation R of an instance of WEAK REALIZABILITY as a conflict graph on the edges of the input graph and to study the complexity subject to structural constraints on this conflict graph.

Our Contributions. In this work, we study the complexity of the SP and SPB Problems. In particular, we show the following results.

1. SPB is $\mathcal{NP}$-complete for all $\omega \geq 2$ when the backbone graph is a spanning tree.
2. There is a constant ω_0 such that SP with window size ω_0 is $\mathcal{NP}$-complete.
3. We give an efficient algorithm with running time $O(n + \omega m)$ for SPB when the backbone graph consists of one 2-connected component plus, possibly, isolated vertices and no stream edge connects two isolated vertices.
4. We give an efficient algorithm for SPB with running time $O(n + m)$ for $\omega = 1$.

It is worth pointing out that the second hardness result shows that WEAK REALIZABILITY is $\mathcal{NP}$-complete even if the conflict graph describing the non-crossing pairs of edges has bounded degree, i.e., every edge may not be crossed only by a constant number of other edges. In particular, this rules out the existence of FPT algorithms with respect to the maximum degree of the conflict graph unless $\mathcal{P} = \mathcal{NP}$.

For the positive results, note that the structural restrictions on the variant for arbitrary values of ω are necessary to overcome the two hardness results and are hence, in a sense, best possible. Moreover, the algorithm for $\omega = 1$ improves the previously best algorithm for PARTIAL PLANARITY by Schaefer [11] (with running time $O((nm)^3)$-time) to linear. Again, since the problem is hard for all $\omega \geq 2$, this result is tight.

For space limitations, some proofs are sketched; refer to [7] for complete proofs.

2   Preliminaries

For standard terminology about graphs, drawings, and embeddings refer to [8].

Given a $(k-1)$-connected graph G with $k \geq 1$, we denote by $k(G)$ the number of its maximal k-connected subgraphs. The maximal 2-connected subgraphs are called *blocks*. Also, a k-connected component is *trivial* if it consists of a single vertex. Further, given a simply connected graph G, that is $1(G) = 1$, the *block-cutvertex tree* T of G is the tree whose nodes are the cutvertices and the blocks of G, and whose edges connect nodes representing cutvertices with nodes representing the blocks they belong to.

Contracting an edge (u, v) in a graph G is the operation of first removing (u, v) from G, then identifying u and v to a new vertex w, and finally removing multi-edges.

Let G be a planar graph and let $\mathcal{E}$ be a planar embedding of G. Further, let H be a subgraph of G. We denote by $\mathcal{E}|_H$ the embedding of H determined by $\mathcal{E}$.

Let $\langle G_i(V, E_i) \rangle_{i=1}^k$ be k planar graphs on the same set V of vertices. A *simultaneous embedding with fixed edges (SEFE)* of graphs $\langle G(V, E_i) \rangle_{i=1}^k$ consists of k planar embeddings $\langle \mathcal{E}_i \rangle_{i=1}^k$ such that $\mathcal{E}_i|_{G_{ij}} = \mathcal{E}_j|_{G_{ij}}$, with $G_{ij} = (V, E_i \cap E_j)$ for $i \neq j$. The SEFE Problem corresponds to the problem of deciding whether the k input graphs admit a *SEFE*. Further, if all graphs share the same set of edges (*sunflower intersection*), that is, the graph $G_\cap = (V, E_i \cap E_j)$ is the same for every i and j, with $1 \leq i < j \leq k$, the problem is called SUNFLOWER SEFE and graph $G_\cap$ is the *common graph*.

In the following, we denote a streamed graph by a triple $\langle G(V, S), E, \Psi \rangle$ such that $G(V, S)$ is a planar graph, called *backbone graph*, $E \subseteq V^2 \setminus S$ is the set of edges of a stream $e_1, e_2, \ldots, e_m$, and $\Psi : E \leftrightarrow \{1, \ldots, m\}$ is a bijective function that encodes the ordering of the edges of the stream.

Given an instance $I = \langle G(V, S), E, \Psi \rangle$, we call graph $G_\cup = (V, S \cup E)$ the *union graph* of I. Observe that, if $G_\cup$ has k connected components, then I can be efficiently decomposed into k independent smaller instances, whose Stream Planarity can be tested independently. Hence, in the following we will only consider streamed graphs with connected union graph. Also, we denote by $\mathcal{Q}$ the set of isolated vertices of G.

Note that, an obvious necessary condition for a streamed graph $\langle G(V, S), E, \Psi \rangle$ to admit an ω-SDB is the existence of a planar combinatorial embedding $\mathcal{E}$ of the backbone graph G such that the endpoints of each edge of the stream lie on the boundary of the same face of $\mathcal{E}$, as otherwise a crossing between an edge of the stream and an edge of G would occur. However, since each edge of the stream must be represented by the same curve at each time, this condition is generally not sufficient, unless $\omega = 1$; see Fig. 1.

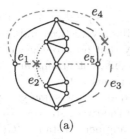
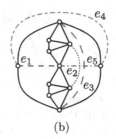

(a)   (b)

Fig. 1. Illustration of an instance $\langle G(V,S), E, \Psi \rangle$ of SPB with $\omega = 2$, where G is a 2-connected graph, $E = \{e_i : 1 \le i \le 5\}$, and $\Psi(e_i) = i$. Solid edges belong to G. (a) and (b) show different embeddings of G and assignments of the edges in E to the faces of such embeddings. (a) determines a 2-SDB of $\langle G(V,S), E, \Psi \rangle$, while (b) does not.

3   Complexity

In the following we study the computational complexity of testing planarity of streamed graphs with and without a backbone graph. First, we show that SPB is $\mathcal{NP}$-complete, even when the backbone graph is a spanning tree and $\omega = 2$. This implies that SUNFLOWER SEFE is $\mathcal{NP}$-complete for an arbitrary number of input graphs, even if every graph contains at most $\xi = 2$ exclusive edges. Second, we show that SP is $\mathcal{NP}$-complete even for a constant window size ω. This also has connections to the fundamental WEAK REALIZABILITY Problem. Namely, Theorem 2 implies the $\mathcal{NP}$-completeness of WEAK REALIZABILITY even for instances $\langle G(V,E), R \rangle$ such that the maximum number of occurrences θ of each edge of E in the pairs of edges in R is bounded by a constant, i.e., for each edge there is only a constant number θ of other edges it may not cross.

These results imply that, unless P=NP, no FPT algorithm with respect to ω, to ξ, or to θ exists for STREAM PLANARITY (WITH BACKBONE), SEFE, and WEAK REALIZABILITY Problems, respectively.

Theorem 1. SPB is $\mathcal{NP}$-complete for $\omega \ge 2$, even when the backbone graph is a tree and the edges of the stream form a matching.

Sketch of Proof. The membership in $\mathcal{NP}$ follows from [12]. The $\mathcal{NP}$-hardness is proved by means of a polynomial-time reduction from problem SUNFLOWER SEFE, which has been proved $\mathcal{NP}$-complete for $k = 3$ graphs, even when the common graph is a tree T and the exclusive edges of each graph only connect leaves of the tree [2]. As shown in [3], we can also assume that the exclusive edges of graphs G_i, with $i \in \{1, 2, 3\}$, form a matching. Given an instance $\langle G_i(V, E_i) \rangle_{i=1}^{3}$ of SUNFLOWER SEFE, we construct a streamed graph $\langle G(V,S), E, \Psi \rangle$ that admits an ω-SDB for $\omega = 2$ if and only if $\langle G_i(V, E_i) \rangle_{i=1}^{3}$ is a positive instance of SUNFLOWER SEFE, as follows. The backbone graph can be obtained from T by adding a set $L(v)$ of $|E_i| - 1$ new leaves as children of each leaf vertex v of T that is incident to an edge in E_i. Then, for each pair of edges (a, b) and (c, d) in E_i, the stream is presented with an edge whose endpoints belong to $L(a)$ and $L(b)$ immediately followed by an edge whose endpoints belong to $L(c)$ and $L(d)$. Further, dummy vertices and dummy edges can be added to $\langle G(V,S), E, \Psi \rangle$, in such a way that no two edges of the stream corresponding to pairs of edges belonging to different graphs are displayed at the same time. Clearly, a crossing between two consecutive edges of the stream corresponds to a crossing between exclusive edges of the same graph, and vice versa. Hence, proving the equivalence between the two instances. To extend the theorem to any value of $\omega \geq 2$, it suffices to augment instance $\langle G(V,S), E, \Psi \rangle$ with additional dummy vertices and dummy edges.   □

Theorem 2. *There is a constant ω_0 such that deciding whether a given streamed graph is ω_0-stream planar is $\mathcal{NP}$-complete.*

Sketch of Proof. The membership in $\mathcal{NP}$ follows from [12]. In the following we describe a reduction that, given a 3-SAT formula φ, produces a streamed graph that is ω_0-stream planar if and only if φ is satisfiable.

For simplicity, we do not describe the stream, but rather important keyframes. Our construction has the property that edges have a FIFO behavior, i.e., if edge e appears before edge f, then e also disappears before f. This, together with the fact that in each key frame only $O(1)$ edges are visible ensures that the construction can indeed be encoded as a stream with $\omega_0 \in O(1)$. The value ω_0 we use is simply the maximum number of visible edges in any of the key frames. Even though we do not take steps to further minimize ω_0, the value produced by the reduction is less than 120. Sometimes, we wish to wait until a certain set of edges has disappeared. In this case we add to the stream sufficiently many isolated edges, i.e., edges whose endpoints are otherwise unused in the construction. Clearly, this does not change the ω_0-planarity of the stream.

We now sketch the construction, which consists of two main pieces. The first is a *cage* providing two faces, called *cells*, one for vertices representing satisfied literals and one for vertices representing unsatisfied literals. We then present a clause stream for each clause of φ. It contains one literal vertex for each literal occurring in the clause and it ensures that these literal vertices are distributed to the two cells of the cage such that at least one goes in the cell for satisfied literals. Throughout we ensure that none of the previously distributed vertices leave their respective cells.

Second, we present a sequence of edges that is ω_0-stream planar if and only if the previously chosen distribution of the literal vertices forms a truth assignment, i.e., any two vertices representing the same literal are in the same cell and any two vertices representing complementary literals of one variable are in distinct cells.

It is clear that, if the constructions work as described, then the resulting streamed graph is ω_0-stream planar if and only if φ is satisfiable. The first part of the stream ensures that from each clause one of the literals must be assigned to the cell containing satisfied literals (i.e. the literal receives the value true). The second part ensures that these choices are consistent over all literals, i.e., these choices actually correspond to a truth assignment of the variables.

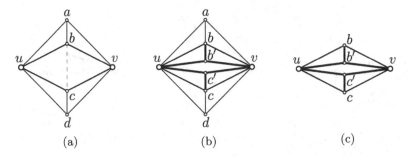

Fig. 2. A persistent edge. The thinner the edges the earlier they leave the window. (a) The initial configuration; the dashed edge bc dissolves first, enforcing a unique planar embedding. (b) New edges incident to b' and c' are introduced. (c) After the edges incident to a and d disappear, the drawing has again the same structure as in (a).

First of all, we need a construction for a gadget that behaves like an edge that persists over time and does not allow vertices to traverse it over time. Such a *persistent edge* is shown in Fig. 2. By including in the stream the dashed edges, which are incident to new vertices, a persistent edge is *renewed*. After the edges incident to a and d leave the sliding window, the gadget still looks the same. Since new edges are only added in the interior of the gadget vertices that are embedded outside the gadget cannot traverse the persistent edge. For simplicity we will not describe in detail when to perform this book keeping. Rather, we just assume that the sliding window is sufficiently large to allow regular book keeping. For example, before each of the steps described later, we might first update all persistent edges, then present the gadget performing one of the steps, then update the persistent edges again, and finally wait for the gadget edges to be removed from the sliding window again.

Next, we describe the cage, which is shown in Fig. 3a, where the gray edges are persistent. The interior faces f^+ and f^- of the two cycles are the positive and negative *literal faces*, respectively. Note that at any point in time only a constant number of edges are necessary for the cage. Before we describe the clause gadget, which is the most involved part of the construction, we sketch how to perform the test for the end of sequence. Namely, assume that we have a set $V' \subseteq V$ of literal vertices, and each of them is contained in one of the two literal faces. To check whether two literal vertices are in the same face, we simply present an edge between them as shown in Fig. 3b. To check whether two literal vertices are in different literal faces, we present edges from them to all three vertices shared by the two literal faces as shown in Fig. 3c. Using this, we can easily check whether all literal vertices corresponding to the same literal are in the same face and whether any two complementary literals are in distinct faces. The clause gadget is illustrated in Fig. 4a. When the stream contains all these edges, the only embedding choice is to decide for each of the literal vertices x, y and z whether they should be embedded in the face closer to the center of the gadget (shaded light gray) or rather closer to the face on

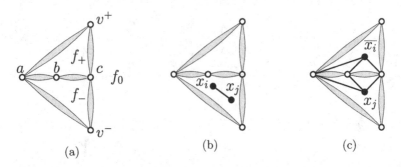

Fig. 3. (a) The cage, the thick gray edges are persistent edges. (b) Edge used to check whether two literal vertices x_i and x_j are in the same face. (c) Edges used to check whether literal vertices $\overline{x_i}$ and x_j are in distinct faces.

the boundary. The former represents truth value `true`, the latter value `false`. Attached to each literal vertex (solid) is a corresponding indicator vertex (empty disk). The edges connecting each indicator to its literal are presented first, so that they also leave the sliding window first and the indicator vertices become isolated and can start traveling to different parts of the clause gadget once some of its edges disappear. The main idea is that, after a short amount of time a triangle on the three indicators is included in the stream. By the time these edges appear in the sliding window, they must hence have managed to meet in one face of the gadget. For this, first the edges incident to the central face of the gadget disappear, allowing indicator vertices whose literals are embedded in the face close to the center to reach the shaded area shown in Fig. 4b. Afterwards, these edges are again presented in the stream. Now the thick dotted edges and afterwards the thick dashed edges of the clause disappear and reappear. This allows indicator vertices that are embedded in the face close to the boundary to travel to the regions shown in Fig 4c. Clearly they can meet in one face if and only if at least one of them was embedded close to the center, i.e., if at least one literal has value `true`. It remains to move the literal vertices to the corresponding faces of the cage, as shown in Fig. 4d,e.   □

4   Algorithms for ω-Streamed Drawings with Backbone

We now describe a linear-time algorithm to solve the SPB Problem for $\omega = 1$ with no restrictions on the backbone graph. The algorithm is based on a subprocedure to decide the SPB Problem for any value of ω in the case in which the backbone graph consists of a 2-connected component plus, possibly, isolated vertices with no edge of the stream connecting any two of them. We call instances satisfying these properties *star instances*, as the isolated vertices are the centers of star subgraphs of the union graph. As proved in Theorem 2, dropping the restriction of the absence of edges of the stream between the isolated vertices of a star instance makes the SPB Problem computationally tough.

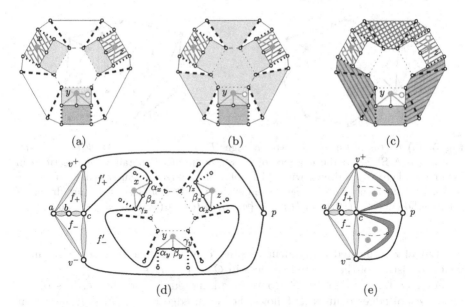

Fig. 4. Illustration of the clause sequence. (a) Initial embedding of the clause. (b), (c) faces indicator vertices can reach if they are embedded in the face close to the center and close to the boundary, respectively. (d) Separating the vertices corresponding to satisfied and unsatisfied literals into two distinct faces. (e) Integrating the now separated literal vertices into the corresponding faces of the cage.

4.1 Star Instances

In this section we describe an efficient algorithm to test the existence of an ω-SDB for star instances (see Fig. 5(a)). The problem is equivalent to finding an embedding $\mathcal{E}$ of the unique non-trivial 2-connected component β of G and an assignment of the edges of the stream and of the isolated vertices of G to the faces of $\mathcal{E}$ that yield an ω-SDB.

Lemma 1. *Let $\langle G(V,S), E, \Psi \rangle$ be a star instance of* SPB *and let ω be a positive integer window size. There exists an equivalent instance $\langle G_i(V, E_i) \rangle_{i=1}^{m+1}$ of* SUN-FLOWER SEFE *such that the common graph $G_\cap$ consists of disjoint 2-connected components. Further, instance $\langle G_i(V, E_i) \rangle_{i=1}^{m+1}$ can be constructed in $O(n + \omega m)$ time.*

Sketch of Proof. Given a star instance $\langle G(V,S), E, \Psi \rangle$ of SPB we construct an instance $\langle G_i(V, E_i) \rangle_{i=1}^{m+1}$ of SUNFLOWER SEFE that admits a SEFE if and only if $\langle G(V,S), E, \Psi \rangle$ admits an ω-SDB, as follows. See Fig 5 for an example of the construction.

Initialize graph $G_\cap$ to the backbone graph G. Also, for every edge $e \in E$, add to $G_\cap$ a set of vertices $D(e) = \{v_i(e) \mid \Psi(e) \leq i < \min(\Psi(e) + \omega, m + 1)\}$. Observe that, since $\langle G(V,S), E, \Psi \rangle$ is a star instance, graph $G_\cap$ contains a single

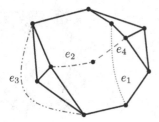
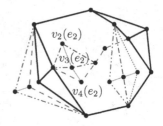

Fig. 5. (a) A star instance with stream edges $E = \{e_i : 1 \leq i \leq 4\}$, $\Psi(e_i) = i$, and $\omega = 3$. (b) A SEFE of the instance of SUNFLOWER SEFE obtained as described in Lemma 1 where $G_\cup$ is drawn with thick solid black edges, exclusive edges of G_i are drawn with the same style as edge e_i and exclusive edges of $G_{m+1} = G_5$ are drawn as yellow solid curves. Vertices in $D(e_2) = \{v_2(e_2), v_3(e_2), v_4(e_2)\}$ are also shown.

non-trivial 2-connected component β, plus a set of trivial 2-connected components consisting of the isolated vertices in $Q \cup \bigcup_{e \in E} D(e)$.

For $i = 1, \dots, m$, graph G_i contains all the edges and the vertices of $G_\cap$ plus a set of edges defined as follows. For each edge $e = (u, v) \in E$ such that $0 \leq i - \Psi(e) < \omega$, add to $E(G_i)$ edges $(u, v_i(e))$ and $(v_i(e), v)$. From a high-level view, graphs G_i, with $i = 1, \dots, m$, are defined in such a way to enforce the same constraints on the possible embeddings of the common graph as the constraints enforced by the edges of the stream on the possible embeddings of the backbone graph.

Finally, graph G_{m+1} contains all the edges and the vertices of $G_\cap$ plus a set of edges defined as follows. For each edge $e \in E$, add to E_{m+1} edges $(v_{\Psi(e)}(e), v_k(e))$, with $\Psi(e) < k < \min(\Psi(e) + \omega, m + 1)$. Observe that, in any planar drawing Γ_{m+1} of G_{m+1}, vertices $v_k(e)$ lie inside the same face of Γ_{m+1}, for any edge $e \in E$. The aim of graph G_{m+1} is to combine the constrains imposed on the embedding of the backbone graph by each graph G_i, with $i = 1, \dots, m$, in such a way that, for each edge $e \in E$, the edges of set $D(e)$ are embedded in the same face of the backbone graph.

We omit the proof of equivalence between the two instances. It is easy to see that instance $\langle G_i(V, E_i) \rangle_{i=1}^{m+1}$ can be constructed in time $O(n + \omega m)$. In fact, the construction of the common graph $G_\cap$ takes $O(n)$-time, since the backbone graph G is planar. Also, each graph G_i can be encoded as the union of a pointer to the encoding of $G_\cap$ and of the encoding of its exclusive edges. Further, each graph G_i, with $i = 1, \dots, m$, has at most ω exclusive edges, and graph G_{m+1} has at most ωm exclusive edges. This concludes the proof of the lemma. $\square$

Lemma 1 provides a straight-forward technique to decide whether a star istance $\langle G(V, S), E, \Psi \rangle$ of SPB admits an ω-SDB. First, transform instance $\langle G(V, S), E, \Psi \rangle$ into an equivalent instance $\langle G_i(V, E_i) \rangle_{i=1}^{m+1}$ of SEFE of $m + 1$ graphs with sunflower intersection and such that the common graph consists of disjoint 2-connected components, by applying the reduction described in the proof of Lemma 1. Then, apply to instance $\langle G_i(V, E_i) \rangle_{i=1}^{m+1}$ the algorithm by

Bläsius *et al.* [5] that tests instances of SEFE with the above properties in linear time. Thus, we obtain the following theorem.

Theorem 3. *Let $\langle G(V,S), E, \Psi \rangle$ be a star instance of SPB. There exists an $O(n + \omega m)$-time algorithm to decide whether $\langle G(V,S), E, \Psi \rangle$ admits an ω-SDB.*

4.2   Unit Window Size

In this section we describe a polynomial-time algorithm to test whether an instance $\langle G(V,S), E, \Psi \rangle$ of SPB admits an ω-SDB for $\omega = 1$. Observe that, in the case in which $\omega = 1$, the SPB Problem equals to the problem of deciding whether an embedding of the backbone graph exists such that the endpoints of each edge of the stream lie on the boundary of the same face of such an embedding.

Lemma 2. *Let $\langle G(V,S), E, \Psi \rangle$ be an instance of SPB. There exists a set of instances $\langle G(V_i, S_i), E_i, \Psi_i \rangle$ whose backbone graph $G(V_i, S_i)$ contains at most one non-trivial connected component $\mathcal{G}_i$ such that $\langle G(V,S), E, \Psi \rangle$ admits an ω-SDB with $\omega = 1$ if and only if all instances $\langle G(V_i, S_i), E_i, \Psi_i \rangle$ admit an ω-SDB with $\omega = 1$. Further, such instances can be constructed in $O(n + m)$ time.*

Sketch of Proof. We construct instances $\langle G(V_i, S_i), E_i, \Psi_i \rangle$ starting from $G_{\cup}$ in two steps. First, for each vertex $v \in V$, we set $l(v) = i$, if $v \in V(\mathcal{G}_i)$. Then, we recursively contract each edge (u, v) of $G_{\cup}$ with $\{u, v\} \subseteq V(\mathcal{G}_i)$ to a vertex w and set $l(w) = i$, for $i = 1, \ldots, 1(G)$. Thus, obtaining an auxiliary graph H on $1(G)$ vertices. Third, we obtain instance $\langle G(V_i, S_i), E_i, \Psi_i \rangle$ from H by recursively uncontracting each vertex w with $l(w) = i$, for $i = 1, \ldots, 1(G)$. Observe that, by construction, $\mathcal{G}_i \subseteq G(V_i, S_i)$. Further, building all instances $\langle G(V_i, S_i), E_i, \Psi_i \rangle$ sums up to $O(n + m)$ time in total.

The necessity is trivial. In order to prove the sufficiency, assume that all instances $\langle G(V_i, S_i), E_i, \Psi_i \rangle$ admit a 1-SDB. Then, a 1-SDB Γ of the original instance can be obtained, starting from a 1-SDB Γ_i of any $\langle G(V_i, S_i), E_i, \Psi_i \rangle$, by recursively replacing the drawing of each isolated vertex $v_j \in \mathcal{Q}_i$ with the 1-SDB Γ_j of $\langle G(V_j, S_j), E_j, \Psi_j \rangle$ (after, possibly, promoting a different face to be the outer face of Γ_j) . The fact that Γ is a 1-SDB of $\langle G(V,S), E, \Psi \rangle$ derives from the fact that each Γ_i is a 1-SDB of $\langle G(V_i, S_i), E_i, \Psi_i \rangle$, that in a 1-SDB crossings among edges in E do not matter, and that, by the connectivity of the union graph, the assignment of the isolated vertices in $\mathcal{Q}_i$ to the faces of the embedding $\mathcal{E}_i$ of $\mathcal{G}_i$ in Γ_i must be such that any two isolated vertices connected by a path of edges of the stream E_i lie inside the same face of $\mathcal{E}_i$. □

By Lemma 2, in the following we only consider the case in which the backbone graph consists of a single non-trivial connected component plus, possibly, isolated vertices. We now present a simple recursive algorithm to test instances with this property.

 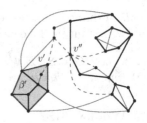

Fig. 6. (a) Instance I and (b) instance I^* obtained in *CASE R2* of Algorithm ALGO-CON. Edges of the backbone graph are black thick curves; edge of the stream are green thin curves; and edges of the stream incident to v' and v'' in I^* are blue dashed curves.

Algorithm **ALGOCON**.

○ *INPUT:* an instace $I = \langle G(V,S), E, \Psi \rangle$ of the SPB Problem with $\omega = 1$ with union graph $G_\cup$ such that G contains at most one non-trivial connected component.

○ *OUTPUT:* YES, if $\langle G(V,S), E, \Psi \rangle$ is positive, or NO, otherwise.

BASE CASE 1: instance I is such that $2(G) = 0$, that is, every connected component of G is an isolated vertex. Return YES, as instances of this kind are trivially positive.

BASE CASE 2: instance I is such that (i) $2(G) = 1$, that is, the backbone graph G consists of a single 2-connected component plus, possibly, isolated vertices and (ii) no edge of the stream connects any two isolated vertices. In this case, apply the algorithm of Theorem 3 to decide I and return YES, if the test succeeds, or NO, otherwise.

RECURSIVE STEP: instance I is such that either (CASE R1) $2(G) = 1$ and there exists edges of the stream between pairs of isolated vertices or (CASE R2) $2(G) > 1$. First, replace instance I with two smaller instances $I^\diamond = \langle G(V_\diamond, S_\diamond), E_\diamond, \Psi_\diamond \rangle$ and $I^\circ = \langle G(V_\circ, S_\circ), E_\circ, \Psi_\circ \rangle$, as described below. Then, return YES, if $ALGOCON(I^\diamond) =$
$ALGOCON(I^\circ) =$ YES, or NO, otherwise.

CASE R1. Instance $I^\diamond$ is obtained from I by recursively contracting every edge (u,v) of $G_\cup$ with $\{u,v\} \not\subseteq V(\mathcal{G})$. Instance I° is obtained from I by recursively contracting every edge (u,v) of $G_\cup$ with $\{u,v\} \subseteq V(\mathcal{G})$.

CASE R2. Let $\mathcal{G}$ be the unique non-trivial connected component of G, let T be the block-cutvertex tree of $\mathcal{G}$ rooted at any block, and let β be any leaf block in T. Also, let v be the parent cutvertex of β in T. We first construct an auxiliary equivalent instance $I^* = \langle G(V_*, S_*), E_*, \Psi_* \rangle$ starting from I and then obtain instances $I^\diamond$ and I° from I^*, as follows. See Fig. 6 for an illustration of the construction of instance I^*. Initialize I^* to I. Replace vertex v in V_* with two vertices v' and v'' and make (i) v' adjacent to all the vertices of β vertex v used to be adjacent to and (ii) v'' adjacent to all the vertices in $V(\mathcal{G}) \setminus V(\beta)$ vertex v used to be adjacent to. Then, replace each edge (v, x) of E^* with edge (v', x), if $x \in V(\beta)$ or if $x \in \mathcal{Q}^*$ and there exists a path composed of edges of the stream connecting x to a vertex $y \neq v \in V(\beta)$, and edge (v'', x), if $x \in V(\mathcal{G}) \setminus V(\beta)$ or if $x \in \mathcal{Q}^*$ and there exists a path composed of edges of the stream connecting x to a vertex $y \neq v \in V(\mathcal{G}) \setminus V(\beta)$. Finally, add edge (v', v'') to E^*. It is easy to see that instances I and I^* are equivalent.

Instance $I^\diamond$ is obtained from I^* by recursively contracting every edge (u, v) of $G_{\cup}^*$ with $u, v \nsubseteq V(\beta)$, where $G_{\cup}^*$ is the union graph of I^*. Instance I° is obtained from I^* by recursively contracting every edge (u, v) of $G_{\cup}^*$ with $\{u, v\} \subseteq V(\beta)$.

Theorem 4. *Let* $\langle G(V, S), E, \Psi \rangle$ *be an instance of* SPB. *There exists an* $O(n + m)$-*time algorithm to decide whether* $\langle G(V, S), E, \Psi \rangle$ *admits an* ω-*SDB for* $\omega = 1$.

Proof. The algorithm runs in two steps, as follows.

- **STEP** 1 applies the reduction illustrated in the proof of Lemma 2 to $\langle G(V, S), E, \Psi \rangle$ to construct $1(G)$ instances $\langle G(V_i, S_i), E_i, \Psi_i \rangle$ such that the backbone graphs $G(V_i, S_i)$ contain at most one non-trivial connected component.
- **STEP** 2 applies Algorithm ALGOCON to every instance $\langle G(V_i, S_i), E_i, \Psi_i \rangle$ and return YES, if all such instances are positive, or NO, otherwise.

Observe that, the correctness of the presented algorithm follows from the correctness of Lemma 2, of Theorem 3, and of Algorithm ALGOCON. We now prove the correctness for Algorithm ALGOCON. Obviously, the fact that instances $I^\diamond$ and I° constructed in CASE R1 and CASE R2 are both positive is a necessary and sufficient condition for instance I to be positive. We prove termination by induction on the number $2(G)$ of blocks of the backbone graph G of instance I, primarily, and on the number of edges of the stream connecting isolated vertices of the backbone graph, secondarily. (i) If $2(G) = 0$, then BASE CASE 1 applies and the algorithm stops; (ii) if $2(G) = 1$ and no two isolated vertices of the backbone graph are connected by an edge of the stream, then BASE CASE 2 applies and the algorithm stops; (iii) if $2(G) = 1$ and there exist edges of the stream between any two isolated vertices of the backbone graph G, then, by CASE R1, instance I is split into (a) an instance $I^\diamond$ with $2(G(V_\diamond, E_\diamond)) = 1$ and no edges of the stream connecting any two isolated vertices of the backbone graph $G(V_\diamond, E_\diamond)$, and (b) an instance I° with $2(G(V_\diamond, E_\diamond)) = 0$; (iv) finally, if $2(G) > 1$, then, by CASE R2, instance I is split into (a) an instance $I^\diamond$ with $2(G(V_\diamond, E_\diamond)) = 1$ and (b) an instance I° with $2(G(V_\diamond, E_\diamond)) = 2(G) - 1$.

The running time easily derives from the fact that all instances $\langle G(V_i, S_i), E_i, \Psi_i \rangle$ can be constructed in $O(n + m)$-time and that the algorithm for star instances described in the proof of Theorem 3 runs in $O(n + \omega m)$-time. This concludes the proof. $\square$

References

1. Angelini, P., Binucci, C., Da Lozzo, G., Didimo, W., Grilli, L., Montecchiani, F., Patrignani, M., Tollis, I.G.: Drawing non-planar graphs with crossing-free subgraphs. In: Wismath, S., Wolff, A. (eds.) GD 2013. LNCS, vol. 8242, pp. 292–303. Springer, Heidelberg (2013)
2. Angelini, P., Da Lozzo, G., Neuwirth, D.: Advancements on SEFE and partitioned book embedding problems. Theor. Comput. Sci. **575**, 71–89 (2015). doi:10.1016/j.tcs.2014.11.016. http://dx.doi.org/10.1016/j.tcs.2014.11.016
3. Angelini, P., Di Battista, G., Frati, F., Patrignani, M., Rutter, I.: Testing the simultaneous embeddability of two graphs whose intersection is a biconnected or a connected graph. J. Discrete Algorithms **14**, 150–172 (2012)
4. Bläsius, T., Kobourov, S.G., Rutter, I.: Simultaneous embedding of planar graphs. In: Tamassia, R. (ed.) Handbook of Graph Drawing and Visualization. CRC Press (2013)
5. Bläsius, T., Karrer, A., Rutter, I.: Simultaneous embedding: edge orderings, relative positions, cutvertices. In: Wismath, S., Wolff, A. (eds.) GD 2013. LNCS, vol. 8242, pp. 220–231. Springer, Heidelberg (2013)
6. Brandes, C.B.U., Di Battista, G., Didimo, W., Gaertler, M., Palladino, P., Patrignani, M., Symvonis, A., Zweig, K.A.: Drawing trees in a streaming model. Inf. Process. Lett. **112**(11), 418–422 (2012)
7. Da Lozzo, G., Rutter, I.: Planarity of Streamed Graphs. ArXiv e-prints 1501. 07106 (2015)
8. Di Battista, G., Eades, P., Tamassia, R., Tollis, I.G.: Graph Drawing. Upper Saddle River, NJ (1999)
9. Di Battista, G., Tamassia, R.: On-line planarity testing. SIAM J. Comput. **25**(5), 956–997 (1996)
10. Gassner, E., Jünger, M., Percan, M., Schaefer, M., Schulz, M.: Simultaneous graph embeddings with fixed edges. In: WG 2006. pp. 325–335 (2006)
11. Schaefer, M.: Picking planar edges; or, drawing a graph with a planar subgraph. In: Duncan, C., Symvonis, A. (eds.) GD 2014. LNCS, vol. 8871, pp. 13–24. Springer, Heidelberg (2014)
12. Schaefer, M., Sedgwick, E., Stefankovic, D.: Recognizing string graphs in NP. J. Comput. Syst. Sci. **67**(2), 365–380 (2003)

Clique-Width of Graph Classes Defined by Two Forbidden Induced Subgraphs

Konrad K. Dabrowski$^{(\boxtimes)}$ and Daniël Paulusma

School of Engineering and Computing Sciences, Durham University Science Laboratories, South Road, Durham DH1 3LE, UK
{konrad.dabrowski,daniel.paulusma}@durham.ac.uk

Abstract. If a graph has no induced subgraph isomorphic to any graph in a finite family $\{H_1, \ldots, H_p\}$, it is said to be $(H_1, \ldots, H_p)$-free. The class of H-free graphs has bounded clique-width if and only if H is an induced subgraph of the 4-vertex path P_4. We study the (un)boundedness of the clique-width of graph classes defined by two forbidden induced subgraphs H_1 and H_2. Prior to our study it was not known whether the number of open cases was finite. We provide a positive answer to this question. To reduce the number of open cases we determine new graph classes of bounded clique-width and new graph classes of unbounded clique-width. For obtaining the latter results we first present a new, generic construction for graph classes of unbounded clique-width. Our results settle the boundedness or unboundedness of the clique-width of the class of (H_1, H_2)-free graphs

(i) for all pairs (H_1, H_2), both of which are connected, except two non-equivalent cases, and

(ii) for all pairs (H_1, H_2), at least one of which is not connected, except 11 non-equivalent cases.

We also consider classes characterized by forbidding a finite family of graphs $\{H_1, \ldots, H_p\}$ as subgraphs, minors and topological minors, respectively, and completely determine which of these classes have bounded clique-width. Finally, we show algorithmic consequences of our results for the graph colouring problem restricted to (H_1, H_2)-free graphs.

Keywords: Clique-width · Forbidden induced subgraph · Graph class

1 Introduction

Clique-width is a well-known graph parameter studied both in a structural and in an algorithmic context; we refer to the surveys of Gurski [24] and Kamiński, Lozin and Milanič [27] for an in-depth study of the properties of clique-width. We are interested in determining whether the clique-width of some given class of

The research in this paper was supported by EPSRC (EP/G043434/1 and EP/K025090/1) and ANR (TODO ANR-09-EMER-010).

© Springer International Publishing Switzerland 2015
V.Th. Paschos and P. Widmayer (Eds.): CIAC 2015, LNCS 9079, pp. 167–181, 2015.
DOI: 10.1007/978-3-319-18173-8_12

graphs is *bounded*, that is, is there a constant c such that every graph from the class has clique-width at most c. For this purpose we study classes of graphs in which one or more specified graphs are forbidden as a "pattern". In particular, we consider classes of graphs that contain no graph from some specified family $\{H_1, \ldots, H_p\}$ as an *induced subgraph*; such classes are said to be $(H_1, \ldots, H_p)$-free. Our research is well embedded in the literature, as there are many papers that determine the clique-width of graph classes characterized by one or more forbidden induced subgraphs; see e.g. [1–10, 14–17, 23, 31–34].

As we show later, it is not difficult to verify that the class of H-free graphs has bounded clique-width if and only if H is an induced subgraph of the 4-vertex path P_4. Hence, it is natural to consider the following problem:

For which pairs (H_1, H_2) does the class of (H_1, H_2)-free graphs have bounded clique-width?

In this paper we address this question by narrowing the gap between the known and open cases significantly; in particular we show that the number of open cases is finite. We emphasise that the *underlying* research question is: what kind of properties of a graph class ensure that its clique-width is bounded? Our paper is to be interpreted as a further step towards this direction, and in our research project (see also [3, 15, 17]) we aim to develop general techniques for attacking a number of the open cases simultaneously.

Algorithmic Motivation. For problems that are NP-complete in general, one naturally seeks to find subclasses of graphs on which they are tractable, and graph classes of bounded clique-width have been studied extensively for this purpose, as we discuss below.

Courcelle, Makowsky and Rotics [13] showed that all MSO_1 graph problems, which are problems definable in Monadic Second Order Logic using quantifiers on vertices but not on edges, can be solved in linear time on graphs with clique-width at most c, provided that a c-expression of the input graph is given. Later, Espelage, Gurski and Wanke [19], Kobler and Rotics [28] and Rao [43] proved the same result for many non-MSO_1 graph problems. Although computing the clique-width of a given graph is NP-hard, as shown by Fellows, Rosamond, Rotics and Szeider [20], it is possible to find an $(8^c - 1)$-expression for any n-vertex graph with clique-width at most c in cubic time. This is a result of Oum [38] after a similar result (with a worse bound and running time) had already been shown by Oum and Seymour [39]. Hence, the NP-complete problems considered in the aforementioned papers [13, 19, 28, 43] are all polynomial-time solvable on any graph class of bounded clique-width even if no c-expression of the input graph is given.

As a consequence of the above, when solving an NP-complete problem on some graph class $\mathcal{G}$, it is natural to try to determine *first* whether the clique-width of $\mathcal{G}$ is bounded. In particular this is the case if we aim to determine the computational complexity of some NP-complete problem when restricted to graph classes characterized by some common type of property. This property may be the absence of a family of forbidden induced subgraphs $H_1, \ldots, H_p$ and we may want to classify for which families of graphs $H_1, \ldots, H_p$ the problem is still NP-hard and for which ones it becomes polynomial-time solvable (in order to increase our understanding of the hardness of the problem in general). We give examples later.

Our Results. In Section 2 we state a number of basic results on clique-width and two results on H-free bipartite graphs that we showed in a very recent paper [17]; we need these results for proving our new results. We then identify a number of new classes of (H_1, H_2)-free graphs of bounded clique-width (Section 3) and unbounded clique-width (Section 4). In particular, the new unbounded cases are obtained from a new, general construction for graph classes of unbounded clique-width. In Section 5, we first observe for which graphs H_1 the class of H_1-free graphs has bounded clique-width. We then present our main theorem that gives a summary of our current knowledge of those pairs (H_1, H_2) for which the class of (H_1, H_2)-free graphs has bounded clique-width and unbounded clique-width, respectively.[1] In this way we are able to narrow the gap to 13 open cases (up to some equivalence relation, which we explain later); when we only consider pairs (H_1, H_2) of connected graphs the number of non-equivalent open cases is only two. In order to present our summary, we will need several results from the papers listed above. We also consider graph classes characterized by forbidding a finite family of graphs $\{H_1, \ldots, H_p\}$ as subgraphs, minors and topological minors, respectively. For these containment relations we are able to completely determine which of these classes have bounded clique-width.

Algorithmic Consequences. Our results are of interest for any NP-complete problem that is solvable in polynomial time on graph classes of bounded clique-width. In Section 6 we give a concrete application of our results by considering the well-known COLOURING problem, which is that of testing whether a graph can be coloured with at most k colours for some given integer k and which is solvable in polynomial time on any graph class of bounded clique-width [28]. The complexity of COLOURING has been studied extensively for (H_1, H_2)-free graphs [14,16,22,29,35,44], but a full classification is still far from being settled. Many of the polynomial-time results follow directly from bounding the clique-width in such classes. As such this forms a direct motivation for our research.

Related Work. We finish this section by briefly discussing one related result. A graph class $\mathcal{G}$ has power-bounded clique-width if there is a constant r so that the class consisting of all r-th powers of all graphs from $\mathcal{G}$ has bounded clique-width. Recently, Bonomo, Grippo, Milanič and Safe [2] determined all pairs of connected graphs H_1, H_2 for which the class of (H_1, H_2)-free graphs has power-bounded clique-width. If a graph class has bounded clique-width, it

---

[1] Before finding the combinatorial proof of our main theorem we first obtained a computer-assisted proof using Sage [46] and the Information System on Graph Classes and their Inclusions [18] (which keeps a record of classes for which boundedness or unboundedness of clique-width is known). In particular, we would like to thank Nathann Cohen and Ernst de Ridder for their help.

has power-bounded clique-width. However, the reverse implication does not hold in general. The latter can be seen as follows. Bonomo et al. [2] showed that the class of H-free graphs has power-bounded clique-width if and only if H is a linear forest (recall that such a class has bounded clique-width if and only if H is an induced subgraph of P_4). Their classification for connected graphs H_1, H_2 is the following. Let $S_{1,i,j}$ be the graph obtained from a 4-vertex star by subdividing one leg $i - 1$ times and another leg $j - 1$ times. Let $T_{1,i,j}$ be the line graph of $S_{1,i,j}$. Then the class of (H_1, H_2)-free graphs has power-bounded clique-width if and only if one of the following two cases applies: (i) one of H_1, H_2 is a path or (ii) one of H_1, H_2 is isomorphic to $S_{1,i,j}$ for some $i, j \geq 1$ and the other one is isomorphic to $T_{1,i',j'}$ for some $i', j' \geq 1$. In particular, the classes of power-unbounded clique-width were already known to have unbounded clique-width.

2   Preliminaries

Below we define the graph terminology used throughout our paper. Let G be a graph. The set $N(u) = \{v \in V(G) \mid uv \in E(G)\}$ is the *(open) neighbourhood* of $u \in V(G)$ and $N[u] = N(u) \cup \{u\}$ is the *closed neighbourhood* of $u \in V(G)$. The *degree* of a vertex in a graph is the size of its neighbourhood. The *maximum degree* of a graph is the maximum vertex degree. For a subset $S \subseteq V(G)$, we let $G[S]$ denote the subgraph of G *induced* by S, which has vertex set S and edge set $\{uv \mid u, v \in S, uv \in E(G)\}$. If $S = \{s_1, \ldots, s_r\}$ then, to simplify notation, we may also write $G[s_1, \ldots, s_r]$ instead of $G[\{s_1, \ldots, s_r\}]$. Let H be another graph. We write $H \subseteq_i G$ to indicate that H is an induced subgraph of G.

Let $\{H_1, \ldots, H_p\}$ be a set of graphs. We say that a graph G is $(H_1, \ldots, H_p)$-*free* if G has no induced subgraph isomorphic to a graph in $\{H_1, \ldots, H_p\}$. If $p = 1$, we may write H_1-free instead of (H_1)-free. The *disjoint union* $G + H$ of two vertex-disjoint graphs G and H is the graph with vertex set $V(G) \cup V(H)$ and edge set $E(G) \cup E(H)$. We denote the disjoint union of r copies of G by rG.

For positive integers s and t, the *Ramsey number* $R(s, t)$ is the smallest number n such that all graphs on n vertices contain an independent set of size s or a clique of size t. Ramsey's Theorem [40] states that such a number exists for all positive integers s and t.

The *clique-width* of a graph G, denoted $\mathrm{cw}(G)$, is the minimum number of labels needed to construct G by using the following four operations:

1. creating a new graph consisting of a single vertex v with label i (denoted by $i(v)$);
2. taking the disjoint union of two labelled graphs G_1 and G_2 (denoted by $G_1 \oplus G_2$);
3. joining each vertex with label i to each vertex with label j ($i \neq j$, denoted by $\eta_{i,j}$);
4. renaming label i to j (denoted by $\rho_{i \to j}$).

An algebraic term that represents such a construction of G and uses at most k labels is said to be a k-*expression* of G (i.e. the clique-width of G is the minimum k for which G has a k-expression). For instance, an induced path on four consecutive vertices a, b, c, d has clique-width equal to 3, and the following 3-expression can be used to construct it:

$$\eta_{3,2}(3(d) \oplus \rho_{3\to2}(\rho_{2\to1}(\eta_{3,2}(3(c) \oplus \eta_{2,1}(2(b) \oplus 1(a)))))).$$

Alternatively, any k-expression for a graph G can be represented by a rooted tree, where the leaves correspond to the operations of vertex creation and the internal nodes correspond to the other three operations. The rooted tree representing the above k-expression is depicted in Fig. 1. A class of graphs $\mathcal{G}$ has *bounded* clique-width if there is a constant c such that the clique-width of every graph in $\mathcal{G}$ is at most c; otherwise the clique-width of $\mathcal{G}$ is *unbounded*.

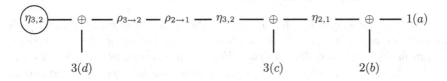

Fig. 1. The rooted tree representing a 3-expression for P_4

Let G be a graph. The *complement* of G, denoted by $\overline{G}$, has vertex set $V(\overline{G}) = V(G)$ and an edge between two distinct vertices if and only if these vertices are not adjacent in G.

Let G be a graph. We define the following five operations. The *contraction* of an edge uv removes u and v from G, and replaces them by a new vertex made adjacent to precisely those vertices that were adjacent to u or v in G. By definition, edge contractions create neither self-loops nor multiple edges. The *subdivision* of an edge uv replaces uv by a new vertex w with edges uw and vw. Let $u \in V(G)$ be a vertex that has exactly two neighbours v, w, and moreover let v and w be non-adjacent. The *vertex dissolution* of u removes u and adds the edge vw. For an induced subgraph $G' \subseteq_i G$, the *subgraph complementation* operation (acting on G with respect to G') replaces every edge present in G' by a non-edge, and vice versa. Similarly, for two disjoint vertex subsets X and Y in G, the *bipartite complementation* operation with respect to X and Y acts on G by replacing every edge with one end-vertex in X and the other one in Y by a non-edge and vice versa.

We now state some useful facts for dealing with clique-width. We will use these facts throughout the paper. Let $k \geq 0$ be a constant and let γ be some graph operation. We say that a graph class $\mathcal{G}'$ is (k, γ)-*obtained* from a graph class $\mathcal{G}$ if the following two conditions hold:

(i) every graph in $\mathcal{G}'$ is obtained from a graph in $\mathcal{G}$ by performing γ at most k times, and

(ii) for every $G \in \mathcal{G}$ there exists at least one graph in $\mathcal{G}'$ obtained from G by performing γ at most k times.

If we do not impose a finite upper bound k on the number of applications of γ then we write that $\mathcal{G}'$ is (∞, γ)-obtained from $\mathcal{G}$.

We say that γ *preserves* boundedness of clique-width if for any finite constant k and any graph class $\mathcal{G}$, any graph class $\mathcal{G}'$ that is (k, γ)-obtained from $\mathcal{G}$ has bounded clique-width if and only if $\mathcal{G}$ has bounded clique-width.

Fact 1. Vertex deletion preserves boundedness of clique-width [31].

Fact 2. Subgraph complementation preserves boundedness of clique-width [27].

Fact 3. Bipartite complementation preserves boundedness of clique-width [27].

Fact 4. For a class of graphs $\mathcal{G}$ of *bounded* maximum degree, let $\mathcal{G}'$ be a class of graphs that is $(\infty, \mathbf{es})$-obtained from $\mathcal{G}$, where $\mathbf{es}$ is the edge subdivision operation. Then $\mathcal{G}$ has bounded clique-width if and only if $\mathcal{G}'$ has bounded clique-width [27].

For $r \geq 1$, the graphs C_r, K_r, P_r denote the cycle, complete graph and path on r vertices, respectively, and the graph $K_{1,r}$ denotes the star on $r+1$ vertices. The graph $K_{1,3}$ is also called the *claw*. For $1 \leq h \leq i \leq j$, let $S_{i,j,k}$ denote the tree that has only one vertex x of degree 3 and that has exactly three leaves, which are of distance i, j and k from x, respectively. Observe that $S_{1,1,1} = K_{1,3}$. A graph $S_{i,j,k}$ is said to be a *subdivided claw*. We let $\mathcal{S}$ be the class of graphs each connected component of which is either a subdivided claw or a path.

The following lemma is well known.

Lemma 1 ([32]). *Let $\{H_1, \ldots, H_p\}$ be a finite set of graphs. If $H_i \notin \mathcal{S}$ for $i = 1, \ldots, p$ then the class of $(H_1, \ldots, H_p)$-free graphs has unbounded clique-width.*

We say that G is *bipartite* if its vertex set can be partitioned into two (possibly empty) independent sets B and W. We say that (B, W) is a *bipartition* of G. Lozin and Volz [33] characterized all bipartite graphs H for which the class of strongly H-free bipartite graphs has bounded clique-width (see [17] for the definition of strongly). Recently, we proved a similar characterization for H-free bipartite graphs; we will use this result in Section 5.

Lemma 2 ([17]). *Let H be a graph. The class of H-free bipartite graphs has bounded clique-width if and only if one of the following cases holds: $H = sP_1$ for some $s \geq 1$, $H \subseteq_i K_{1,3} + 3P_1$, $H \subseteq_i K_{1,3} + P_2$, $H \subseteq_i P_1 + S_{1,1,3}$, or $H \subseteq_i S_{1,2,3}$.*

From the same paper we will also need the following lemma.

Lemma 3 ([17]). *Let $H \in \mathcal{S}$. Then H is $(2P_1 + 2P_2, 2P_1 + P_4, 4P_1 + P_2, 3P_2, 2P_3)$-free if and only if $H = sP_1$ for some integer $s \geq 1$ or H is an induced subgraph of one of the graphs in $\{K_{1,3} + 3P_1, K_{1,3} + P_2, P_1 + S_{1,1,3}, S_{1,2,3}\}$.*

We say that a graph G is *complete multipartite* if $V(G)$ can be partitioned into k independent sets $V_1, \ldots, V_k$ for some integer k, such that two vertices are adjacent if and only if they belong to two different sets V_i and V_j. The next result is due to Olariu [37] (the graph $\overline{P_1 + P_3}$ is also called the *paw*).

Lemma 4 ([37]). *Every connected $(\overline{P_1 + P_3})$-free graph is either complete multipartite or K_3-free.*

Every complete multipartite graph has clique-width at most 2. Also, the definition of clique-width directly implies that the clique-width of any graph is equal to the maximum clique-width of its connected components. Hence, Lemma 4 immediately implies the following (well-known) result.

Lemma 5. *For any graph H, the class of $(\overline{P_1 + P_3}, H)$-free graphs has bounded clique-width if and only if the class of (K_3, H)-free graphs has bounded clique-width.*

Kratsch and Schweitzer [30] proved that the GRAPH ISOMORPHISM problem is graph-isomorphism-complete for the class of $(K_4, P_1 + P_4)$-free graphs. It is a straightforward exercise to simplify their construction and use analogous arguments to prove that the class of $(K_4, P_1 + P_4)$-free graphs has unbounded clique-width. Very recently, Schweitzer [45] showed that any graph class that allows a so-called simple path encoding has unbounded clique-width, implying this result as a direct consequence.

Lemma 6 ([45]). *The class of $(K_4, P_1 + P_4)$-free graphs has unbounded clique-width.*

3   New Classes of Bounded Clique-Width

In this section we identify two new graph classes that have bounded clique-width, namely the classes of $(\overline{P_1 + P_3}, P_1 + S_{1,1,2})$-free graphs and $(\overline{P_1 + P_3}, K_{1,3} + 3P_1)$-free graphs. We omit the proofs of both these results. The proof of the first result uses a similar approach to that used by Dabrowski, Lozin, Raman and Ries [16] to prove that the classes of $(K_3, S_{1,1,3})$-free and $(K_3, K_{1,3} + P_2)$-free graphs have bounded clique-width.

Theorem 1. *The class of $(\overline{P_1 + P_3}, P_1 + S_{1,1,2})$-free graphs has bounded clique-width.*

Theorem 2. *The class of $(\overline{P_1 + P_3}, K_{1,3} + 3P_1)$-free graphs has bounded clique-width.*

4   New Classes of Unbounded Clique-Width

In order to prove our results, we first present a general construction for obtaining graph classes of unbounded clique-width. We then use our construction to obtain two new classes of unbounded clique-width. Our construction generalizes the constructions used by Golumbic and Rotics [23],[2] Brandstädt et al. [4] and Lozin and Volz [33] to prove that the classes of square grids, K_4-free co-chordal graphs and $2P_3$-free graphs, respectively, have unbounded clique-width.

---

[2] The class of (square) grids was first shown to have unbounded clique-width by Makowsky and Rotics [34]. The construction of [23] determines the exact clique-width of square grids and narrows the clique-width of non-square grids to two values.

Theorem 3. *For $m \geq 0$ and $n > m+1$ the clique-width of a graph G is at least $\lfloor \frac{n-1}{m+1} \rfloor + 1$ if $V(G)$ has a partition into sets $V_{i,j}(i,j \in \{0,\ldots,n\})$ with the following properties:*

1. $|V_{i,0}| \leq 1$ *for all* $i \geq 1$.
2. $|V_{0,j}| \leq 1$ *for all* $j \geq 1$.
3. $|V_{i,j}| \geq 1$ *for all* $i,j \geq 1$.
4. $G[\cup_{j=0}^{n} V_{i,j}]$ *is connected for all* $i \geq 1$.
5. $G[\cup_{i=0}^{n} V_{i,j}]$ *is connected for all* $j \geq 1$.
6. *For* $i,j,k \geq 1$, *if a vertex of* $V_{k,0}$ *is adjacent to a vertex of* $V_{i,j}$ *then* $i \leq k$.
7. *For* $i,j,k \geq 1$, *if a vertex of* $V_{0,k}$ *is adjacent to a vertex of* $V_{i,j}$ *then* $j \leq k$.
8. *For* $i,j,k,\ell \geq 1$, *if a vertex of* $V_{i,j}$ *is adjacent to a vertex of* $V_{k,\ell}$ *then* $|k-i| \leq m$ *and* $|\ell - j| \leq m$.

Proof. Fix integers n,m with $m \geq 0$ and $n > m+1$, and let G be a graph with a partition as described above. For $i > 0$ we let $R_i = \cup_{j=0}^{n} V_{i,j}$ be a *row* of G and for $j > 0$ we let $C_j = \cup_{i=0}^{n} V_{i,j}$ be a *column* of G. Note that $G[R_i]$ and $G[C_j]$ are non-empty by Property 3. They are connected graphs by Properties 4 and 5, respectively.

Consider a k-expression for G. We will show that $k \geq \lfloor \frac{n-1}{m+1} \rfloor + 1$. As stated in Section 2, this k-expression can be represented by a rooted tree T, whose leaves correspond to the operations of vertex creation and whose internal nodes correspond to the other three operations (see Fig. 1 for an example). We denote the subgraph of G that corresponds to the subtree of T rooted at node x by $G(x)$. Note that $G(x)$ may not be an induced subgraph of G as missing edges can be added by operations corresponding to $\eta_{i,j}$ nodes higher up in T.

Let x be a deepest (i.e. furthest from the root) $\oplus$ node in T such that $G(x)$ contains an entire row or an entire column of G (the node x may not be unique). Let y and z be the children of x in T. Colour all vertices in $G(y)$ blue and all vertices in $G(z)$ red. Colour all remaining vertices of G yellow. Note that a vertex of G appears in $G(x)$ if and only if it is coloured either red or blue and that there is no edge in $G(x)$ between a red and a blue vertex. Due to our choice of x, G contains a row or a column none of whose vertices are yellow, but no row or column of G is entirely blue or entirely red. Without loss of generality, assume that G contains a non-yellow column.

Because G contains a non-yellow column, each row of G contains a non-yellow vertex, by Property 3. Since no row is entirely red or entirely blue, every row of G is therefore coloured with at least two colours. Let R_i be an arbitrary row. Since $G[R_i]$ is connected, there must be two adjacent vertices $v_i, w_i \in R_i$ in G, such that v_i is either red or blue and w_i has a different colour than v_i. Note that v_i and w_i are therefore not adjacent in $G(x)$ (recall that if w_i is yellow then it is not even present as a vertex of $G(x)$).

Now consider indices $i,k \geq 1$ with $k > i+m$. By Properties 6 and 8, no vertex of R_i is adjacent to a vertex of $R_k \setminus V_{k,0}$ in G. Therefore, since $|V_{k,0}| \leq 1$ by Property 1, we conclude that either v_i and w_i are not adjacent to v_k in G, or v_i and w_i

are not adjacent to w_k in G. In particular, this implies that w_i is not adjacent to v_k in G or that w_k is not adjacent to v_i in G. Recall that v_i and w_i are adjacent in G but not in $G(x)$, and the same holds for v_k and w_k. Hence, a $\eta_{i,j}$ node higher up in the tree, makes w_i adjacent to v_i but not to v_k, or makes w_k adjacent to v_k but not to v_i. This means that v_i and v_k must have different labels in $G(x)$. We conclude that $v_1, v_{(m+1)+1}, v_{2(m+1)+1}, v_{3(m+1)+1}, \ldots, v_{(\lfloor \frac{n-1}{m+1} \rfloor)(m+1)+1}$ must all have different labels in $G(x)$. Hence, the k-expression of G uses at least $\lfloor \frac{n-1}{m+1} \rfloor + 1$ labels. □

We now use Theorem 3 to determine two new graph classes that have unbounded clique-width. We omit the proofs.

Theorem 4. *The class of* $(P_6, \overline{2P_1 + P_2})$*-free graphs has unbounded clique-width.*

Theorem 5. *The class of* $(3P_2, P_2 + P_4, P_6, \overline{P_1 + P_4})$*-free graphs has unbounded clique-width.*

5 Classifying Classes of (H_1, H_2)-Free Graphs

In this section we study the boundedness of clique-width of classes of graphs defined by two forbidden induced subgraphs. Recall that this study is partially motivated by the fact that it is easy to obtain a full classification for the boundedness of clique-width of graph classes defined by one forbidden induced subgraph, as shown in the next theorem (we omit the proof). This classification does not seem to have previously been explicitly stated in the literature.

Theorem 6. *Let H be a graph. The class of H-free graphs has bounded clique-width if and only if H is an induced subgraph of P_4.*

We are now ready to study classes of graphs defined by two forbidden induced subgraphs. Given four graphs H_1, H_2, H_3, H_4, we say that the class of (H_1, H_2)-free graphs and the class of (H_3, H_4)-free graphs are *equivalent* if the unordered pair H_3, H_4 can be obtained from the unordered pair H_1, H_2 by some combination of the following operations:

1. complementing both graphs in the pair;
2. if one of the graphs in the pair is K_3, replacing it with $\overline{P_1 + P_3}$ or vice versa.

By Fact 2 and Lemma 5, if two classes are equivalent then one has bounded clique-width if and only if the other one does. Given this definition, we can now classify all classes defined by two forbidden induced subgraphs for which it is known whether or not the clique-width is bounded. This includes both the already-known results and our new results. We will later show that (up to equivalence) this leaves only 13 open cases.

Theorem 7. *Let $\mathcal{G}$ be a class of graphs defined by two forbidden induced subgraphs. Then:*

(i) $\mathcal{G}$ *has bounded clique-width if it is equivalent to a class of* (H_1, H_2)*-free graphs such that one of the following holds:*

1. H_1 *or* $H_2 \subseteq_i P_4$;
2. $H_1 = sP_1$ *and* $H_2 = K_t$ *for some* s, t;
3. $H_1 \subseteq_i P_1 + P_3$ *and* $\overline{H_2} \subseteq_i K_{1,3} + 3P_1$, $K_{1,3} + P_2$, $P_1 + S_{1,1,2}$, P_6 *or* $S_{1,1,3}$;
4. $H_1 \subseteq_i 2P_1 + P_2$ *and* $\overline{H_2} \subseteq_i 2P_1 + P_3$, $3P_1 + P_2$ *or* $P_2 + P_3$;
5. $H_1 \subseteq_i P_1 + P_4$ *and* $\overline{H_2} \subseteq_i P_1 + P_4$ *or* P_5;
6. $H_1 \subseteq_i 4P_1$ *and* $\overline{H_2} \subseteq_i 2P_1 + P_3$;
7. $H_1, \overline{H_2} \subseteq_i K_{1,3}$.

(ii) $\mathcal{G}$ *has unbounded clique-width if it is equivalent to a class of* (H_1, H_2)*-free graphs such that one of the following holds:*

1. $H_1 \notin \mathcal{S}$ *and* $H_2 \notin \mathcal{S}$;
2. $\overline{H_1} \notin \mathcal{S}$ *and* $\overline{H_2} \notin \mathcal{S}$;
3. $H_1 \supseteq_i K_{1,3}$ *or* $2P_2$ *and* $\overline{H_2} \supseteq_i 4P_1$ *or* $2P_2$;
4. $H_1 \supseteq_i P_1 + P_4$ *and* $\overline{H_2} \supseteq_i P_2 + P_4$;
5. $H_1 \supseteq_i 2P_1 + P_2$ *and* $\overline{H_2} \supseteq_i K_{1,3}$, $5P_1$, $P_2 + P_4$ *or* P_6;
6. $H_1 \supseteq_i 3P_1$ *and* $\overline{H_2} \supseteq_i 2P_1 + 2P_2$, $2P_1 + P_4$, $4P_1 + P_2$, $3P_2$ *or* $2P_3$;
7. $H_1 \supseteq_i 4P_1$ *and* $\overline{H_2} \supseteq_i P_1 + P_4$ *or* $3P_1 + P_2$.

Proof. We first consider the bounded cases. Statement (i).1 follows from Theorem 6. To prove Statement (i).2 note that if $H_1 = sP_1$ and $H_2 = K_t$ for some s, t then by Ramsey's Theorem, all graphs in the class of (H_1, H_2)-free graphs have a bounded number of vertices and therefore the clique-width of graphs in this class is bounded. By the definition of equivalence, when proving Statement (i).3, we may assume that $H_1 = K_3$. Then Statement (i).3 follows from Fact 2 combined with the fact that (K_3, H)-free graphs have bounded clique-width if H is $K_{1,3} + 3P_1$ (Theorem 2), $K_{1,3} + P_2$ [16], $P_1 + S_{1,1,2}$ (Theorem 1), P_6 [5] or $S_{1,1,3}$ [16]. Statement (i).4 follows from Fact 2 and the fact that $(\overline{2P_1 + P_2}, 2P_1 + P_3)$-free, $(\overline{2P_1 + P_2}, 3P_1 + P_2)$-free and $(\overline{2P_1 + P_2}, P_2 + P_3)$-free graphs have bounded clique-width [15]. Statement (i).5 follows from Fact 2 and the fact that both $(P_1 + P_4, \overline{P_1 + P_4})$-free graphs [7] and $(P_5, \overline{P_1 + P_4})$-free graphs [8] have bounded clique-width. Statement (i).6 follows from Fact 2 and the fact that $(2P_1 + P_3, K_4)$-free graphs have bounded clique-width [3]. Statement (i).7 follows from the fact that $(K_{1,3}, \overline{K_{1,3}})$-free graphs have bounded clique-width [1,9].

We now consider the unbounded cases. Statements (ii).1 and (ii).2 follow from Lemma 1 and Fact 2. Statement (ii).3 follows from the fact that the classes of $(C_4, K_{1,3}, K_4, \overline{2P_1 + P_2})$-free [4], $(K_4, 2P_2)$-free [4] and $(C_4, C_5, 2P_2)$-free graphs (or equivalently, split graphs) [34] have unbounded clique-width. Statement (ii).4 follows from Fact 2 and the fact that the class of $(P_2 + P_4, 3P_2, P_6, \overline{P_1 + P_4})$-free (Theorem 5) graphs have unbounded clique-width. Statement (ii).5 follows from Fact 2 and the fact that $(C_4, K_{1,3}, K_4, \overline{2P_1 + P_2})$-free [4], $(5P_1, \overline{2P_1 + P_2})$-free [14], $(\overline{2P_1 + P_2}, P_2 + P_4)$-free (see arXiv version of [15]) and $(P_6, \overline{2P_1 + P_2})$-free (Theorem 4) graphs have unbounded clique-width. To prove Statement (ii).6, suppose $H_1 \supseteq_i 3P_1$ and $\overline{H_2} \supseteq_i 2P_1 + 2P_2, 2P_1 + P_4, 4P_1 + P_2, 3P_2$ or $2P_3$. Then

$\overline{H_1} \notin \mathcal{S}$, so $\overline{H_2} \in \mathcal{S}$, otherwise we are done by Statement (ii).2. By Lemma 3, $\overline{H_2}$ is not an induced subgraph of any graph in $\{K_{1,3}+3P_1, K_{1,3}+P_2, P_1+S_{1,1,3}, S_{1,2,3}\}$. The class of (H_1, H_2)-free graphs contains the class of complements of $\overline{H_2}$-free bipartite graphs. By Fact 2 and Lemma 2, this latter class has unbounded clique-width. Statement (ii).7 follows from the Fact 2 and the fact that the classes of $(K_4, P_1 + P_4)$-free graphs (Lemma 6) and $(4P_1, \overline{3P_1 + P_2})$-free graphs [14] have unbounded clique-width.   □

As we will prove in Theorem 8, the above classification leaves exactly 13 open cases (up to equivalence).

Open Problem 1. *Does the class of (H_1, H_2)-free graphs have bounded clique-width when:*

1. $H_1 = 3P_1, \overline{H_2} \in \{P_1 + P_2 + P_3, P_1 + 2P_2, P_1 + P_5, P_1 + S_{1,1,3}, P_2 + P_4, S_{1,2,2}, S_{1,2,3}\}$;
2. $H_1 = 2P_1 + P_2, \overline{H_2} \in \{P_1 + P_2 + P_3, P_1 + 2P_2, P_1 + P_5\}$;
3. $H_1 = P_1 + P_4, \overline{H_2} \in \{P_1 + 2P_2, P_2 + P_3\}$ *or*
4. $H_1 = \overline{H_2} = 2P_1 + P_3$.

Note that the two pairs $(3P_1, \overline{S_{1,1,2}})$ and $(3P_1, \overline{S_{1,2,3}})$, or equivalently, the two pairs $(K_3, S_{1,2,2})$ and $(K_3, S_{1,2,3})$ are the only pairs that correspond to open cases in which both H_1 and H_2 are connected. We also observe the following. Let $H_2 \in \{P_1+P_2+P_3, P_1+2P_2, P_1+P_5, P_1+S_{1,1,3}, P_2+P_4, S_{1,2,2}, S_{1,2,3}\}$. Lemma 2 shows that all bipartite H_2-free graphs have bounded clique-width. Moreover, the graph $P_1 + 2P_2$ is an induced subgraph of H_2. Hence, for investigating whether the boundedness of the clique-width of bipartite H_2-free graphs can be extended to (K_3, H_2)-free graphs, the $H_2 = P_1 + 2P_2$ case is the starting case.

Theorem 8. *Let $\mathcal{G}$ be a class of graphs defined by two forbidden induced subgraphs. Then $\mathcal{G}$ is not equivalent to any of the classes listed in Theorem 7 if and only if it is equivalent to one of the 13 cases listed in Open Problem 1.*

A graph G is $(H_1, \ldots, H_p)$-*subgraph-free* if G has no subgraph isomorphic to a graph in $\{H_1, \ldots, H_p\}$. Let G and H be graphs. Then G contains H as a *minor* or *topological minor* if G can be modified into H by a sequence that consists of edge contractions, edge deletions and vertex deletions, or by a sequence that consists of vertex dissolutions, edge deletions and vertex deletions, respectively. If G does not contain any of the graphs $H_1, \ldots, H_p$ as a (topological) minor, we say that G is $(H_1, \ldots, H_p)$-*(topological-)minor-free*. We omit the proof of the following result, which completely characterizes which of these graph classes have bounded clique-width.

Theorem 9. *Let $\{H_1, \ldots, H_p\}$ be a finite set of graphs. Then the following statements hold:*

(i) *The class of $(H_1, \ldots, H_p)$-subgraph-free graphs has bounded clique-width if and only if $H_i \in \mathcal{S}$ for some $1 \le i \le p$.*

(ii) *The class of* $(H_1, \ldots, H_p)$-*minor-free graphs has bounded clique-width if and only if* H_i *is planar for some* $1 \leq i \leq p$.

(iii) *The class of* $(H_1, \ldots, H_p)$-*topological-minor-free graphs has bounded clique-width if and only if* H_i *is planar and has maximum degree at most 3 for some* $1 \leq i \leq p$.

6   Consequences for Colouring

One of the motivations of our research was to further the study of the computational complexity of the COLOURING problem for (H_1, H_2)-free graphs. Recall that COLOURING is polynomial-time solvable on any graph class of bounded clique-width by combining results of Kobler and Rotics [28] and Oum [38]. By combining a number of known results [11,12,16,22,29,35,41,42,44] with new results, Dabrowski, Golovach and Paulusma [14] presented a summary of known results for COLOURING restricted to (H_1, H_2)-free graphs. Combining Theorem 7 with the results of Kobler and Rotics [28] and Oum [38] and incorporating a number of recent results [25,26,36] leads to an updated summary. This updated summary (and a proof of it) can be found in the recent survey paper of Golovach, Johnson, Paulusma and Song [21].

From this summary we note that not only the case when $H_1 = P_4$ or $H_2 = P_4$ but thirteen other maximal classes of (H_1, H_2)-free graphs for which COLOURING is known to be polynomial-time solvable can be obtained by combining Theorem 7 with the results of Kobler and Rotics [28] and Oum [38] (see also [21]). One of these thirteen classes is one that we obtained in this paper (Theorem 2), namely the class of $(K_{1,3} + 3P_1, \overline{P_1 + P_3})$-free graphs, for which COLOURING was not previously known to be polynomial-time solvable. Note that Dabrowski, Lozin, Raman and Ries [16] already showed that COLOURING is polynomial-time solvable for $(\overline{P_1 + P_3}, P_1 + S_{1,1,2})$-free graphs, but in Theorem 1 we strengthened their result by showing that the clique-width of this class is also bounded.

Theorem 8 shows that there are 13 classes of (H_1, H_2)-free graphs (up to equivalence) for which we do not know whether their clique-width is bounded. These classes correspond to 28+6+4+1=39 distinct classes of (H_1, H_2)-free graphs. The complexity of COLOURING is unknown for only 15 of these classes. We list these cases below:

1. $\overline{H_1} \in \{3P_1, P_1 + P_3\}$ and $H_2 \in \{P_1 + S_{1,1,3}, S_{1,2,3}\}$;
2. $H_1 = 2P_1 + P_2$ and $\overline{H_2} \in \{P_1 + P_2 + P_3, P_1 + 2P_2, P_1 + P_5\}$;
3. $H_1 = \overline{2P_1 + P_2}$ and $H_2 \in \{P_1 + P_2 + P_3, P_1 + 2P_2, P_1 + P_5\}$;
4. $H_1 = P_1 + P_4$ and $\overline{H_2} \in \{P_1 + 2P_2, P_2 + P_3\}$;
5. $\overline{H_1} = P_1 + P_4$ and $H_2 \in \{P_1 + 2P_2, P_2 + P_3\}$;
6. $H_1 = \overline{H_2} = 2P_1 + P_3$.

Note that Case 1 above reduces to two subcases by Lemma 4. All classes of (H_1, H_2)-free graphs, for which the complexity of COLOURING is still open and which are not listed above have unbounded clique-width. Hence, new techniques will need to be developed to deal with these classes.

References

1. Boliac, R., Lozin, V.V.: On the Clique-Width of Graphs in Hereditary Classes. In: Bose, P., Morin, P. (eds.) ISAAC 2002. LNCS, vol. 2518, pp. 44–54. Springer, Heidelberg (2002)
2. Bonomo, F., Grippo, L.N., Milanič, M., Safe, M.D.: Graphs of power-bounded clique-width. arXiv, abs/1402.2135 (2014)
3. Brandstädt, A., Dabrowski, K.K., Huang, S., Paulusma, D.: Bounding the clique-width of H-free chordal graphs. CoRR, abs/1502.06948 (2015)
4. Brandstädt, A., Engelfriet, J., Le, H.-O., Lozin, V.V.: Clique-width for 4-vertex forbidden subgraphs. Theory of Computing Systems **39**(4), 561–590 (2006)
5. Brandstädt, A., Klembt, T., Mahfud, S.: P_6- and triangle-free graphs revisited: structure and bounded clique-width. Discrete Mathematics and Theoretical Computer Science **8**(1), 173–188 (2006)
6. Brandstädt, A., Kratsch, D.: On the structure of (P_5, gem)-free graphs. Discrete Applied Mathematics **145**(2), 155–166 (2005)
7. Brandstädt, A., Le, H.-O., Mosca, R.: Gem- and co-gem-free graphs have bounded clique-width. International Journal of Foundations of Computer Science **15**(1), 163–185 (2004)
8. Brandstädt, A., Le, H.-O., Mosca, R.: Chordal co-gem-free and (P_5, gem)-free graphs have bounded clique-width. Discrete Applied Mathematics **145**(2), 232–241 (2005)
9. Brandstädt, A., Mahfud, S.: Maximum weight stable set on graphs without claw and co-claw (and similar graph classes) can be solved in linear time. Information Processing Letters **84**(5), 251–259 (2002)
10. Brandstädt, A., Mosca, R.: On variations of P_4-sparse graphs. Discrete Applied Mathematics **129**(2–3), 521–532 (2003)
11. Broersma, H., Golovach, P.A., Paulusma, D., Song, J.: Determining the chromatic number of triangle-free $2P_3$-free graphs in polynomial time. Theoretical Computer Science **423**, 1–10 (2012)
12. Broersma, H., Golovach, P.A., Paulusma, D., Song, J.: Updating the complexity status of coloring graphs without a fixed induced linear forest. Theoretical Computer Science **414**(1), 9–19 (2012)
13. Courcelle, B., Makowsky, J.A., Rotics, U.: Linear time solvable optimization problems on graphs of bounded clique-width. Theory of Computing Systems **33**(2), 125–150 (2000)
14. Dabrowski, K.K., Golovach, P.A., Paulusma, D.: Colouring of graphs with Ramsey-type forbidden subgraphs. Theoretical Computer Science **522**, 34–43 (2014)
15. Dabrowski, K.K., Huang, S., Paulusma, D.: Bounding Clique-Width via Perfect Graphs. In: Dediu, A.-H., Formenti, E., Martín-Vide, C., Truthe, B. (eds.) LATA 2015. LNCS, vol. 8977, pp. 676–688. Springer, Heidelberg (2015). Full version: arXiv CoRR abs/1406.6298
16. Dabrowski, K.K., Lozin, V.V., Raman, R., Ries, B.: Colouring vertices of triangle-free graphs without forests. Discrete Mathematics **312**(7), 1372–1385 (2012)
17. Dabrowski, K.K., Paulusma, D.: Classifying the Clique-Width of H-Free Bipartite Graphs. In: Cai, Z., Zelikovsky, A., Bourgeois, A. (eds.) COCOON 2014. LNCS, vol. 8591, pp. 489–500. Springer, Heidelberg (2014)
18. de Ridder, H.N., et al. Information System on Graph Classes and their Inclusions, 2001-2013. http://www.graphclasses.org

19. Espelage, W., Gurski, F., Wanke, E.: How to Solve NP-hard Graph Problems on Clique-Width Bounded Graphs in Polynomial Time. In: Brandstädt, A., Le, V.B. (eds.) WG 2001. LNCS, vol. 2204, pp. 117–128. Springer, Heidelberg (2001)
20. Fellows, M.R., Rosamond, F.A., Rotics, U., Szeider, S.: Clique-width is NP-Complete. SIAM Journal on Discrete Mathematics **23**(2), 909–939 (2009)
21. Golovach, P.A., Johnson, M., Paulusma, D., Song, J.: A survey on the computational complexity of colouring graphs with forbidden subgraphs. CoRR, abs/1407.1482 (2014)
22. Golovach, P.A., Paulusma, D.: List coloring in the absence of two subgraphs. Discrete Applied Mathematics **166**, 123–130 (2014)
23. Golumbic, M.C., Rotics, U.: On the clique-width of some perfect graph classes. International Journal of Foundations of Computer Science **11**(03), 423–443 (2000)
24. Gurski, F.: Graph operations on clique-width bounded graphs. CoRR, abs/cs/0701185 (2007)
25. Hoàng, C.T., Lazzarato, D.A.: Polynomial-time algorithms for minimum weighted colorings of $(P_5, \overline{P_5})$-free graphs and related graph classes. Discrete Applied Mathematics 186(0166–218X), 106–111 (2015). doi:http://dx.doi.org/10.1016/j.dam.2015.01.022. http://www.sciencedirect.com/science/article/pii/S0166218X15000244
26. Huang, S., Johnson, M., Paulusma, D.: Narrowing the complexity gap for colouring (C_s, P_t)-free graphs. In: Gu, Q., Hell, P., Yang, B. (eds.) AAIM 2014. LNCS, vol. 8546, pp. 162–173. Springer, Heidelberg (2014)
27. Kamiński, M., Lozin, V.V., Milanič, M.: Recent developments on graphs of bounded clique-width. Discrete Applied Mathematics **157**(12), 2747–2761 (2009)
28. Kobler, D., Rotics, U.: Edge dominating set and colorings on graphs with fixed clique-width. Discrete Applied Mathematics **126**(2–3), 197–221 (2003)
29. Král', D., Kratochvíl, J., Tuza, Z., Woeginger, G.J.: Complexity of Coloring Graphs without Forbidden Induced Subgraphs. In: Brandstädt, A., Le, V.B. (eds.) WG 2001. LNCS, vol. 2204, pp. 254–262. Springer, Heidelberg (2001)
30. Kratsch, S., Schweitzer, P.: Graph Isomorphism for Graph Classes Characterized by Two Forbidden Induced Subgraphs. In: Golumbic, M.C., Stern, M., Levy, A., Morgenstern, G. (eds.) WG 2012. LNCS, vol. 7551, pp. 34–45. Springer, Heidelberg (2012)
31. Lozin, V.V., Rautenbach, D.: On the band-, tree-, and clique-width of graphs with bounded vertex degree. SIAM Journal on Discrete Mathematics **18**(1), 195–206 (2004)
32. Lozin, V.V., Rautenbach, D.: The tree- and clique-width of bipartite graphs in special classes. Australasian Journal of Combinatorics **34**, 57–67 (2006)
33. Lozin, V.V., Volz, J.: The clique-width of bipartite graphs in monogenic classes. International Journal of Foundations of Computer Science **19**(02), 477–494 (2008)
34. Makowsky, J.A., Rotics, U.: On the clique-width of graphs with few P_4's. International Journal of Foundations of Computer Science **10**(03), 329–348 (1999)
35. Malyshev, D.S.: The coloring problem for classes with two small obstructions. Optimization Letters **8**(8), 2261–2270 (2014)
36. Malyshev, D.S.: Two cases of polynomial-time solvability for the coloring problem. Journal of Combinatorial Optimization (in press)
37. Olariu, S.: Paw-free graphs. Information Processing Letters **28**(1), 53–54 (1988)
38. Oum, S.-I.: Approximating rank-width and clique-width quickly. ACM Transactions on Algorithms **5**(1), 10 (2008)
39. Oum, S.-I., Seymour, P.D.: Approximating clique-width and branch-width. Journal of Combinatorial Theory, Series B **96**(4), 514–528 (2006)

40. Ramsey, F.P.: On a problem of formal logic. Proceedings of the London Mathematical Society s2-30(1), 264–286 (1930)
41. Randerath, B.: 3-colorability and forbidden subgraphs. I: Characterizing pairs. Discrete Mathematics **276**(1–3), 313–325 (2004)
42. Randerath, B., Schiermeyer, I.: A note on Brooks' theorem for triangle-free graphs. Australasian Journal of Combinatorics **26**, 3–9 (2002)
43. Rao, M.: MSOL partitioning problems on graphs of bounded treewidth and clique-width. Theoretical Computer Science **377**(1–3), 260–267 (2007)
44. Schindl, D.: Some new hereditary classes where graph coloring remains NP-hard. Discrete Mathematics **295**(1–3), 197–202 (2005)
45. Schweitzer, P.: Towards an isomorphism dichotomy for hereditary graph classes. In: Mayr, E.W., Ollinger, N. (eds.) 32nd International Symposium on Theoretical Aspects of Computer Science (STACS 2015). LIPIcs, vol. 30, pp. 689–702. Schloss Dagstuhl - Leibniz-Zentrum für Informatik, Dagstuhl (2015). doi:http://dx.doi.org/10.4230/LIPIcs.STACS.2015.689. http://drops.dagstuhl.de/opus/volltexte/2015/4951
46. Stein, W.A., et al.: Sage Mathematics Software (Version 5.9). The Sage Development Team (2013). http://www.sagemath.org

Randomized Adaptive Test Cover

Peter Damaschke[✉]

Department of Computer Science and Engineering,
Chalmers University, 41296 Göteborg, Sweden
ptr@chalmers.se

Abstract. In a general combinatorial search problem with binary tests we are given a set of elements and a hypergraph of possible tests, and the goal is to find an unknown target element using a minimum number of tests. We explore the expected test number of randomized strategies. We obtain several general results on the ratio of the expected and worst-case deterministic test number, as well as complexity results for hypergraphs of small rank, and we state some open problems.

Keywords: Combinatorial search · Randomization · Game theory · LP duality · Fractional graph theory

1 Introduction

A hypergraph $\mathcal{H}$ is a set U of n elements (vertices) equipped with a family of subsets called the *edges*. We consider a search problem on hypergraphs, therefore we also call the edges *tests*: One unknown element $u \in U$ is the *target*. A searcher can apply any tests T from $\mathcal{H}$. The test answers *positive* if $u \in T$, and *negative* else. The searcher aims to identify u efficiently from the outcomes of some tests from $\mathcal{H}$. The primary goal is to minimize the number of tests. We will silently assume that $\mathcal{H}$ is *separating*, that is, no two targets cause the same outcomes of all tests (otherwise it would be impossible to distinguish them).

Combinatorial search has applications, e.g., in biological testing [2,3]. A number of more specific, classic combinatorial search problems can be formulated in the above way, perhaps the foremost example is the group testing problem [7] which found various applications. Note that even problems like sorting by comparisons fit in this framework. (There the target is a sorted sequence of numbers, the elements are all permutations, and a test finds out the relation between two numbers.) One may also think of $\mathcal{H}$ as a system of binary attributes of objects, and then an efficient test strategy is also a compact *classification system*.

A search strategy can work in *rounds*, where all tests in a round are done in parallel, without waiting for each other's outcomes. An *adaptive* strategy performs only one test per round. In a *deterministic* strategy, the choice of tests for each round is uniquely determined by the outcomes of earlier tests. In a *randomized* strategy, the choice of tests for each round can, additionally, depend on random decisions. We stress that the tests still behave deterministically; here

© Springer International Publishing Switzerland 2015
V.Th. Paschos and P. Widmayer (Eds.): CIAC 2015, LNCS 9079, pp. 182–193, 2015.
DOI: 10.1007/978-3-319-18173-8_13

we assume neither random errors nor target probability distributions, only the searcher's choice of tests may be randomized. Also note that deterministic strategies are by definition a special case of randomized strategies, they just do not use the possibility to take random decisions. A deterministic strategy is *optimal* among all deterministic strategies if it minimizes the worst-case number of tests, and a randomized strategy is optimal if it minimizes the worst-case *expected* number of tests, where the worst case refers to maximization over all targets. For clarity we give formal definitions of the test numbers.

Definition 1. *For a separating hypergraph $\mathcal{H}$ let $Det(\mathcal{H})$ and $Rand(\mathcal{H})$ denote the set of all deterministic and randomized search strategies, respectively, on $\mathcal{H}$. Note that $Det(\mathcal{H}) \subset Rand(\mathcal{H})$. For $A \in Rand(\mathcal{H})$ and an element $u \in U$, let $t_A(\mathcal{H}, u)$ denote the expected number of tests done by A if u is the target. (If $A \in Det(\mathcal{H})$, this "expected" number is simply the deterministic number of tests.) The test number of A is $t_A(\mathcal{H}) := \max_{u \in U} t_A(\mathcal{H}, u)$. We define the optimum test numbers as $t_{rand}(\mathcal{H}) := \min_{A \in Rand(\mathcal{H})} t_A(\mathcal{H})$ and $t_{det}(\mathcal{H}) := \min_{A \in Det(\mathcal{H})} t_A(\mathcal{H})$. We may also limit the number r of rounds of strategies A, in which case we write $t_{rand,r}(\mathcal{H})$ and $t_{det,r}(\mathcal{H})$.*

Trivially we have $t_{rand,r}(\mathcal{H}) \leq t_{det,r}(\mathcal{H})$. As we will see, the randomized test number can be significantly smaller, however, note that $t_{rand,1}(\mathcal{H}) = t_{det,1}(\mathcal{H})$, since $t_A(\mathcal{H}, u)$ is the same for all $u \in U$ if A has only one round. In other words, meaningful randomized strategies need at least two rounds.

The case $r = 1$ (thus restricted to the deterministic setting) is well known as the *test cover* problem and can be rephrased as follows: Given a separating hypergraph, find a smallest subset of the edges of $\mathcal{H}$ that still form a separating hypergraph. The complexity of the test cover problem has been intensively studied [1,3,5,6,8,10], whereas very little is known for $r > 1$; see [12] for some combinatorial results. The *rank* of a hypergraph $\mathcal{H}$ is the maximum size of its edges. The test cover problem with fixed rank found independent interest, since already this restricted case has practical relevance [2,4,9].

To the best of our knowledge, there is no study of the search problem *in full generality and in the randomized setting* so far, although randomization has well been used in many specific search problems. E.g., Quicksort is a famous sorting algorithm, and randomized constructions have been extensively studied for group testing (however we cannot possibly give a survey here).

Contributions. We focus on adaptive testing. An obvious question is how much the test number can benefit from randomization. That is, we study the ratio $t_{det}(\mathcal{H})/t_{rand}(\mathcal{H})$. As a preparation we express the search problem as a matrix game and linear program, which implies some simple lower bounds on $t_{rand}(\mathcal{H})$. We use them to show that $t_{det}(\mathcal{H})/t_{rand}(\mathcal{H})$ is, essentially, at least 2 for large hypergraphs of fixed rank, and the ratio can be up to 4 in certain hypergraphs. By a certain composition of small hypergraphs we get examples where the ratio is away from 1 even when the test number is logarithmic, i.e., close to the information-theoretic lower bound. The largest possible ratio remains an open problem. For rank-2 hypergraphs we also relate the search problem to some standard graph problems. They do not exactly correspond to the search problem but enable a $\frac{7}{6}$-approximation algorithm for the randomized test number. This approximation is based on the primal-dual method and certain half-integral solutions to the minimum fractional edge cover problem. While the actual complexity of computing $t_{rand}(\mathcal{H})$ remains open even in the rank-2 case, we show that finding optimal deterministic strategies is NP-hard already for rank 3.

2   Game-Theoretic Interpretation and Lower Bounds

Any randomized search strategy on a hypergraph $\mathcal{H}$ can be viewed as a probability distribution on the (finite) set of the deterministic strategies on $\mathcal{H}$, also called a *mixed strategy*. This allows us to apply the classic theory of zero-sum games. The searcher applies a mixed strategy as above, and an adversary plays a mixed strategy, too, which is a probability distribution on U specifying the probability of each element being the target. We also refer to it as a *target distribution*. By von Neumann's minimax theorem, a special case of LP strong duality, we get:

Proposition 1. *There exists a pair of optimal mixed strategies such that the player does an expected number of at most $t_{rand}(\mathcal{H})$ tests whatever the target is, and every deterministic strategy needs an expected number of at least $t_{rand}(\mathcal{H})$ tests. Only deterministic strategies that attain this expected value can be involved with nonzero probability in an optimal mixed search strategy.*

The optimal strategies can be computed by a linear program (LP), with the caveat that the number of deterministic strategies to consider may not be polynomial in n. In the following note that any tests T and $U \setminus T$ are equivalent.

Proposition 2. *The following lower bounds hold.*
Averaging bound: $t_{rand}(\mathcal{H}) \geq \min_{A \in Det(\mathcal{H})} \frac{1}{n} \sum_{u \in U} t_A(\mathcal{H}, u)$; *in words: the best average deterministic test number lower-bounds the randomized test number.*
Set cover lower bound: *For any fixed $u \in U$, replace all tests $T \ni u$ with their complements. Then $t_{rand}(\mathcal{H})$ is at least the size of a smallest set cover of $U \setminus \{u\}$.*

Proof. By Proposition 1, any target distribution yields a lower bound on $t_{rand}(\mathcal{H})$. The uniform target distribution yields the averaging bound. The target distribution that concentrates probability 1 on some fixed $u \in U$ yields the set cover lower bound, since every non-target must occur in some negative test.   □

Definition 2. *For some arbitrary but fixed order of U, we call the vector of the n values $t_A(\mathcal{H}, u)$ the* test number vector *of strategy A on $\mathcal{H}$.*

Hence the test number vector of any randomized strategy is a convex linear combination of test number vectors of some deterministic strategies.

A deterministic strategy can be viewed as a *strategy tree* with n leaves for the n elements. Every inner node is marked with the set of tests applied in a round. Depending on their outcomes, the searcher is sent to a child node for the next round. The depth of the tree is the maximum number of rounds.

Proposition 3. *Let $\mathcal{H}$ be a hypergraph on $n = 2^k + m$ elements, where $2^k \leq n$ is the largest power of 2 not exceeding n. Then we have:*
$$t_{rand}(\mathcal{H}) \geq \tfrac{1}{n}((n - 2m)k + 2m(k + 1)) = k + \tfrac{2m}{n}.$$

Proof. (Sketch.) Since parallel tests could also be done sequentially, the test number vector of any deterministic strategy is the vector of distances of the n leaves to the root in some *binary* tree. Due to simple exchange arguments, among the binary trees with a fixed number n of leaves, the average test number is minimized if $n - 2m$ numbers are k and $2m$ numbers are $k + 1$. Together with the averaging bound in Proposition 2 this yields the assertion. □

This bound is particularly useful in practical computations of optimal strategies for small instances $\mathcal{H}$: First we check whether $\mathcal{H}$ allows deterministic strategies A with $k + \tfrac{2m}{n}$ tests on average, since (by Proposition 1) only such A can build a randomized strategy with $k + \tfrac{2m}{n}$ expected tests. For small $\mathcal{H}$ there are not so many possible strategy trees. We enumerate them and solve the LP that balances the expected test numbers for all targets. If the LP has no balanced solution, then we know $t_{rand}(\mathcal{H}) > k + \tfrac{2m}{n}$. Thus we would next include deterministic strategies A where $\tfrac{1}{n} \sum_{u \in U} t_A(\mathcal{H}, u)$ is by $\tfrac{1}{n}$ larger, solve the extended LP, and so on. (Calculation examples are omitted due to space limitations.)

3   Some Structural Lemmas

Lemma 1. *Let $\mathcal{H}$ be a hypergraph and u an element such that $\{u\}$ is not an edge. Let $\mathcal{H}_u$ be the hypergraph $\mathcal{H}$ with the edge $\{u\}$ inserted. Then $t_{det}(\mathcal{H}_u) = t_{det}(\mathcal{H})$.*

Proof. Trivially, $t_{det}(\mathcal{H}_u) \leq t_{det}(\mathcal{H})$. To show that $t_{det}(\mathcal{H})$ is not strictly larger, we consider an optimal deterministic strategy A_u for $\mathcal{H}_u$. Let A be the strategy for $\mathcal{H}$ that mimics A_u, and whenever A_u wants to test $\{u\}$, strategy A skips this non-existing test and continues as A_u would do if this test were negative. Strategy A can be incomplete in the sense that it may not find the target. Below we discuss how to complete it, using $t_{det}(\mathcal{H}_u)$ tests in the worst case.

Consider any leaf ℓ of the strategy tree of A. If the strategy has not skipped $\{u\}$ on the path to ℓ, then A has behaved as A_u, thus a target is identified. The other case is that A has skipped $\{u\}$ on the path to ℓ. The path to ℓ is identical to the corresponding path in A_u, except that test $\{u\}$ is not done and a negative outcome assumed. Moreover, A_u would have identified some target v on this path. The only missing information is now that u is in fact negative. It follows that at most two candidates, u and v, remain at ℓ. Since $\mathcal{H}$ is separating, we can append any test to distinguish u and v. Since one test was skipped, the path to ℓ in A is strictly shorter than the one in A_u, hence adding a test at the end does not increase the test number compared to A_u. □

Lemma 2. *Let $\mathcal{H}$ be a hypergraph containing an edge $\{u\}$. Against any deterministic searcher, there is an optimal adversary strategy (enforcing at least $t_{det}(\mathcal{H})$ tests) that gives a negative answer whenever $\{u\}$ is tested.*

Proof. When the searcher tests $\{u\}$ and the answer is positive, then u is identified as the target, and $\{u\}$ was the last test. Hence the adversary cannot miss a longest path in the strategy tree by giving a negative answer instead.   □

Definition 3. *We define the* composition $\mathcal{H} \triangleright \mathcal{J}$ *of hypergraphs* $\mathcal{H}$ *and* $\mathcal{J}$ *as follows. Let* $U = \{u_1, \ldots, u_n\}$ *and* V *be the vertex sets of* $\mathcal{H}$ *and* $\mathcal{J}$, *respectively. We replace every element* $u_i \in U$ *with a copy of* $\mathcal{J}$, *denoted by* $\mathcal{J}_i$, *on a vertex set* V_i. *Every edge* E *of* $\mathcal{H}$ *is kept in* $\mathcal{H} \triangleright \mathcal{J}$, *by replacing the elements: We define* E *in* $\mathcal{H} \triangleright \mathcal{J}$ *as the union of those* V_i *with* $u_i \in E$ *in* $\mathcal{H}$.

Informally, we substitute $\mathcal{J}$ into every element of $\mathcal{H}$, or $\mathcal{H} \triangleright \mathcal{J}$ models a hierarchical classification where $\mathcal{H}$ gives a coarse classification refined by $\mathcal{J}$. While the following Lemma is not too surprising, the proof becomes somewhat tricky. Besides the previous Lemmas it uses a technique that often simplifies lower-bound proofs for search problems: An adversary may reveal more information than the searcher has asked for. Since this only helps the searcher, any lower bound obtained in this way is also a lower bound for the original search problem.

Lemma 3. *The deterministic test number of the composition of any two hypergraphs is additive:* $t_{det}(\mathcal{H} \triangleright \mathcal{J}) = t_{det}(\mathcal{H}) + t_{det}(\mathcal{J})$.

Proof. Subadditivity is easy to see: In order to search $\mathcal{H} \triangleright \mathcal{J}$ one can first run an optimal strategy on $\mathcal{H}$ to determine the copy of $\mathcal{J}$ containing the target, followed by an optimal strategy for $\mathcal{J}$ applied to this copy. The reverse $t_{det}(\mathcal{H} \triangleright \mathcal{J}) \geq t_{det}(\mathcal{H}) + t_{det}(\mathcal{J})$ is much less obvious, as the searcher may interleave tests in $\mathcal{H}$ and in the copies of $\mathcal{J}$. Define $\mathcal{H}_1$ as the hypergraph obtained from $\mathcal{H}$ by inserting all singleton edges $\{u\}$ that are not yet in $\mathcal{H}$. We claim:

$$t_{det}(\mathcal{H} \triangleright \mathcal{J}) \geq t_{det}(\mathcal{H}_1 \triangleright \mathcal{J}) \geq t_{det}(\mathcal{H}_1) + t_{det}(\mathcal{J}) = t_{det}(\mathcal{H}) + t_{det}(\mathcal{J}).$$

The first inequality is trivial, and the equation holds due to Lemma 1. To show $t_{det}(\mathcal{H}_1 \triangleright \mathcal{J}) \geq t_{det}(\mathcal{H}_1) + t_{det}(\mathcal{J})$ we describe an adversary on $\mathcal{H}_1 \triangleright \mathcal{J}$ forcing the searcher to do at least $t_{det}(\mathcal{H}_1) + t_{det}(\mathcal{J})$ tests. We follow an optimal adversary strategy on $\mathcal{H}_1$ with the property from Lemma 2, as long as the searcher keeps on testing edges of $\mathcal{H}_1$. Whenever the searcher tests an edge from a copy of $\mathcal{J}$, say $\mathcal{J}_i$, although the copy containing the target is not yet identified, we answer negatively and also reveal that the target is not in V_i at all. This gives the searcher the same information as if she had tested V_i, which equals the edge $\{u_i\}$ of $\mathcal{H}_1$, and received a negative answer. Moreover, this still complies with our optimal adversary strategy on $\mathcal{H}_1$. Thus we have "indirectly forced" the searcher to test only edges of $\mathcal{H}_1$ until the set V_k with the target is identified. Since the adversary has run an optimal strategy, at least $t_{det}(\mathcal{H}_1)$ tests have been done so far. At this moment there still remains the set V_k of candidates, and no tests have been performed in $\mathcal{J}_k$. From this it is clear that the searcher needs $t_{det}(\mathcal{J})$ further tests in the worst case.   □

4   The Case of Singleton Tests

The *rank* of a hypergraph $\mathcal{H}$ is the maximum number of elements in an edge. First we briefly settle the case of rank 1. Since $\mathcal{H}$ is separating, at least $n-1$ of its n elements must be singleton edges.

Proposition 4. *Let $\mathcal{H}$ be a hypergraph of rank 1. If $\mathcal{H}$ has exactly $n-1$ edges then $t_{rand}(\mathcal{H}) = n-1$. If $\mathcal{H}$ has n edges then $t_{rand}(\mathcal{H}) = \frac{1}{n}(\sum_{i=1}^{n} i-1) = \frac{n+1}{2} - \frac{1}{n}$.*

Proof. The assertion for $n-1$ edges follows from the set cover lower bound in Proposition 2 and the trivial fact that $n-1$ tests suffice. If $\mathcal{H}$ has n edges then the available test number vectors are all permutations of the following vector: $(1, 2, 3, \ldots, n-3.n-2, n-1.n-1)$. We take one of them and its cyclic shifts, each with probability $\frac{1}{n}$. This yields the claimed test number. Optimality is seen by assigning the target probability $\frac{1}{n}$ to every element. $\square$

The above strategy needs $\log_2 n$ random bits. Using only one random bit we can combine the test number vector $(1, 2, 3, \ldots, n-3.n-2, n-1.n-1)$ and its reverse, each with probability $\frac{1}{2}$. Then the balance is not perfect, still the inner elements have an expected test number $\frac{n+1}{2}$ which is only slightly worse. Note that $t_{det}(\mathcal{H})/t_{rand}(\mathcal{H})$ tends to 2 as n grows. This raises the question is how large $t_{det}(\mathcal{H})/t_{rand}(\mathcal{H})$ can ever be for general hypergraphs. We will address it later. Similarly we ask how large $t_{det,r}(\mathcal{H})/t_{rand,r}(\mathcal{H})$ can ever be.

Back to rank 1, next suppose that only r rounds are permitted. For ease of presentation we state only the asymptotic result. In particular, we assume that r divides n, and the nth test must be done even if $n-1$ tests were negative.

Proposition 5. *Let $\mathcal{H}$ be a hypergraph of rank 1, with all n edges. Then we have $t_{rand}(\mathcal{H}) = \frac{r(r+1)}{2} \cdot \frac{n}{r} \cdot \frac{1}{r} = \frac{n}{2}(1 + \frac{1}{r})$ subject to lower-order terms.*

Proof. (Sketch.) We divide U into r bins of $\frac{n}{r}$ elements, arrange the bins in a cycle, and in each round we test all elements of one bin, following the cyclic order and starting at a random bin. Then every element is tested in each round with the same probability $1/r$, and the expected number of tests is as claimed. To show optimality we take again the uniform target distribution. The expected test number of any deterministic r-round strategy is then uniquely determined by the numbers x_i of elements tested in rounds $i = 1, \ldots, r$, and it amounts to $\frac{1}{n}\sum_{j=1}^{r} x_j(\sum_{i=1}^{j} x_i)$, where $\sum_{i=1}^{r} x_i = n$. By standard methods for multivariate extremal problems with constraints, this expression is minimized if all x_i equal $\frac{n}{r}$, and then it becomes $\frac{r(r+1)}{2} \cdot \frac{n^2}{r^2} \cdot \frac{1}{n} = \frac{n}{2}(1 + \frac{1}{r})$ again. $\square$

5   Deterministic vs. Randomized Test Number

The rank-1 case suggests that $t_{det}(\mathcal{H})/t_{rand}(\mathcal{H})$ might typically be around 2 if the rank is small compared to n. In the following we study this question. We consider hypergraphs $\mathcal{H}$ of arbitrary but fixed rank. Let $\nu := \nu(\mathcal{H})$ be the size of a minimum *edge cover* in $\mathcal{H}$, that is, a subset of edges that covers all elements in U. Since $\mathcal{H}$ is separating, at most one element $u \in U$ is in no edge, and if such u exists, we define ν to be the size of a minimum edge cover of $U \setminus \{u\}$.

Theorem 1. *Consider all hypergraphs $\mathcal{H}$ of any fixed rank h. There we have $t_{det}(\mathcal{H}) = \nu \pm O(1)$. If the edges of $\mathcal{H}$ cover all elements, then we also have $t_{rand}(\mathcal{H}) < \frac{\nu}{2} + O(1)$. Thus, for any $\epsilon > 0$, all large enough $\mathcal{H}$ without uncovered elements satisfy $t_{det}(\mathcal{H})/t_{rand}(\mathcal{H}) > 2 - \epsilon$.*

Proof. We can test the edges of any fixed minimum edge cover $\mathcal{E}$ in one round. For any test outcomes there remain at most h candidates for the target, and we can trivially distinguish them by fewer than h more tests in a second round. This shows $t_{det,2}(\mathcal{H}) < \nu + h$. Next we argue that even an adaptive deterministic strategy cannot be essentially better in the worst case: All test outcomes could be negative until the tested edges cover all elements. Then we have to do at least $\nu - 1$ tests. In short, $t_{det}(\mathcal{H}) \geq \nu - 1$. From the set cover lower bound in Proposition 2 we also get $t_{rand}(\mathcal{H}) \geq \nu$ if some element is in no edge. Thus, in the following we consider only hypergraphs where the edges cover all elements.

Now, an obvious randomized strategy is to test the edges of $\mathcal{E}$ in random order. More precisely, we may take all cyclic shifts of some fixed order or two reverse orders, with equal probability, just as in the case of rank 1. We argue that every fixed target is found after $\frac{\nu}{2} + h$ expected tests. We assign every element u arbitrarily to some edge $E \in \mathcal{E}$ with $u \in E$, called the designated edge of u. Due to the random order of $\mathcal{E}$, the designated edge of any target u appears after at most $\frac{\nu}{2} + 1$ expected tests, and then fewer than further tests identify u as the target. This yields the assertions and finishes the proof.   □

By modifying the proof for r rounds and using the randomized strategy from Section 4, we similarly obtain $t_{det,r}(\mathcal{H})/t_{rand,r}(\mathcal{H}) > \frac{2r}{r+1} - \epsilon$.

One might conjecture that shuffling disjoint edges is already the best use of randomization, which would imply that ratio 2 is the best. But in the end of the proof of Theorem 1 we notice that element u may appear earlier, in some non-designated edge, hence $t_{rand}(\mathcal{H})$ might be smaller in a specific hypergraph. The following example demonstrates that, indeed, the ratio can be up to 4.

Proposition 6. *For any fixed h and $\epsilon > 0$ there exist hypergraphs $\mathcal{H}$ of rank h where $t_{det}(\mathcal{H})/t_{rand}(\mathcal{H}) > \frac{4(h+1)}{(h+3)} - \epsilon$.*

Proof. An h-simplex is the set of all $h + 1$ possible edges of size h in a set of $h + 1$ elements Let $\mathcal{H}$ consist of k disjoint h-simplices. Clearly, $\nu = 2k$. Now we arrange the h-simplices in random order and first test one random edge from every h-simplex. With probability $\frac{h}{h+1}$ we hit the target, and fewer than h further tests identify it. In this case we do an expected number of $\frac{k}{2} + O(1)$ tests (remember that h is fixed). With probability $\frac{1}{h+1}$ we miss the target, but there remains only one candidate in every h-simplex. In this case we test a second edge from every h-simplex to find the target. Thus we do an expected number of $k + \frac{k}{2} + O(1)$ tests in total. Hence the overall expected test number is $\frac{h}{h+1} \cdot \frac{k}{2} + \frac{1}{h+1} \cdot \frac{3k}{2} + O(1) = \frac{h+3}{2(h+1)}k + O(1)$. Our ratio tends to $\frac{2k \cdot 2(h+1)}{(h+3)k} = \frac{4(h+1)}{(h+3)}$ for large k.   □

Problem: Is $t_{det}(\mathcal{H})/t_{rand}(\mathcal{H}) < 4$ for all $\mathcal{H}$? Is it bounded by any constant?

Next we look into a different direction. One may conjecture that randomization significantly helps "bad" instances only, where linearly many tests are needed. However we show that $t_{det}(\mathcal{H})/t_{rand}(\mathcal{H})$ can be away from 1 even for "good" hypergraphs where $O(\log n)$ tests are enough.

Theorem 2. *Let $\mathcal{H}$ be any hypergraph with g elements, such that $t_{det}(\mathcal{H}) = t$, and $t_{det}(\mathcal{H})/t_{rand}(\mathcal{H}) \geq c$. Then there exist hypergraphs with arbitrarily large numbers n of elements, such that $t_{det}(\mathcal{L})/t_{rand}(\mathcal{L}) \geq c$ as well, and $t_{det}(\mathcal{L}) = \frac{t}{\log_2 g} \log_2 n$.*

Proof. Let $\mathcal{L}$ be the k-fold composition of $\mathcal{H}$ with itself: $\mathcal{L} := \mathcal{H} \triangleright \ldots \triangleright \mathcal{H}$ (k terms). One may think of $\mathcal{L}$ as a "tree of hypergraphs $\mathcal{H}$" with degree g and depth k. Clearly $\mathcal{L}$ has $n := g^k$ elements. An obvious randomized strategy that goes recursively down this tree needs at most $k \cdot t_{rand}(\mathcal{H})$ expected tests for every target, by linearity of expectation. Hence $t_{rand}(\mathcal{L}) \leq k \cdot t_{rand}(\mathcal{H})$. Lemma 3 inductively applied to k factors yields $t_{det}(\mathcal{L}) = k \cdot t_{det}(\mathcal{H})$. Furthermore, note that $k = \frac{\log_2 n}{\log_2 g}$. Together these bounds imply the assertions. □

To give an example, let $\mathcal{H}$ consist of three elements and singleton tests. Then $g = 3$, $t = t_{det}(\mathcal{H}) = 2$, $t_{rand}(\mathcal{H}) = \frac{5}{3}$, thus we get hypergraphs $\mathcal{L}$ with ratio $c = \frac{6}{5}$ and $t_{det}(\mathcal{L}) = \frac{3}{\log_2 5 - \log_2 3} \log_2 n$. It would be interesting to figure out the largest possible ratio c for any given factor of $\log_2 n$.

6   The Case of Graphs

Next we consider hypergraphs $\mathcal{H}$ of rank 2. For convenience we look at usual graphs only, where all edges have exactly two elements. This is not a severe restriction, due to the following reasoning. Small instances could be solved exhaustively, and for large instances, where also t_{det} is large, we do not have to care about one test more or less. Now, let $\{u\}$ be any singleton edge. Unless u is isolated, it also belongs to some edge $\{u, v\}$. Any given strategy can be changed in the way that it tests $\{u, v\}$ rather than $\{u\}$. In the negative case this is even more efficient. In the positive case, either one further test of an edge $\{u, w\}$ or $\{v, w\}$ can distinguish u and v, or u and v form a connected component. In the latter case we modify the instance by removing v and keeping the edge $\{u\}$ only. Altogether, subject to differences of at most 1 in the test numbers, it suffices to study usual graphs $G = (V, E)$ with edges of size 2, except that isolated vertices are also considered as edges.

A *fractional independent set* assigns a non-negative weight to every vertex, such that every edge has a total weight at most 1. In particular, isolated vertices can get the weight 1.

Proposition 7. *If $\mathcal{H}$ has a fractional independent set of total weight α, then $t_{rand}(\mathcal{H}) \geq \frac{\alpha}{2} - O(1)$.*

Proof. We normalize the assumed fractional independent set, that is, divide all weights by α, such that the weights sum up to 1 and hence form a probability distribution on V. It suffices to show that every deterministic strategy A needs an expected number of at least $\frac{\alpha}{2} - O(1)$ tests on this target distribution. As long as all tests are negative, strategy A tests the edges in some specific order. Since every edge has a probability mass of at most $\frac{1}{\alpha}$, the first i tests cover elements of total probability mass at most $\frac{i}{\alpha}$, for every i. By a simple exchange argument, the expected number of tests until A hits the target is minimized if every tested edge covers new elements whose total probability mass is exactly $\frac{1}{\alpha}$. This yields the assertion.   □

A *fractional edge cover* assigns a non-negative weight x_e to every edge $e \in E$, such that every vertex is incident to edges with a total weight at least 1. In particular, isolated vertices (recall that we count them as edges) must get the weight 1. Finding a maximum-weight fractional independent set and a minimum-weight fractional edge cover are natural LP relaxations of the corresponding integral optimization problems, and they form a well-known pair of dual LPs, hence due to LP strong duality they achieve the same optimal value on a given graph.

A function with range $[0, 1]$ is called *half-integral* if it attains only the values $0, \frac{1}{2}, 1$. Although half-integrality is well studied for several optimization problems and is a tool for approximation algorithms (e.g., in [11]), we are not aware of an earlier proof of the following structural result that we will use to approximate the randomized test number.

Theorem 3. *In every graph, the fractional edge cover problem has an optimal solution that is half-integral, with the additional property that the edges e with $x_e = \frac{1}{2}$ form vertex-disjoint odd cycles. Moreover, this solution can be obtained in polynomial time.*

Proof. Consider any optimal fractional edge cover. Observe $x_e \leq 1$ for all edges, since otherwise we could reduce x_e to 1 and get a better valid solution. We call an edge e fractional if $0 < x_e < 1$. A tour is an sequence of edges that starts and ends in the same vertex; note that edges may appear several times in a tour and be traversed in arbitrary directions. The length of a tour is the number of edges, where repeatedly traversed edges are counted again.

Let C be a tour of fractional edges, and assume that C has even length. We may change the weights of the edges alternatingly by some amount $+\epsilon$ and $-\epsilon$. This changes neither the sum of weights incident to any vertex, nor the total weight. As for multiple edges e in C, note that the net effect on x_e is the sum of all changes applied there, and it may be zero. In the latter case we call e an inert edge. Suppose that C has some non-inert edge. Then, by increasing ϵ (starting from $\epsilon := 0$) we reach a value where some x_e becomes 0 or 1 while all other x_e are still in $[0, 1]$. Altogether we get an optimal solution with one fractional edge less. Iterating the procedure we can destroy all even tours that do not consist of inert edges only.

Now, in the graph F induced by the fractional edges, every connected component has at most one cycle, and it has odd length. Namely, any even cycle is obviously an even tour of non-inert edges, any two intersecting odd cycles also contain an even cycle, and any two disjoint odd cycles connected by a path can be merged into one even tour where not all edges are inert. Next, F cannot have any vertex of degree 1, since the only incident edge would have weight 1 and thus not be fractional. It follows that every connected component of F is merely an odd cycle. Finally, any cycle of length k has a fractional edge cover size of at least $\frac{k}{2}$ (just sum up all constraints for the k vertices), and giving each edge in the cycle the weight $\frac{1}{2}$ is a valid solution.

An optimal fractional vertex cover is obtained by solving an LP, and the above transformations of x_e values in even tours can be done in polynomial time.   □

Now we devise an algorithm to compute a randomized search strategy on a given graph G. Note that, in the case of graphs, a search strategy is completely specified by a (random) sequence of edges to be tested until the first positive test occurs. If two target candidates remain at this moment, then one more test identifies the target. Hence, for any target u, the expected test number equals (up to one test) the expected position of the first edge where u occurs in the sequence.

First we compute a fractional edge cover according to Theorem 3. From every odd cycle we randomly choose one vertex, called the left-over vertex. Let M be the set of all edges e with $x_e = 1$ (note that this includes all isolated vertices) plus disjoint edges that cover the odd cycles except their left-over vertices. (Clearly, any path with an even number of vertices is covered by a unique set of disjoint edges.) Let L be a further set of edges, one for each left-over vertex. That is, every edge in L contains a left-over vertex and some arbitrary neighbor. It is important to notice that $M \cup L$ is still an edge cover, thus every vertex occurs in some edge of $M \cup L$.

Let α be the total weight of our fractional edge cover, and let $\ell := |L|$. Observe that $|M| = \alpha - \frac{\ell}{2}$. Let c be the length of a shortest odd cycle C in our fractional edge cover. Our algorithm chooses among two strategies as follows.

Strategy 1 tests the edges of $M \cup L$ in random order. Since $|M| + |L| = \alpha + \frac{\ell}{2}$, the expected test number is at most $\frac{\alpha}{2} + \frac{\ell}{4}$.

Strategy 2 tests the edges of M in random order, followed by the edges of L in random order. Then any fixed vertex occurs already in M with probability at least $\frac{c-1}{c}$, and it occurs only in L with probability at most $\frac{1}{c}$. Hence the expected position of its first occurence is at most

$$\frac{c-1}{c} \cdot \frac{|M|}{2} + \frac{1}{c} \cdot \left(|M| + \frac{\ell}{2}\right) = \frac{c-1}{c} \cdot \left(\frac{\alpha}{2} - \frac{\ell}{4}\right) + \frac{1}{c} \cdot \alpha = \frac{c+1}{2c} \cdot \alpha - \frac{c-1}{c} \cdot \frac{\ell}{4}.$$

For any fixed ℓ this is maximized when c has the smallest possible value, $c = 3$, and then it becomes $\frac{2}{3}(\alpha - \frac{\ell}{4})$. (Actually we have already used the idea of Strategy 2 in the example that proves Proposition 6.)

Numbers α and ℓ are known from the fractional edge cover. We choose Strategy 1 or 2 with the smaller bound on the expected test number. The cut-off point is $\ell = \frac{\alpha}{3}$, hence the bound is at most $\frac{7}{12}\alpha$. Together with the lower bound $\frac{1}{2}\alpha$ from Proposition 7 we finally obtain:

Theorem 4. *For hypergraphs $\mathcal{H}$ with rank 2 we can approximate $t_{rand}(\mathcal{H})$ within a factor $\frac{7}{6}$ in polynomial time.*

7 The P-NP Borderline

The algorithm from Theorem 4 could be refined, e.g., by treating odd cycles of different length differently and analyzing the expecting target position more carefully. The lower bound from Proposition 7 can be raised, too. However, our main point in Theorem 4 was to achieve a rather good approximation ratio. Before making more efforts, one would like to know the real complexity status of computing $t_{rand}(\mathcal{H})$ for hypergraphs of rank 2. As we have seen, the problem is closely related to some standard fractional optimization problems. Moreover, we can solve it in polynomial time for special graphs. For instance, for graphs possessing a perfect matching it is not hard to show that it is optimal to test the matching edges in random order. The difficulty in general graphs is to schedule the tests that cover further, single vertices. Still it seems quite possible that the problem is polynomial-time solvable in graphs. We state this as an open problem. For rank 3 we have the following hardness result.

Theorem 5. *The problem of computing $t_{det}(\mathcal{H})$ for hypergraphs $\mathcal{H}$ of rank 3 is NP-complete.*

Proof. We give a reduction from the NP-complete Exact 3-Cover problem. An instance of Exact 3-Cover consists of a set U of $3k$ elements and a family of triples, i.e., subsets of U with 3 elements. The problem is to cover U by k of them. We construct a hypergraph $\mathcal{H}$ on $U \cup \{x, y, z\}$ where x, y, z are new elements. The edges of U are the given triples and all singletons. Trivially, the construction is polynomial. We claim that the Exact 3-Cover instance is affirmative if and only if $t_{det}(\mathcal{H}) \leq k + 2$.

If we can cover U by k triples, then we can test these k edges and are left with three target candidates, for any possible test outcomes. Then two further tests are sufficient (and necessary) to spot the target. Hence $t_{det}(\mathcal{H}) \leq k + 2$.

For the reverse direction, suppose that we cannot cover U by k triples. Then any k edges within U cover at most $3k-1$ elements. It follows that any $k-1$ edges within U cover at most $3k-2$ elements. Consider any deterministic strategy and an adversary giving negative answers to the first $k-1$ edges in U that are tested. These edges do not cover x, y, z and at least two more elements of U. Let $e \subset U$ be the kth test in U. Note that e still leaves some $u \in U$ uncovered. Regardless when and in which order $\{x\}, \{y\}, \{z\}$ and e are tested, after the first three of them we still have at least two target candidates. Namely, if e is among the first three tests, then some elements both in U and outside U are yet uncovered, and if e is the last of these four tests, then two elements in U are yet uncovered. Altogether, $(k-1) + 3 = k + 2$ tests do not determine the target.   □

It appears natural to conjecture that computing $t_{rand}(\mathcal{H})$ for hypergraphs $\mathcal{H}$ of rank 3 is NP-complete as well. An attempt would be to combine the same reduction idea with the averaging bound. However this looks more challenging. We remark that $t_{det,2}(\mathcal{H})$ is computable in polynomial time if $\mathcal{H}$ has rank 2: It is optimal to test some maximum matching in round 1. (Details are omitted due to space limitations.) So we conclude with the following

Problem: What is the complexity status of computing $t_{rand}(\mathcal{H})$ for hypergraphs $\mathcal{H}$ of any fixed rank?

References

1. Basavaraju, M., Francis, M.C., Ramanujan, M.S., Saurabh, S.: Partially Polynomial Kernels for Set Cover and Test Cover. In: Seth, A., Vishnoi, N.K. (Eds.) FSTTCS 2013. LIPIcs, vol. 24, pp. 67–78. Dagstuhl (2013)
2. de Bontridder, K.M.J., Halldórsson, B.V., Halldórsson, M.M., Hurkens, C.A.J., Lenstra, J.K., Ravi, R., Stougie, L.: Approximation Algorithms for the Test Cover Problem. Math. Progr. Series B **98**, 477–491 (2003)
3. De Bontridder, K.M.J., Lageweg, B.J., Lenstra, J.K., Orlin, J.B., Stougie, L.: Branch-and-bound algorithms for the test cover problem. In: Möhring, R.H., Raman, R. (eds.) ESA 2002. LNCS, vol. 2461, pp. 223–233. Springer, Heidelberg (2002)
4. Crowston, R., Gutin, G., Jones, M., Muciaccia, G., Yeo, A.: Parameterizations of Test Cover with Bounded Test Sizes. CoRR abs/1209.6528 (2012)
5. Crowston, R., Gutin, G., Jones, M., Saurabh, S., Yeo, A.: Parameterized study of the test cover problem. In: Rovan, B., Sassone, V., Widmayer, P. (eds.) MFCS 2012. LNCS, vol. 7464, pp. 283–295. Springer, Heidelberg (2012)
6. Cui, P.: A tighter analysis of set cover greedy algorithm for test set. In: Chen, B., Paterson, M., Zhang, G. (eds.) ESCAPE 2007. LNCS, vol. 4614, pp. 24–35. Springer, Heidelberg (2007)
7. Du, D.Z., Hwang, F.K.: Combinatorial Group Testing and Its Applications. Series on Appl. Math. vol. 3. World Scientific (2000)
8. Fahle, T., Tiemann, K.: A Faster Branch-and-Bound Algorithm for the Test-Cover Problem Based on Set-Covering Techniques. ACM J. Experim. Algor. 11 (2006)
9. Fernau, H., Raible, D.: A parameterized perspective on packing paths of length two. In: Yang, B., Du, D.-Z., Wang, C.A. (eds.) COCOA 2008. LNCS, vol. 5165, pp. 54–63. Springer, Heidelberg (2008)
10. Gutin, G., Muciaccia, G., Yeo, A.: (Non-)existence of Polynomial Kernels for the Test Cover Problem. Info. Proc. Letters **113**, 123–126 (2013)
11. Hochbaum, D.S.: Solving Integer Programs over Monotone Inequalities in Three Variables: A Framework for Half Integrality and Good Approximations. Eur. J. Operational Res. **140**, 291–321 (2002)
12. Wiener, G.: Rounds in Combinatorial Search. Algorithmica **67**, 315–323 (2013)

Contraction Blockers for Graphs
with Forbidden Induced Paths

Öznur Yaşar Diner[1]([✉]), Daniël Paulusma[2], Christophe Picouleau[3],
and Bernard Ries[4]

[1] Computer Engineering Department, Kadir Has University, Istanbul, Turkey
`oznur.yasar@khas.edu.tr`
[2] School of Engineering and Computing Sciences, Durham University, Durham, UK
`daniel.paulusma@durham.ac.uk`
[3] Laboratoire CEDRIC, CNAM, Paris, France
`christophe.picouleau@cnam.fr`
[4] PSL, Université Paris-Dauphine and CNRS, LAMSADE UMR 7243, Paris, France
`bernard.ries@dauphine.fr`

Abstract. We consider the following problem: can a certain graph
parameter of some given graph be reduced by at least d for some integer d
via at most k edge contractions for some given integer k? We examine
three graph parameters: the chromatic number, clique number and independence number. For each of these graph parameters we show that,
when d is part of the input, this problem is polynomial-time solvable on
P_4-free graphs and NP-complete as well as W[1]-hard, with parameter
d, for split graphs. As split graphs form a subclass of P_5-free graphs,
both results together give a complete complexity classification for P_ℓ-
free graphs. The W[1]-hardness result implies that it is unlikely that the
problem is fixed-parameter tractable for split graphs with parameter d.
But we do show, on the positive side, that the problem is polynomial-
time solvable, for each parameter, on split graphs if d is fixed, i.e., not
part of the input. We also initiate a study into other subclasses of perfect
graphs, namely cobipartite graphs and interval graphs.

1   Introduction

A graph modification problem is usually defined as follows. We fix a graph class $\mathcal{G}$
and a set S of one or more graph operations. The input consists of a graph G
and an integer k. The question is whether G can be modified into a graph $H \in \mathcal{G}$
by using at most k operations from S. Now, instead of fixing a particular graph
class $\mathcal{G}$, one may want to fix a *graph parameter* π instead. Then the question
becomes whether G can be modified, by using at most k operations from S, into
a graph H with $\pi(H) \leq \pi(G) - d$ for some *threshold* d, which is a nonnegative
integer that can either be fixed or be part of the input. These problems have

Öznur Yaşar Diner—Supported partially by Marie Curie International Reintegration
Grant PIRG07/GA/2010/268322.
Daniël Paulusma—Supported by EPSRC EP/K025090/1.

© Springer International Publishing Switzerland 2015
V.Th. Paschos and P. Widmayer (Eds.): CIAC 2015, LNCS 9079, pp. 194–207, 2015.
DOI: 10.1007/978-3-319-18173-8_14

been studied in a number of papers [2–4, 11, 21–23], where the graph parameters that were considered are the chromatic number, clique number, independence number, matching number, and the vertex cover number, while the set S was a singleton consisting of a vertex deletion, edge deletion or edge addition. In this paper we focus on another graph operation: the *edge contraction*, for which the graph modification problem has been studied for fixed graph classes already in the early eighties [24,25] but not yet for fixed graph parameters.

Let G be a finite undirected graph with no self-loops and no multiple edges. The *contraction* of an edge uv of G removes the vertices u and v from G, and replaces them by a new vertex made adjacent to precisely those vertices that were adjacent to u or v in G. We say that a graph G can be k-*contracted* into a graph H if G can be modified into H by a sequence of at most k edge contractions.

We consider the following generic problem, where we fix the graph parameter π and the threshold d (that is, they are not part of the input):

d-CONTRACTION BLOCKER(π)
Instance: a graph $G = (V, E)$ and a nonnegative integer k.
Question: can G be k-contracted into a graph H with $\pi(H) \leq \pi(G) - d$?

We also consider the following version of the above problem where d is part of the input (thus only π is fixed):

CONTRACTION BLOCKER(π)
Instance: a graph $G = (V, E)$ and two nonnegative integers d, k.
Question: can G be k-contracted into a graph H with $\pi(H) \leq \pi(G) - d$?

These problems have been studied implicitly in the literature already in various settings. For instance, Belmonte et al. [5] proved that 1-CONTRACTION BLOCKER(Δ), where Δ denotes the maximum vertex-degree, is NP-complete even for split graphs. In this paper we consider the following graph parameters: the chromatic number χ, the clique number ω and the independence number α of a graph. The following two results follow directly from known results.

First, 1-CONTRACTION BLOCKER(χ) is NP-complete even for graphs of chromatic number 3. This can be seen as follows. Consider the problem BIPARTITE CONTRACTION, which is that of testing whether a graph can be made bipartite by at most k edge contractions. It is readily seen that 1-CONTRACTION BLOCKER(χ) and BIPARTITE CONTRACTION are equivalent for graphs of chromatic number 3. Heggernes, van't Hof, Lokshtanov and Paul [18] observed that BIPARTITE CONTRACTION is NP-complete by reducing from the NP-complete problem EDGE BIPARTIZATION, which is that of testing whether a graph can be made bipartite by deleting at most k edges. Given an instance (G, k) of EDGE BIPARTIZATION, they obtain an instance (G', k') of BIPARTITE CONTRACTION by replacing every edge in G by a path of sufficiently large odd length. Note that the resulting graph G' has chromatic number 3.

Second, 1-CONTRACTION BLOCKER(α) is NP-complete even for graphs with independence number 2. This can be seen as follows. Golovach, Heggernes, van't Hof and Paul [15] considered the s-CLUB CONTRACTION problem, which takes as input a graph G and an integer k and asks whether G can be k-contracted into a graph with diameter at most s for some fixed integer s. They showed that 1-CLUB CONTRACTION is NP-complete even for cobipartite graphs. Graphs of diameter 1 are complete graphs, that is, graphs with independence number 1, whereas cobipartite graphs have independence number at most 2.

Our Results. In Section 2 we first introduce some definitions and notations. In the same section we show that 1-CONTRACTION BLOCKER(ω) is NP-complete even for graphs with clique number 3. In Section 3 we prove that CONTRAC-TION BLOCKER(π) is polynomial-time solvable on cographs for $\pi \in \{\alpha, \chi, \omega)$. Cographs are also known as P_4-free graphs (a graph is P_ℓ-free if it has no induced path on ℓ vertices).

Our result generalizes a recent result of Golovach et al. [15] who proved that the HADWIGER NUMBER problem is polynomial time solvable on cographs. This problem is to test whether a graph contains the complete graph K_r as a minor (or equivalently as a contraction) for some given integer r, which is equivalent to the CONTRACTION BLOCKER(α) problem restricted to instances (G, d, k) where $d = \alpha(G) - 1$ and $k = |V(G)| - r$. Our result can be viewed as best possible as in Section 4 we show that for $\pi \in \{\alpha, \chi, \omega)$ the CONTRACTION BLOCKER(π) problem is NP-complete for split graphs, which form a subclass of P_5-free graphs. We show that the same hardness reduction can also be used to prove that the three problems, restricted to split graphs, are W[1]-hard when parameterized by d. The latter result means that for split graphs these problems are unlikely to be fixed-parameter tractable with parameter d. We complement the hardness results for split graphs by proving in the same section that, for all (fixed) $d \geq 1$, the d-CONTRACTION BLOCKER(π) problem is polynomial-time solvable for split graphs if $\pi \in \{\alpha, \chi, \omega\}$. See Table 1 for an overview of these results.

Cographs and split graphs are subclasses of perfect graphs. Section 5 contains, besides a number of directions for future work, some initial results for other subclasses of perfect graphs, namely for interval graphs and cobipartite graphs.

Table 1. Our results from Sections 3 and 4 for CONTRACTION BLOCKER(π) with $\pi \in \{\alpha, \chi, \omega\}$ (recall that, when d is fixed, we denote the problem by d-CONTRACTION BLOCKER(π)). Here, NP-c stands for NP-complete.

	general graphs	cographs	split graphs
d fixed	NP-c even if $d = 1$	P	P
d part of input	NP-c	P	NP-c and W[1]-hard with parameter d

2   Preliminaries

We denote a graph by $G = (V(G), E(G))$, where $V(G)$ is the vertex set and $E(G)$ is the edge set. We may write $G = (V, E)$ if no confusion is possible. All graphs considered are finite, undirected and without self-loops or multiple edges. Let $G = (V, E)$ be a graph. The *complement* of G is the graph $\overline{G} = (V, \overline{E})$ with vertex set V and an edge between two vertices u and v if and only if $uv \notin E$. For a subset $S \subseteq V$, we let $G[S]$ denote the *induced* subgraph of G, which has vertex set S and edge set $\{uv \in E \mid u, v \in S\}$. A set $I \subseteq V$ is an *independent set* of G if no two vertices in I are adjacent to each other. The *independence number* $\alpha(G)$ is the number of vertices in a maximum independent set of G. A subset $C \subseteq V$ is called a *clique* of G if any two vertices in C are adjacent to each other. The *clique number* $\omega(G)$ is the number of vertices in a maximum clique of G. For a positive integer k, a *k-coloring* of G is a mapping $c : V \to \{1, 2, \ldots, k\}$ such that $c(u) \neq c(v)$ whenever $uv \in E$. The *chromatic number* $\chi(G)$ is the smallest number k for which G has a k-coloring. Recall that the contraction of an edge $uv \in E$ removes the vertices u and v from G, and replaces them by a new vertex made adjacent to precisely those vertices that were adjacent to u or v in G (so neither self-loops nor multiple edges are created). We may also say that a vertex u is *contracted onto* v, and we use v to denote the new vertex resulting from the edge contraction.

Let G be a graph and let $\{H_1, \ldots, H_p\}$ be a set of graphs. We say that G is $(H_1, \ldots, H_p)$-*free* if G has no induced subgraph isomorphic to a graph in $\{H_1, \ldots, H_p\}$. If $p = 1$ we may write H_1-free instead of (H_1)-free. For $n \geq 1$, the graph P_n denotes the *path* on n vertices, that is, $V(P_n) = \{u_1, \ldots, u_n\}$ and $E(P_n) = \{u_i u_{i+1} \mid 1 \leq i \leq n - 1\}$. For $n \geq 3$, the graph C_n denotes the *cycle* on n vertices, that is, $V(C_n) = \{u_1, \ldots, u_n\}$ and $E(C_n) = \{u_i u_{i+1} \mid 1 \leq i \leq n - 1\} \cup \{u_n u_1\}$.

A graph $G = (V, E)$ is a *split graph* if G has a *split partition*, which is a partition of its vertex set into a clique K and an independent set I. A split partition (K, I) of a graph G is called *maximal* if $K \cup \{u\}$ is not a clique for all $u \in I$. A split partition (K, I) of a graph G is called *minimal* if $I \cup \{v\}$ is not an independent set for all $v \in K$. Split graphs coincide with $(2P_2, C_4, C_5)$-free graphs [12] (where $2P_2$ is the disjoint union of two copies of P_2). A split graph is *chordal*, that is, contains no induced cycle on four or more vertices. A graph is *cobipartite* if it is the complement of a *bipartite* graph, which is a graph whose vertex set can be split into two non-empty subsets A and B such that any edge is between a vertex of A and a vertex of B. A graph is an *interval graph* if it is the intersection graph of a set of closed intervals on the real line, i.e., its vertices correspond to the intervals and two vertices are adjacent in G if and only if their intervals have at least one point in common. A P_4-free graph is also called a *cograph*. A graph is *perfect* if the chromatic number of every induced subgraph equals the size of a largest clique in that subgraph. Chordal graphs, cobipartite graphs, cographs, interval graphs and split graphs all form subclasses of perfect graphs.

We finish this section by showing the following general result which motivated our study of special graph classes. Note that it is trivial to solve 1-CONTRACTION BLOCKER(χ) in polynomial-time on graphs with chromatic number 2 as well as 1-CONTRACTION BLOCKER(ω) on graphs with clique number 2.[1]

---

[1] We omitted the proofs of some results due to space constraints. These results are marked by ♠.

Theorem 1 (♠). 1-CONTRACTION BLOCKER(π) *is* NP-*complete for*

(i) *graphs with independence number 2 if* $\pi = \alpha$;
(ii) *graphs with chromatic number 3 if* $\pi = \chi$;
(iii) *graphs with clique number 3 if* $\pi = \omega$.

3   Cographs

Before presenting our results on cographs we first give some additional terminology. Let G_1 and G_2 be two vertex-disjoint graphs. The *join* operation $\otimes$ adds an edge between every vertex of G_1 and every vertex of G_2. The *union* operation $\oplus$ creates the disjoint union of G_1 and G_2 which is the graph with vertex set $V(G_1) \cup V(G_2)$ and edge set $E(G_1) \cup E(G_2)$. We denote the disjoint union of G_1 and G_2 by $G_1 \oplus G_2$. We denote the disjoint union of r copies of a graph G by rG.

It is well known (see, for example, [7]) that a graph G is a cograph if and only if G can be generated from K_1 by a sequence of operations, where each operation is either a join or a union. Such a sequence corresponds to a decomposition tree T, which has the following properties:

1. its root r corresponds to the graph $G_r = G$;
2. every leaf x of T corresponds to exactly one vertex of G, and vice versa, implying that x corresponds to a unique single-vertex graph G_x;
3. every internal node x of T has at least two children, is either labeled $\oplus$ or $\otimes$, and corresponds to an induced subgraph G_x of G defined as follows:
   - if x is a $\oplus$-node, then G_x is the disjoint union of all graphs G_y where y is a child of x;
   - if x is a $\otimes$-node, then G_x is the join of all graphs G_y where y is a child of x.

A cograph G may have more than one such tree but has exactly one unique tree [9], called the *cotree* T_G of G, if the following additional property is required:

4. Labels of internal nodes on the (unique) path from any leaf to r alternate between $\oplus$ and $\otimes$.

Note that T_G has $O(n)$ vertices. For our purposes we must modify T_G by applying the following known procedure (see e.g. [6]). Whenever an internal node x of T_G has more than two children y_1 and y_2, we remove the edges xy_1 and xy_2 and add a new vertex x' with edges xx', $x'y_1$ and $x'y_2$. If x is a $\oplus$-node, then x' is a $\oplus$-node, and if x is a $\otimes$-node, then x' is a $\otimes$-node. Applying this rule exhaustively yields a tree in which each internal node has exactly two children. We denote this tree by T'_G. Because T_G has $O(n)$ vertices, modifying T_G into T'_G takes linear time.

Corneil, Perl and Stewart [10] proved that the problem of deciding whether a graph with n vertices and m edges is a cograph can be solved in time $O(n+m)$. They also showed that in the same time it is possible to construct its cotree (if it exists). As modifying T_G into T'_G takes $O(n+m)$ time, we obtain the following lemma.

Lemma 1. *Let G be a graph with n vertices and m edges. Deciding if G is a cograph and constructing T_G' (if it exists) can be done in time $O(n + m)$.*

For two integers k and l we say that a graph G can be (k, l)-*contracted* into a graph H if G can be modified into H by a sequence containing k edge contractions and l vertex deletions. Note that cographs are closed under edge contraction and under vertex deletion. In fact, to prove our results for cographs, we will prove that the problem whether a cograph G can be (k, l)-contracted into a cograph H with $\pi(H) \leq \pi(G) - d$ is polynomial-time solvable for all given integers d, k, l and for all $\pi \in \{\alpha, \chi, \omega\}$.

Theorem 2. *For $\pi \in \{\alpha, \chi, \omega\}$, the* CONTRACTION BLOCKER(π) *problem can be solved in $O(n^2 + mn + k^3 n)$ time on cographs with n vertices and m edges.*

Proof. First consider $\pi = \alpha$. Let G be a cograph with n vertices and m edges that together with an integer k forms an instance of CONTRACTION BLOCKER(α). We first construct T_G'. We then consider each node of T_G' by following a bottom-up approach starting at the leaves of T_G' and ending in its root r.

Let x be a node of T_G'. Recall that G_x is the subgraph of G induced by all vertices that corresponds to leaves in the subtree of T_G' rooted at x. We associate a table with x that records the following data: for each pair of integers $i, j \geq 0$ with $i + j \leq k$ we compute the largest integer d such that G_x can be (i, j)-contracted into a graph H_x with $\alpha(H_x) \leq \alpha(G_x) - d$. We denote this integer d by $d(i, j, x)$. Let $i, j \geq 0$ with $i + j \leq k$.

Case 1. x is a leaf.
Then G_x is a 1-vertex graph meaning that $d(i, j, x) = 0$ if $j = 0$, whereas $d(i, j, x) = 1$ if $j \geq 1$.

Case 2. x is a $\oplus$-node.
Let y and z be the two children of x. Then, as G_x is the disjoint union of G_y and G_z, we find that $\alpha(G_x) = \alpha(G_y) + \alpha(G_z)$. Hence, we have

$$d(i, j, x) = \max \{\alpha(G_x) - (\alpha(G_y) - d(a, b, y) + \alpha(G_z) - d(i - a, j - b, z)) \mid$$
$$0 \leq a \leq i, 0 \leq b \leq j\}$$
$$= \max \{d(a, b, y) + d(i - a, j - b, z) \mid 0 \leq a \leq i, 0 \leq b \leq j\}.$$

Case 3. x is a $\otimes$-node.
Since x is a $\otimes$-node, G_x is connected and as such has a spanning tree T. If $i + j \geq |V(G_x)|$ and $j \geq 1$, then we can contract i edges of T in the graph G_x followed by j vertex deletions. As each operation will reduce G_x by exactly one vertex, this results in the empty graph. Hence, $d(i, j, x) = \alpha(G_x)$. From now on assume that $i + j < |V(G_x)|$ or $j = 0$. As such, any graph we can obtain from G_x by using i edge contractions and j vertex deletions is non-empty and hence has independence number at least 1.

Let y and z be the two children of x. Then, as G_x is the join of G_y and G_z, we find that $\alpha(G_x) = \max\{\alpha(G_y), \alpha(G_z)\}$. In order to determine $d(i, j, x)$ we must do some further analysis. Let S be a sequence that consists of i edge contractions and j vertex deletions of G_x such that applying S on G_x results in a graph H_x with $\alpha(H_x) = \alpha(G_x) - d(i,j,x)$. We partition S into five sets S_y^e, S_z^e, S_{yz}^e, S_y^v, S_z^v, respectively, as follows. Let S_y^e and S_z^e be the set of contractions of edges with both end-vertices in G_y and with both end-vertices in G_z, respectively. Let S_{yz}^e be the the set of contractions of edges with one end-vertex in G_y and the other one in G_z. Let $a_y = |S_y^e|$ and let $a_z = |S_z^e|$. Then $|S_{yz}^e| = i - a_y - a_z$. Let S_y^v and S_z^v be the set of deletions of vertices in G_y and G_z, respectively. Let $b = |S_y^v|$. Then $|S_z^v| = j - b$. We distinguish between two cases.

First assume that $S_{yz}^e = \emptyset$. Then $a_y + a_z = i$. Let H_y be the graph obtained from G_y after applying the subsequence of S, consisting of operations in $S_y^e \cup S_y^v$, on G_y. Let H_z be defined analogously. Then we have

$$
\begin{aligned}
\alpha(H_x) &= \max\{\alpha(H_y), \alpha(H_z)\} \\
&= \max\{\alpha(G_y) - d(a_y, b, y), \alpha(G_z) - d(a_z, j - b, z)\} \\
&= \max\{\alpha(G_y) - d(a_y, b, y), \alpha(G_z) - d(i - a_y, j - b, z)\},
\end{aligned}
$$

where the second equality follows from the definition of S.

Now assume that $S_{yz}^e \neq \emptyset$. Recall that $i + j < |V(G_x)|$ or $j = 0$. Hence $\alpha(H_x) \geq 1$. Our approach is based on the following observations.

First, contracting an edge with one end-vertex in G_y and the other one in G_z is equivalent to removing these two end-vertices and introducing a new vertex that is adjacent to all other vertices of G_x (such a vertex is said to be *universal*).

Second, assume that G_y contains two distinct vertices u and u' and that G_z contains two distinct vertices v and v'. Suppose that we are to contract two edges from $\{uv, uv', u'v, u'v'\}$. Contracting two edges of this set that have a common end-vertex, say edges uv and uv', is equivalent to deleting u, v, v' from G_x and introducing a new universal vertex. Contracting two edges with no common end-vertex, say uv and $u'v'$, is equivalent to deleting all four vertices u, u', v, v' from G_x and introducing two new universal vertices. Because the two new universal vertices in the latter choice are adjacent, whereas the vertex u' may not be universal after making the former choice, the latter choice decreases the independence number by the same or a larger value than the former choice. Hence, we may assume without loss of generality that the latter choice happened. More generally, the contracted edges with one end-vertex in G_y and the other one in G_z can be assumed to form a matching. We also note that introducing a new universal vertex to a graph does not introduce any new independent set other than the singleton set containing the vertex itself.

We conclude that each edge contraction in S_{yz}^e may be considered to be equivalent to deleting one vertex from G_y and one from G_z and introducing a new universal vertex. If one of the two graphs G_y or G_z becomes empty in this way, then an edge contraction in S_{yz}^e can be considered to be equivalent to the deletion of a vertex of the other one. Finally, if both sets G_y and G_z become empty, then we can stop as in that case H_x has independence number 1 (which we assumed was the smallest value of $\alpha(H_x)$).

By the above observations and the definition of S we find that

$$\alpha(H_x) = \max\{1, \alpha(G_y) - d(a_y, b+i-a_y-a_z, y), \alpha(G_z) - d(a_z, j-b+i-a_y-a_z, z)\}.$$

Hence we can do as follows. We consider all tuples (a_y, b) with $0 \leq a_y \leq i$ and $0 \leq b \leq j$ and compute $\max\{\alpha(G_y) - d(a_y, b, y), \alpha(G_z) - d(i - a_y, j - b, z)\}$. Let α'_x be the minimum value over all values found. We then consider all tuples (a_y, a_z, b) with $a_y \geq 0$, $a_z \geq 0$, $a_y + a_z \leq i$ and $0 \leq b \leq j$ and compute $\max\{1, \alpha(G_y) - d(a_y, b+i-a_y-a_z, y), \alpha(G_z) - d(a_z, j-b+i-a_y-a_z, z)\}$. Let α''_x be the minimum value over all values found. Then $d(i, j, x) = \alpha(G_x) - \min\{\alpha'_x, \alpha''_x\}$.

After reaching the root r, we let our algorithm return the integer $d(k, 0, r)$. By construction, $d(k, 0, r)$ is the largest integer such that $G = G_r$ can be k-contracted into a graph H with $\alpha(H) \leq \alpha(G) - d(k, 0, r)$. We are left to analyze the running time.

Constructing T'_G can be done in $O(n + m)$ time by Lemma 1. We now determine the time it takes to compute one entry $d(i, j, x)$ in the table associated with a node x. It takes linear time to compute the independence number of a cograph[2]. The total number of tuples (a_y, b) and (a_y, a_z, b) that we need to consider is $O(k^3)$. Note that the table associated with a node x has $O(k^2)$ entries but that we only have to compute $\alpha(G_x)$ once. Hence, it takes $O(n + m + k^3)$ time to construct a table for a node. As $T_{G'}$ has $O(n)$ vertices, the total running time is $O(n + m) + O(n(n + m + k^3)) = O(n^2 + mn + k^3 n)$.

Now consider $\pi = \chi$. Note that we cannot consider the complement of a cograph (which is a cograph) because an edge contraction in a graph does not correspond to an edge contraction in its complement. However, we can re-use the previous proof after making a few modifications. Let G be a cograph with n vertices and m edges that together with an integer k forms an instance of CONTRACTION BLOCKER(χ). We follow the same approach as in the proof for $n = \alpha$. We only have to swap Cases 2 and 3 after observing that $\chi(G_x) = \max\{\chi(G_y), \chi(G_z)\}$ if x is a $\oplus$-node with y and z as its two children and $\chi(G_x) = \chi(G_y) + \chi(G_z)$ if x is a $\otimes$-node. We can use the same arguments as used in the proof for $n = \alpha$ for the running time analysis as well; we only have to observe that it takes $O(n+m)$ time to compute the chromatic number of a cograph (using the same arguments as before or another algorithm of [8]).

Finally consider $\pi = \omega$. As cographs are perfect and closed under edge contractions, the proof follows immediately from the corresponding result for $\pi = \chi$. □

Remark. As can be seen from the proofs of our results, our algorithms for solving CONTRACTION BLOCKER(π) on cographs for $\pi \in \{\alpha, \chi, \omega\}$ in fact determine the largest integer d for which the input graph G can be k-contracted into a graph H with $\pi(H) \leq \pi(G) - d$.

---

[2] For a cograph G, compute T'_G and use the formula $\alpha(G_x) = \alpha(G_y) + \alpha(G_z)$ if x is a $\oplus$-node with children y and z and $\alpha(G_x) = \max\{\alpha(G_y), \alpha(G_z)\}$ otherwise. Alternatively, see for example [8] for a linear-time algorithm on a superclass of cographs.

4 Split Graphs

We first show the following result.

Theorem 3. *Let $\pi \in \{\alpha, \chi, \omega\}$. For any fixed $d \geq 0$, the d-CONTRACTION BLOCKER(π) problem is polynomial-time solvable on split graphs.*

Proof. First consider $\pi = \alpha$. Let (G, k) be an instance of d-CONTRACTION BLOCKER(α) where $G = (V, E)$ is a split graph. Let (K, I) be a minimal split partition of G. Let I' be the set of vertices in I that have at least one neighbor in K, and let $I'' = I \setminus I'$. Because G is a split graph, all vertices of I' belong to the same connected component D of G. Moreover, we have $\alpha(G) = |I| = |I'| + |I''| = \alpha(D) + |I''|$.

First suppose that $|I'| \leq d$. For (G, k) to be a yes-instance, G must be contracted into a graph G' with $\alpha(G') \leq \alpha(G) - d = |I'| + |I''| - d \leq |I''|$. This means that we must contract D into the empty graph, which is not possible. Hence, (G, k) is a no-instance in this case. Hence, we may assume without loss of generality that $|I'| \geq d + 1$.

Suppose that $k \geq d + 1$. If $k \geq |I'|$, then we contract every vertex of I' onto a neighbor in K. In this way we have k-contracted G into a graph G' with $\alpha(G') = |I''| + 1 \leq |I'| + |I''| - (|I'| - 1) \leq |I'| + |I''| - d = \alpha(G) - d$. So, (G, k) is a yes-instance in this case. If $k \leq |I'| - 1$, we contract each vertex of an arbitrary subset of k vertices of I' onto a neighbor in K. In this way we have k-contracted G into a graph G' with $\alpha(G') \leq |I'| - k + 1 + |I''| \leq |I'| + |I''| - d = \alpha(G) - d$. So, (G, k) is a yes-instance in this case as well.

If $k \leq d + 1$, we consider all possible sequences of at most k edge contractions. This takes time $O(|E(G)|^k)$, which is polynomial as d, and consequently k, is fixed. For every such sequence we check in polynomial time whether the resulting graph has chromatic number at most $\chi(G) - d$. As split graphs are closed under edge contraction and moreover are chordal graphs, the latter can be verified in linear time (see [16]).

Now let $\pi = \chi$. Let (G, k) be an instance of d-CONTRACTION BLOCKER(χ) where $G = (V, E)$ is a split graph.

Case 1. $\chi(G) \leq d$.
For (G, k) to be a yes-instance, G must be contracted into a graph G' with $\chi(G') \leq \chi(G) - d \leq 0$. The only graph with chromatic number at most 0, is the empty graph. However, a non-empty graph cannot be contracted to an empty graph. Hence, (G, k) is a no-instance in this case.

Case 2. $\chi(G) = d + 1$.
For (G, k) to be a yes-instance, G must be contracted into a graph G' with $\chi(G') \leq \chi(G) - d = 1$. Hence, every connected component of G' must consist of exactly one vertex. If G has no connected components with edges, then (G, k) is a yes-instance. Otherwise, because G is a split graph, G has exactly one connected component D containing one or more edges. In that case, (G, k) is a yes-instance if and only if $k \geq |V(D)| - 1$; this can be checked in constant time.

Case 3. $\chi(G) \geq d + 2$.

First, assume that $k < d$. Because every edge contraction reduces the chromatic number by at most 1, (G, k) is a no-instance.

Second, assume that $k = d$. We consider all possible sequences of at most k edge contractions. This takes time $O(|E(G)|^k)$, which is polynomial as d, and consequently k, is fixed. For every such sequence we check in polynomial time whether the resulting graph has chromatic number at most $\chi(G) - d$. As split graphs are closed under edge contractions and moreover are chordal graphs, the latter can be verified in polynomial time (see [16]).

Third, assume that $k > d$. We claim that (G, k) is a yes-instance. This can be seen as follows. Let (K, I) be a maximal split partition of G.

If $k < |K|$, then we contract k arbitrary edges of K. The resulting graph G' has a split partition (K', I) with $|K'| = |K| - k \leq |K| - d - 1$. Hence $\chi(G') \leq |K'| + 1 \leq |K| - d = \chi(G) - d$. Note that the latter equality follows from our assumption that (K, I) is maximal. Now suppose that $k \geq |K|$. We contract $|K|$ arbitrary edges of K. The resulting graph G' has chromatic number $2 \leq \chi(G) - d$. Hence, in both cases, we conclude that (G, k) is a yes-instance.

Finally consider $\pi = \omega$. We use the previous result combined with the fact that split graphs are perfect and closed under edge contractions. □

In our next theorem we give two hardness results which, as explained in Section 1, show that Theorem 3 can be seen as best possible. In their proofs we will reduce from the RED-BLUE DOMINATING SET problem. This problem takes as input a bipartite graph $G = (R \cup B, E)$ and an integer k, and asks whether there exists a *red-blue dominating set* of size at most k, that is, a subset $D \subseteq B$ of at most k vertices such that every vertex in R has at least one neighbor in D. This problem is NP-complete, because it is equivalent to the NP-complete problems SET COVER and HITTING SET [14]. The RED-BLUE DOMINATING SET problem is also W[1]-complete when parameterized by $|B| - k$ [17]. Belmonte et al. [5] reduced from the same problem for showing that 1-CONTRACTION BLOCKER(Δ) is NP-complete and W[2]-hard (with parameter k) for split graphs, but the arguments we use to prove our results are quite different from the ones they used.

Theorem 4. *For $\pi \in \{\alpha, \chi, \omega\}$, the* CONTRACTION BLOCKER(π) *problem, restricted to split graphs, is* NP-*complete as well as* W[1]-*hard when parameterized by d.*

Proof. The problem is readily seen to be in NP for $\pi \in \{\alpha, \chi, \omega\}$. Recall that we reduce from RED BLUE DOMINATING SET in order to show NP-hardness and W[1]-hardness with parameter d.

First consider $\pi = \alpha$. Let $G = (R \cup B, E)$ be a bipartite graph that together with an integer k forms an instance of RED-BLUE DOMINATING SET. We may assume without loss of generality that $k \leq |B|$. Moreover, we may assume that every vertex of R is adjacent to at least one vertex of B. We add all possible edges between vertices in R. This yields a split graph G^* with a split partition (R, B).

Because every vertex in R is assumed to be adjacent to at least one vertex of B in G, we find that (R, B) is a minimal split partition of G^*.

Because RED-BLUE DOMINATING SET problem is NP-complete [14] and W[1]-complete when parameterized by $|B| - k$ [17], it suffices to prove that G has a red-blue dominating set of size at most k if and only if $(G^*, |B| - k)$ is a yes-instance of $(|B| - k)$-CONTRACTION BLOCKER(α). We prove this claim below.

First suppose that G has a red-blue dominating set D of size at most k. Because $k \leq |B|$, we may assume without loss of generality that $|D| = k$ (otherwise we would just add some vertices from $B \setminus D$ to D).

In G^* we contract every $u \in B \setminus D$ onto a neighbor in R. In this way we $(|B| - k)$-contracted G^* into a graph G'. Note that G' is a split graph that has a split partition (R, D). Because every vertex in R is adjacent to at least one vertex of D in G by definition of D, it is adjacent to at least one vertex of D in G^*. The latter statement is still true for G', as contracting an edge incident to a vertex $u \in B$ is equivalent to deleting u. Hence, (R, D) is a minimal split partition of G', so $\alpha(G') = |D|$. Because (R, B) is a minimal split partition of G^*, we have $\alpha(G^*) = |B|$. This means that $\alpha(G') = |D| = |B| - (|B| - |D|) = \alpha(G^*) - (|B| - k)$. We conclude that $(G^*, |B| - k)$ is a yes-instance of $(|B| - k)$-CONTRACTION BLOCKER(α).

Now suppose that $(G^*, |B| - k)$ is a yes-instance of $(|B| - k)$-BLOCKER(α), that is, G^* can be $(|B| - k)$-contracted into a graph G' such that $\alpha(G') \leq \alpha(G^*) - (|B| - k)$. Recall that $\alpha(G^*) = |B|$. Hence, $\alpha(G') \leq k$. Let p be the number of contractions of edges with one end-vertex in B. Note that any such contraction decreases the size of the independent set B by exactly one. If $p < |B| - k$, then G' contains an independent set of size $|B| - p > k$, which would mean that $\alpha(G') > k$, a contradiction. Hence, $p \geq |B| - k$, which implies that $p = |B| - k$ as we performed no more than $|B| - k$ contractions in total. Let D denote the independent set obtained from B after all edge contractions. Then we find that $k = |B| - (|B| - k) = |B| - p = |D| \leq \alpha(G') \leq \alpha(G^*) - (|B| - k) = |B| - (|B| - k) = k$. Hence, $|D| = \alpha(G')$, which means that (D, R) is a minimal split partition of G'. This means that every vertex of R is adjacent to at least one vertex of D in G'. Because all our contractions were performed on edges with one end-vertex in B, we have only removed vertices from G^*, that is, G' is an induced subgraph of G^*. Hence, every vertex of R is adjacent to at least one vertex of D in G'. Consequently, D is a red-blue dominating set of G with size $|D| = k$.

We omit the proof for $\pi = \chi$. As split graphs are perfect and closed under edge contractions, the case $\pi = \omega$ follows directly from the case $\pi = \chi$. □

5 Conclusions

Because split graphs are $(2P_2, C_4, C_5)$-free [12], they are P_5-free. This means that Theorem 2, combined with Theorem 4, has the following consequence.

Corollary 1. *Let* $\pi \in \{\alpha, \chi, \omega\}$. *Then* CONTRACTION BLOCKER(π) *restricted to* P_ℓ-*free graphs is polynomial-time solvable if* $\ell \leq 4$ *and* NP-*complete if* $\ell \geq 5$.

Recently, Lokshtanov, Vatshelle, and Villanger [20] proved that the independence number of a P_5-free graph can be computed polynomial time (thereby solving a long-standing open problem). In contrast, already 1-CONTRACTION BLOCKER(α) is NP-complete for P_5-free graphs (recall that it is NP-complete even for cobipartite graphs, as explained in Section 1). The problems of determining the chromatic number [19] and the clique number [1] are NP-hard for P_5-free graphs. One might be able to use these two results to prove NP-hardness of d-CONTRACTION BLOCKER(π) for $\pi \in \{\chi, \omega\}$ and $d \geq 1$.

The classes of cographs and split graphs are subclasses of the class of perfect graphs. Thus, it is interesting to study CONTRACTION BLOCKER(π) for other subclasses of perfect graphs, such as interval graphs or cobipartite graphs with $\pi \in \{\alpha, \chi\}$ (since for perfect graphs CONTRACTION BLOCKER(ω) and CONTRACTION BLOCKER(χ) are equivalent). For interval graphs we can show the following result.

Theorem 5 (♠). *Let* $\pi \in \{\chi, \omega\}$. *Then* CONTRACTION BLOCKER(π) *can be solved in polynomial time on interval graphs.*

Whether the same result holds for CONTRACTION BLOCKER(α) is not clear and left as future work.

Cobipartite graphs have independence number at most 2, that is, are $3P_1$-free. We can show the following.

Theorem 6 (♠). *For any fixed* $d \geq 0$, *the* d-CONTRACTION BLOCKER(χ) *problem can be solved in polynomial time on* $3P_1$-*free graphs.*

Whether Theorem 6 can be generalized to the class of $4P_1$-free graphs is an open problem. Its proof cannot be translated to $4P_1$-free graphs, because computing the chromatic number is NP-hard for $4P_1$-free graphs [19]. Also, determining the complexity of CONTRACTION BLOCKER(χ) for the class of cobipartite graphs and its superclass of $3P_1$-free graphs is still open. Moreover, we do not know the complexity of d-CONTRACTION BLOCKER(ω) for $3P_1$-free graphs and $d \geq 1$ (whereas, for $\pi = \alpha$ this problem is NP-complete already for $d = 1$ even for cobipartite graphs, as we recalled earlier).

Finally, we note that a similar table as Table 1 is not complete for the other variants of the blocker problem where the operation permitted is the edge addition, edge deletion or vertex deletion, respectively. For edge deletions the problem, for $\pi = \chi$, is known [2] to be NP-hard for general graphs even if $d = 1$, polynomial-time solvable on threshold graphs (which form a proper subclass of P_4-free graphs) if d is part of the input and polynomial-time solvable on split graphs but only if d is fixed. For edge additions the problem, for $\pi = \alpha$, is known [2] to be NP-hard for general graphs even if $d = 1$ and polynomial-time solvable on split graphs if d is fixed. For vertex deletions the problem, for $\pi = \omega$, is known to be NP-complete for general graphs [21] and, for $\pi = \alpha$, polynomial-time solvable for cographs if d is part of the input [4]. It would be interesting to complete these results in the way we have done for edge contractions.

References

1. Alekseev, V.E.: On easy and hard hereditary classes of graphs with respect to the independent set problem. Discrete Applied Math. **132**, 17–26 (2004)
2. Bazgan, C., Bentz, C., Picouleau, C., Ries, B.: Blockers for the stability number and the chromatic number. Graphs and Combinatorics **31**, 73–90 (2015)
3. Bazgan, C., Toubaline, S., Tuza, Z.: Complexity of Most Vital Nodes for Independent Set in Graphs Related to Tree Structures. In: Iliopoulos, C.S., Smyth, W.F. (eds.) IWOCA 2010. LNCS, vol. 6460, pp. 154–166. Springer, Heidelberg (2011)
4. Bazgan, C., Toubaline, S., Tuza, Z.: The most vital nodes with respect to independent set and vertex cover. Discrete Applied Mathematics **159**(17), 1933–1946 (2011)
5. Belmonte, R., Golovach, P.A., van' t Hof, P.: Parameterized complexity of three edge contraction problems with degree constraints. Acta Informatica **51**, 473–497 (2014)
6. Bodlaender, H.L., Möhring, R.H.: The pathwidth and treewidth of cographs. SIAM J. Discrete Math. **6**, 181–188 (1993)
7. Brandstädt, A., Le, V.B., Spinrad, J.: Graph Classes: A Survey. SIAM Monographs on Discrete Mathematics and Applications (1999)
8. Chvátal, V., Hoàng, C.T., Mahadev, N.V.R., de Werra, D.: Four classes of perfectly orderable graphs. J. Graph Theory **11**, 481–495 (1987)
9. Corneil, D.G., Lerchs, H., Stewart Burlingham, L.: Complement reducible graphs. Discrete Applied Mathematics **3**, 163–174 (1981)
10. Corneil, D.G., Perl, Y., Stewart, L.K.: A linear recognition algorithm for cographs. SIAM J. Comput. **14**, 926–934 (1985)
11. Costa, M.-C., de Werra, D., Picouleau, C.: Minimum d-blockers and d-transversals in graphs. Journal of Combinatorial Optimization **22**, 857–872 (2011)
12. Földes, S., Hammer, P.L.: Split graphs. In: 8th South-Eastern Conf. on Combinatorics, Graph Theory and Computing, Congressus Numerantium. vol. 19, pp. 311–315 (1977)
13. Fulkerson, D., Gross, O.: Incidence matrices and interval graphs. Pacific Journal of Mathematics **15**, 835–855 (1965)
14. Garey, M.R., Johnson, D.S.: Computers and Intractability: A Guide to the Theory of NP-Completeness. Freeman (1979)
15. Golovach, P.A., Heggernes, P., van 't Hof, P., Paul, C.: Hadwiger number of graphs with small chordality. In: Kratsch, D., Todinca, I. (eds.) WG 2014. LNCS, vol. 8747, pp. 201–213. Springer, Heidelberg (2014)
16. Golumbic, M.C.: Algorithmic Graph Theory and Perfect Graphs. Academic Press, New York (1980)
17. Gutin, G., Jones, M., Yeo, A.: Kernels for Below-Upper-Bound Parameterizations of the Hitting Set and Directed Dominating Set Problems. Theor. Comput. Sci. **412**, 5744–5751 (2011)
18. Heggernes, P., van't Hof, P., Lokshtanov, D., Paul, C.: Obtaining a Bipartite Graph by Contracting Few Edges. SIAM Journal on Discrete Mathematics **27**, 2143–2156 (2013)
19. Král', D., Kratochvíl, J., Tuza, Z., Woeginger, G.J.: Complexity of coloring graphs without forbidden induced subgraphs. In: Brandstädt, A., Le, V.B. (eds.) WG 2001. LNCS, vol. 2204, p. 254. Springer, Heidelberg (2001)
20. Lokshtanov, D., Vatshelle, M., Villanger, Y.: Independent Set in P5-Free Graphs in Polynomial Time. In: Proc. SODA, pp. 570–581 (2014)

21. Pajouh, F.M., Boginski, V., Pasiliao, E.L.: Minimum vertex blocker clique problem. Networks **64**, 48–64 (2014)
22. Ries, B., Bentz, C., Picouleau, C., de Werra, D., Costa, M.-C., Zenklusen, R.: Blockers and Transversals in some subclasses of bipartite graphs : when caterpillars are dancing on a grid. Discrete Mathematics **310**, 132–146 (2010)
23. Toubaline, S.: Détermination des éléments les plus vitaux pour des problèmes de graphes, Ph. D. thesis, Université Paris-Dauphine (2010)
24. Watanabe, T., Tadashi, A.E., Nakamura, A.: On the removal of forbidden graphs by edge-deletion or by edge-contraction. Discrete Applied Mathematics **3**, 151–153 (1981)
25. Watanabe, T., Tadashi, A.E., Nakamura, A.: On the NP-hardness of edge-deletion and -contraction problems. Discrete Applied Mathematics **6**, 63–78 (1983)

On the Complexity
of Wafer-to-Wafer Integration

Guillerme Duvillié[1]($\boxtimes$), Marin Bougeret[1], Vincent Boudet[1], Trivikram Dokka[2],
and Rodolphe Giroudeau[1]

[1] LIRMM, Université Montpellier 2, Montpellier, France
{guillerme.duvillie,marin.bougeret,vincent.boudet,
rodolphe.giroudeau}@lirmm.fr
[2] Department of Management Science, Lancaster University, Lancaster, UK
t.dokka@lancaster.ac.uk

Abstract. In this paper we consider the Wafer-to-Wafer Integration problem. A wafer is a p-dimensional binary vector. The input of this problem is described by m disjoints sets (called "lots"), where each set contains n wafers. The output of the problem is a set of n disjoint stacks, where a stack is a set of m wafers (one wafer from each lot). To each stack we associate a p-dimensional binary vector corresponding to the bit-wise AND operation of the wafers of the stack. The objective is to maximize the total number of "1" in the n stacks. We provide $O(m^{1-\epsilon})$ and $O(p^{1-\epsilon})$ non-approximability results even for $n = 2$, as well as a $\frac{p}{r}$-approximation algorithm for any constant r. Finally, we show that the problem is **FPT** when parameterized by p, and we use this **FPT** algorithm to improve the running time of the $\frac{p}{r}$-approximation algorithm.

1 Introduction

1.1 Problem Definition

In this paper we consider Wafer-to-Wafer Integration problems. In these problems, we are given m disjoint sets $V^1 \ldots V^m$, where each set V^i contains n binary p-dimensional vectors. For any $j \in [n]_1$[1], and any $i \in [m]_1$, we denote by v_j^i the jth vector of set V^i, and for any $k \in [p]_1$ we denote by $v_j^i(k) \in \{0,1\}$ the kth component of v_j^i.

Let us now define the output. A stack $s = (v_1^s, \ldots, v_m^s)$ is an $m-tuple$ of vectors such that $v_i^s \in V^i$, for any $i \in [m]_1$. An output of the problem is a set $S = \{s_1, \ldots, s_n\}$ of n stacks such that for any i and j, vector v_j^i is contained exactly in one stack. An example of input and output is depicted in Figure 1.

These problems are motivated by an application in IC manufacturing in semiconductor industry, see [10] for more details about this application. A wafer can be seen as a string of bad dies (0) and good dies (1). Integrating two wafers corresponds to superimposing the two corresponding strings. In this operation, a position in the merged string is only 'good' when the two corresponding dies

---

[1] The notation $[n]_j$ stands for $\{j, \ldots, n\}$.

© Springer International Publishing Switzerland 2015
V.Th. Paschos and P. Widmayer (Eds.): CIAC 2015, LNCS 9079, pp. 208–220, 2015.
DOI: 10.1007/978-3-319-18173-8_15

are good, otherwise it is 'bad'. The objective of Wafer-to-Wafer Integration is to form n stacks, while maximizing the overall quality of the stacks (depending on the objective function).

Let us now define several objective functions, and the corresponding optimization problems. We consider the operator $\wedge$ which maps two p-dimensional vectors to another one by performing the logical *and* operation on each component of entry vectors. More formally, given two p-dimensional vectors u and v, we define $u \wedge v = (u(1) \wedge v(1), u(2) \wedge v(2), \ldots, u(p) \wedge v(p))$. We associate to any stack $s = (v_1^s, \ldots, v_m^s)$ a binary p-dimensional vector $v_s = \bigwedge_{i=1}^{m} v_i^s$. Then, the profit of a stack s is given by $c(v_s)$, where $c(v) = \sum_{k=1}^{p} v(k)$. Roughly speaking, the profit of a stack is the number of good bits in the stack, where a good bit (in position k) survives iff all the vectors of the stack have a good bit in position k.

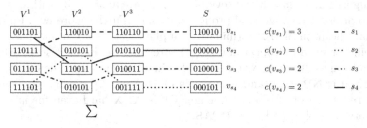

Fig. 1. Example of max $\sum 1$ instance with $m = 3, n = 4, p = 6$ and of a feasible solution S of profit $f_{\Sigma 1}(S) = 7$

We are now ready to define two following optimization problems:

Set of problems 1. max $\sum 1$ and min $\sum 0$

Input	m sets of n binary p-dimensional vectors
Output	a set S of n disjoint stacks
Objective functions	max $\sum 1$: maximize $f_{\Sigma 1}(S) = \sum_{j=1}^{n} c(v_{s_j})$, the total number of good bits
	min $\sum 0$: minimize $f_{\Sigma 0}(S) = np - \sum_{j=1}^{n} c(v_{s_j})$, the number of bad bits

---

Instance of these problems will be denoted by $I[m, n, p]$. The notation $f(S)$ (instead of $f_{\Sigma 0}(S), f_{\Sigma 1}(S), \ldots$) will be used when the context is non ambiguous.

1.2   Related Work

In this paper we consider results in the framework of approximation and fixed parameter tractability theory. We only briefly recall the definitions here and refer the reader to [9,11] for more information. For any $\rho > 1$, a ρ-approximation algorithm A (for a maximization problem) is such that for any instance I, $A(I) \geq \frac{Opt(I)}{\rho}$, where $Opt(I)$ denotes the optimal value. The input of a parameterized (decision) problem Π is a couple (X, κ), where $X \subseteq \Sigma^*$ is a classical decision problem, and $\kappa : \Sigma^* \longrightarrow \mathbb{N}$ is a parameterization. Deciding Π requires to determine for any instance $I \in \Sigma^*$ if $I \in X$. Finally, we say that an algorithm

A decides Π in **FPT** time (or that Π is **FPT** parameterized by κ) iff there exists a computable function f and a constant c such that for any I, $A(I)$ runs in $\mathcal{O}(f(\kappa(I))|I|^c)$.

The $\max \sum 1$ problem was originally defined in [10] as the "yield maximization problem in wafer-to-wafer 3-D integration technology". Authors of [10] point out that "the classical **NP**-hard 3-D matching problem is reducible to the $\max \sum 1$ problem". However, they do not provide the reduction and they only conclude that $\max \sum 1$ is **NP**-hard without stating consequences on the approximability. They also notice that $\max \sum 1$ is polynomial for $m = 2$ (as it reduces to finding a maximum profit perfect matching in a bipartite graph, solved by Hungarian Method), and design the "iterative matching heuristic" (IMH) that computes a solution based on (2D) matchings.

In [3] and [4] we investigated the $\min \sum 0$ problem by providing a $\frac{4}{3}$- approximation algorithm for $m = 3$ and several $f(m)$-approximation algorithms for arbitrary m (and for a more general profit function c). Furthermore, we also noticed in [3] that the natural ILP formulation implied that $\min \sum 0$ and $\max \sum 1$ are polynomial for fixed p. Concerning negative results, the implicit straightforward reduction from k-DIMENSIONAL MATCHING in [3] and made explicit in [5], shows that $\min \sum 0$ is **NP**-hard, and $\max \sum 1$ is $\mathcal{O}(\frac{m}{\ln m})$ non-approximable. The more complex reduction of [4] shows that $\min \sum 0$ is **APX**-hard even for $m = 3$, and thus is very unlikely to admit a **PTAS**.

Table 1. Overview of results on Wafer-to-Wafer Integration

	[10]	[3], [4]	This paper
$\max \sum 1$	**NP**-hard	$\mathcal{O}(\frac{m}{\ln m})$ inapproximability polynomial for fixed p	for any ε, $\mathcal{O}(p^{1-\varepsilon})$ and $\mathcal{O}(m^{1-\varepsilon})$ inapproximability (even for $n = 2$) $\frac{p}{r}$-approximation in $\mathcal{O}(\hat{f}(r)poly(m+n+p))$ **FPT**/p
$\min \sum 0$		for $m = 3$: $\frac{4}{3}$-approximation, **APX**-hard $f(m)$-approximation for general m polynomial for fixed p	**FPT**/p

1.3 Contributions

In this paper we mainly study the $\max \sum 1$ problem, with a particular focus on parameter p. We prove in Subsection 2.2 that even for $n = 2$, for any ϵ, there is no $\rho(m, p)$-approximation algorithm for $\max \sum 1$ such that $\rho(x, x) = \mathcal{O}(x^{1-\epsilon})$ unless **P** = **NP** (this implies in particular no $\mathcal{O}(p^{1-\epsilon})$ and no $\mathcal{O}(m^{1-\epsilon})$ ratios). These negative results show that the simple p-approximation presented in Section 3.1 is somehow the best ratio we can hope for. Nevertheless, looking for better positive results we focus on $\frac{p}{r}$-approximation algorithm for any constant r. It turns out that any $\mathcal{O}(f(n, m, p))$ exact algorithm for $\max \sum 1$ can be used to derive a $\frac{p}{r}$-approximation in $\mathcal{O}(p \times f(n, m, r))$. This motivates our main result: determining the complexity of the $\max \sum 1$ problem when parameterized by p. The natural ILP in [4] implied that $\max \sum 1$ (and $\min \sum 0$) is polynomial for fixed p. In Section 3.2, we improve this result by showing that $\max \sum 1$ (and $\min \sum 0$) are **FPT** parameterized by p.

2   Negative Results

In order to obtain negative results for $\max\sum 1$, let us first introduce two related problems defined in the table Set of Problems 2.

Set of problems 2. $\max\max 1$ and $\max_{\neq 0}$

Input	m sets of n binary p-dimensional vectors
Output	a set S of n disjoint stacks
Objective functions	$\max\max 1$: maximize $f_{\max 1}(S) = \max_{j\in[n]_1} c(v_{s_j})$, the profit of the best stack
	$\max_{\neq 0}$: maximize $f_{\neq 0}(S) = |\{j\,|\,c(v_{s_j}) \geq 1\}|$, the number of non null stacks

---

Roughly speaking, we will see that approximating $\max\sum 1$ is *harder* than approximating these two problems, and that these problems are themselves non-approximable.

To show that approximability is preserved we will provide strict reductions [1]. Indeed, if there is a strict reduction from Π_1 to Π_2, then any polynomial ρ-approximation for Π_2 yields to a ρ-approximation for Π_1. Notice that in the following we will rather provide reductions as defined in the following property.

Property 1. *Let Π_1 and Π_2 be two maximization problems with their given objective functions m_1 and m_2. Let f be a polynomial function that given any instance x of Π_1 associate an instance $f(x)$ of Π_2. Let g be a polynomial function that given any instance x of Π_1, and feasible solution S_2 of $f(x)$, associates a feasible solution $g(x, S_2)$ of Π_1. If f and g verify the two following conditions:*

1. *$Opt(x) = Opt(f(x))$*
2. *$m_1(g(x, S_2)) \geq m_2(f(x))$*

then (f, g) is a strict reduction.

2.1   Relation Between $\max_{\neq 0}$, $\max\max 1$ and $\max\sum 1$

Observation 1. *There exists a strict reduction from $\max\max 1$ to $\max\sum 1$.*

Proof. Let us construct (f, g) as in Property 1. Consider an instance $I'[m', n', p']$ of $\max\max 1$. We construct an instance $I[m, n, p]$ of $\max\sum 1$ as follows: we set $p = p'$, $n = n'$, $m = m' + 1$. The m' sets of $I'[m', n', p']$ remain unchanged in $I[m, n, p]$: $\forall i \in [m']_1$, $V^i = V'^i$ and the last set $V^{m'+1}$ contains $(n - 1)$ "zero vectors" (*i.e.* vectors having only 0) and one "one vector" (*i.e.* vector having only 1).

Informally, the set $V^{m'+1}$ of I behaves like a selecting mask: since all stacks except one are turned into zero stacks when assigning the vectors of last set, the unique "one vector" of set $V^{m'+1}$ must be assigned to the best stack, and maximizing the sum of the stacks is equivalent to maximizing the best stack.

More precisely, it is straightforward to see that the following statement is true: $\forall x$, $\exists$ *solution* S' of $\max\max 1$ *of value* $f_{\max 1}(S') = x \Leftrightarrow \exists$ *solution* S of $\max\sum 1$ *of value* $f_{\sum 1}(S) = x$. Thus, we get $Opt_{\max\max 1}(I') = Opt_{\max\sum 1}(I)$. As the previous reduction is polynomial, and a solution of I' can be deduced from a solution of I in polynomial time, we get the desired result.   □

Observation 2. *There exists a strict reduction from* $\max_{\neq 0}$ *to* $\max\sum 1$.

We refer the reader to [5] for the reduction proving this lemma.

According to Observations 1 and 2, any non-approximability result for $\max_{\neq 0}$ or $\max\max 1$ will transfer to $\max\sum 1$. This motivates the next section.

2.2   Hardness of $\max_{\neq 0}$ and $\max\max 1$

The reduction from k-DIMENSIONAL MATCHING (kDM) provided in [3] can be adapted to $\max\sum 1$ instead of $\min\sum 0$ as shown in [5]. Unlike the case of $\min\sum 0$, the reduction preserves approximability:

Theorem 1 (implicit in [3]). *There is a strict reduction from kDM to* $\max_{\neq 0}$.

As it is **NP**-hard to approximate kDM to a factor $\mathcal{O}(\frac{k}{ln(k)})$ [6], we get the following corollary:

Corollary 1. *It is* **NP**-*hard to approximate* $\max_{\neq 0}$ *within a factor* $\mathcal{O}(\frac{m}{ln(m)})$.

We can also notice that any $\frac{m}{r}$-approximation ratio (for a constant $r \geq 3$) for $\max_{\neq 0}$ or $\max\sum 1$ would improve the currently best known ratio for kDM set to $\frac{k+1+\epsilon}{3}$ in [2].

Let us now consider a new reduction which provides results for $n = 2$ and according to parameter p.

Theorem 2. *There is a strict reduction from* MAXIMUM CLIQUE PROBLEM *to* $\max\max 1$ *for* $n = 2$.

Proof. Let us construct (f, g) as in Property 1. Let us consider an instance $G = (V, E)$ of the MAXIMUM CLIQUE PROBLEM. The corresponding instance of $\max\max 1$ is constructed as follows. We consider $m = |V|$ sets, each having two vectors. All the vectors have $p = |V|$ bits. For each vertex i of V, we create the set $V^i = (v_1^i, v_2^i)$. For any i, we define $v_1^i = (v_1^i(1), v_1^i(2), \ldots, v_1^i(p))$, where $v_1^i(k) = 1$ iff $\{i, k\} \in E$ or $i = k$, and $v_2^i = (v_2^i(1), v_2^i(2), \ldots, v_2^i(p))$, where $v_2^i(k) = 1$ iff $i \neq k$. In other words, v_1^i corresponds to the ith row of the adjacency matrix of G, with a self loop.

The idea is that selecting v_1^i corresponds to selecting vertex i in graph, and selecting v_2^i will turn the ith component to 0, which corresponds to a penalty for not choosing vertex i.

We first need to state an intermediate lemma. For any stack $s = \{v_1^s, \ldots, v_m^s\}$, let $X_s = \{i | v_i^s = v_1^i\}$ be the associated set of vertices in G. Recall that v_s is the p dimensional vector representing s.

Lemma 1. $\forall i \in [p]_1, \ v_s(i) = 1 \Leftrightarrow ((i \in X_s) \ and \ (\forall x \in X_s \setminus i, \ \{x, i\} \in E)).$

$\triangle$ Let us first prove Lemma 1. Suppose i^{th} component of v_s is 1. This implies that $v_1^i \in s$, and thus $i \in X_s$. Now, suppose by contradiction that $\exists x \in X_s \setminus i$ such that $\{x, i\} \notin E$. $x \in X_s$ implies that $v_1^x \in s$. Moreover, $v_s(i) = 1$ implies that $v_1^x(i) = 1$, and thus $\{x, i\} \in E$, which leads to a contradiction. Suppose now that $i \in X_s$, and $\forall x \in X_s \setminus i, \ \{x, i\} \in E$. Let us prove that $\forall i', \ v_{i'}^s(i) = 1$. Notice first that for $i' = i$ we have $v_{i'}^s(i) = v_1^i(i) = 1$. Moreover, $\forall i' \neq i$ such that $i' \notin X_s$ we have $v_{i'}^s(i) = v_2^{i'}(i) = 1$. Finally, $\forall i' \neq i$ such that $i' \in X_s$, we have $v_{i'}^s(i) = v_1^{i'}(i) = 1$ as $\{i', i\} \in E$. $\triangle$

It is now straightforward to prove that $\forall x$, "$\exists$ solution S for max max 1 of value $f_{\max 1}(S) = x \Leftrightarrow \exists$ a clique X in G of size x." Indeed, suppose first that we have a solution S such that $f_{\max 1}(S) = x$. Let $s = (v_1^s, \ldots, v_m^s)$ be the stack in S of value x, and let $G_s = \{k | v_s(k) = 1\}$ be the set of good bits of s. We immediately get that the vertices corresponding to G_s form a clique in G, as $\forall i$ and $j \in G_s$ the previous property implies that $i \in X_s$, $j \in X_s$, and thus $\{i, j\} \in E$. Suppose now that there is a clique X^* in G, and let s be a stack such that $X_s = X^*$. The previous property implies that $\forall i \in X_s, \ v_s(i) = 1$.

Thus, $Opt_{\max \max 1}(I)$ is equal to the size of the maximum clique in G. As the previous reduction is polynomial, and as a solution of S of I can be translated back in polynomial time into the corresponding clique in G (of same size), we get the desired result. $\square$

As for any ϵ there is no $\mathcal{O}(|V|^{1-\epsilon})$-approximation for MAXIMUM CLIQUE PROBLEM (with set of vertices V) unless **P=NP** [12], we get the following corollary:

Corollary 2. *Even for $n = 2$, for any ϵ, there is no $\rho(m, p)$-approximation such that $\rho(x, x) = \mathcal{O}(x^{1-\epsilon})$ for max max 1 and thus for $\max \sum 1$. Notice that in particular, $\mathcal{O}(p^{1-\epsilon})$ and $\mathcal{O}(m^{1-\epsilon})$ are not possible, but for example $(pm)^{\frac{1}{2}}$ is not excluded.*

To summarize, the main negative results for $\max \sum 1$ are no $\mathcal{O}(p^{1-\epsilon})$-approximation and no $\mathcal{O}(m^{1-\epsilon})$ approximation for $n = 2$, and no $\mathcal{O}(\frac{m}{\ln m})$-approximation for arbitrary n (using the reduction from k-DIMENSIONAL MATCHING of [4]). Notice that it does not seem obvious to adapt the previous reductions to provide the same non-approximability results for $\min \sum 0$. Thus, the question of improving the $f(m)$ ratios provided in [4] is still open.

3   Positive Results

In this Section, we develop some polynomial-time approximation algorithm for $\max \sum 1$. Then, we show that $\max \sum 1$ and $\min \sum 0$ are **FPT** parameterized by p.

3.1 $\frac{p}{r}$-approximation

Given the previous negative results, it seems natural to look for ratio $\frac{p}{r}$, where r is a constant. Let us first see how to achieve a ratio p with Algorithm 1.

Algorithm 1. p-approximation for $\max \sum 1$

$x = 0$;

while $\exists k$ *such that it is possible to create a stack s such that $v_s(k) = 1$* **do**

    Add s to the solution;

    $x = x + 1$;

if $x < n$ **then**

    Add $n - x$ arbitrary (null) stacks to the solution;

Property 1. Algorithm 1 is a p-approximation algorithm for $\max \sum 1$.

Proof. Let $S = S_{\neq 0} \bigcup S_0$ be the solution computed by the algorithm, where $S_{\neq 0}$ is the set of non zero stacks, and S_0 is the set of the remaining null stacks. Since $S_{\neq 0}$ and S_0 are disjoint, we have $S_0 = S \setminus S_{\neq 0}$. Let $n_1 = |S_{\neq 0}|$, and $\forall i$, let $V'^i = V^i \bigcap S_0$. Let $n_2 = |S_0| = |V'^i|$ (all the V'^i have the same size). Notice that $n = n_1 + n_2$.

As the algorithm cannot create any non null stack at the end of the loop, we know that for any position $k \in [p]_1$, there is a set $i(k)$ such for any vector $w \in V'^{i(k)}$, $w(k) = 0$. In other words, we can say that there is a column of n_2 zeros in set $V'^{i(k)}$. Notice there may be several columns of zeros in a given set. Thus, we deduce that there are at least p columns (of n_2 zeros) in the vectors of $V'^{i(k)}$. Moreover, as none of these zeros can be matched in a solution, we know that these $n_2 p$ zeros will appear in any solution.

Thus, given S^* an optimal solution, we have $f(S^*) \leq np - n_2 p = n_1 p$. As $f(S) \geq n_1$, we get the desired result. $\qquad \square$

Given a fixed integer r (and targeting a ratio $\frac{p}{r}$), a natural way to extend Algorithm 1 is to first look for r t-tuples (*i.e.* find $(k_1, \ldots, k_r)$ such that it is possible to create s such that $v_s(k_1) = \cdots = v_s(k_r) = 1$)), then $(r-1)$ t-tuple, *etc.* However, even for $r = 2$ this algorithm is not sufficient to get a ratio $\frac{p}{2}$, as shown by the example depicted in Figure2.

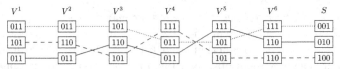

Fig. 2. Counter-example showing that Algorithm 1 for $r = 2$ remains a p-approximation. The depicted stacks correspond to an optimal solution of profit 3 whereas the algorithm outputs a solution of profit 1.

In this example it is not possible to create any stack of value strictly greater than 1 since set V^1 kills positions $\{1, 2\}$ (we say that a set kills positions $\{k_1, k_2\}$ iff there is no vector in the set such that $w(k_1) = w(k_2) = 1$), set V^2 kills positions $\{1, 3\}$, and set V^3 kills positions $\{2, 3\}$.

Thus, in this case (and more generally when no stack of value greater than 1 can be created), the solution computed by the algorithm for $r = 2$ is the same as one computed by Algorithm 1. In the worst case, the algorithm creates only one stack of value 1 (by choosing the first vector of each set). However, as depicted in Figure 2, the optimal value is 3, and thus the ratio $\frac{p}{2}$ is not verified. In other words, knowing that no stack of profit 2 can be created does not provide better results for Algorithm 1. This motivates the different approach we follow hereafter.

Property 2. Suppose that there exists an exact algorithm for $\max \sum 1$ running in $f(n, m, p)$. Then, for any $r \in [p]_1$ we have a $\frac{p}{r}$-approximation running in $\mathcal{O}(p \times f(n, m, r))$.

Proof. The idea is to use a classical "shifting technique" by guessing the subset of the r most valuable consecutive positions in the optimal solution, and run the exact algorithm on these r positions.

Let S^* be an optimal solution for $\max \sum 1$. Let us write $f(S^*) = \sum_{k=1}^{p} a_k$, where $a_k = |\{s \in S^* | v_s(k) = 1\}|$ is the number of stacks in S^* that save position k. $\forall k \in [p-1]_0$, let $X_k = \{k, \ldots, (k+r-1) \bmod p\}$, and $\sigma_k = \sum_{t \in X_k} a_t$. Notice that we have $\sum_{k=1}^{p} \sigma_k = r \sum_{k=1}^{p} a_k = rf(S^*)$, as each value a_k appears exactly r times in $\sum_{k=1}^{p} \sigma_k$. This implies $\max_k \sigma_k \geq \frac{r}{p} f(S^*)$.

For any k, let I_k be the restricted instance where all the vectors are truncated to only keep positions in X_k (there are still nm vectors in I_k, but each vector is now a r dimensional vector). By running the exact algorithm on all the I_k and keeping the best solution, we get a $\frac{p}{r}$-approximation running in $\mathcal{O}(pf(n, m, r))$. □

The previous lemma motivates the exact resolution of $\max \sum 1$ in polynomial-time for fixed p. It is already proved in [4] that $\min \sum 0$ can be solved in $\mathcal{O}(m(n^{2^p}))$. As this result also apply to $\max \sum 1$, we get a $\frac{p}{r}$-approximation running in $\mathcal{O}(pm(n^{2^r}))$, for any $r \in [p]_1$. Our objective is now to improve this running time by showing that $\max \sum 1$ (and $\min \sum 0$) are even **FPT** parameterized by p (and not only polynomial for fixed p).

3.2   Faster Algorithm for Fixed p for $\max \sum 1$

Definition 1. *For any $t \in [2^p - 1]_0$, we define configuration t as B_t: the p-dimensional binary vector that represents t in binary. We say that a p-dimensional vector v is in configuration t iff $v = B_t$.*

First ideas to get an **FPT** *algorithm*
Let us first recall our previous algorithm in [4] for fixed p. This result is obtained using an integer linear programming formulation of the following form. The objective function is $\min \sum_{t=0}^{2^p-1} x_t \bar{c}_t$ (recall that in [4] the considered objective function is $\min \sum 0$), where $x_t \in [n]_0$ is an integer variable representing the number of stacks in configuration t, and $\bar{c}_t \in [p]_0$ is the number of 0 in configuration t.

This is a good starting point to get an **FPT** algorithm. Indeed, if we note n_{var} (resp. m_{ctr}) the number of variables (resp. number of constraints) of an ILP, for any $A \in \mathbb{Q}^{n_{var} \times m_{ctr}}, b \in \mathbb{Q}^{m_{ctr}}$, the famous algorithm of Lenstra [8] allows us to decide the feasibility of an ILP, under the form $\exists? x \in \mathbb{Z}^{n_{var}} | Ax \leq b$, in time $\mathcal{O}(C^{n_{var}^3} m_{ctr}^{O(1)})$ (this running time is given in [7]), where C is a constant. Thus, to get an **FPT** algorithm parameterized by p, it is sufficient to write $\min \sum 0$ (and $\max \sum 1$) as an ILP using $f(p)$ variables.

However, it remains now to add constraints that represent the $\min \sum 0$ problem. In [4], these constraints are added using z_{jt}^i variables (for $i \in [m]_1, j \in [n]_1, t \in [2^p - 1]_0$)), where $z_{jt}^i = 1$ iff v_j^i is assigned to a stack of type t. Nevertheless these new $\mathcal{O}(mn2^p)$ variables prevent us to use [8]. Thus, we now come back to the $\max \sum 1$ problem, and our objective is to express the constraints using only the $\{x_t\}$ variables.

Presentation of the new ILP for $\max \sum 1$

For any $t \in [2^p - 1]_0$, we define an integer variable $x_t \in [n]_0$ representing the number of stacks in configuration t. Let also $c_t \in [p]_1 = c(B_t)$ be the number of 1 in configuration t.

Definition 2. *A profile is a tuple* $P = \{x_0, \ldots, x_{2^p-1}\}$ *such that* $\sum_{t=0}^{2^p-1} x_t = n$.

Definition 3. *The profile* $Pr(S) = \{x_0, \ldots, x_{2^p-1}\}$ *of a solution* $S = \{s_1, \ldots, s_n\}$ *is defined by* $x_t = |\{i|v_{s_i} \text{ is in configuration } t\}|$, *for* $t \in [2^p - 1]_0$.

Definition 4. *Given a profile* P, *an associated solution* S *is a solution such that* $Pr(S) = P$. *We say that a profile* P *is feasible iff there exists an associated solution* S *that is feasible.*

Notice that the definition of associated solutions also applies to a non feasible profile. In this case, any associated solution will also be non feasible.

Obviously, the $\max \sum 1$ problem can be formulated using the following ILP:

$$\max \sum_{t=0}^{2^p-1} x_t c_t \quad \text{subject to} \quad \sum_{t=0}^{2^p-1} x_t = n$$

$$\forall 0 \leq t < 2^p, \ x_t \in \mathbb{N}$$

$$P = \{x_t\} \text{ is feasible}$$

Our objective is now to express the feasibility of a profile by using only these 2^p variables. Roughly speaking, the idea to ensure the feasibility is the following. Let us suppose (with $p = 2$ and $n = 4$ for example) that there exists a feasible solution of fixed profile $x_0 = 0, x_1 = 1, x_2 = 2, x_3 = 1$. Suppose also that the first set is as depicted in Figure 3. To create a feasible solution with this profile, we have to "satisfy" (for each set V^i) the demands x_t for all configurations t. For example in set 1, the demand x_2 can be satisfied by using one vector in configuration 2 and one vector of configuration 3, and the demand 3 can be satisfied using the remaining vector of 3 (the demand x_0 is clearly satisfied). Notice that a demand of a given configuration (e.g. configuration 2 here) can be satisfied using a vector that "dominates" this configuration (e.g. configuration 3 here). The notion of domination will be introduced in Definition 5. Thus, a feasible profile implies that for any set i there exists a perfect matching between the vectors of V^i and the profile $\{x_t\}$.

Let us now define more formally the previous ideas.

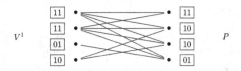

Fig. 3. Example showing that satisfying demands of profile P with set 1 requires to find a perfect matching. Edges represent domination between configuration.

Definition 5 (Domination).

A p-dimensional vector v_1 dominates a p-dimensional vector v_2 (denoted by $v_1 \gg v_2$) iff $\forall k \in [p]_1$, $v_2(k) = 1 \Rightarrow v_1(k) = 1$.

A configuration $t_1 \in [2^p - 1]_0$ dominates a configuration $t_2 \in [2^p - 1]_0$ (denoted by $t_1 \gg t_2$) iff $B_{t_1} \gg B_{t_2}$ (recall that B_t is the p-dimensional binary representation of t).

A solution S' dominates a solution S (denoted by $S' \gg S$) iff $\exists$ a bijection $\phi : [n]_1 \to [n]_1$ such that for any $i \in [n]_1$, $v_{s'_i} \gg v_{s_{\phi(i)}}$ (in other word, there is a one to one domination between stacks of S' and stacks of S).

A profile P' dominates a profile P (denoted by $P' \gg P$) iff there exists solutions S' and S such that $Pr(S') = P'$, $Pr(S) = P$ and $S' \gg S$.

Definition 6. For any $i \in [m]_1$ and any $t \in [2^p - 1]_0$, let b_t^i be the number of vectors of set V^i in configuration t.

Definition 7 (Graph G_P^i).

Let P be a profile not necessarily feasible. Let $G_P^i = ((\Delta_P, \Lambda^i), E_\gg)$, where $\Lambda^i = \{\lambda_t^{i,l}, 0 \leq t \leq 2^p - 1, 1 \leq l \leq b_t^i\}$, and $\Delta_P = \{\delta_t^l, 0 \leq t \leq 2^p - 1, 1 \leq l \leq x_t\}$. Let us fix a bijection $f : \Delta_P \cup \Lambda^i \mapsto [2^p - 1]_0$, that associates to each vertex $\lambda_t^{i,l}$ and to each vertex δ_t^l the vector in configuration t. Λ^i (resp. Δ_P) represents the set of vectors of V^i (resp. the demands of profile P) grouped according to their configurations. Notice that $|\Lambda^i| = |\Delta_P| = n$. Finally, we set $E_\gg = \{\{a, b\} | a \in \Delta_P, b \in \Lambda^i, f(a) \ll f(b)\}$.

We are now ready to show the following proposition.

Proposition 1. For any profile $P = \{x_0, \ldots, x_{2^p-1}\}$,

$$(\exists P' \text{ feasible, with } P' \gg P) \Leftrightarrow \forall i \in [m]_1, \exists a \text{ matching of size } n \text{ in } G_P^i$$

Before starting the proof, notice that the simpler proposition "for any P, P feasible $\Leftrightarrow \forall i \in [m]_1$, there is a matching of size n in G_P^i" does not hold. Indeed, $\Rightarrow$ is correct, but $\Leftarrow$ is not: consider P with $x_0 = n$ (recall that configuration 0 is the null vector), and an instance with nm "1 vectors" (containing only 1). In this case, there is a matching of size n in all the G_P^i, but P is not feasible. This explains the formulation of Proposition 1. An example of the correction formulation is depicted Figure 4.

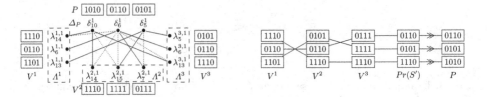

Fig. 4. Illustration of Proposition 1 with $m = n = 3$ and $p = 4$. Left: The three G_P^i graphs (edges are depicted by solid and dotted lines), and three matchings (in solid lines) corresponding to S'. Right: Solution S' s.t. $Pr(S') \gg P$.

Proof. Let P be a profile.

($\Rightarrow$) Let P' be a feasible profile that dominates P. Let $S = \{s_1, \ldots, s_n\}$ and $S' = \{s'_1, \ldots, s'_n\}$ two solutions such that S' is feasible, $Pr(S) = P$, $Pr(S') = P'$ (notice that S and P are not necessarily feasible), and $S' \gg S$. Without loss of generality, let us assume that $\forall j$, $s'_j \gg s_j$ (*i.e.* the bijection ϕ of Definition 5 is the identity), and let us assume that for any j, $s'_j = (v_j^1, \ldots, v_j^m)$. Since $v_j^i \in s'_j$, then for any i, we know that $v_j^i \gg s'_j \gg s_j$, $\forall j \in [n]_1$. This implies a matching of size n in all the graphs G_P^i.

($\Leftarrow$) Let us suppose that $\forall i \in [m]_1$, there is a matching $\mathcal{M}^i$ of size n in G_P^i.

W.l.o.g. let us rename $\{\delta_1, \ldots, \delta_n\}$ the vertices of Δ_P, and $\{\lambda_1^i, \ldots, \lambda_n^i\}$ the vertices of Λ^i such that for any i, $\mathcal{M}^i = \{\{\lambda_1^i, \delta_1\}, \ldots, \{\lambda_n^i, \delta_n\}\}$. This implies $f(\lambda_1^i) \gg f(\delta_1)$, ..., $f(\lambda_n^i) \gg f(\delta_n)$. Let us define $S = \{s_1, \ldots, s_n\}$, where $\forall j \in [n]_1$, $s_j = (f(\lambda_j^1), \ldots, f(\lambda_j^m))$. Notice that for any j, $s_j \gg f(\delta_j)$, as all the $f(\lambda_j^i) \gg f(\delta_j)$, and combining two vectors $f(\lambda_j^{i_1}) \gg f(\delta_j)$ and $f(\lambda_j^{i_2}) \gg f(\delta_j)$ creates another vector that dominates $f(\delta_j)$. Thus, S is feasible, and $Pr(S) \gg P$, and we set $P' = Pr(S)$.   □

Now, we can use the famous Hall's Theorem to express the existence of a matching in every set.

Theorem 3 (Hall's Theorem). *Let $G = ((V^1, V^2), E)$ a bipartite graph with $|V^1| = |V^2| = n$. There is a matching of size n in G iff $\forall \sigma \subseteq V^1$, $|\sigma| \leq |\Gamma(\sigma)|$, where $\Gamma(\sigma) = \{v_2 \in V^2 | \exists v_1 \in \sigma$ such that $\{v_1, v_2\} \in E\}$.*

Remark 1. Notice that we cannot use Hall's Theorem directly on graphs G_P^i, as we would have to add the 2^n constraints of the form $\forall S \subseteq V^i$. However, we will reduce the number of constraints to a function $f(p)$ by exploiting the particular structure of G_P^i.

Proposition 2 (Matching in G_P^i).

$\forall i \in [m]_1$, $\forall P = \{x_0, \ldots, x_{2^p - 1}\}$:
$(\forall \sigma \subseteq \Delta_P, |\sigma| \leq |\Gamma(\sigma)|) \Leftrightarrow (\forall \sigma_{cfg} \subseteq [2^p - 1]_0, \sum_{t \in \sigma_{cfg}} x_t \leq \sum_{t \in dom(\sigma_{cfg})} b_t^i)$ *where* $dom(\sigma_{cfg}) = \{t' | \exists t \in \sigma_{cfg}$ such that $t' \gg t\}$ *is the set of configurations that dominate* σ_{cfg}.

Proof. ($\Rightarrow$) Let $\sigma_{cfg} = \{t_1, \ldots, t_\alpha\}$. Let $\sigma = \{\delta_{t_i}^l, 1 \leq i \leq \alpha, 1 \leq l \leq x_{t_i}\}$ be the vertices of Δ_P corresponding to the demands in σ_{cfg}. Observe that $\sum_{t \in \sigma_{cfg}} x_t = |\sigma|$.

Notice also that $\Gamma(\sigma) = \{\lambda_t^{i,l}, t \in dom(\sigma), 1 \leq l \leq b_t^i\}$ by construction. Thus, $|\sigma| \leq |\Gamma(\sigma)|$ implies $\sum_{t \in \sigma_{cfg}} x_t \leq \sum_{t \in dom(\sigma_{cfg})} b_t^i$.

($\Leftarrow$) Let $\sigma \subseteq \Delta_P$. $\forall t \in [2^p - 1]_0$, let $X_t = \{\delta_t^l, 1 \leq l \leq x_t\}$, let $\sigma_t = \sigma \cap X_t$. Let $\sigma_{cfg} = \{t_1, \ldots, t_\alpha\} = \{t | \sigma_t \neq \emptyset\}$. Let $X = \bigcup_{t \in \sigma_{cfg}} \{X_t\}$. Notice that $|\sigma| \leq |X| = \sum_{t \in \sigma_{cfg}} x_t$.

Let us first prove that $\Gamma(\sigma) = \Gamma(X)$. $\Gamma(\sigma) \subseteq \Gamma(X)$ is obvious. Now, if there is a $\lambda_{t'}^{i,l'} \in \Gamma(X)$, it means that there is a $t \in \sigma_{cfg}$ such that $\lambda_{t'}^{i,l'} \in \Gamma(X_t)$, and thus there exists l such that $\{\delta_t^l, \lambda_{t'}^{i,l'}\} \in E$ (which implies that $t' \gg t$). As $\sigma_t \neq \emptyset$, there exists l' such that $\delta_t^{l'} \in \sigma_t$, and $\{\delta_t^{l'}, \lambda_{t'}^{i,l'}\} \in E$ as $t' \gg t$.

Finally, the hypothesis with our set σ_{cfg} leads to

$$|\sigma| \leq |X| = \sum_{t \in \sigma_{cfg}} x_t \leq \sum_{t \in dom(\sigma_{cfg})} b_t^i = |\Gamma(X)| = |\Gamma(\sigma)|$$

$\square$

Using Propositions 1 and 2, we can now write that for any profile $P = \{x_0, \ldots, x_{2^p-1}\}$:

$$\exists P' \text{ feasible, with } P' \gg P \Leftrightarrow \forall i, \forall \sigma_{cfg} \subseteq [2^p - 1]_0, \sum_{t \in \sigma_{cfg}} x_t \leq \sum_{t \in dom(\sigma_{cfg})} b_t^i.$$

Thus, we use now the following ILP to describe the $\max \sum 1$ problem:

$$\max \quad \sum_{t=0}^{2^p-1} x_t c_t$$

subject to $\quad \forall i \in [m]_1 : \forall \sigma_{cfg} \subseteq [2^p - 1]_0, \sum_{t \in \sigma_{cfg}} x_t \leq \sum_{t \in dom(\sigma_{cfg})} b_t^i$

$$\forall t \in [2^p - 1]_0, x_t \in \mathbb{N}$$

This linear program has 2^p variables and $(m2^{2^p} + 2^p)$ constraints. Thus, we can solve it using [8] in time $f(p)poly(n+m)$, we get that $\max \sum 1$ and $\min \sum 0$ are **FPT** parameterized by p. Using Property 2 this ILP leads to a $\frac{p}{r}$-approximation algorithm for $\max \sum 1$ running in time $f(r)poly(n + m + p)$.

4 Conclusion

In this article, we established that $\max \sum 1$ is $\mathcal{O}(m^{1-\varepsilon})$ and $\mathcal{O}(p^{1-\varepsilon})$ non-approximable for $n = 2$. On the positive side, we provided a **FPT** algorithm for $\max \sum 1$ leading to a $\frac{p}{r}$-approximation algorithm running in $\mathcal{O}(f(r)poly(m + n + p))$, which is the best we can hope for. The existence of a $\rho(m)$-approximation (typically m) algorithm remains open.

References

1. Crescenzi, P., Kann, V., Silvestri, R., Trevisan, L.: Structure in approximation classes. SIAM Journal on Computing **28**(5), 1759–1782 (1999)
2. Cygan, M.: Improved approximation for 3-dimensional matching via bounded pathwidth local search. In: 2013 IEEE 54th Annual Symposium on Foundations of Computer Science (FOCS), pp. 509–518. IEEE (2013)
3. Dokka, T., Bougeret, M., Boudet, V., Giroudeau, R., Spieksma, F.C.R.: Approximation algorithms for the wafer to wafer integration problem. In: Erlebach, T., Persiano, G. (eds.) WAOA 2012. LNCS, vol. 7846, pp. 286–297. Springer, Heidelberg (2013)
4. Dokka, T., Crama, Y., Spieksma, F.C.R.: Multi-dimensional vector assignment problems. Discrete Optimization **14**, 111–125 (2014)
5. Duvillié, G., Bougeret, M., Boudet, V., Dokka, T., Giroudeau, R.: On the complexity of Wafer-to-Wafer Integration. Research report, Lirmm; UM II montpellier, Faculté des Sciences et Techniques du Languedoc, January 2015. HAL id:lirmm-01110027
6. Hazan, E., Safra, S., Schwartz, O.: On the complexity of approximating k-dimensional matching. In: Arora, S., Jansen, K., Rolim, J.D.P., Sahai, A. (eds.) RANDOM 2003 and APPROX 2003. LNCS, vol. 2764, pp. 83–97. Springer, Heidelberg (2003)
7. Kratsch, S.: On polynomial kernels for integer linear programs: covering, packing and feasibility. In: Bodlaender, H.L., Italiano, G.F. (eds.) ESA 2013. LNCS, vol. 8125, pp. 647–658. Springer, Heidelberg (2013)
8. Lenstra Jr., H.W.: Integer programming with a fixed number of variables. Mathematics of Operations Research **8**(4), 538–548 (1983)
9. Niedermeier, R.: Invitation to fixed-parameter algorithms (2006)
10. Reda, S., Smith, G., Smith, L.: Maximizing the functional yield of wafer-to-wafer 3-d integration. IEEE Transactions on Very Large Scale Integration (VLSI) Systems **17**(9), 1357–1362 (2009)
11. Williamson, D.P., Shmoys, D.B.: The design of approximation algorithms. Cambridge University Press (2011)
12. Zuckerman, D.: Linear degree extractors and the inapproximability of max clique and chromatic number. In: Proceedings of the Thirty-Eighth Annual ACM Symposium on Theory of Computing, pp. 681–690. ACM (2006)

Label Placement in Road Maps

Andreas Gemsa, Benjamin Niedermann$^{(\boxtimes)}$, and Martin Nöllenburg

Institute of Theoretical Informatics, Karlsruhe Institute of Technology,
Karlsruhe, Germany
{gemsa,niedermann,noellenburg}@kit.edu

Abstract. A road map can be interpreted as a graph embedded in the
plane, in which each vertex corresponds to a road junction and each
edge to a particular road section. We consider the cartographic problem
to place non-overlapping road labels along the edges so that as many
road sections as possible are identified by their name, i.e., covered by a
label. We show that this is NP-hard in general, but the problem can be
solved in polynomial time if the road map is an embedded tree.

1 Introduction

Map labeling is a well-known cartographic problem in computational geometry [12, Chapter 58.3.1],[14]. Depending on the type of map features, one can
distinguish labeling of *points*, *lines*, and *areas*. Common cartographic quality criteria are that labels must be disjoint and clearly identify their respective map
features [8]. Most of the previous work concerns point labeling, while labeling line
and area features received considerably less attention. In this paper we address
labeling linear features, namely roads in a road map.

Geometrically, a *road map* is the representation of a *road graph* G as an
arrangement of fat curves in the plane $\mathbb{R}^2$. Each *road* is a connected subgraph
of G (typically a simple path) and each edge belongs to exactly one road. Roads
may intersect each other in *junctions*, the vertices of G, and we denote an edge
connecting two junctions as a *road section*. In road labeling the task is to place
the road names inside the fat curves so that the road sections are identified
unambiguously, see Fig. 1.

Chirié [1] presented a set of rules and quality criteria for label placement
in road maps based on interviews with cartographers. This includes that (C1)
labels are placed inside and parallel to the road shapes, (C2) every road section
between two junctions should be clearly identified, and (C3) no two road labels
may intersect. Further, he gave a mathematical description for labeling a single
road and introduced a heuristic for sequentially labeling all roads in the map.
Imhof's foundational cartographic work on label positioning in maps lists very
similar quality criteria [4]. Edmondson et al. [2] took an algorithmic perspective
on labeling a single linear feature (such as a river). While Edmondson et al.
considered *non-bent* labels, Wolff et al. [13] introduced an algorithm for single
linear feature that places labels following the curvature of the linear feature.

© Springer International Publishing Switzerland 2015
V.Th. Paschos and P. Widmayer (Eds.): CIAC 2015, LNCS 9079, pp. 221–234, 2015.
DOI: 10.1007/978-3-319-18173-8_16

Strijk [10] considered static road labeling with embedded labels and presented a heuristic for selecting non-overlapping labels out of a set of label candidates. Seibert and Unger [9] considered grid-shaped road networks. They showed that in those networks it is NP-complete and APX-hard to decide whether for every road at least one label can be placed. Yet, Neyer and Wagner [7] introduced a practically efficient algorithm that finds such a grid labeling if possible. Maass and Döllner [6] presented a heuristic for labeling the roads of an interactive 3D map with objects (such as buildings). Apart from label-label overlaps, they also resolve label-object occlusions. Vaaraniem et al. [11] used a force-based labeling algorithm for 2D and 3D scenes including road label placement.

Contribution. While in grid-shaped road networks it is sufficient to place a single label per road to clearly identify all its road sections, this is not the case in general road networks. Consider the example in Fig. 1. In Fig. 1a), it is not obvious whether the orange road section in the center belongs to *Knuth St.* or to *Turing St.* Simply maximizing the number of placed labels, as often done for labeling point features, can cause undesired effects like unnamed roads or clumsy label placements (e.g., around *Dijkstra St.* and *Hamming St.* in Fig. 1a)). Therefore, in contrast to Seibert and Unger [9], we aim for maximizing the number of *identified* road sections, i.e., road sections that can be clearly assigned to labels; see Fig. 1b).

Based on criteria (C1)–(C3) we introduce a new and versatile model for road labeling in Section 2. In Section 3 we show that the problem of maximizing the number of identified road sections is NP-hard for general road graphs, even if each road is a path. For the special case that the road graph is a tree, we present a polynomial-time algorithm in Section 4. This special case is not only of theoretical interest, but our algorithm in fact provides a very useful subroutine in exact or heuristic algorithms for labeling general road graphs. Our initial experiments, sketched in the full version [3], show that real-world road networks decompose into small subgraphs, a large fraction of which (more than 85.1%) are actually trees, and thus can be labeled optimally by our algorithm.

2 Preliminaries

As argued above, a road map is a collection of fat curves in the plane, each representing a particular piece of a named road. If two (or more) such curves intersect, they form junctions. A *road label* is again a fat curve (the bounding shape of the road name) that is contained in and parallel to the fat curve representing its road. We observe that labels of different roads can intersect only within junctions and that the actual width of the curves is irrelevant, except for defining the shape and size of the junctions. These observations allow us to define the following more abstract but equivalent road map model.

A *road map* $\mathcal{M}$ is a planar *road graph* $G = (V, E)$ together with a planar embedding $\mathsf{E}(G)$, which can be thought of as the geometric representation of the road axes as thin curves; see Fig. 1c). We denote the number of vertices of G

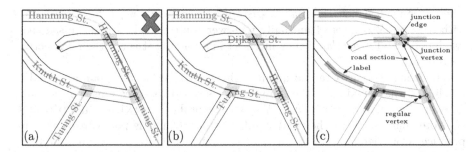

Fig. 1. a–b): Two ways to label the same road network. Each road section has its own color. Junctions are marked gray. Fig. b) identifies all road sections. c) Illustration of the road graph and relevant terms.

by n, and the number of edges by m. Observe that since G is planar $m = O(n)$. Each edge $e \in E$ is either a *road section*, which is not part of a junction, or a *junction edge*, which is part of a junction. Each vertex $v \in V$ is either a *junction vertex* incident only to junction edges, or a *regular vertex* incident to one road section and at most one junction edge, which implies that each regular vertex has degree at most two. A junction vertex v and its incident edges are denoted as a *junction*. The edge set E decomposes into a set $\mathcal{R}$ of edge-disjoint *roads*, where each road $R \in \mathcal{R}$ induces a connected subgraph of G. Without loss of generality we assume no two road sections G are incident to the same vertex. Thus, a road decomposes into road sections, separated by junction vertices and their incident junction edges. In realistic road networks the number of roads connected passing through a junction is small and does not depend on the size of the road network. We therefore assume that each vertex in G has constant degree. We assume that each road $R \in \mathcal{R}$ has a name whose length we denote by $\lambda(R)$.

For simplicity, we identify the embedding $\mathsf{E}(G)$ with the points in the plane covered by $\mathsf{E}(G)$, i.e. $\mathsf{E}(G) \subseteq \mathbb{R}^2$. We also use $\mathsf{E}(v)$, $\mathsf{E}(e)$, and $\mathsf{E}(R)$ to denote the embeddings of a vertex v, an edge e, and a road R.

We model a label as a simple open curve $\ell\colon [0,1] \to \mathsf{E}(G)$ in $\mathsf{E}(G)$. Unless mentioned otherwise, we consider a curve ℓ always to be simple and open, i.e., ℓ has no self-intersections and its end points do not coincide. In order to ease the description, we identify a curve ℓ in $\mathsf{E}(G)$ with its image, i.e., ℓ denotes the set $\{\ell(t) \in \mathsf{E}(G) \mid t \in [0,1]\}$. The start point of ℓ is denoted as the *head* $h(\ell)$ and the endpoint as the *tail* $t(\ell)$. The length of ℓ is denoted by $\mathrm{len}(\ell)$. The curve ℓ *identifies* a road section r if $\ell \cap \mathsf{E}(r) \neq \emptyset$. For a set $\mathcal{L}$ of curves $\omega(\mathcal{L})$ is the number of road sections that are identified by the curves in $\mathcal{L}$. For a single curve ℓ we use $\omega(\ell)$ instead of $\omega(\{\ell\})$. For two curves ℓ_1 and ℓ_2 it is not necessarily true that $\omega(\{\ell_1, \ell_2\}) = \omega(\ell_1) + \omega(\ell_2)$, because they may identify the same road section twice.

A *label* ℓ for a road R is a curve $\ell \subseteq \mathsf{E}(R)$ of length $\lambda(R)$ whose endpoints must lie on road sections and not on junction edges or junction vertices. Requiring that labels end on road sections avoids ambiguous placement of labels in junctions where it is unclear how the road passes through it. A *labeling* $\mathcal{L}$ for a road map with road set $\mathcal{R}$ is a set of mutually non-overlapping labels, where we say that two labels ℓ and ℓ' *overlap* if they intersect in a point that is not their respective head or tail.

Following the cartographic quality criteria (C1)–(C3), our goal is to find a labeling $\mathcal{L}$ that maximizes the number of identified road sections, i.e., for any labeling $\mathcal{L}'$ we have $\omega(\mathcal{L}') \leq \omega(\mathcal{L})$. We call this problem MAXIDENTIFIEDROADS.

Note that assuming the road graph G to be planar is not a restriction in practice. Consider for example a road section r that overpasses another road section r', i.e., r is a bridge over r', or r' is a tunnel underneath r. In order to avoid overlaps between labels placed on r and r', we either can model the intersection of r and r' as a regular crossing of two roads or we split r' in smaller road sections that do not cross r. In both cases the corresponding road graph becomes planar. In the latter case we may obtain more independent roads created by chopping r' into smaller pieces.

3   Computational Complexity

We first study the computational complexity of road labeling and prove NP-hardness of MAXIDENTIFIEDROADS in the following sense.

Theorem 1. *For a given road map $\mathcal{M}$ and an integer K it is NP-hard to decide if in total at least K road sections can be identified.*

Proof. We perform a reduction from the NP-complete PLANAR MONOTONE 3-SAT problem [5]. An instance of PLANAR MONOTONE 3-SAT is a Boolean formula φ with n variables and m clauses (disjunctions of at most three literals) that satisfies the following additional requirements: (i) φ is *monotone*, i.e., every clause contains either only positive literals or only negative literals and (ii) the induced variable-clause graph H_φ of φ is planar and can be embedded in the plane with all variable vertices on a horizontal line, all positive clause vertices on one side of the line, all negative clauses on the other side of the line, and the edges drawn as rectilinear curves connecting clauses and contained variables on their respective side of the line. We construct a road map $\mathcal{M}_\varphi$ that mimics the shape of the above embedding of H_φ by defining variable and clause gadgets, which simulate the assignment of truth values to variables and the evaluation of the clauses. We refer to Fig. 2 for a sketch of the construction.

Chain Gadget. The basic building block is the *chain gadget*, which consists of an alternating sequence of equally long horizontal and vertical roads with identical label lengths that intersect their respective neighbors in the sequence and form junctions with them as indicated in Fig. 2c). Assume that the chain consists of $k \geq 3$ roads. Then each road except the first and last one decomposes into three road sections split by two junctions, a longer central section and two short end sections; the first and last road consist of only two road sections, a short one and a long one, separated by one junction. (These two roads will later be connected to other gadgets; indicated by dotted squares in Fig. 2c).) The label

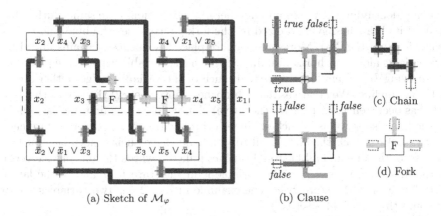

Fig. 2. Illustration of NP-hardness proof. (a) 3-Sat formula $\varphi = (x_4 \vee x_1 \vee x_5) \wedge (x_2 \vee x_4 \vee x_3) \wedge (\bar{x}_2 \vee \bar{x}_1 \vee \bar{x}_3) \wedge (\bar{x}_3 \vee \bar{x}_5 \vee \bar{x}_4)$ represented as road graph $\mathcal{M}_\varphi$. Truth assignment is $x_1 = true$, $x_2 = true$, $x_3 = false$, $x_4 = false$ and $x_5 = false$. (b) Clause gadget in two states. (c) The chain is the basic building block for the proof. (d) Schematized fork gadget.

length and distance between junctions is chosen so that for each road either the central and one end section is identified, or no section at all is identified. For the first and last road, both sections are identified if the junction is covered and otherwise only the long section can be identified. We have k roads and $k - 1$ junctions. Each label must block a junction, if it identifies two sections. So the best possible configuration blocks all junctions and identifies $2(k-1)+1 = 2k-1$ road sections.

The chain gadget has exactly two states, in which $2k - 1$ road sections are identified. Either the label of the first road does not block a junction and identifies a single section and all subsequent roads have their label cover the junction with the preceding road in the sequence, or the label of the last road does not block a junction and all other roads have their label cover the junction with the successive road in the sequence. In any other configuration there is at least one road without any identified section and thus at most $2k - 2$ sections are identified. We use the two optimal states of the gadget to represent and transmit the values *true* and *false* from one end to the other.

Fork Gadget. The *fork gadget* allows to split the value represented in one chain into two chains, which is needed to transmit the truth value of a variable into multiple clauses. To that end it connects to an end road of three chain gadgets by sharing junctions. The detailed description of the fork gadget is found in the full version[3].

Variable Gadget. We define the *variable gadgets* simply by connecting chain and fork gadgets into a connected component of intersecting roads. This construction already has the functionality of a variable gadget: it represents (in a labeling identifying the maximum number of road sections) the same truth value in all of its branches, synchronized by the fork gadgets, see the blue chains and yellow forks in Fig. 2a). More precisely, we place a sequence of chains linked by fork gadgets along the horizontal line on which the variable vertices are placed in the drawing H_φ. Each fork creates a branch of the variable gadget either above or below the line. We create as many branches above (below) the line as the variable has occurrences in positive (negative) clauses in φ. The first and last chain on the line also serve as branches. The synchronization of the different branches via the forks is such that either all top branches have their road labels pushed away from the line and all bottom branches pulled towards the line or vice versa. In the first case, we say that the variable is in the state *false* and in the latter case that it is in the state *true*. The example in Fig. 2 has two variables set to *true* and three variables set to *false*.

Clause Gadget. Finally, we need to create the clause gadget, which links three branches of different variables. The core of the gadget is a single road that consists of three sub-paths meeting in one junction. Each sub-path of that road shares another junction with one of the three incoming variable branches. Beyond each of these three junctions the final road sections are just long enough so that a label can be placed on the section. However, the section between the central junction of the clause road and the junctions with the literal roads is shorter than the label length. The road of the clause gadget has six sections in total and we argue that the six sections can only be identified if at least one incoming literal evaluates to *true*. Otherwise at most five sections can be identified. By construction, each road in the chain of a false literal has its label pushed towards the clause, i.e., it blocks the junction with the clause road. As long as at least one of these three junctions is not blocked, all sections can be identified; see Fig. 2b). But if all three junctions are blocked, then only two of the three inner sections of the clause road can be identified and the third one remains unlabeled; see Fig. 2b).

Reduction. Obviously, the size of the instance $\mathcal{M}_\varphi$ is polynomial in n and m. If we have a satisfying variable assignment for φ, we can construct the corresponding road labeling and the number of identified road sections is six per clause and a fixed constant number K' of sections in the variable gadgets, i.e., at least $K = K' + 6m$. On the other hand, if we have a road labeling with at least K identified sections, each variable gadget is in one of its two maximum configurations and each clause road has at least one label that covers a junction with a literal road, meaning that the corresponding truth value assignment of the variables is indeed a satisfying one. This concludes the reduction.

Since MAXIDENTIFIEDROADS is an optimization problem, we only present the NP-hardness proof. Still, one can argue that the corresponding decision problem is NP-complete by guessing which junctions are covered by which label and then using linear programming for computing the label positions. We omit the technical details. Further, most roads in the reduction are paths, except for the central road in each clause gadget, which is a degree-3 star. In fact, we can strengthen Theorem 1 by using a more complex clause gadget instead that uses only paths; see full version [3].

4   An Efficient Algorithm for Tree-Shaped Road Maps

In this section we assume that the underlying road graph of the road map is a tree $T = (V, E)$. In Section 4.1 we present a polynomial-time algorithm to optimally solve MaxIdentifiedRoads for trees;

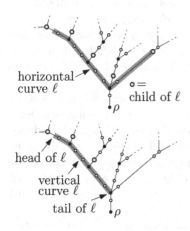

Fig. 3. Basic definitions

Section 4.2 shows how to improve its running time and space consumption. Our approach uses the basic idea that removing the vertices, whose embeddings lie in a curve $c \subseteq \mathsf{E}(T)$, splits the tree into independent parts. In particular this is true for labels. We assume that T is rooted at an arbitrary leaf ρ and that its edges are directed away from ρ; see Fig. 3. For two points $p, q \in \mathsf{E}(T)$ we define $d(p, q)$ as the length of the shortest curve in $\mathsf{E}(T)$ that connects p and q. For two vertices u and v of T we also write $d(u, v)$ instead of $d(\mathsf{E}(u), \mathsf{E}(v))$. For a point $p \in \mathsf{E}(T)$ we abbreviate the distance $d(p, \rho)$ to the root ρ by d_p. For a curve ℓ in $\mathsf{E}(T)$, we call $p \in \ell$ the *lowest point* of ℓ if $d_p \leq d_q$, for any $q \in \ell$. As T is a tree, p is unique. We distinguish two types of curves in $\mathsf{E}(T)$. A curve ℓ is *vertical* if $h(\ell)$ or $t(\ell)$ is the lowest point of ℓ; otherwise we call ℓ *horizontal* (see Fig. 3). Without loss of generality we assume that the lowest point of each vertical curve ℓ is its tail $t(\ell)$. Since labels are modeled as curves, they are also either vertical or horizontal. For a vertex $u \in V$ let T_u denote the subtree rooted at u.

4.1   Basic Approach

We first determine a finite set of candidate positions for the heads and tails of labels, and transform T into a tree $T' = (V', E')$ by subdividing some of T's edges so that it contains a vertex for every candidate position. To that end we construct for each regular vertex $v \in V$ a chain of tightly packed vertical labels that starts at $\mathsf{E}(v)$, is directed towards ρ, and ends when either the road ends, or adding the next label does not increase the number of identified road sections. More specifically, we place a first vertical label ℓ_1 such that $h(\ell_1) = \mathsf{E}(v)$. For $i = 2, 3, \ldots$ we add a new vertical label ℓ_i with $h(\ell_i) = t(\ell_{i-1})$, as long as $h(\ell_i)$ and $t(\ell_i)$ do not lie on the same road section and none of ℓ_i's endpoints lie on a junction edge. We use the tails of all those labels to subdivide the tree T. Doing this for all regular vertices of T we obtain the tree T', which we call the *subdivision tree* of T. The vertices in $V' \setminus V$ are neither junction vertices nor regular vertices. Since each chain consists of $O(n)$ labels the cardinality of V' is $O(n^2)$. We call an optimal labeling $\mathcal{L}$ of T an *canonical labeling* if for each label $\ell \in \mathcal{L}'$ there exists a vertex v in T' with $\mathsf{E}(v) = h(\ell)$ or $\mathsf{E}(v) = t(\ell)$. The next lemma proves that is sufficient to consider canonical labelings.

Lemma 1. *For any road graph T that is a tree, there exists a canonical labeling $\mathcal{L}$.*

The idea behind the proof is to *push* the labels of an optimal labeling $\mathcal{L}$ as far as possible towards the leaves of T without changing the identified road sections; see Fig. 4. The labels then start or end either at a leaf, a regular vertex or at an endpoint of another label, which yields a canonical labeling. For the full proof see the full version [3].

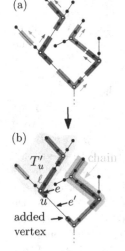

Fig. 4.   Canonical labeling

We now explain how to construct such a canonical labeling. To that end we first introduce some notations. For a vertex $u \in V'$ let $\mathcal{L}(u)$ denote a labeling that identifies a maximum number of road sections in T only using valid labels in $\mathsf{E}(T'_u)$, where T'_u denotes the subtree of T' rooted at u. Note that those labels also may identify the incoming road section of u, e.g., label ℓ in Fig. 4b) identifies the edge e'.

Further, the children of a vertex $u \in V'$ are denoted by the set $N(u)$; we explicitly exclude the parent of u from $N(u)$. Further, consider an arbitrary curve ℓ in $\mathsf{E}(T)$ and let $\ell' = \ell \setminus \{t(\ell), h(\ell)\}$. We observe that removing all vertices of T' contained in ℓ' together with their incident outgoing edges creates several independent subtrees. We call the roots of these subtrees (except the one containing ρ) *children* of ℓ (see Fig. 3). If no vertex of T' lies in ℓ', the curve is contained in a single edge $(u, v) \in E'$. In that case v is the only child of ℓ. We denote the set of all children of ℓ as $N(\ell)$.

For each vertex u in T' we introduce a set $C(u)$ of *candidates*, which model potential labels with lowest point $\mathsf{E}(u)$. If u is a regular vertex of T or $u \in V' \setminus V$, the set $C(u)$ contains all vertical labels ℓ with lowest point $\mathsf{E}(u)$. If u is a junction vertex, $C(u)$ contains all horizontal labels that start or end at a vertex of T' and whose lowest point is $\mathsf{E}(u)$. In both cases we assume that $C(u)$ also contains the degenerated curve $\perp_u = \mathsf{E}(u)$, which is the *dummy label* of u. We set $N(\perp_u) = N(u)$ and $\omega(\perp_u) = 0$.

For a curve ℓ we define $\mathcal{L}(\ell) = \bigcup_{v \in N(\ell)} \mathcal{L}(v) \cup \{\ell\}$. Thus, $\mathcal{L}(\ell)$ is a labeling comprising ℓ and the labels of its children's optimal labelings. We call a label $\bar{\ell} \in C(u)$ with $\bar{\ell} = \text{argmax}\{\omega(\mathcal{L}(\ell)) \mid \ell \in C(u)\}$ an *optimal candidate* of u. Next, we prove that it is sufficient to consider optimal candidates to construct a canonical labeling.

Lemma 2. *Given a vertex u of T' and an optimal labeling $\mathcal{L}(u)$ and let $\bar{\ell}$ be an optimal candidate of u, then it is true that $\omega(\mathcal{L}(u)) = \omega(\mathcal{L}(\bar{\ell}))$.*

Proof. First note that $\omega(\mathcal{L}(u)) \geq \omega(\mathcal{L}(\bar{\ell}))$ because both labelings $\mathcal{L}(u)$ and $\mathcal{L}(\bar{\ell})$ only contain labels that are embedded in $\mathsf{E}(T'_u)$. By Lemma 1 we can assume without loss of generality that $\mathcal{L}(u)$ is a canonical labeling. Let ℓ be the label of $\mathcal{L}(u)$ with $\mathsf{E}(u)$ as the lowest point of ℓ (if it exists).

If ℓ exists, then the vertices in $N(\ell)$ are roots of independent subtrees, which directly yields $\omega(\mathcal{L}(u)) = \omega(\mathcal{L}(\ell))$. By construction of $C(u)$ we further know that ℓ is contained in $C(u)$. Hence, ℓ is an optimal candidate of u, which implies $\omega(\ell) = \omega(\bar{\ell})$.

If ℓ does not exist, then we have

$$\omega(\mathcal{L}(u)) = \omega\left(\bigcup_{v \in N(u)} \mathcal{L}(v) \right) \overset{(1)}{=} \omega\left(\bigcup_{v \in N(\perp_u)} \mathcal{L}(v) \cup \{\perp_u\} \right) = \omega(\mathcal{L}(\perp_u)).$$

Equality (1) follows from $N(\perp_u) = N(u)$ and the definition that $\perp_u$ does not identify any road section. Since $\perp_u$ is contained in $C(u)$, the dummy label $\perp_u$ is the optimal candidate $\bar{\ell}$. $\square$

Algorithm 1 first constructs the subdivision tree $T' = (V', E')$ from T. Then starting with the leaves of T' and going to the root ρ of T', it computes an optimal candidate $\bar{\ell} = \texttt{OptCandidate}\ (u)$ for each vertex $u \in V'$ in a bottom-up fashion. By Lemma 2 the labeling $\mathcal{L}(\bar{\ell})$ is an optimal labeling of T'_u. In particular $\mathcal{L}(\rho)$ is the optimal labeling of T.

Due to the size of the subdivision tree T' we consider $O(n^2)$ vertices. Implementing $\texttt{OptCandidate}(u)$, which computes an optimal candidate $\bar{\ell}$ for u, naively, creates $C(u)$ explicitly. We observe that if u is a junction vertex, $C(u)$ may contain $O(n^2)$ labels; $O(n^2)$ pairs of road sections of different subtrees of u can be connected by horizontal labels. Each label can be constructed in $O(n)$ time using a breadth-first search. Thus, for each vertex u the procedure $\texttt{OptCandidate}$ needs in a naive implementation $O(n^3)$ time, which yields $O(n^5)$ running time in total. Further, we need $O(n^2)$ storage to store T'. Note that we do not need to store $\mathcal{L}(u)$ for each vertex u of T', but by Lemma 2 we can reconstruct it using $\mathcal{L}(\bar{\ell})$, where $\bar{\ell}$ is the optimal candidate of u. To that end we store for each vertex of T' its optimal candidate $\bar{\ell}$ and $w(\mathcal{L}(\bar{\ell}))$.

Algorithm 1. Computing an optimal labeling of T.

Input: Road graph T, where T is a tree with root ρ.
Output: Optimal labeling $\mathcal{L}(\rho)$ of T.
1 $T' \leftarrow$ compute subdivision tree of T
2 **for** *each leaf v of T'* **do** $\mathcal{L}(v) \leftarrow \emptyset$
3 **for** *each vertex u of T' considered in a bottom-up traversal of T'* **do**
4 $\quad \lfloor\ \mathcal{L}(u) \leftarrow \mathcal{L}(\texttt{OptCandidate}(u))$
5 **return** $\mathcal{L}(\rho)$

Theorem 2. *For a road map with a tree as underlying road graph,* MAXIDEN-
TIFIEDROADS *can be solved in $O(n^5)$ time using $O(n^2)$ space.*

In case that all roads are paths, Algorithm 1 runs in $O(n^4)$ time, because
for each $u \in V'$ the set $C(u)$ contains $O(n)$ labels. Further, besides the *primary
objective* to identify a maximum number of road sections, Chirié [1] also suggested several additional *secondary objectives*, e.g., labels should overlap as few
junctions as possible. Our approach allows us to easily incorporate secondary
objectives by changing the weight function ω appropriately.

4.2   Improvements on Running Time

In this part we describe how the running time of Algorithm 1 can be improved
to $O(n^3)$ time by speeding up OptCandidate(u) to $O(n)$ time.

For an edge $e = (u, v) \in E \cup E'$ we call a vertical curve $\ell \subseteq \mathsf{E}(T)$ an e-
rooted curve, if $t(\ell) = \mathsf{E}(u)$, $h(\ell)$ lies on a road section, and $\mathrm{len}(\mathsf{E}(e) \cap \ell) =
\min\{\mathrm{len}(\ell), \mathrm{len}(\mathsf{E}(e))\}$, i.e., ℓ emanates from $\mathsf{E}(u)$ passing through e; for example
the red label in Fig. 4b) is an e-rooted curve. An e-rooted curve ℓ is *maximal* if
there is no other e-rooted curve ℓ' with $\mathrm{len}(\ell) = \mathrm{len}(\ell')$ and $\omega(\mathcal{L}(\ell')) > \omega(\mathcal{L}(\ell))$.
We observe that in any canonical labeling each vertical label ℓ is a (u, v)-rooted
curve with $(u, v) \in E'$, and each horizontal label ℓ can be composed of a
(u, v_1)-rooted curve ℓ_1 and a (u, v_2)-rooted curve ℓ_2 with $(u, v_1), (u, v_2) \in E'$
and $\mathsf{E}(u)$ is the lowest point of ℓ; see Fig. 6 and Fig. 7, respectively. Further,
for a vertical curve c in $\mathsf{E}(T)$ its
distance interval $I(c)$ is $[\mathrm{d}_{t(c)}, \mathrm{d}_{h(c)}]$.
Since T is a tree, for every point p
of c we have $\mathrm{d}_p \in I(c)$. Two vertical
curves c and c' *superpose* each other
if $I(c) \cap I(c') \neq \emptyset$; see Fig 5.

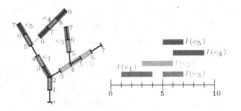

Next, we introduce a data structure that encodes for each edge (u, v)
of T all maximal (u, v)-rooted curves   **Fig. 5.** Superposing curves, e.g., c_1 and c_2
as $O(n)$ superposition-free curves in   superpose each other, while c_1 and c_5 do
$\mathsf{E}(T_u)$. In particular, each of those   not. The tree is annotated with distance
curves lies on a single road section such   marks.
that all (u, v)-rooted curves ending on
that curve are maximal and identify the same number of road sections. We define
this data structure as follows.

Definition 1 (Linearization). *Let $e = (u, v)$ be an edge of T. A tuple $(L, \overline{\omega})$ is
called a* linearization *of e, if L is a set of superposition-free curves and $\overline{\omega} \colon L \to \mathbb{R}$
such that*
(1) for each curve $c \in L$ there is a road section e' in T_u with $c \subseteq \mathsf{E}(e')$,
(2) for each e-rooted curve ℓ there is a curve $c \in L$ with $\mathrm{len}(\ell) + \mathrm{d}_u \in I(c)$,
(3) for each point p of each curve $c \in L$ there is a maximal e-rooted curve ℓ with
    $h(\ell) = p$ and $\overline{\omega}(c) = \omega(\mathcal{L}(\ell))$.

Assume that we apply Algorithm 1 on T' and that we currently consider the vertex u of T'. Hence, we can assume that for each vertex $v \neq u$ of T'_u its optimal candidate and $\omega(\mathcal{L}(v))$ is given. We first explain how to speed up OptCandidate using linearizations. Afterwards, we present the construction of linearizations.

Application of linearizations. Here we assume that the linearizations are given for the edges of T. Concerning the type of u we describe how to compute its optimal candidate.

Case 1, u is regular. If u is a leaf, the set $C(u)$ contains only $\perp_u$. Hence, assume that u has one outgoing edge $e = (u, v) \in E'$, which belongs to a road R. Let P be the longest path of vertices in T'_u that starts at u and does not contain any junction vertex. Note that the path must be unique. Further, by construction of T' the last vertex w of P must be a regular vertex in V, but not in $V' \setminus V$. We consider two cases; see Fig 6.

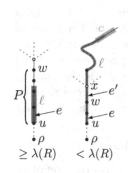

If $d(u, w) \geq \lambda(R)$, the optimal candidate is either $\perp_u$ or the e-rooted curve ℓ of length $\lambda(R)$ that ends on $\mathsf{E}(P)$. By assumption and due to $\omega(\mathcal{L}(\perp_u)) = \omega(\mathcal{L}(v))$, we decide in $O(1)$ time whether $\omega(\mathcal{L}(\perp_u)) \geq \omega(\mathcal{L}(\ell))$, obtaining the optimal candidate.

If $d(u, w) < \lambda(R)$, the optimal candidate is either $\perp_u$ or goes through a junction. Since w is regular, it has only one outgoing edge $e' = (w, x)$. Further, by the choice of P the edge e' is a junction edge in T; therefore the linearization $(L, \overline{\omega})$ of e' is given. In linear time we search for the curve $c \in L$ such that there is an e-rooted curve ℓ of length $\lambda(R)$ with its head on c. To that end we consider for each curve $c \in L$ its distance interval $I(c)$ and check whether there is $t \in I(c)$ with $t - d_u = \lambda(R)$. Note that using a binary search tree for finding c speeds this procedure up to $O(\log n)$ time, however, this does not asymptotically improve the total running time. The e-rooted curve ℓ then can be easily constructed in $O(n)$ time by walking from c to u in $\mathsf{E}(T)$.

Fig. 6. Case 1

If such a curve c exist, by definition of a linearization the optimal candidate is either $\perp_u$ or ℓ, which we can decide in $O(1)$ time by checking $\omega(\mathcal{L}(\perp_u)) \geq \omega(\mathcal{L}(\ell))$.

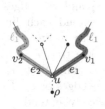

Note that we have $\omega(\mathcal{L}(\perp_u)) = \omega(\mathcal{L}(v))$ and $\omega(\mathcal{L}(\ell)) = \overline{\omega}(c)$. If c does not exist, again by definition of a linearization there is no vertical label $\ell \in C(u)$ and $\perp_u$ is the optimal candidate.

Case 2, u is a junction vertex. The set $C(u)$ contains horizontal labels. Let ℓ be such a label and let $e_1 = (u, v_1)$ and $e_2 = (u, v_2)$ be two junction edges in E covered by ℓ; see Fig. 7. Then there is an e_1-rooted curve ℓ_1 and an e_2-rooted curve ℓ_2 whose composition is ℓ. Further, we have

Fig. 7. Case 2

$\omega(\mathcal{L}(\ell)) = \omega(\mathcal{L}(\ell_1) \cup \mathcal{L}(\ell_2)) + \sum_{v \in N(u) \setminus \{v_1, v_2\}} \omega(\mathcal{L}(v))$. We use this as follows.

Let e_1 and e_2 be two outgoing edges of u that belong to the same road R, and let $(L_1, \overline{\omega}_1)$ and $(L_2, \overline{\omega}_2)$ be the linearizations of e_1 and e_2, respectively.

We define for e_1 and e_2 and their linearizations the operation opt-cand(L_1, L_2) that finds an optimal candidate of u restricted to labels identifying e_1 and e_2.

For $i = 1, 2$ let $d_i = \max\{d_u \mid u$ is vertex of $T_{v_i}\}$ and let $f_u(t) = d_u - (t - d_u) = 2\,d_u - t$ be the function that "mirrors" the point $t \in \mathbb{R}^2$ at d_u. Applying $f_u(t)$ on the boundaries of the distance intervals of the curves in L_1, we first mirror these intervals such that they are contained in the interval $[2\,d_u - d_1, d_u]$; see Fig. 8. Thus, the curves in $L_1 \cup L_2$ are mutually superposition-free such that their distance intervals lie in $J = [2\,d_u - d_1, d_2]$.

We call an interval $[x, y] \subseteq J$ a *window*, if it has length $\lambda(R)$, $d_u \in [x, y]$ and there are curves $c_1 \in L_1$ and $c_2 \in L_2$ with $x \in I(c_1)$ and $y \in I(c_2)$; see Fig. 8. By the definition of a linearization there is a maximal e_1-rooted curve ℓ_1 ending on c_1 and a maximal e_2-rooted curve ℓ_2 ending on c_2 such that $\text{len}(\ell_1) + \text{len}(\ell_2) = \lambda(R)$. Consequently, the composition of ℓ_1 and ℓ_2 forms a horizontal label ℓ with $\omega(\mathcal{L}(\ell)) = \omega(\mathcal{L}(\ell_1) \cup \mathcal{L}(\ell_2)) + \sum_{v \in N(u) \setminus \{v_1, v_2\}} \mathcal{L}(v)$; we call $\omega(\mathcal{L}(\ell))$ the *value* of the window. Using a simple sweep from left to right we compute for the distance interval $I(c)$ of each curve $c \in L_1 \cup L_2$ a window $[x, y]$ that starts or ends in $I(c)$ (if such a window exists). The result of opt-cand(L_1, L_2) is then the label ℓ of the window with maximum value. For each pair e_1 and e_2 of outgoing edges we apply opt-cand(L_1, L_2) computing a label ℓ. By construction either the label ℓ with maximum $\omega(\ell)$ or $\perp_u$ is the optimal candidate for u, which we can check in $O(1)$ time. Later on we prove that we consider only linearizations of linear size. Since each vertex of T' has constant degree, we obtain the next lemma.

Lemma 3. *For each $u \in V'$ the optimal candidate can be found in $O(n)$ time.*

Construction of linearizations. It remains to show that a linearization of an edge $e = (u, v)$ can be constructed in $O(n)$ time assuming that the linearizations

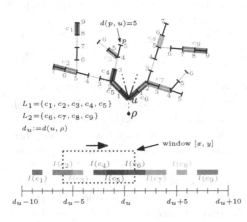

$(L_1, \overline{\omega}_1), \ldots, (L_k, \overline{\omega}_k)$ of the outgoing edges $e_1 = (v, w_1), \ldots, e_k = (v, w_k)$ of v are given. Due to space restrictions we only give a rough sketch; the details can be found in the full version [3]. For $1 \leq i \leq k$ let T_i be the tree induced by the edges e, e_i and the edges of T_{w_i}. We first extend each linearization $(L_i, \overline{\omega}_i)$ to a linearization restricted to the tree T_i, i.e., conceptually we assume that T_u only consists of the edges of T_i. To that end we basically extend L_i by disjoint sub-curves of $\mathsf{E}(e)$ if e is a road section and update $\overline{\omega}_i$. Afterwards we merge those constructed linearizations to one linearization $(L, \overline{\omega})$ of e without any

Fig. 8. Constructing the optimal candidate of u based on the linearizations $(L_1, \overline{\omega}_1)$ and $(L_2, \overline{\omega}_2)$. The tree is annotated with distance marks.

restrictions on T_u. In particular we enforce that L is a set of superposition-free curves, which we achieve by splitting superposing curves c and c' into three superposition-free curves c_1, c_2, c_3 such that each of them is either contained in c or c' and $\bigcup_{i=1}^{3} I(c_i) = I(c) \cup I(c')$. The choice of c_1, c_2 and c_3 depends on the number of road sections identified by an e-rooted curve that ends on c and c', respectively. We prove that this merging runs in $O(n)$ time per edge $e \in E$ and $(L, \overline{\omega})$ has size $O(n)$. This and Lemma 3 yield the next proposition.

Proposition 1. *For a road map $\mathcal{M}$ with a tree T as underlying road graph,* MAXIDENTIFIEDROADS *can be solved in $O(n^3)$ time.*

Since T' contains $O(n^2)$ vertices, the algorithm needs $O(n^2)$ space. This can be improved to $O(n)$ space. To that end T' is constructed *on the fly* while executing Algorithm 1. Parts of T' that become unnecessary are discarded. In the full version [3] we prove that it is sufficient to store $O(n)$ vertices of T' at any time such that the optimal labeling can still be constructed. We summarize in the following theorem.

Theorem 3. *For a road map $\mathcal{M}$ with a tree T as underlying road graph,* MAX-IDENTIFIEDROADS *can be solved in $O(n^3)$ time using $O(n)$ space.*

5 Conclusions and Outlook

In this paper we investigated the problem of maximizing the number of identified road sections in a labeling of a road map; we showed that it is NP-hard in general, but can be solved in $O(n^3)$ time and linear space for the special case of trees.

The underlying road graphs of real-world road maps are rarely trees. Initial experimental evidence indicates, however, that road maps can be decomposed into a large number of subgraphs by placing trivially optimal road labels and removing the corresponding edges from the graph. It turns out that between 85.1% and 97.7% of the resulting subgraphs are actually trees, which we can label optimally by our proposed algorithm. As a consequence, this means that a large fraction (between 88.6% and 96.1%) of all road sections in our real-world road graphs can be labeled optimally by combining this simple preprocessing strategy with the tree labeling algorithm. We are investigating further heuristic and exact approaches for labeling the remaining non-tree subgraphs (e.g., by finding suitable spanning trees and forests) for a separate companion paper.

References

1. Chirié, F.: Automated name placement with high cartographic quality: City street maps. Cartography and Geo. Inf. Science **27**(2), 101–110 (2000)
2. Edmondson, S., Christensen, J., Marks, J., Shieber, S.M.: A general cartographic labelling algorithm. Cartographica **33**(4), 13–24 (1996)
3. Gemsa, A., Niedermann, B., Nöllenburg, M.: Label placement in road maps. CoRR, abs/1501.07188 (2015)

4. Imhof, E.: Positioning names on maps. Amer. Cartogr., 128–144 (1975)
5. Lichtenstein, D.: Planar formulae and their uses. SIAM J. Comput. **11**(2), 329–343 (1982)
6. Maass, S., Döllner, J.: Embedded labels for line features in interactive 3d virtual environments. In: Proc. 5th Int. Conf. Computer Graphics, Virtual Reality, Visualisation and Interaction in Africa, AFRIGRAPH 2007, pp. 53–59. ACM (2007)
7. Neyer, G., Wagner, F.: Labeling downtown. In: Bongiovanni, G., Petreschi, R., Gambosi, G. (eds.) CIAC 2000. LNCS, vol. 1767, pp. 113–124. Springer, Heidelberg (2000)
8. Reimer, A., Rylov, M.: Point-feature lettering of high cartographic quality: a multi-criteria model with practical implementation. In: European Workshop on Computational Geometry (EuroCG 2014), Ein-Gedi, Israel (2014)
9. Seibert, S., Unger, W.: The hardness of placing street names in a Manhattan type map. Theor. Comp. Sci. **285**, 89–99 (2002)
10. Strijk, T.: Geometric Algorithms for Cartographic Label Placement. Dissertation - Utrecht University (2001)
11. Vaaraniemi, M., Treib, M., Westermann, R.: Temporally coherent real-time labeling of dynamic scenes. In: Proc. 3rd Int. Conf. Comput. Geospatial Research Appl., COM.Geo 2012, pp. 17:1–17:10. ACM (2012)
12. van Kreveld, M.: Geographic information systems. In: Handbook of Discrete and Computational Geometry, Second Edition, chap. 58, pp. 1293–1314. CRC Press (2010)
13. Wolff, A., Knipping, L., van Kreveld, M., Strijk, T., Agarwal, P. K.: A simple and efficient algorithm for high-quality line labeling. In: Innovations in GIS VII: GeoComputation, chap. 11, pp. 147–159. Taylor & Francis (2000)
14. Wolff, A., Strijk, T.: The map labeling bibliography (2009). http://liinwww.ira.uka. de/bibliography/Theory/map.labeling.html

Discrete Stochastic Submodular Maximization: Adaptive vs. Non-adaptive vs. Offline

Lisa Hellerstein, Devorah Kletenik, and Patrick Lin[✉]

Department of Computer Science and Engineering,
Polytechnic School of Engineering, New York University, New York, NY, USA
{lisa.hellerstein,dkletenik,patrick.lin}@nyu.edu

Abstract. We consider the problem of stochastic monotone submodular function maximization, subject to constraints. We give results on adaptivity gaps, and on the gap between the optimal offline and online solutions. We present a procedure that transforms a decision tree (adaptive algorithm) into a non-adaptive chain. We prove that this chain achieves at least τ times the utility of the decision tree, over a product distribution and binary state space, where $\tau = \min_{i,j} \Pr[x_i = j]$. This proves an adaptivity gap of $\frac{1}{\tau}$ (which is 2 in the case of a uniform distribution) for the problem of stochastic monotone submodular maximization subject to state-independent constraints. For a cardinality constraint, we prove that a simple adaptive greedy algorithm achieves an approximation factor of $(1 - \frac{1}{e^\tau})$ with respect to the optimal offline solution; previously, it has been proven that the algorithm achieves an approximation factor of $(1 - \frac{1}{e})$ with respect to the optimal adaptive online solution. Finally, we show that there exists a non-adaptive solution for the stochastic max coverage problem that is within a factor $(1 - \frac{1}{e})$ of the optimal adaptive solution and within a factor of $\tau(1 - \frac{1}{e})$ of the optimal offline solution.

1 Introduction

We consider stochastic submodular function maximization, subject to constraints. This problem is motivated by problems in application areas such as machine learning, social networks, and recommendation systems.

In traditional (non-stochastic) submodular function maximization, the goal is to find a subset of "items" with maximum utility, as measured by a submodular utility function assigning a real value to each possible subset of items. In *stochastic* submodular function maximization, items have *states*. For example, if each item is a sensor, the item might be either working or broken. The utility of a subset of items depends not only on which items are in the subset, but also on their states. The state of each item is initially unknown, and can only be determined by performing a "test" on the item.

Algorithms for stochastic submodular maximization work in an on-line setting, sequentially choosing which item to test next. The choice can be adaptive, depending on the outcomes of previous tests. The state of each item is an independent random variable. The goal is to maximize the expected utility of the tested items. Previous work has sought to determine the adaptivity gap, which

© Springer International Publishing Switzerland 2015
V.Th. Paschos and P. Widmayer (Eds.): CIAC 2015, LNCS 9079, pp. 235–248, 2015.
DOI: 10.1007/978-3-319-18173-8_17

is the ratio between the optimal adaptive and non-adaptive solutions. In this paper we present new adaptivity gap results for discrete monotone submodular functions. We also consider another type of gap that has not been previously explored in the context of stochastic submodular maximization: the ratio between the optimal offline solution and the optimal adaptive solution.

Our main result is an adaptivity gap of 2 for all *state-independent* constraints, when the state set is binary and the item state distribution is uniform. More generally, for arbitrary product distributions, we prove an adaptivity gap of $\frac{1}{\tau}$. Here τ is the minimum value of $p_{i,j}$, where $p_{i,j}$ is the probability that item i is in state j. We say that a constraint is state-independent if the restriction on the items tested does not depend on their states. (A constraint requiring testing to stop when an item is found to be in state 1 is not state-independent.) A standard knapsack constraint is state-independent, and this is the first adaptivity gap for knapsack constraints. We prove the gap using a simple, bottom-up procedure that transforms a decision tree (adaptive algorithm) into a single non-adaptive chain corresponding to a root-leaf path in the tree.

Asadpour and Nazerzadeh previously showed an adaptivity gap of $\frac{e}{e-1}$ for a matroid constraint, using a stronger monotonicity condition than the one we use here (their results also apply to continuous states) [3]. For a cardinality constraint, we show that the simple adaptive greedy algorithm gives a $(1 - \frac{1}{e^\tau})$-approximation with respect to the optimal *offline* solution, and that a dependence on τ in the approximation factor is necessary.

Finally, we consider the discrete stochastic version of the maximum coverage problem, which is a special case of submodular maximization subject to a cardinality constraint. We modify an approximation algorithm for the deterministic version of this problem, due to Ageev and Sviridenko [1,2], to prove that the optimal non-adaptive solution for this problem is within a factor of $1 - \frac{1}{e}$ of the optimal adaptive solution. We also show that the optimal non-adaptive solution for this problem is within a factor of $\tau(1 - \frac{1}{e})$ with respect to the optimal offline solution.

2   Preliminaries and Definitions

Let $\mathbb{S} = \{0, \ldots, \ell - 1\}$, and $\hat{\mathbb{S}} = \mathbb{S} \cup \{*\}$. A partial assignment is a vector $b \in \hat{\mathbb{S}}^n$. Hence b can be viewed as an assignment to variables $x_1, \ldots, x_n$. We will use these partial assignments to represent the outcomes of tests giving the states of n items, where each item can be in one of ℓ states. We write $b_i = s$ to indicate that item i has been tested and found to be in state s, and $b_i = *$ to indicate that the state of item i is unknown. We assume that the states of different items are independent.

If $b', b \in \hat{\mathbb{S}}^n$ and $b'_i = b_i$ for all $b_i \neq *$, then we call b' an *extension* of b, which we will write as $b' \succ b$. We will use $b_{j \leftarrow s}$ to denote the extension of b setting the j-th bit of b to s.

As is standard in the literature, given a set $N = \{1, \ldots, n\}$, we say a function $g : 2^N \to \mathbb{R}_{\geq 0}$ is a *utility function*. We will use the notation $g_S(j)$ to denote $g(S \cup \{j\}) - g(S)$.

We extend the notion of a utility function to the stochastic setting, wherein we have $g : \hat{\mathbb{S}}^n \to \mathbb{R}_{\geq 0}$ defined on partial assignments. In this case, we will write $g(S, b) := g(b')$ where b' is a partial assignment consistent with b on all entries i where $i \in S$, and $b_i = *$ whenever $i \notin S$. The notation $g_{S,b}(j)$ will denote $g(S \cup \{j\}, b) - g(S, b)$. If S' is a set, then $g_S(S')$ will mean $g(S \cup S') - g(S)$.

Utility function $g : \hat{\mathbb{S}}^n \to \mathbb{R}_{\geq 0}$ is called *submodular* if $g(b_{i \leftarrow s}) - g(b) \geq g(b'_{i \leftarrow s}) - g(b')$ when $b' \succ b$, $b'_i = b_i = *$, and $s \in \mathbb{S}$. We say g is *monotone* if $g(b_{i \leftarrow s}) \geq g(b)$ when $b_i = *$. That is, testing a bit can only increase the utility.

We will work with product distributions over the vectors $\mathbb{S}^n$: for $i \in N$, $j \in \mathbb{S}$, we use $p_{i,j}$ to mean the probability of the i-th coordinate being j (so that for each i, $\sum_j p_{i,j} = 1$). Many of our results are with respect to $\tau = \min_{i,j} p_{i,j}$. We use $\mathbb{E}[g_{S,b}(j)]$ to denote the expected increase in utility from testing the j-th bit.

We define the *Stochastic Submodular Maximization* problem as the problem of maximizing a monotone submodular function, in the stochastic setting with a discrete state space, subject to one or more constraints. More specifically, in this problem, we are given as input a monotone submodular $g : \hat{\mathbb{S}}^n \to \mathbb{R}_{\geq 0}$, the constraints, and the parameters of a product distribution over $\hat{\mathbb{S}}^n$.

Solving a Stochastic Submodular Maximization problem entails finding an *adaptive solution* that builds a set $Q \subseteq N$ item by item, testing each item after selecting it, that maximizes $\mathbb{E}[g(Q, b)]$ subject to the constraints. This is effectively a decision tree, whose nodes are labeled with $j \in N$, and we branch depending on the outcome of b_j. A *chain* is a balanced tree such that the labels are the same for all nodes at the same level. A chain is a *non-adaptive procedure*. We also consider the so-called *offline solution*, which knows a priori the outcomes of the random bits, and takes the optimal set with respect to said outcome.

A knapsack constraint has the form $\sum_{j \in Q} c_j \leq B$ where each $c_j \geq 0$, and $B \geq 0$. The c_j are called costs, and B is a budget. A special case is when $c_j = 1$ for all j and B is an integer; this is a cardinality constraint. (A cardinality constraint is also a special case of a matroid constraint.)

The *Stochastic Max Coverage* problem is a special case of the Stochastic Submodular Maximization problem with a cardinality constraint. The utility function $g : \hat{\mathbb{S}} \to \mathbb{R}_{\geq 0}$ in the Stochastic Max Coverage problem is defined as follows: let $E = \{e_1, \ldots, e_m\}$ be a ground set of elements, and for $i \in N$ and $r \in \mathbb{S}$, $S_{i,r} \subseteq E$. For convenience of notation, we will also write $S_{i,a}$ to mean S_{i,a_i} where $a \in \mathbb{S}^n$. Then $g(S, b) = |\bigcup_{i \in S} S_{i,b}|$. If b is a partial assignment, then we say e_j is *covered* with respect to b if $e_j \in \bigcup_{i : b_i \neq *} S_{i,b}$. This function g is clearly submodular and monotone.

The expected values of the optimal adaptive, non-adaptive, and offline solutions will be denoted by ADAPT, NONADAPT, and OFFLINE. We are interested in the *adaptivity gap* $\frac{\text{ADAPT}}{\text{NONADAPT}}$, as well as the ratios $\frac{\text{OFFLINE}}{\text{ADAPT}}$ and $\frac{\text{OFFLINE}}{\text{NONADAPT}}$.

3   Related Work

The submodular maximization problems studied in this paper were all initially studied in the deterministic setting. Feige showed that for all of these problems, under the assumption of P $\neq$ NP, no polynomial time algorithm can achieve an approximation factor better than $(1 - \frac{1}{e})$ [5]. For the problem of maximizing a monotone submodular function subject to a cardinality constraint, Nemhauser et al. showed that the natural greedy algorithm achieves an approximation factor of $(1 - \frac{1}{e})$ [8]. For the problem with a knapsack constraint, Sviridenko subsequently showed that an algorithm of Khuller et al. also achieves an approximation factor of $(1 - \frac{1}{e})$ [7,10]. The results of Golovin and Krause achieve the same approximation factor for a cardinality constraint in the stochastic setting [6].

As mentioned earlier, Asadpour and Nazerzadeh showed an adaptivity gap of $\frac{e}{e-1}$ for Stochastic Submodular Maximization with a matroid constraint, using a stronger definition of monotonicity. In their definition of monotonicity, for any partial assignment b and $s \in \mathbb{S}$, they require that $g(b_{i \leftarrow s}) \geq g(b)$ if either $b_i = *$ or $s \geq b_i$, whereas we only require that $g(b_{i \leftarrow s}) \geq g(b)$ if $b_i = *$. Their proof is based on Poisson clocks and pipage rounding [3], and applies to continuous state spaces. Our adaptivity gap results do not apply to continuous state spaces, but our proofs are combinatorial.

Chan and Farias studied the related Stochastic Depletion problem and gave a $\frac{1}{2}$-approximation for the problem with respect to what they call the offline solution in their model [4]; in our model, their algorithm translates to a $\frac{1}{2}$-approximation with respect to ADAPT for Stochastic Submodular Maximization with a cardinality constraint.

Table 1. Bounds for Stochastic Submodular Maximization

	Knapsack Constraint	Cardinality Constraint	Max Coverage
Deterministic	$(1 - \frac{1}{e})$OPT [10]	$(1 - \frac{1}{e})$OPT [9]	$(1 - \frac{1}{e})$OPT [7]
Stochastic: Adaptive	OPEN	$(1 - \frac{1}{e})$ADAPT [3,6] $(1 - \frac{1}{e^\tau})$**OFFLINE**	$(1 - \frac{1}{e})$ADAPT [3,6] $(1 - \frac{1}{e^\tau})$**OFFLINE**
Stochastic: Non-Adaptive	τ**ADAPT**	$(1 - \frac{1}{e})$ADAPT[†] [3] τ**ADAPT**	$(1 - \frac{1}{e})$ADAPT[†] [3] $(1 - \frac{1}{e})$**ADAPT** $\tau(1 - \frac{1}{e})$**OFFLINE**

Table 1 summarizes approximation bounds for stochastic discrete monotone submodular function maximization with a knapsack constraint, a cardinality constraint, and for max-coverage. Results from this paper are in bold. We denote with a † bounds relying on the stronger definition of monotonicity of [3].

The entries in the first row give the best bounds known for polynomial-time algorithms solving the deterministic versions of the problems, assuming oracle access to the utility function g. The bounds are given in terms of OPT, the optimal solution to the deterministic problem. As noted above, these are the best bounds possible, assuming P $\neq$ NP [5].

The entries in the second row refer to the best nontrivial bounds achieved by polynomial-time algorithms for the stochastic versions of the problems, assuming polynomial-time access to g. Both bounds are achieved by the Adaptive Greedy algorithm (described in Section 5). We note that Golovin and Krause give a randomized version of the algorithm for the knapsack constraint achieving an approximation factor of $(1 - \frac{1}{e})$, but with the relaxation that the budget only needs to be met in expectation [6]. The last row refers to the bounds achieved by the respective best possible non-adaptive solutions for the problems in the stochastic setting (irrespective of running time).

4   An Adaptivity Gap for State-Independent Constraints

In this section we present an adaptivity gap for Stochastic Submodular Maximization with state-independent constraints. We use a technique that takes a decision tree and outputs a root-leaf path by collapsing the tree bottom up in a greedy manner; at each step, one child chain of a node replaces the other, leaving a single longer chain.

We show that under a product distribution over $\{0,1\}^n$, this gives a non-adaptive procedure that is a τ-approximation of expected utility of the original tree, where $\tau = \min_{i,j} p_{i,j}$. This gives a bound on the adaptivity gap for the problem: $\frac{\text{ADAPT}}{\text{NONADAPT}} \leq \frac{1}{\tau}$ for binary states.

Theorem 1. *For binary states, the Stochastic Submodular Maximization problem with state-independent constraints has an adaptivity gap of at most $\frac{1}{\tau}$.*

Proof. Let T be a decision tree corresponding to a solution to an instance of Stochastic Submodular Maximization with state-independent constraints, and binary states. We show that if T achieves expected utility U, then there exists a chain, corresponding to a root-leaf path in T, that achieves expected utility $\tau \cdot U$. Since the constraints are state-independent, this root-leaf path must obey the constraints, and the theorem follows.

We use a recursive procedure to turn T into a chain. At any intermediate step, we have a subtree consisting of a parent node whose child subtrees are chains. We show that when performing the procedure on this subtree, the loss incurred in expected utility is at most $1 - \tau$ times the expected utility contributed by the parent node. Since the expected utility of a decision tree is a weighted sum of the expected utility contributed by its nodes, the total loss from the entire procedure is at most $1 - \tau$ times the expected utility of the whole tree.

Hence without loss of generality, suppose T is a tree with a root node labeled x_i for some $i \in N$ whose two child subtrees are chains, as shown in Figure 1. For convenience of notation, assume the nodes on the left child chain are labeled x_{l_1} through x_{l_c}, and the nodes on the right child chain are labeled x_{r_1} through x_{r_d}.

Let $L = \{l_m\}_{m=1}^c$ and $R = \{r_m\}_{m=1}^d$. For $D \in \{L, R\}$ and fixed assignment b, monotonicity gives $g_{\varnothing,b}(\{i\} \cup D) \geq g_{\varnothing,b}(D)$, and submodularity gives $g_{\varnothing,b}(D) \geq$

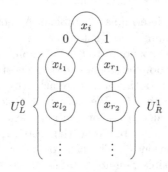

Fig. 1. T

$g_{\{i\},b}(D)$. Thus, in expectation,

$$\mathbb{E}[g_{\varnothing,b}(\{i\} \cup L) \mid b_i = 1] \geq \mathbb{E}[g_{\varnothing,b}(L) \mid b_i = 1]$$
$$= \mathbb{E}[g_{\varnothing,b}(L) \mid b_i = 0] \geq \mathbb{E}[g_{\{i\},b}(L) \mid b_i = 0] \quad (1)$$

and similarly

$$\mathbb{E}[g_{\varnothing,b}(\{i\} \cup R) \mid b_i = 0] \geq \mathbb{E}[g_{\{i\},b}(R) \mid b_i = 1]. \quad (2)$$

We introduce the following notation: Set $U_i^0 = \mathbb{E}[g_{\varnothing,b}(i) \mid b_i = 0]$, $U_i^1 = \mathbb{E}[g_{\varnothing,b}(i) \mid b_i = 1]$ (note that $g_{\varnothing,b}(i)$ is actually constant on all b with the same value for b_i), and $U_i = \mathbb{E}[g_{\varnothing,b}(i)] = p_{i,0}U_i^0 + p_{i,1}U_i^1$. So U_i is the expected utility contributed by the root node x_i.

Set $U_L^0 = \mathbb{E}[g_{\{i\},b}(L) \mid b_i = 0]$, that is, the expected increase in utility from testing $x_{l_1}, \dots, x_{l_c}$ after testing x_i and getting $b_i = 0$. Similarly, we set $U_L^1 = \mathbb{E}[g_{\{i\},b}(L) \mid b_i = 1]$, $U_R^0 = \mathbb{E}[g_{\{i\},b}(R) \mid b_i = 0]$ and $U_R^1 = \mathbb{E}[g_{\{i\},b}(R) \mid b_i = 1]$.

With this notation in mind, we can rewrite, respectively, (1) and (2) as $U_i^1 \geq U_L^0 - U_L^1$ and $U_i^0 \geq U_R^1 - U_R^0$, that is,

$$U_i \geq p_{i,0}(U_R^1 - U_R^0) + p_{i,1}(U_L^0 - U_L^1). \quad (3)$$

Consider the two variants T_L and T_R of T as follows: T_L is the result of replacing the left child chain of T with the right child chain, i.e, both the left and right child chains are labeled x_{l_1} through x_{l_d}; T_R is instead the result of replacing the right child chain with the left. T_L and T_R are shown in Figures 2a and 2b, respectively. Note that they are both fully non-adaptive, and thus can just as easily be represented by pure chains.

Without loss of generality, we will assume

$$p_{i,0}(U_L^0 - U_R^0) \leq p_{i,1}(U_R^1 - U_L^1) \quad (4)$$

and substitute T_R for T by replacing the left child chain of T with its right child chain (in the symmetric setting, we have $p_{i,0}(U_L^0 - U_R^0) \geq p_{i,1}(U_R^1 - U_L^1)$ and

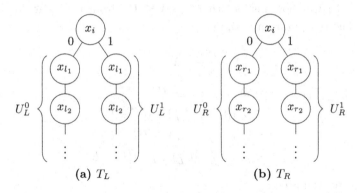

(a) T_L **(b)** T_R

Fig. 2. Variants of T

substitute T_L for T). Set

$$S(b) = \begin{cases} \{i\} \cup L & \text{when } b_i = 0 \\ \{i\} \cup R & \text{when } b_i = 1 \end{cases}.$$

Let Δ be the expected loss in utility. We want to show that $\tau U_i \geq \Delta$. We compute Δ:

$$\begin{aligned}
\Delta &= \mathbb{E}[g(S(b), b)] - \mathbb{E}[g(\{i\} \cup R, b)] \\
&= [p_{i,0}(U_i^0 + U_L^0) + p_{i,1}(U_i^1 + U_R^1)] - [p_{i,0}(U_i^0 + U_R^0) + p_{i,1}(U_i^1 + U_R^1)] \\
&= p_{i,0}(U_L^0 - U_R^0).
\end{aligned}$$

There are two cases: (i) $p_{i,0} \leq p_{i,1}$ and (ii) $p_{i,0} > p_{i,1}$.

In case (i) we show that $\frac{1}{p_{i,1}} \Delta \leq U_i$, from which it follows that $\Delta \leq \tau U_i$ since $\tau \leq p_{i,1}$. Indeed, using $p_{i,1} = 1 - p_{i,0}$ we have

$$\begin{aligned}
\frac{1}{p_{i,1}} \Delta &= \left(1 + \frac{p_{i,0}}{p_{i,1}}\right) \Delta \\
&\leq \Delta + p_{i,0}(U_R^1 - U_L^1) \\
&= p_{i,0}[(U_L^0 - U_R^0) + (U_R^1 - U_L^1)] \\
&= p_{i,0}[(U_L^0 - U_L^1) + (U_R^1 - U_R^0)] \\
&\leq p_{i,0}(U_L^0 - U_L^1) + p_{i,1}(U_R^1 - U_R^0) \\
&\leq U_i
\end{aligned}$$

where the first inequality follows by (4), the second inequality since $p_{i,0} \leq p_{i,1}$, and the last inequality from (3).

In case (ii) we show instead that $\frac{1}{p_{i,0}}\Delta \leq U_i$, from which again we have $\Delta \leq \tau U_i$. The computation is similar:

$$
\begin{aligned}
\frac{1}{p_{i,0}}\Delta &= \left(1 + \frac{p_{i,1}}{p_{i,0}}\right)\Delta \\
&\leq \Delta + p_{i,1}(U_L^0 - U_R^0) \\
&\leq p_{i,1}[(U_R^1 - U_L^1) + (U_L^0 - U_R^0)] \\
&= p_{i,1}[(U_L^0 - U_L^1) + (U_R^1 - U_R^0)] \\
&< p_{i,0}(U_L^0 - U_L^1) + p_{i,1}(U_R^1 - U_R^0).
\end{aligned}
$$

This completes the proof.   □

Our bottom-up greedy approach cannot achieve a constant-factor approximation; for every threshold $\theta \in [0, \frac{1}{2}]$, it is possible to construct a simple example where $\tau < \theta$ in which the procedure used in the proof loses more than a $1 - \theta$ fraction of the original utility. Some examples can be found in the appendix. Other methods may be able to achieve better bounds.

5 The Gap Between **ADAPT** and **OFFLINE**

In this section, we consider the gap $\frac{\text{OFFLINE}}{\text{ADAPT}}$ for Stochastic Submodular Maximization with a cardinality constraint.

We first define some notation. We use k to denote the number of items allowed by the cardinality constraint. For consistency, any variant of b will refer to a partial assignment, and any variant of a will refer to only a full assignment, that is, $a \in \mathbb{S}^n$. For the sake of clarity, in this section we will explicitly specify the assignments over which we are taking expectations, except with the shorthand that when b is a fixed partial assignment, $\mathbb{E}_{a \succ b}[\,\cdot\,]$ will mean $\mathbb{E}_a[\,\cdot \mid a \succ b]$. If we have a sequence $\{S^t\}_{t=0}^k$, we write $g_t(j)$ in place of $g_{S^t}(j)$.

We use AGREEDY to denote the expected utility of the *Adaptive Greedy* algorithm of Golovin and Krause [6] (also called *Adaptive Myopic* by Asadpour and Nazerzadeh [3]), which, starting with $Q^0 = \varnothing$, at each step t, based on the partial assignment b^{t-1} of bits tested so far, adaptively picks i_t satisfying

$$
i_t = \underset{i \in N \setminus Q^{t-1}}{\arg\max} \; \underset{a \succ b^{t-1}}{\mathbb{E}}\left[g_{t-1,a}(i)\right],
$$

sets $Q^t = Q^{t-1} \cup \{i_t\}$, and tests b_{i_t} to get b^t. This implicitly forms a decision tree of depth k that branches based on the outcome of b_{i_t}, and outputs Q^k.

It is clear that AGREEDY $\leq$ ADAPT $\leq$ OFFLINE. We show:

Theorem 2. *AGREEDY* $\geq (1 - \frac{1}{e^\tau}) \cdot$ *OFFLINE for Stochastic Submodular Maximization with a cardinality constraint.*

In other words, $\frac{\text{OFFLINE}}{\text{ADAPT}} \leq \frac{e^\tau}{e^\tau - 1}$. Although this is bound is weak for small τ, we observe that some dependence on τ is unfortunately unavoidable:

Proposition 1. *ADAPT cannot achieve an approximation bound better than τ relative to OFFLINE.*

The proof of Proposition 1 is by example, and is given in the appendix. We will note, however, that τ and $1 - \frac{1}{e^\tau}$ are close: for $0 < \tau \leq \frac{1}{2}$ the difference between τ and $1 - \frac{1}{e^\tau}$ is at most ~ 0.107, which is achieved at $\tau = \frac{1}{2}$.

We now prove Theorem 2. We use the following lemma due to Wolsey:

Lemma 1 ([11]). *Let k be a positive integer, and $s > 0$, $\rho_1, ..., \rho_k \geq 0$ be reals. Then*

$$\frac{\sum_{i=1}^{k} \rho_i}{\min_{t \in \{1,...,k\}} \left(s\rho_t + \sum_{i=1}^{t-1} \rho_i \right)} \geq 1 - \left(1 - \frac{1}{s} \right)^k \geq 1 - \frac{1}{e^{k/s}}.$$

Next, let Q_a^* be the optimal offline solution on assignment a, $\tau = \min_{i,j} p_{i,j}$, and (Q^t, b^t) be the collection Q^t given by the Adaptive Greedy algorithm at step t corresponding to the partial assignment b^t. Then:

Lemma 2. *For $t = 1, 2, \ldots, k$,*

$$\mathbb{E}_a \left[g(Q_a^*, a) \right] \leq \mathbb{E}_a \left[g(Q^{t-1}, a) \right] + \frac{k}{\tau} \cdot \mathbb{E}_a \left[g(Q^t, a) - g(Q^{t-1}, a) \right]$$

Proof. Suppose Greedy chooses i_t at step t, that is, $Q^t \setminus Q^{t-1} = \{i_t\}$. By definition we have

$$\mathbb{E}_{a' \succ b^{t-1}} \left[g_{t-1,a'}(i_t) \right] = \sum_{m=1}^{\ell} p_{i_t,m} \left[g_{t-1,b_{i_t \leftarrow m}^{t-1}}(i_t) \right].$$

Next, suppose $j \in Q_{a'}^* \setminus Q^{t-1}$ where a' is a full assignment and $a' \succ b^{t-1}$. We can write $\Pr[a'] = \prod_i \psi_i$ where $\psi_i \in \{p_{i,0}, \ldots, p_{i,\ell-1}\}$. If we then let $a'_{j \leftarrow m}$ be the assignment a' with the j-th bit set to m, we have

$$g_{t-1,a'}(j) \leq \frac{1}{\psi_j} \left(\sum_{m=0}^{\ell-1} p_{j,m} \left[g_{t-1,a'_{j \leftarrow m}}(j) \right] \right)$$

$$\leq \frac{1}{\tau} \left(\sum_{m=0}^{\ell-1} p_{i_t,m} \left[g_{t-1,b_{i_t \leftarrow m}^{t-1}}(i_t) \right] \right)$$

$$= \frac{1}{\tau} \cdot \mathbb{E}_{a' \succ b^{t-1}} \left[g_{t-1,a'}(i_t) \right]$$

where the first inequality follows from the fact that $\psi_j = p_{j,m}$ for some m, and the second from the definitions of τ and i_t.

Then since $|Q_{a'}^* \setminus Q^{t-1}| \leq k$ we get

$$\sum_{j \in Q_{a'}^* \setminus Q^{t-1}} g_{t-1,a'}(j) \leq \frac{k}{\tau} \cdot \mathbb{E}_{a' \succ b^{t-1}} \left[g_{t,a'}(i_t) \right]. \tag{5}$$

Next, we have

$$
\begin{aligned}
\mathbb{E}_a \left[g(Q_a^*, a) \right] &= \mathbb{E}_{b^{t-1}} \left[\mathbb{E}_{a' \succ b^{t-1}} \left[g(Q_{a'}^*, a') \right] \right] \\
&\leq \mathbb{E}_{b^{t-1}} \left[\mathbb{E}_{a' \succ b^{t-1}} \left[g(Q_{a'}^* \cup Q^{t-1}, a') \right] \right] \\
&\leq \mathbb{E}_{b^{t-1}} \left[\mathbb{E}_{a' \succ b^{t-1}} \left[g(Q^{t-1}, a') + \sum_{j \in Q_{a'}^* \setminus Q^{t-1}} g_{t-1,a'}(j) \right] \right] \\
&\leq \mathbb{E}_a \left[g(Q^{t-1}, a) \right] + \mathbb{E}_{b^{t-1}} \left[\frac{k}{\tau} \cdot \mathbb{E}_{a' \succ b^{t-1}} \left[g_{t-1,a'}(i_t) \right] \right] \\
&= \mathbb{E}_a \left[g(Q^{t-1}, a) \right] + \frac{k}{\tau} \cdot \mathbb{E}_a \left[g_{t-1,a}(i_t) \right]
\end{aligned}
$$

where the first inequality follows from monotonicity, the second from submodularity, and the third from (5). □

In light of Lemma 1 we see that Theorem 2 follows from Lemma 2:

Proof (of Theorem 2). Let $\rho_i = \mathbb{E}_a[g(Q^i, a) - g(Q^{i-1}, a)]$, then clearly we have $\mathbb{E}_a[g(Q^{t-1}, a)] = \sum_{i=1}^{t-1} \rho_i$. Since the Adaptive Greedy algorithm outputs Q^k,

$$
\begin{aligned}
\frac{\text{AGREEDY}}{\text{OFFLINE}} &= \frac{\sum_{i=1}^{k} \rho_i}{\mathbb{E}_a[g(Q_a^*, a)]} \\
&\geq \frac{\sum_{i=1}^{k} \rho_i}{\min_{t \in \{1,\dots,k\}} \left(\frac{k}{\tau} \rho_t + \sum_{i=1}^{t-1} \rho_i \right)} \geq 1 - \frac{1}{e^\tau}
\end{aligned}
$$

as desired. □

6   Gaps for Stochastic Max Coverage

In this section we consider the special case of Stochastic Max Coverage. We still achieve an adaptivity gap of $\frac{e}{e-1}$, which is tight by an example given by Asadpour and Nazerzadeh [3, §3.1]. Furthermore, we obtain a bound of $\frac{e}{\tau(e-1)}$ for $\frac{\text{OFFLINE}}{\text{NONADAPT}}$.

We use of the following fact, which is easily seen by inspection or by calculus.

Lemma 3. *For all $x \geq 0$, $1 - (1 - \frac{x}{k})^k \geq x(1 - \frac{1}{e})$.*

We give a combinatorial proof of the following result, which is adapted from work of Ageev and Sviridenko on the deterministic version of the problem [1,2].

Theorem 3. *For the Stochastic Max Coverage problem, there exists a non-adaptive solution that achieves coverage $(1 - \frac{1}{e})$ADAPT.*

Proof. For simplicity, we give the analysis for the uniform distribution over binary states; the analysis is similar for product distributions over $\ell > 2$ states.

We make use of the *neighbor property*, which every adaptive algorithm satisfies by definition: given two assignments a, a' differing only in bit j, either b_j is tested for both assignments, or for neither. We denote the expected value of the optimal offline solution satisfying the neighbor property by NBR. Clearly, NONADAPT $\leq$ ADAPT $\leq$ NBR $\leq$ OFFLINE. We will prove that NONADAPT $\geq$ $(1 - \frac{1}{e})$NBR, thus achieving the desired adaptivity gap. Let $\mathcal{X}$ denote the procedure giving the optimal offline solution satisfying the neighbor property.

For each $i \in N$ and assignment a we assign a variable $x_{i,a}$ such that $x_{i,a} = 1$ if and only if $S_{i,a}$ is included in Q_a, the subcollection given by $\mathcal{X}$. Correspondingly we assign to each $e_j \in E$ a variable $y_{j,a}$ such that $y_{j,a} = 1$ if and only if $j \in \bigcup_{i \in Q_a} S_{i,a}$. Since NBR denotes the expected number of ground elements by the solution given by $\mathcal{X}$, NBR $= \sum_j \mathbb{E}_a[y_{j,a}]$.

Consider the following (randomized) algorithm for producing a non-adaptive solution from the solution given by $\mathcal{X}$: randomly pick k sets according to the following probability distribution: pick i with probability $\frac{1}{k} \sum_a \Pr[a] \sum_{i=1}^n x_{i,a} = \frac{1}{k2^n} \sum_a \sum_{i=1}^n x_{i,a}$.

For each e_j we say i is a *promising cover* for j if either $e_j \in S_{i,0}$ or $e_j \in S_{i,1}$. We divide the promising covers into two categories: i is of type B if $e_j \in S_{i,0} \cap S_{i,1}$ and i is of type A otherwise. We abuse notation slightly and let $A = \frac{1}{2^n} \sum_{i \text{ of type } A} \sum_{a:e_j \in S_{i,a}} x_{i,a}$, and similarly $B = \frac{1}{2^n} \sum_{i \text{ of type } B} \sum_a x_{i,a}$. By definition $A + B \geq \mathbb{E}_a[y_{j,a}]$.

In expectation, the probability that a random i produces a promising cover of type A for e_j is $\frac{2A}{k}$; hence the probability that, in expectation, at least one of the chosen i is a promising cover of type A for e_j is at least $1 - (1 - \frac{2A}{k})^k \geq 2A(1 - \frac{1}{e})$. A promising cover of type A covers e_j with probability $\frac{1}{2}$, so e_j is covered by at least one promising cover of type A with probability at least $A(1 - \frac{1}{e})$.

Similarly, picking a random set produces a promising cover of type B for j with probability $\frac{B}{k}$, so at least one of the chosen i is a promising cover of type B for e_j is at least $1 - (1 - \frac{B}{k})^k \geq B(1 - \frac{1}{e})$, and a promising cover of type B covers e_j with probability 1, so e_j is covered by at least one promising cover of type B with probability at least $B(1 - \frac{1}{e})$. Since the promising covers of type A and B are disjoint, we conclude that e_j is covered with probability at least $(1 - \frac{1}{e})(A + B) \geq (1 - \frac{1}{e}) \mathbb{E}_a[y_{j,a}]$.

Before finishing the proof, we note that for product distributions, we will actually have many more such categories: for each $i \in N$ we will have a different set of categories. Furthermore, when there are $\ell > 2$ states, there is a larger number of possible ways of covering each e_j and hence a larger number of categories. However, the analysis will be similar, so that it will still the case that e_j is covered with probability $\geq (1 - \frac{1}{e}) \mathbb{E}_a[y_{j,a}]$.

We return to the proof at hand. By linearity of expectation, the expected number of elements covered by this non-adaptive solution is equal to the sum of the probabilities that each j is covered, in other words,

$$\mathbb{E}[\# \text{ elements covered}] = \sum_j \Pr[j \text{ is covered}]$$

$$\geq \left(1 - \frac{1}{e}\right) \sum_j \sum_a \mathbb{E}[y_{j,a}] = \left(1 - \frac{1}{e}\right) \text{NBR}.$$

Since this randomized procedure achieves at least $(1 - \frac{1}{e})$NBR in expectation, then there must exist a non-adaptive solution that achieves at least $(1-\frac{1}{e})$NBR $\geq$ $(1 - \frac{1}{e})$ADAPT, thus completing the proof.   □

A similar analysis gives the following result:

Theorem 4. *For the Stochastic Max Coverage problem, there exists a non-adaptive solution that achieves coverage* $\tau(1 - \frac{1}{e})$*OFFLINE.*

Proof. The proof is similar to the proof of Theorem 3. Again we use variables $x_{i,a}$ such that $x_{i,a} = 1$ if and only if $S_{i,a}$ is included in the optimal subcollection Q_a for a, and $y_{j,a}$ such that $y_{j,a} = 1$ if and only if $e_j \in \bigcup_{i \in Q_a} S_{i,a}$. Clearly $\sum_{i:e_j \in S_{i,a}} x_{i,a} \geq y_{j,a}$ for each assignment a and OFFLINE $= \sum_j \mathbb{E}_a[y_{j,a}]$.

Again we randomly pick k sets according to the following probability distribution: pick i with probability $\frac{1}{k} \sum_a \Pr[a] \sum_{i=1}^n x_{i,a}$.

In expectation, the probability that picking a random i produces a promising cover for e_j is

$$\frac{1}{k} \sum_a \Pr[a] \sum_{i:e_j \in S_{i,a}} x_{i,a} \geq \frac{1}{k} \sum_a \Pr[a] y_{j,a} = \frac{\mathbb{E}_a[y_{j,a}]}{k}$$

hence the probability that at least one promising cover for j is chosen is

$$1 - \left(1 - \frac{\mathbb{E}_a[y_{j,a}]}{k}\right)^k \geq \left(1 - \frac{1}{e}\right) \mathbb{E}_a[y_{j,a}].$$

Since each promising cover covers j with probability $\geq \tau$, the probability that j is actually covered is at least $\tau(1 - \frac{1}{e}) \mathbb{E}_a[y_{j,a}]$. The rest of the proof follows from analysis similar to the analysis used in the proof of Theorem 3.   □

Acknowledgements. Patrick Lin was partially supported by NSF Grants 1217968 and 1319648. Devorah Kletenik and Lisa Hellerstein were partially supported by NSF Grants 1217968 and 0917153.

A   Appendix: Counterexamples

A.1   Counterexamples for Section 4

Without loss of generality we will assume $p_{i,1}(U_R^1 - U_L^1) \geq p_{i,0}(U_L^0 - U_R^0)$. This is the situation in which we substitute T_R for T. The expected loss in utility is $\Delta = \mathbb{E}[g(S(b), b)] - \mathbb{E}[g(\{i\} \cup R, b)] = p_{i,0}(U_L^0 - U_R^0)$.

We give a systematic way of finding counterexamples roughly showing that for $\theta \in [0, \frac{1}{2}]$, by picking $\tau < \theta$ we can achieve $\Delta \geq \theta \cdot U_i$ where $U_i = p_{i,0}U_i^0 + p_{i,1}U_i^1$.

Example 1. Suppose $\theta = 1 - 10^{-m}$, eg. if $m = 3$ then $\theta = 0.999$. Then let $p_{i,0} = 10^{-m-1}$ (so $p_{i,1} = 1 - 10^{-m-1}$), $U_R^1 = 10^{m+1}$, $U_R^0 = 5$, $U_L^1 = 10^{m+1} - 1$, $U_L^0 = 10^{m+1} - 2$, $U_i^0 = 10^{m+1} - 4$, and $U_i^1 = 3 \cdot 10^{-m-1}$. Then

$$\Delta = 10^{-m-1}(10^{m+1} - 2 - 5) = 1 - 7 \cdot 10^{-m-1}$$

and

$$\theta \cdot U_i = (1 - 10^{-m})[3 \cdot 10^{-m-1}(1 - 10^{-m-1}) + (10^{m+1} - 4) \cdot 10^{-m-1}]$$
$$\approx 1 - 2 \cdot 10^{-m}$$

as desired. It is easy to verify that $p_{i,1}(U_R^1 - U_L^1) \geq p_{i,0}(U_L^0 - U_R^0)$. Further, the utility values given here can be shown to be consistent with a monotone, submodular utility function g. □

A.2 Proof of Proposition 1

We construct a counterexample for which ADAPT $\approx (\tau - \varepsilon)$OFFLINE.

Example 2. Consider an instance of the stochastic submodular coverage with a cardinality constraint problem as follows: say $\ell = 2$ (so this is over binary states), let $t = \frac{1}{\tau}$ and set $n = 1 + t^2$. Let $p_{i,1} = \tau$ and $p_{i,0} = 1 - \tau$ for all $i \in N$. Let $B = 1$, so the problem is to maximize the expected utility of picking a single bit.

Consider the following stochastic monotone submodular function g defined over $\mathbb{S}$: for all subcollections $Q \subseteq N$ such that $i \notin Q$, let

$$g_{Q,a}(i) = \begin{cases} 1 & \text{if } i = 1 \\ t - \varepsilon & \text{if } i \in \{2, \ldots, n\} \text{ and } a_i = 1 \\ 0 & \text{if } i \in \{2, \ldots, n\} \text{ and } a_i = 0 \end{cases}$$

where the monotonicity and submodularity of g follow from the fact that g is an additive utility function.

For $i \notin Q$, $\mathbb{E}[g_{Q,b}(i)] = 1$ if $i = 1$ and $\mathbb{E}[g_{Q,b}(i)] = \frac{t-\varepsilon}{t}$ if $i \in \{2, \ldots, n\}$. Due to the low probability of any variable from the second group having a value of 1, no adaptive tree outperforms the greedy choice of picking the first variable. Thus ADAPT $= 1$.

The optimal offline procedure, on the other hand, will pick any variable of the second group that has a value of 1. With probability $1 - (1 - \frac{1}{t})^{t^2}$, at least one variable of the second group will evaluate to 1; hence, OFFLINE $= \left(1 - (1 - \frac{1}{t})^{t^2}\right) \cdot (t - \varepsilon) + (1 - \frac{1}{t})^{t^2} \cdot 1$.

It follows that for sufficiently large t, ADAPT $\approx (\tau - \varepsilon)$OFFLINE. □

References

1. Ageev, A.A., Sviridenko, M.I.: Approximation algorithms for maximum coverage and max cut with given sizes of parts. In: Cornuéjols, G., Burkard, R.E., Woeginger, G.J. (eds.) IPCO 1999. LNCS, vol. 1610, pp. 17–30. Springer, Heidelberg (1999)
2. Ageev, A.A., Sviridenko, M.I.: Pipage rounding: A new method of constructing algorithms with proven performance guarantee. J. Comb. Optim. **8**(3), 307–328 (2004)
3. Asadpour, A., Nazerzadeh, H.: Maximizing stochastic monotone submodular functions (2014). arXiv preprint arXiv:0908.2788v2
4. Chan, C.W., Farias, V.F.: Stochastic depletion problems: Effective myopic policies for a class of dynamic optimization problems. Math. Oper. Res. **34**(2), 333–350 (2009)
5. Feige, U.: A threshold of ln n for approximating set cover. J. ACM **45**(4), 634–652 (1998)
6. Golovin, D., Krause, A.: Adaptive submodularity: Theory and applications in active learning and stochastic optimization. J. Artif. Intell. Res. (JAIR) **42**, 427–486 (2011)
7. Khuller, S., Moss, A., Naor, J.: The budgeted maximum coverage problem. Inf. Process. Lett. **70**(1), 39–45 (1999)
8. Nemhauser, G.L., Wolsey, L.A.: Best algorithms for approximating the maximum of a submodular set function. Math. Oper. Res. **3**(3) (1978)
9. Nemhauser, G.L., Wolsey, L.A., Fisher, M.L.: An analysis of approximations for maximizing submodular set functions - I. Math. Program. **14**(1), 265–294 (1978)
10. Sviridenko, M.I.: A note on maximizing a submodular set function subject to a knapsack constraint. Oper. Res. Lett. **32**(1), 41–43 (2004)
11. Wolsey, L.A.: Maximising real-valued submodular functions: Primal and dual heuristics for location problems. Math. Oper. Res. **7**(3), 410–425 (1982)

Parameterized Algorithms and Kernels for 3-Hitting Set with Parity Constraints

Vikram Kamat[1] and Neeldhara Misra[2]([✉])

[1] University of Warsaw, Warsaw, Poland
vkamat@mimuw.edu.pl
[2] Indian Institute of Science, Bangalore, India
neeldhara@csa.iisc.ernet.in

Abstract. The 3-HITTING SET problem involves a family of subsets $\mathcal{F}$ of size at most three over an universe $\mathcal{U}$. The goal is to find a subset of $\mathcal{U}$ of the smallest possible size that intersects every set in $\mathcal{F}$. The version of the problem with parity constraints asks for a subset S of size at most k that, in addition to being a hitting set, also satisfies certain parity constraints on the sizes of the intersections of S with each set in the family $\mathcal{F}$. In particular, an odd (even) set is a hitting set that hits every set at either one or three (two) elements, and a perfect code is a hitting set that intersects every set at exactly one element. These questions are of fundamental interest in many contexts for general set systems. Just as for Hitting Set, we find these questions to be interesting for the case of families consisting of sets of size at most three. In this work, we initiate an algorithmic study of these problems in this special case, focusing on a parameterized analysis. We show, for each problem, efficient fixed-parameter tractable algorithms using search trees that are tailor-made to the constraints in question, and also polynomial kernels using sunflower-like arguments in a manner that accounts for equivalence under the additional parity constraints.

1 Introduction

The 3-HITTING SET problem involves a family of subsets $\mathcal{F}$ of size at most three over an universe $\mathcal{U}$. The goal is to find a subset of $\mathcal{U}$ of the smallest possible size that intersects every set in $\mathcal{F}$. This is a fundamental NP-complete problem, and has been extensively studied from an algorithmic perspective. In particular, consider the decision version of the problem: here we are given, in addition to $\mathcal{U}$ and $\mathcal{F}$, a positive integer k, and the question is if $\mathcal{F}$ admits a hitting set of size at most k. Using a standard exhaustive search, analyzed using a depth-bounded search tree, this question can be answered in time $3^k n^{\mathcal{O}(1)}$.

V. Kamat—Supported by a postdoctoral fellowship from the Warsaw Center of Mathematics and Computer Science
N. Misra—Supported by the INSPIRE Faculty Scheme, DST India (project DSTO-1209).

© Springer International Publishing Switzerland 2015
V.Th. Paschos and P. Widmayer (Eds.): CIAC 2015, LNCS 9079, pp. 249–260, 2015.
DOI: 10.1007/978-3-319-18173-8_18

Such algorithms are said to be *fixed-parameter tractable* with respect to k, since the exponential complexity is contained in k alone, and the algorithm is efficient for small values of k. Such an analysis belongs naturally to the framework of parameterized complexity: here, in addition to the overall input size n, one studies how a secondary measurement (called the *parameter*), that captures additional relevant information, affects the computational complexity of the problem in question. Parameterized decision problems are defined by specifying the input, the parameter, and the question to be answered. The two-dimensional analogue of the class P is decidability within a time bound of $f(k)n^c$, where n is the total input size, k is the parameter, f is some computable function and c is a constant that does not depend on k or n. A parameterized problem that can be decided in such a time-bound is termed *fixed-parameter tractable* (FPT). For general background on the theory of fixed-parameter tractability, see [3], [4], and [10].

A parameterized problem is said to admit a *polynomial kernel* if every instance (I, k) can be reduced in polynomial time to an equivalent instance with both size and parameter value bounded by a polynomial in k. The study of kernelization is a major research frontier of parameterized complexity and many important recent advances in the area are on kernelization. For overviews of kernelization we refer to surveys [1,5] and to the corresponding chapters in books on parameterized complexity [4,10].

As it turns out, 3-HITTING SET is known to admit an instance kernel of size $\mathcal{O}(k^3)$, with an universe of size $\mathcal{O}(k^2)$ [11]. The earlier arguments for kernels relied on sunflowers, while the argument in [11] relies on a linear vertex-kernel for Vertex Cover. Further, unless the polynomial hierarchy collapses, it is also known that there are no kernels with $\mathcal{O}(k^{d-\varepsilon})$ sets [2]. In this work, we analyze the 3-Hitting Set problem with parity constraints. The constraints we study correspond to the well known problems ODD SET, EVEN HITTING SET and PERFECT CODE when restricted to families with sets of size at most three:

3-Odd Set Given a family $\mathcal{F}$ of sets of size at most three over an universe $\mathcal{U}$, and an integer k, is there a subset S of $\mathcal{U}$ of size at most k such that $|S \cap X|$ is odd for every $X \in \mathcal{F}$?

3-Perfect Code Given a family $\mathcal{F}$ of sets of size at most three over an universe $\mathcal{U}$, and an integer k, is there a non-empty subset S of $\mathcal{U}$ of size at most k such that $|S \cap X| = 1$ for every $X \in \mathcal{F}$?

3-Even Hitting Set Given a family $\mathcal{F}$ of sets of size at most three over an universe $\mathcal{U}$, and an integer k, is there a hitting set S of $\mathcal{U}$ of size at most k such that $|S \cap X|$ is even for every $X \in \mathcal{F}$?

All of these problems have been studied closely in the setting of families that involve sets of unbounded size. They are of fundamental interest in several contexts, including matroids and coding theory. Surprisingly, the natural question of understanding these problems on families where the set sizes are bounded appears to be relatively unexplored, although HITTING SET is very well-understood in the specialized context (for example, see the algorithm in [9]). In this work, we study the parameterized complexity of each of these problems.

We note that each of these problems are NP-complete even when restricted to families that have sets of size at most three, due to the characterization in [7]. For each of these problems, we provide fixed-parameter tractable algorithms and polynomial kernels. We remark that all the problems above can be reduced to the MIN-ONES 3-SAT problem by an appropriate formulation of the constraints. Therefore, by the results in [12], all of these problems are FPT with running time $2.85^k n^{\mathcal{O}(1)}$. We improve upon this running time significantly in each case by exploiting branching rules that are tailor-made to the constraints in question. In particular, for 3-Odd Set and 3-Perfect Code, we obtain a running time of $2.56^k n^{\mathcal{O}(1)}$; while for 3-Even Hitting Set, we obtain a running time of $1.73^k n^{\mathcal{O}(1)}$.

The approaches that have been successful for 3-HITTING SET are not easy to employ directly for the versions of the problem with parity constraints. The fundamental operations of deleting "irrelevant" elements from the instance or incorporating "forced" elements in our solution are now non-trivial to implement. For instance, notice that when an element x is *forced*, it is typically safe to delete all sets containing x from the family — and this is essential to proving kernel bounds. For, say, ODD SET on the other hand, it may not be safe to delete these sets without a mechanism for remembering the additional constraints imposed on them. In particular, let $S := \{x, y, z\}$ be a set containing x. If x belongs to a solution, then we are obliged to "remember" that the elements y and z are, in some sense, coupled to each other — they must now either belong together or completely avoid any odd set that contains S.

We develop a natural auxiliary constraint graph to keep track of these additional implications. With a more refined notion of forcing vertices, a standard branching argument works. For the kernels, similarly, using these carefully defined operations, we are able to apply sunflower-style reduction rules to reduce the size of the instance. To the best of our knowledge, this is the first explicit attempt to obtain FPT algorithms that are faster than that suggested by the exhaustive branching strategy, or from the known algorithms for MIN-ONES 3-SAT.

The work of [8] has a very intricate framework for applying reduction rules involving sunflowers to solve a much more general problem, namely, MIN-ONES CSPs. As a corollary of this work, we already know that all three problems that we are discussing admit kernels where the instance size is bounded by $\mathcal{O}(k^4)$. However, with the help of a slightly specialized analysis, we obtain kernels that have instance sizes bounded $\mathcal{O}(k^3)$ for all the problems considered. It would be very interesting to improve these to quadratic or even linear bounds on the number of vertices in the universe. We also note that we are technically describing *bikernels*, which are instances of "annotated" versions of the original problems. However, in each case, there are simple reductions that will generate an instance of the original problem that has the same size bounds asymptotically.

The remainder of this paper is organized as follows. After introducing the terminology and notation in Section 2, we address the 3-ODD SET problem. We the describe both the branching algorithm and reduction rules towards kernelization. The details for 3-PERFECT CODE are quite similar to 3-ODD SET.

On the other hand, 3-EVEN HITTING SET requires significantly different definitions of the two basic operations of deletion and incorporation, the details of which are deferred to a full version due to space constraints.

2   Preliminaries

A parameterized problem is denoted by a pair $(Q, k) \subseteq \Sigma^* \times \mathbb{N}$. The first component Q is a classical language, and the number k is called the parameter. Such a problem is *fixed–parameter tractable* (FPT) if there exists an algorithm that decides it in time $\mathcal{O}(f(k)n^{\mathcal{O}(1)})$ on instances of size n. A *kernelization algorithm* takes an instance (x, k) of the parameterized problem as input, and in time polynomial in $|x|$ and k, produces an equivalent instance (x', k') such that both $|x'|$ and k' are functions purely of k. The output x' is called the kernel of the problem. A kernel is said to be a *polynomial kernel* if its size $|x'|$ is polynomial in the parameter k. We refer the reader to standard text books on the subject [3,10] for more details on the notion of fixed-parameter tractability.

The MIN-ONES 3-SAT problem is the following: Given a 3-CNF SAT formula ϕ, is there a satisfying assignment for ϕ that sets at most k variables to TRUE? A standard branching algorithm for this problem would branch on all clauses with no negated literals, attempting to set each variable in turn to TRUE in turn, and the instances at the leaf nodes of the search tree can be satisfied by setting all variables to FALSE. This algorithm has running time $\mathcal{O}(3^k)$. This has since been improved in [12] to $\mathcal{O}^*(2.85^k)$. We note that all the hitting set problems with parity constraints can be reduced to MIN-ONES 3-SAT in a manner that preserves the size of the solution. We refer the reader to the appendix for a sketch of the proof. Therefore, an immediate corollary is that each of these problems admit algorithms with running time $\mathcal{O}^*(2.85^k)$.

We will use the notion of a sunflower frequently during our description of kernelization algorithms. A *sunflower* with k petals and a core Y is a collection of sets $\{S_1, \ldots, S_k\}$ such that $S_i \cap S_j = Y$ for all $i \neq j$; the sets $S_i \setminus Y$ are *petals*, and we require that none of them is empty. Note that a family of pairwise disjoint sets is a sunflower (with an empty core).

Lemma 1 (The Sunflower Lemma[6]).   *Let $\mathcal{F}$ be family of sets each of cardinality s. If $|\mathcal{F}| > s!(k-1)^s$ then $\mathcal{F}$ contains a sunflower with k petals.*

We use standard notation for branching vectors, focusing only on the drop on the measures in question and ignoring polynomial factors in n. We refer the reader to [10] for an overview of this notation. We also use the $\mathcal{O}^*()$ notation to suppress running times that are polynomial in n, focusing on the function of the parameter.

3   A Polynomial Case: Two-Sized Sets

A subroutine we will need in the subsequent sections is the case of ODD SET restricted to families where all sets have size two. We show that not only is this polynomially solvable, but that any solution has a special structure. Recall that the problem would require us to find a subset of the universe of size at most k that chooses *exactly* one element from any set of size two. First, it is easy to see that the graphs corresponding to YES-instances must be bipartite.

Observation 1 ($\star$). *Let $(\mathcal{U}, \mathcal{F}, k)$ be an instance of* ODD SET, *where every set in $\mathcal{F}$ has size at most two. If this is a* YES-*instance, then the graph given by $\mathcal{Z} := (\mathcal{U}, \mathcal{F})$ is bipartite.*

Now, we show that any solution must pick exactly one of the partitions of every bipartite component of $\mathcal{Z}$.

Lemma 2 ($\star$). *Let $(\mathcal{U}, \mathcal{F}, k)$ be a* YES-*instance of* ODD SET, *where every set in $\mathcal{F}$ has size at most two, and there are no isolated vertices. Let $\mathcal{Z} := (\mathcal{U}, \mathcal{F})$. Further, let $(A_1, B_1), \ldots, (A_t, B_t)$ be the connected components of $\mathcal{Z}$. Then, for any odd set $O \subseteq \mathcal{U}$, we have that for all $1 \leq i \leq t$, either $A_i \subseteq O$ and $B_i \cap O = \emptyset$, or $B_i \subseteq O$ and $A_i \cap O = \emptyset$.*

Due to space constraints, a proof of the lemmas above is deferred to a full version of this work.

4   3-Odd Set

An exhaustive branching algorithm for ODD SET would require us to branch on four possibilities for a given set $S = \{x, y, z\}$ in the input family — either the set S is hit at exactly one element (leading to three different cases), or all of S belongs to the odd set. However, in each of these branches, we are required to execute the operation of "incorporating" an element in our solution and solving an equivalent, recursive instance. incorporating an element $x \in U$ in the solution is straightforward for Hitting Set — we may simply delete all sets that contain x and decrease the parameter by one. For Odd Set, on the other hand, we have to additionally remember for every three-element set S that x appears in, the elements $S \setminus \{x\}$ are no longer independent — they must either both be chosen in the future or not be chosen at all. To remember these constraints, we adopt the use of an auxiliary graph on the universe, adding an edge between elements that must be chosen.

Formally, we solve the following auxiliary problem:

Constrained Odd Set
   Input: A family $\mathcal{F}$ of sets of size at most three over an universe $\mathcal{U}$,
       a graph $G = (\mathcal{U}, E)$, and a positive integer k.
Parameter: k
   Question: Is there a subset $X \subseteq \mathcal{U}$ of size at most k such that for every
       $S \in \mathcal{F}$, $|S \cap X|$ is odd, and further, for every connected
       component $C \in G$, either $C \subseteq X$ or $C \cap X = \emptyset$?

Notice that an instance of ODD SET is a special case of CONSTRAINED ODD SET, where the constraint graph is the edgeless graph on $|U|$ vertices. We define special operations that mimic the process of "incorporating" an element into a solution, and "deleting" an element from the instance. Let $\mathcal{Z} := (\mathcal{U}, \mathcal{F}, G, k)$ be an instance of CONSTRAINED ODD SET, let x be an element from $\mathcal{U}$. Let $C(x)$ be the set of elements in the connected component containing x, in G. We now have the following.

Delete x from the instance. If there is any set $S \in \mathcal{F}$ such that $S \subseteq C(x)$, then the operation of deleting x from $\mathcal{Z}$ results in a trivial NO instance. Otherwise, let $\mathcal{F}' := \{S \setminus C(x) \mid S \in \mathcal{F}\}$. The instance that results from deleting x from $\mathcal{Z}$ is given by $(\mathcal{U} \setminus C(x), \mathcal{F}', G \setminus C(x), k)$.

Incorporate x in the solution. Consider the set $D(x)$ obtained as follows: to begin with, $D(x) = C(x)$. As long as there exists a set $S \in \mathcal{F}$ of size three such that $|S \cap D(x)| = 2$, we include the element $S \setminus D(x)$ in $D(x)$. At any stage, if the number of elements in $D(x)$ is more than k, or if there is a two-sized set $S \in \mathcal{F}$ that is contained in $D(x)$, then the operation of incorporating x from $\mathcal{Z}$ results in a trivial NO instance. Otherwise, we have the following. Let $\mathcal{F}_1'$ be the collection of sets of size two in $\mathcal{F}$ that intersect $D(x)$ at exactly one element, and let $\mathcal{F}_2'$ be the collection of sets of size three in $\mathcal{F}$ that intersect $D(x)$ at exactly one element. That is,

$$\mathcal{F}_1' := \{S \mid S \in \mathcal{F}, |S| = 2, |S \cap D(x)| = 1\},$$

$$\mathcal{F}_2' := \{S \mid S \in \mathcal{F}, |S| = 3, |S \cap D(x)| = 1\},$$

Now, let G' be the graph obtained from G by adding all possible edges among the vertices of $D(x)$. Then, the process of incorporating x entails the following:

1. Let G^* be obtained from G' by adding the edges corresponding to the pairs $(S \setminus D(x))$ for all $S \in \mathcal{F}_2'$.
2. For all $S \in \mathcal{F}_1'$, delete $S \setminus D(x)$ from the instance $(\mathcal{U}, \mathcal{F}, G^*, k)$. If any of these operations result in a trivial NO instance, then abort and return a NO instance. Otherwise, let the resulting instance be $(\mathcal{U}', \mathcal{F}', H, k)$.
3. Return the instance $(\mathcal{U} \setminus D(x), \mathcal{F}'', H \setminus D(x), k - |D(x)|)$, where $\mathcal{F}''$ is the family obtained from $\mathcal{F}'$ after deleting all sets that intersect $D(x)$.

We prove the correctness of these operations in the following lemmas.

Lemma 3 ($\star$). *Let $(\mathcal{U}, \mathcal{F}, \mathcal{G}, k)$ be an instance of CONSTRAINED ODD SET and let $x \in \mathcal{U}$ be such that any odd set of size at most k for $(\mathcal{U}, \mathcal{F})$ respecting the constraints in G must contain x. Let $(\mathcal{U}', \mathcal{F}', \mathcal{G}', k')$ be the instance obtained by applying the operations associated with incorporating x in the solution. Then, $(\mathcal{U}, \mathcal{F}, \mathcal{G}, k)$ is a YES instance if, and only if, $(\mathcal{U}', \mathcal{F}', \mathcal{G}', k')$ is a YES instance.*

Lemma 4 ($\star$). *Let $(\mathcal{U}, \mathcal{F}, \mathcal{G}, k)$ be an instance of CONSTRAINED ODD SET and let $x \in \mathcal{U}$ be such that any odd set of size at most k for $(\mathcal{U}, \mathcal{F})$ respecting the constraints in G cannot contain x. Let $(\mathcal{U}', \mathcal{F}', \mathcal{G}', k')$ be the instance* *obtained by applying the operations associated with deleting x in the solution. Then, $(\mathcal{U}, \mathcal{F}, \mathcal{G}, k)$ is a* YES *instance if, and only if, $(\mathcal{U}', \mathcal{F}', \mathcal{G}', k')$ is a* YES *instance.*

We are now ready to present our branching and kernelization algorithms.

4.1   A Branching Algorithm for 3-Odd Set

We are now ready to describe the branching algorithm. Before branching, we always incorporate the elements of any singleton sets. For hitting set, a further simplification that is often exploited is the *domination strategy*: if there are elements x and y such that x appears in all sets that y appears in, then it is safe to delete x from the instance, by a standard pushing argument — a solution containing x can be replaced with a solution that contains y instead of x. However, consider the following instance of CONSTRAINED ODD SET:

$$(\{x, y, z\}, \{y, w\}, \{w, x_1\}, \cdots, \{w, x_{k+1}\}, k)$$

Note that although z is dominated by y; a solution containing z cannot be morphed into a solution containing y — because any constrained odd set is forced to contain w, which in turn forbids us from picking y. This rule, when valid, is useful in generating sets of size two, which can be branched on more efficiently. However, since we are unable to apply the domination rule in the stated form, several arguments used across different cases of the 3-Hitting Set algorithm in [9] cannot be adapted directly to our scenario. This motivates the need for a somewhat different branching strategy. Fortunately, it turns out that an elegant branching strategy leads us to an algorithm with running time $\mathcal{O}^*(2.56^k)$. We first consider the case when we have at least two sets of size three that overlap; and then consider the case when all the three-sized sets are disjoint and some sets have size two. We use the standard measure, which is the size of the odd set sought.

Case A. There exist a pair of three-sized sets that have at least one element in common. Let these sets be:

$$\{x, a_1, b_1\}, \{x, a_2, b_2\}$$

Case A1. Suppose $a_1 = a_2 = y$. Then we have the following branches:
- Incorporate x and y. This forces us to incorporate b_1 and b_2, and the measure drops by four.
- Incorporate neither x nor y. This forces us to incorporate b_1, and b_2, and the measure drops by two.
- Incorporate x but not y. This leads to a drop of one in the measure.
- Incorporate y but not x. This leads to a drop of one in the measure.

The overall branch vector, therefore, is $(4, 2, 1, 1)$.

Case A2. Without loss of generality, we may assume here that $a_1 \neq a_2$ and $b_1 \neq b_2$. Here, we branch exhaustively on these sets as follows:

- Incorporate x. This leads to a drop of one in the measure.
- Delete x. Incorporate a_1 and b_2, deleting b_1 and a_2. This is leads to a drop of two in the measure.
- Delete x. Incorporate a_1 and a_2, deleting b_1 and b_2. This is leads to a drop of two in the measure.
- Delete x. Incorporate b_1 and a_2, deleting a_1 and b_2. This is leads to a drop of two in the measure.
- Delete x. Incorporate b_1 and b_2, deleting b_1 and a_2. This is leads to a drop of two in the measure.

The overall branch vector, therefore, is $(2, 2, 2, 2, 1)$.

Case B. All the three-sized sets are disjoint. Let $\mathcal{H} \subseteq \mathcal{F}$ be all the two-sized sets in the family $\mathcal{F}$, and let $V \subseteq U$ be the set of all elements $\mathcal{U}$ that appear in some set of $\mathcal{H}$. Let the graph $\mathcal{Z}$ be defined as $(V, \mathcal{H})$. If the graph $\mathcal{Z}$ is not bipartite, then return No. Otherwise, let the connected components of $\mathcal{Z}$ be $Z_1 := (A_1, B_1), \cdots, Z_i := (A_i, B_i), \cdots, Z_\ell := (A_\ell, B_\ell)$, where (A_i, B_i) are the bipartitions of the component Z_i. Recall from Lemma 2, any odd set S for $(\mathcal{U}, \mathcal{F})$ must contain, for all $1 \leq i \leq \ell$, exactly one of A_i or B_i. In other words, for all $1 \leq i \leq \ell$, either $S \cap Z_i = A_i$ or $S \cap C_i = B_i$.

On the other hand, let $C_1, \ldots, C_t$ denote the connected components of the constraint graph G. Now we have a few rules of simplification, applied in the order stated below, based on this observation. The correctness of these rules are easily established.

- If there exists $1 \leq i \leq \ell$ for which both A_i and B_i contain the vertices of the same component C_i in the constraint graph, then return No.
- Let S be a three-sized set in $\mathcal{F}$. If there exists $1 \leq i \leq \ell$ such that $S \subseteq A_i$, then incorporate A_i and delete B_i. If there exists $1 \leq i \leq \ell$ such that $S \subseteq B_i$, then incorporate B_i and delete A_i.
- Let S be a three-sized set in $\mathcal{F}$. If there exists $1 \leq i \leq \ell$ such that $|S \cap B_i| = 2$ and $|S \cap A_i| = 1$, then incorporate A_i and delete B_i. Similarly, there exists $1 \leq i \leq \ell$ such that $|S \cap A_i| = 2$ and $|S \cap B_i| = 1$, then incorporate B_i and delete A_i.

We are now ready to branch as follows: for all $1 \leq i \leq \ell$, we incorporate all of A_i in one branch (while deleting B_i) and all of B_i in the other (while deleting A_i), with the measure dropping by at least one in both branches. In the base case, we are left with a collection of three-sized sets $\{S_1, S_2, \ldots, S_t\}$ that are mutually disjoint. Although tempting at this stage, it may not be correct to pick one element from each set, because different elements have different "costs" of incorporation, corresponding to the sizes of their connected components in the conflict graph. Further, not all elements of a set are equal in the context of the conflict graph; different elements may lead us to different paths of further incorporations; some which may be of greater benefit than others. To resolve such instances, we use the following strategy:

1. If there is some S_i that contains an element a with $|C(a)| > 1$, then branch exhaustively on S_i. Formally, if $S_i := \{a, b, c\}$, then we branch as follows:
   - Incorporate a, b and c. This forces us to incorporate $C(a)$, and since $|C(a)| > 1$, the measure drops by at least four.
   - Incorporate a, delete b and c;. This forces us to incorporate $C(a)$, and since $|C(a)| > 1$, the measure drops by at least two.
   - Incorporate b, delete a and c; this leads to a drop of one in the measure.
   - Incorporate c, delete a and b; this leads to a drop of one in the measure.

   The overall branch vector, therefore, is $(4, 2, 1, 1)$.
2. Now, every S_i contains elements that belong to singleton components in the conflict graph. Here, we may arbitrarily choose an element from each set to be incorporated in the final solution.

The running time of the algorithm that we have just described is bounded by the following recurrence:

$$T(k) \leq \max\{T(k-4) + T(k-2) + T(k-1) + T(k-1),$$
$$4T(k-2) + T(k-1),$$
$$T(k-1) + T(k-1)\}.$$

Notice that correctness follows from Lemmas 3 and 4; along with the exhaustive nature of the branching strategy. Using standard techniques to bound the recurrence above, we have the following theorem.

Theorem 1. *The* CONSTRAINED ODD SET *problem admits a* FPT *algorithm with running time* $\mathcal{O}^*(2.56^k)$.

4.2 A Polynomial Kernel for 3-Odd Set

We now turn to the kernelization algorithm for CONSTRAINED ODD SET. For this discussion, we work with a weighted version of CONSTRAINED ODD SET, where all elements of the universe $\mathcal{U}$ have positive integer weights, and we are looking for a solution of total weight at most k. The delete and incorporate operations are adapted in the natural way, to use the weights of the elements when decreasing the budget, rather than the number of elements. To begin with, we have some simple pre-processing rules for an instance $(\mathcal{U}, \mathcal{F}, G, k)$:

Reduction Rule 1. If $\{x\} \in \mathcal{F}$, then we incorporate x into the solution.

Reduction Rule 2. If $\{x\} \in \mathcal{U}$, and the weight of x is greater than k, then delete x.

Reduction Rule 3. If $\{x, y\} \in \mathcal{F}$, and x and y are in the same connected component of G, then we return a trivial No-instance.

Reduction Rule 4. Let C be a connected component in G such that the sum of weights of all vertices in C is greater than k, and let $x \in C$. We delete x from the instance.

Reduction Rule 5. If $k \leq 0$ but $\mathcal{F}$ is non-empty then return a trivial No-instance.

We call Reduction Rules 1—5 the *basic reduction rules*. The correctness of these rules are easy to verify, and the arguments are omitted here due to space constraints. An instance is always reduced with respect to the basic reduction rules before the application of any further reduction rules. As with the algorithm for 3-HITTING SET, we will also appeal to the Sunflower Lemma (Lemma 1).

We now consider all the three-sized sets in $\mathcal{F}$, let us use $\mathcal{H}$ to denote this sub-family of $\mathcal{F}$. If $|\mathcal{H}| \geq 6k^3$, then $\mathcal{H}$ contains a sunflower with at least $(k+1)$ petals. It is easy to construct the sub-family corresponding to the sunflower in polynomial time. If the sunflower has an empty core, then we abort and return a trivial No-instance. Otherwise, let the petals of the sunflower be $P_1, \ldots, P_t$, where $t \geq k + 1$. We have the following cases.

Reduction Rule 6. (The core has one element.) Let $\bigcap_{i=1}^{t} P_i := \{x\}$. We then incorporate x in the solution.

Reduction Rule 7. (The core has two elements.) Let $\bigcap_{i=1}^{t} P_i := \{x, y\}$.

Let $P_i := \{x, y, z_i\}$. For all $1 \leq i \leq t$, we delete z_i from the instance. Let $(\mathcal{U}', \mathcal{F}', G', k)$ be the resulting instance. We now have the following cases:

1. If $\mathcal{U}'$ does not contain either x or y, then we return a trivial No-instance.
2. If exactly one of x or y is present in $\mathcal{U}'$, then we incorporate the element $\mathcal{U}' \cap \{x, y\}$ in our solution, and return the resulting instance.
3. If both x and y are present in $\mathcal{U}'$, and x and y are in the same connected component of G', then we return a trivial No instance.
4. Otherwise, we return the instance $(\mathcal{U}', \mathcal{F}' \cup \{x, y\}, G', k)$.

We now claim the correctness of the two rules above.

Lemma 5 ($\star$). *Reduction rules 6 and 7 are safe.*

We now consider the collection of two-sized sets in $\mathcal{F}$, let us use $\mathcal{J}$ to denote this sub-family of $\mathcal{F}$. If $|\mathcal{J}| \geq 2k^2$, then $\mathcal{J}$ contains a sunflower with at least $(k+1)$ petals. Again, if the sunflower has an empty core, then we abort and return a trivial No-instance. Otherwise, let the petals of the sunflower be $P_1, \ldots, P_t$, where $t \geq k + 1$. This sunflower necessarily has a core of size one; and this leads us to the following reduction rule:

Reduction Rule 8. Let $P_i := \{x, y_i\}$. For all $1 \leq i \leq t$, delete y_i from the instance. If the resulting instance is a trivial No-instance, then return a trivial No-instance. Otherwise, return the instance obtained by incorporate x in the solution.

Lemma 6 ($\star$). *Reduction rule 8 is safe.*

Notice that once the reduction rules are exhaustively applied, we are left with either a trivial No-instance, or an instance with at most $6k^3$ sets of size three, at most $2k^2$ sets of size two, and no singletons. Let us denote a reduced instance by $(\mathcal{U}^R, \mathcal{F}^R, G^R, k^R)$. We call an element of $\mathcal{U}^R$ *useful* if it appears in some set of $\mathcal{F}^R$. We now have two final reduction rules, whose correctness is self-evident, and which we apply only once on a reduced instance:

Reduction Rule 9. Let C be a connected component in G with no useful vertices, and let $x \in C$. We delete x from the instance.

Reduction Rule 10. Let C be a connected component in G with at least one useful vertex, and ℓ useless vertices of total weight w. We replace all the useless vertices of C by a dummy vertex of weight w and make it adjacent to all vertices in the component C.

Since it is evident that elements of the universe that do not appear in any sets do not belong to any minimal solutions, the correctness of Reduction Rule 9 follows from Lemma 4.

Let us denote an instance reduced with respect to the reduction rule 9 by $(\mathcal{U}^\dagger, \mathcal{F}^\dagger, G^\dagger, k^\dagger)$. Since $|\mathcal{U}^\dagger| = |V(G^\dagger)|$, note that a bound on the universe of the kernel will follow from a bound on the number of vertices in $G^\dagger$. Observe that, by Reduction Rules 4 and 9, every connected component of $G^\dagger$ has size at most k, and there are at most as many components as there are useful vertices. Further, every component contains at most one vertex that is not useful, and this leads to a bound on the number of elements in the universe, as we argue in the following lemma.

Lemma 7 ($\star$). *Let $(\mathcal{U}, \mathcal{F}, G, k)$ be an instance that is recursively subjected to the basic reduction rules and the sunflower reduction rules (in that order), and finally to Reduction Rule 9. Let the resulting instance be denoted by $(\mathcal{U}^\dagger, \mathcal{F}^\dagger, G^\dagger, k^\dagger)$. Then, $|\mathcal{U}^\dagger| = \mathcal{O}(k^3)$.*

Lemmata 4—7 lead to the following theorem, which also concludes the objective of this section.

Theorem 2. *The* CONSTRAINED ODD SET *problem has an instance kernel of size $\mathcal{O}(k^3)$.*

References

1. Bodlaender, H.L.: Kernelization: new upper and lower bound techniques. In: Chen, J., Fomin, F.V. (eds.) IWPEC 2009. LNCS, vol. 5917, pp. 17–37. Springer, Heidelberg (2009)
2. Dell, H., van Melkebeek, D.: Satisfiability Allows No Nontrivial Sparsification unless the Polynomial-Time Hierarchy Collapses. J. ACM **61**(4), 23 (2014)
3. Downey, R.G., Fellows, M.R.: Parameterized Complexity. Springer (1999)

4. Flum, J., Grohe, M.: Parameterized Complexity Theory. Springer-Verlag New York Inc (2006)
5. Guo, J., Niedermeier, R.: Invitation to data reduction and problem kernelization. SIGACT news **38**(1), 31–45 (2007)
6. Jukna, S.: Extremal combinatorics - with applications in computer science, pp. I-XVII, 1–375. Springer (2001)
7. Khanna, et al.: The Approximability of Constraint Satisfaction Problems. SICOMP: SIAM Journal on Computing **30** (2001)
8. Kratsch, S., Wahlström, M.: Preprocessing of min ones problems: a dichotomy. In: Abramsky, S., Gavoille, C., Kirchner, C., Meyer auf der Heide, F., Spirakis, P.G. (eds.) ICALP 2010. LNCS, vol. 6198, pp. 653–665. Springer, Heidelberg (2010)
9. Niedermeier, Rossmanith: An Efficient Fixed-Parameter Algorithm for 3-Hitting Set. Journal of Discrete Algorithms, Elsevier (unseen by me), **1** (2003)
10. Niedermeier, R.: Invitation to Fixed Parameter Algorithms (Oxford Lecture Series in Mathematics and Its Applications). Oxford University Press, USA (2006)
11. Nishimura, N., Ragde, P., Thilikos, D.M.: Smaller kernels for hitting set problems of constant arity. In: Downey, R.G., Fellows, M.R., Dehne, F. (eds.) IWPEC 2004. LNCS, vol. 3162, pp. 121–126. Springer, Heidelberg (2004)
12. Raman, V., Shankar, B.S.: Improved fixed-parameter algorithm for the minimum weight 3-SAT problem. In: Ghosh, S.K., Tokuyama, T. (eds.) WALCOM 2013. LNCS, vol. 7748, pp. 265–273. Springer, Heidelberg (2013)

Simple Strategies Versus Optimal Schedules in Multi-agent Patrolling

Akitoshi Kawamura$^{(\boxtimes)}$ and Makoto Soejima

University of Tokyo, Tokyo, Japan
kawamura@graco.c.u-tokyo.ac.jp, msoejima@is.s.u-tokyo.ac.jp

Abstract. Suppose that we want to patrol a fence (line segment) using k mobile agents with given speeds $v_1, \ldots, v_k$ so that every point on the fence is visited by an agent at least once in every unit time period. A simple strategy where the ith agent moves back and forth in a segment of length $v_i/2$ patrols the length $(v_1 + \cdots + v_k)/2$, but it has been shown recently that this is not always optimal. Thus a natural question is to determine the smallest c such that a fence of length $c(v_1 + \cdots + v_k)/2$ cannot be patrolled. We give an example showing $c \geq 4/3$ (and conjecture that this is the best possible).

We also consider a variant of this problem where we want to patrol a circle and the agents can move only clockwise. We can patrol a circle of perimeter rv_r by a simple strategy where the r fastest agents move at the same speed, but it has been shown recently that this is not always optimal. We conjecture that this is not even a constant-approximation strategy. To tackle this conjecture, we relate it to what we call *constant gap families*. Using this relation, we give another example where the simple strategy is not optimal.

We propose another variant where we want to patrol a single point under the constraint that for each $i = 1, \ldots, k$, the time between two consecutive visits of agent i should be a_i or longer. This problem can be reduced to the discretized version where the a_i are integers and the goal is to visit the point at every integer time. It is easy to see that this discretized patrolling is impossible if $1/a_1 + \cdots + 1/a_k < 1$, and that there is a simple strategy if $1/a_1 + \cdots + 1/a_k \geq 2$. Thus we are interested in the smallest c such that patrolling is always possible if $1/a_1 + \cdots + 1/a_k \geq c$. We prove that $\alpha \leq c < 1.546$, where $\alpha = 1.264\ldots$ (we conjecture that $c = \alpha$). We also discuss the computational complexity of related problems.

1 Introduction

In *patrolling problems*, a set of mobile agents are deployed in order to protect or supervise a given area, and the goal is to leave no point unattended for a long period of time. Besides being a well-studied task in robotics and distributed algorithms, patrolling raises interesting theoretical questions [4]. Recent studies [2,3,6] have shown that finding an optimal strategy is not at all straightforward, even when the terrain to be patrolled is as simple as it could be. We

© Springer International Publishing Switzerland 2015
V.Th. Paschos and P. Widmayer (Eds.): CIAC 2015, LNCS 9079, pp. 261–273, 2015.
DOI: 10.1007/978-3-319-18173-8_19

continue this line of research in three basic settings: patrolling a line segment, a circle, and a point. We will be particularly interested in the ratio by which the best schedule could outperform the simple strategy for each problem.

1.1 Fence Patrolling

In 2011, Czyzowicz et al. [2] proposed the following problem:

Fence Patrolling Problem. We want to patrol a fence (line segment) using k mobile agents. We are given the speed limits of the agents $v_1, \ldots, v_k$ and the *idle time* $T > 0$. For each point x on the fence and time $t \in \mathbb{R}$, there must be an agent who visits the point x during the interval $[t, t + T)$. How long can the fence be?

Formally, a fence is an interval $[0, L]$, and a *schedule* is a k-tuple $(a_1, \ldots, a_k)$ of functions, where each $a_i \colon \mathbb{R} \to \mathbb{R}$ satisfies $|a_i(s) - a_i(t)| \le v_i \cdot |s - t|$ for all s, $t \in \mathbb{R}$. It *patrols* the fence with idle time T if for any time $t \in \mathbb{R}$ and any location $x \in [0, L]$, there are an agent i and a time $t' \in [t, t + T)$ such that $a_i(t') = x$.

Note that if we can patrol a fence of length L with idle time T, we can patrol a fence of length αL with idle time αT by scaling, for any $\alpha > 0$. Thus, we are only interested in the ratio of L and T. Unless stated otherwise, we fix the idle time to $T = 1$.

In Section 2, we will prove that any schedule can be approximated arbitrarily closely by a periodic schedule. Thus, for any $\varepsilon > 0$, we can find in finite time (though not efficiently) a schedule that is $1 - \varepsilon$ times as good as any schedule.

Czyzowicz et al. [2] discussed the following simple strategy that patrols a fence of length $(v_1 + \cdots + v_k)/2$ (with idle time 1), and proved that no schedule can patrol more than twice as long a fence as this strategy:

Partition-Based Strategy. Divide the fence into k segments, the ith of which has length $v_i/2$. The agent i moves back and forth in the ith segment.

They conjectured that this gives the optimal schedule. However, Kawamura and Kobayashi [6] exhibited a setting of speed limits $v_1, \ldots, v_k$ and a schedule that patrols a fence slightly longer than the partition-based strategy. Thus, the following natural question arises: what is the biggest ratio between the optimal schedule and partition-based strategy? Formally, we want to determine the smallest constant c such that no schedule can patrol a fence that is c times as long as the partition-based strategy does.

Czyzowicz et al.'s result [2] says that $1 \le c \le 2$, and their conjecture was that $c = 1$. Kawamura and Kobayashi's example shows that $c \ge 42/41$. Later this lower bound was improved to $25/24$ [1,3]. In Section 3, we will further improve the lower bound to $4/3$. We conjecture that $c = 4/3$. However, we have not been able to prove even $c < 2$.

1.2 Unidirectional Circle Patrolling

In Section 4, we will discuss another problem proposed by Czyzowicz et al. [2]:

Unidirectional Circle Patrolling Problem. We want to patrol a circle using k mobile agents. We are given the speed limits $v_1, \ldots, v_k$ of the agents. For each point x on the circle and time $t \in \mathbb{R}$, there must be an agent who visits the point x during the interval $[t, t + 1)$. Each agent i is allowed to move along the circle in clockwise direction with arbitrary speed between 0 and its speed limit v_i, but it is not allowed to move in the opposite direction. How long can the perimeter of the circle be?

They conjectured that the following strategy is optimal:

Runners Strategy. Without loss of generality, we can assume that $v_1 \geq \cdots \geq v_k$. If all the fastest r agents move at constant speed v_r and placed equidistantly, we can patrol a perimeter of length rv_r. By choosing the optimal r, we can achieve the perimeter $\max_r rv_r$.

However, Dumitrescu et al. [3] constructed an example where this strategy is not optimal.

We conjecture that the Runners Strategy is not even a constant-ratio approximation strategy. Formally, we conjecture that for any constant c, there exist $v_1, \ldots, v_k$ such that we can patrol a perimeter of $c \max_r rv_r$.

To tackle this conjecture, we will define *constant gap families*, and show a relation between them and Circle Patrolling Problem. Also, using this relation, we will construct another example where the Runners Strategy is not optimal.

1.3   Point Patrolling

In Section 5, we propose a new problem that we call Point Patrolling Problem. In a sense, this is a simplification of the Fence Patrolling Problem. In this problem, agents patrol a single point instead of a fence. In this case, it is natural to set a lower bound on the intervals between two consecutive visits by an agent instead of restricting its speed. Formally, we study the following problem:

Point Patrolling Problem. We want to patrol a point using k mobile agents. We are given the lower bounds $a_1, \ldots, a_k$ on the intervals between two consecutive visits of the agents. A *schedule* is a k-tuple of sets $S_1, \ldots, S_k \subseteq \mathbb{R}$, where S_i means the set of times at which the ith agent visits the point. Thus, if t_1 and t_2 are two distinct elements of S_i, they must satisfy $|t_1 - t_2| \geq a_i$. This schedule *patrols* the point with *idle time* T if for any time $t \in \mathbb{R}$, there are an agent i and a time $t' \in [t, t + T)$ such that $t' \in S_i$. How small can the idle time be?

It turns out that this problem can be reduced to a decision problem that asks whether it is possible to visit the point at each integer time under the constraint that for each $i = 1, \ldots, k$, the time between two consecutive visits of agent i should be $a_i \in \mathbb{N}$ or longer. We will see the relation between the amount $1/a_1 + \cdots + 1/a_k$ and this problem.

In Section 6, we will analyze the complexity of problems that are related to this discretized problem.

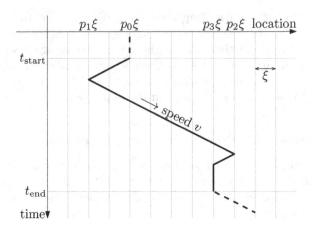

Fig. 1. A (v, ξ)-zigzag movement on a time interval $[t_{\text{start}}, t_{\text{end}}]$

2 Zigzag Schedules for Fence Patrolling

In the following two sections, we will discuss the Fence Patrolling Problem.

In this section, we prove that, for the purpose of discussing upper limit of the length of the fence, we may restrict attention to periodic schedules.

The movement of an agent during time interval $[t_{\text{start}}, t_{\text{end}}] \subseteq \mathbb{R}$ is represented by a function $a \colon [t_{\text{start}}, t_{\text{end}}] \to \mathbb{R}$. This function is called a (v, ξ)-*zigzag movement*, for $v, \xi > 0$ (Figure 1), if there are integers $p_0, p_1, p_2, p_3 \in \mathbb{Z}$ such that the agent

- starts at time t_{start} at location $p_0 \xi$,
- moves at speed v until it reaches $p_1 \xi$,
- moves at speed v until it reaches $p_2 \xi$,
- moves at speed v until it reaches $p_3 \xi$,
- and then stays there until time t_{end}.

For this movement to be possible, the entire route must be short enough to be travelled with speed v; that is,

$$|p_0 - p_1|\xi + |p_1 - p_2|\xi + |p_2 - p_3|\xi \leq \tau v, \tag{1}$$

where $\tau := t_{\text{end}} - t_{\text{start}}$ is the length of the time interval.

We prove in the next lemma (see the full version [7] for a proof) that any schedule can be converted into one that consists of zigzag movements without deteriorating the idle time too much. For positive $\xi, \tau > 0$, a schedule $(a_1, \ldots, a_k)$ (for k agents with speed limits $v_1, \ldots, v_k$) is called a (ξ, τ)-*zigzag schedule* if the movement of each agent $i = 1, \ldots, k$ during each time interval $[m\tau, (m+1)\tau]$, $m \in \mathbb{Z}$, is a (v_i, ξ)-zigzag.

Lemma 1. *For any $\varepsilon > 0$ and speeds $v_1, \ldots, v_k > 0$, there are $\xi > 0$ and $\tau' > 0$ satisfying the following. Suppose that there is a schedule for a set of agents with speed limits $v_1, \ldots, v_k$ that patrols a fence with some idle time $T > 0$. Then there is a (ξ, τ')-zigzag schedule for the same set of agents that patrols the same fence with idle time $T(1 + \varepsilon)$.*

The next lemma says that a zigzag schedule can be made periodic without changing the idle time (see the full version [7] for a proof).

Lemma 2. *Suppose that there is a (ξ, τ)-zigzag schedule for a set of agents that patrols a fence with some idle time. Then there is a periodic (ξ, τ)-zigzag schedule for the same agents that patrols the same fence with the same idle time.*

Using the above lemmas, we obtain an algorithm that solves the Fence Patrolling Problem with arbitrarily high precision in the following sense.

Theorem 3. *There exists an algorithm that, given $v_1, \ldots, v_k$, T and $\varepsilon > 0$, finds a schedule that patrols a fence of length at least $1 - \varepsilon$ times the length of the fence patrolled by the same agents using any schedule.*

Proof. Suppose that there is a schedule that patrols a fence of length L with idle time T using these agents. By Lemma 1, there is a (ξ, τ)-zigzag schedule that patrols a fence of length $(1 - \varepsilon)L$, for some ξ, $\tau > 0$ determined by the inputs ε and $v_1, \ldots, v_k$. By Lemma 2, there is a (ξ, τ)-zigzag schedule with period p that patrols the same length $(1 - \varepsilon)L$, for some $p > 0$ determined by the inputs. Since there are only finitely many such schedules, we can check all of them in a finite amount of time. □

In previous work [2,6], a schedule was defined as functions on the halfline $[0, +\infty)$ (instead of $\mathbb{R}$) and the requirement for patrolling was that each location be visited in every length-T time interval contained in this halfline. Note that the argument for Lemmas 1 and 2 in this section stays valid when we start with a patrolling schedule on $[0, +\infty)$ in this sense. In particular, a patrolling schedule on $[0, +\infty)$ can be converted to a (periodic) schedule on $\mathbb{R}$ without essentially worsening the idle time. Therefore, the ratio bound that we are interested in (the constant c in Section 1.1) is not affected by our slight deviation in the definition.

3   A Schedule Patrolling a Long Fence

In this section, we will prove that for any $c < 4/3$, there exists a schedule that patrols a fence c times as long as the partition-based strategy. This improves the same claim for $c < 25/24$ established previously [1,3].

Theorem 4. *For any $c < 4/3$, there are settings of speed limits $v_1, \ldots, v_k$ and a schedule that patrols a fence of length $c(v_1 + \cdots + v_k)/2$ (with idle time 1).*

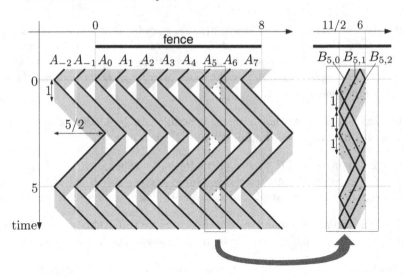

Fig. 2. The strategy in the proof of Theorem 4 when $n = 3$ and $L = 8$. The trajectories of the agents are the thick solid lines, and the regions they cover are shown shaded. The $n + L - 1$ faster agents $A_{-n+1}, \ldots, A_{L-1}$ (left) move back and forth with period $2n - 1$, but leave some triangular regions (dotted) uncovered. These regions are covered by the nL slow agents $B_{0,0}, \ldots, B_{n-1,L-1}$ (right; scaled up horizontally for clarity).

Proof. We construct, for any positive integers n and L, a schedule that patrols a fence of length L with idle time 1 using $n + L - 1$ agents with speed 1 and nL agents with speed $1/(2n - 1)$. Note that with the partition-based strategy, the same set of agents would patrol (with idle time 1) a fence of length $\frac{1}{2}(n + L - 1 + nL/(2n - 1))$. The ratio between L and this approaches $4/3$ when $1 \ll n \ll L$, and hence we have the theorem. The schedule that proves our claim is as follows:

- Each of the $n + L - 1$ agents A_i $(-n < i < L)$ with speed 1 visits the locations i and $i + n - 1/2$ alternately (at its maximal speed); it is at location i at time 0. (This means that some agents occasionally step out of the fence $[0, L]$; to avoid this, we could simply modify the schedule so that they stay at the end of the fence for a while.)
- Each of the nL agents $B_{i,j}$ $(0 \le i < L, 0 \le j < n)$ with speed $1/(2n - 1)$ visits the locations $i + 1/2$ and $i + 1$ alternately (at its maximal speed); it is at location $i + 1/2$ at time $j + 1/2$.

See Figure 2, where we say that an agent *covers* a point $(x, t) \in [0, L] \times \mathbb{R}$ if it visits the location x during the time interval $[t - 1, t]$. It is routine to verify that with the above schedule, the entire region is covered by at least one agent.   □

Note that the intersection of the line segment $[0, L] \times \{1/2\}$ with the region not covered by the agents A_i (Figure 2, left) has total length $L/2$. In order to cover this intersection, the other agents must have total speed of at least $L/2$, because an agent with speed v can only cover a subsegment of length at most v out of any horizontal line. Thus, as long as we use the above schedule for A_i (whose total speed is $n + L - 1 \geq L$), the total speed of all agents must be at least $3L/2$, no matter what other agents we have instead of the $B_{i,j}$. Since the schedule for A_i looks reasonably efficient, we conjecture that the above construction is optimal:

Conjecture 5. No schedule can patrol a fence that is more than $4/3$ times as long as the partition-based strategy.

4   Circle Patrolling

We start by defining constant gap families. As mentioned in the introduction, they are closely related to the Circle Patrolling Problem as we will show in Lemma 6. For a real number $c > 1$ and a positive integer k, a (c, k)-*constant gap family* (henceforth a (c, k)-*family*) is a k-tuple of sets $S_1, \ldots, S_k \subseteq \mathbb{R}$ with $S_1 \cup \cdots \cup S_k = \mathbb{R}$ such that for each i,

1. the set S_i is a union of non-overlapping intervals $S_i = \bigcup_{j \in \mathbb{Z}} [a_{i,j}, b_{i,j}]$;
2. the length of each interval in S_i is at most $1/(ci-1)$, i.e., $b_{i,j} - a_{i,j} \leq 1/(ci-1)$;
3. the distance between two consecutive intervals in S_i is *exactly* 1, i.e., $a_{i,j+1} - b_{i,j} = 1$.

Lemma 6. *Let $c > 1$.*

1. *If k agents with speed limits 1, $1/2$, $\ldots$, $1/k$ can patrol a circle of perimeter c, then there is a (c, k)-family.*
2. *If there is a (c, k)-family, then k agents with speed limits 1, $1/2$, $\ldots$, $1/k$ can patrol a circle of perimeter $c/2$.*

Proof.   Consider a circle with perimeter L. We say an agent *covers* $t \in \mathbb{R}$ if it visits the point $ct \bmod L$ at least once during the time interval $[t, t + L/c]$ (Figure 3). It is straightforward to show that for any possible movement of an agent with speed limit $v < c$, the set of $t \in \mathbb{R}$ covered by this agent is a union of disjoint intervals $\bigcup_{i \in \mathbb{Z}} [a_i, b_i]$ such that

$$b_i - a_i \leq \frac{vL}{c(c-v)}, \qquad a_{i+1} - b_i = \frac{L}{c}. \tag{2}$$

These are also sufficient conditions in the sense that for any a_i and b_i satisfying (2), there is a movement of an agent with speed limit v that covers $\bigcup_{i \in \mathbb{Z}} [a_i, b_i]$; that is, for each $i \in \mathbb{Z}$, the agent is at the point $ca_i \bmod L$, and during the time interval $[a_i, a_{i+1}]$, the agent travels the distance $c(a_{i+1} - a_i) - L$. This movement is possible because $c(a_{i+1} - a_i) - L \leq v(a_{i+1} - a_i)$. We now prove the claims.

1. Define S_i as the set of numbers t that are covered by the agent with speed limit $1/i$. Then, $S_1, \ldots, S_k$ is a (c, k)-family.

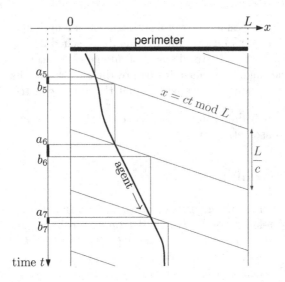

Fig. 3. Time intervals $[a_i, b_i]$ that are covered by an agent

2. Let $S_1, \ldots, S_k$ be a (c, k)-family. Then, we can define a schedule on a circle of perimeter $c/2$ such that the agent of speed limit $1/i$ covers $t/2$ for all $t \in S_i$. This can be verified by checking the sufficient conditions (2). By the definition of "cover", for each t, at least one agent visits ct during the time interval $[t, t + 1/2]$. Note that ct and $ct + nL$ refer to the same point for each integer n. Thus ct is visited at least once during the time interval $[(ct + nL)/c, (ct + nL)/c + 1/2] = [t + n/2, t + n/2 + 1/2]$ for each integer n. This implies that in each unit time interval each point is visited by an agent. □

Theorem 7. *There exist $v_1, \ldots, v_k$ and a schedule that patrols a circle with perimeter $1.05 \max_r r v_r$.*

Proof. By Lemma 6, it suffices to prove the existence of a $(2.1, k)$-family for some k. Using a computer program, we have found a $(2.1, 122)$-family $(S_1, \ldots, S_{122})$, which can be found in the full version [7]. □

In fact, we believe that the constant 1.05 can be replaced by an arbitrary big constant:

Conjecture 8. The runners strategy does not have a constant approximation ratio. Formally, for any constant c, there exist $v_1, \ldots, v_k$ and a schedule that patrols a circle with perimeter $c \max_r r v_r$.

By Lemma 6, the conjecture is correct if and only if for any constant c, there exists k such that a (c, k)-family exists. It looks easier to prove the existence of (c, k)-families than to prove the conjecture directly.

In Theorem 7, we have found a $(2.1, 122)$-family, despite the fact that the existence of (c, k)-families is nontrivial already when $c > 1$. We feel that this supports our Conjecture 8.

5   Point Patrolling

In this section, we will discuss Point Patrolling Problem. First, let's see that this problem can be reduced to a problem in which time is also discrete. Consider a decision version of this problem. That is, you are given T, and you need to decide whether the idle time can be at most T. We can reduce the original problem to this decision problem by binary search. This decision problem can be discretized in the following way:

Discretized Point Patrolling Problem. There are k agents and they want to patrol a point. We are given positive integers $a_1, \ldots, a_k$. The interval between two consecutive visits by the ith agent must be at least a_i. A schedule is called *good* if at each integer time the point is visited by at least one agent. Determine whether there exists a good schedule.

For simplicity, we call a tuple of integers $(a_1, \ldots, a_k)$ *good* if there exists a good strategy in Discretized Point Patrolling Problem, and otherwise call it *bad*.

Theorem 9. *Agents with intervals* $(a_1, \ldots, a_k)$ *(here a_i are real numbers) can achieve the idle time of T for the (non-discretized) Point Patrolling Problem if and only if $(\lceil a_1/T \rceil, \ldots, \lceil a_k/T \rceil)$ is good.*

See the full version [7] for a proof. This discretized problem can be solved in $O(k \prod_{i=1}^{k} a_i)$ time. Construct a graph with $\prod_{i=1}^{k} a_i$ vertices. Each vertex of the graph is labeled with a sequence of integers $(b_1, \ldots, b_k)$ such that $0 \le b_i < a_i$ for all i. This vertex means that $\min\{(\text{current time}) - (\text{the last visit by agent } i), a_i - 1\} = b_i$. If $b_r = a_r - 1$, add an edge from $(b_1, \ldots, b_k)$ to $(\min\{b_1 + 1, a_1 - 1\}, \ldots, \min\{b_{r-1}+1, a_{r-1}-1\}, 0, \min\{b_{r+1}+1, a_{r+1}-1\}, \ldots, \min\{b_k+1, a_k-1\})$. A valid schedule corresponds to an infinite path in this graph. Thus, $(a_1, \ldots, a_k)$ is good if and only if this graph contains an infinite path. Since this graph is finite, this can be checked by finding a cycle in the graph.

However, this algorithm is slow. In order to design a fast approximation algorithm, first we give a sufficient condition for $(a_1, \ldots, a_k)$ to be bad:

Theorem 10. *If $\sum_{i=1}^{k} 1/a_i < 1$, $(a_1, \ldots, a_k)$ is bad.*

Proof. Let M be a sufficiently big integer. Out of any consecutive M integer times, the ith agent can visit the point at most $\lceil M/a_i \rceil$ times. If $(a_1, \ldots, a_k)$ is good, the sum of $\lceil M/a_i \rceil$ must be at least M, but this contradicts $\sum_{i=1}^{k} 1/a_i < 1$ when M is sufficiently big. $\qquad\square$

On the other hand, the following gives a sufficient for $(a_1, \ldots, a_k)$ to be good when $a_1, \ldots, a_k$ are powers of 2:

Lemma 11. *If* $\sum_{i=1}^{k} 1/2^{b_i} \geq 1$, $(2^{b_1}, \ldots, 2^{b_k})$ *is good.*

Proof. We prove the lemma by induction of k. Since $\sum_{i=1}^{k} 1/2^{b_i} \geq 1$, at least one of the following conditions hold:

- For some i, $b_i = 0$. In this case, $(2^{b_1}, \ldots, 2^{b_k})$ is obviously good.
- There exist distinct i, j such that $b_i = b_j = t$. Let S be a set of integers. If an agent with interval $2d$ can visit the point at all elements in S, there exists a schedule of two agents with intervals d such that for each element in S, at least one agent visits the point. Thus, we can replace two agents with intervals 2^t with an agent with interval 2^{t-1}. This replacement doesn't change the inverse sum of intervals, and by the assumption of the induction $(2^{b_1}, \ldots, 2^{b_k})$ is good. □

We can design a polynomial time 2-approximation algorithm for the (non-discretized) Point Patrolling Problem using the previous two lemmas.

Let $a_1, \ldots, a_k \in \mathbb{R}$ be the input and x be the optimal idle time. As we noted above, $(\lceil a_1/x \rceil, \ldots, \lceil a_k/x \rceil)$ is good. Thus, by theorem 10, $x/a_1 + \cdots + x/a_k \geq 1$.

Let y be a number that satisfies $y/a_1 + \cdots + y/a_k = 1$. Let b_i be an integer that satisfies $a_i/2y \leq 2^{b_i} \leq a_i/y$. Since $1/2^{b_1} + \cdots + 1/2^{b_k} \geq y/a_1 + \cdots + y/a_k = 1$, $(2^{b_1}, \ldots, 2^{b_k})$ is good. Since $a_i/2y \leq 2^{b_i}$ for each i, $(\lceil a_1/2y \rceil, \ldots, \lceil a_k/2y \rceil)$ is also good and we can achieve the idle time of $2y$. This is at most twice bigger than the optimal idle time x.

In the remaining part of this section, we focus on the relation between Discretized Point Patrolling Problem and the amount $\sum_{i=1}^{k} 1/a_i$.

Theorem 12. *If* $\sum_{i=1}^{k} 1/a_i \geq 2$, $(a_1, \ldots, a_k)$ *is good.*

Proof. Let b_i be an integer that satisfies $a_i \leq 2^{b_i} < 2a_i$. Since $\sum_{i=1}^{k} \frac{1}{2^{b_i}} \geq \sum_{i=1}^{k} \frac{1}{2a_i} \geq 1$, by lemma 11, $(2^{b_1}, \ldots, 2^{b_k})$ is good. Therefore, $(a_1, \ldots, a_k)$ is also good. □

This constant 2 can be improved, as shown in Theorem 14 below.

Lemma 13. *If* $(a_1, \ldots, a_k)$ *is bad and* $\sum_{i=1}^{k} \frac{1}{a_i} = t$ *and* $a_i \leq 2M$ *for all* i, *there exists a bad* $(b_1, \ldots, b_m)$ *such that* $\sum_{i=1}^{m} \frac{1}{b_i} \geq \frac{M+1}{M+2} t - \frac{1}{M+2}$ *and* $b_i \leq M$ *for all* i.

Proof. Without loss of generality, we can assume that $a_1 \leq \cdots \leq a_k$. Let r be an integer that satisfies $a_r \leq M < a_{r+1}$. First, define $c := (c_1, \ldots, c_k)$ as follows:

- If $i \leq r$ or $i - r$ is even, $c_i = a_i$.
- Otherwise, $c_i = a_{i+1}$.

For all i, $c_i \geq a_i$, so c is bad. Also, we can bound the inverse sum of c: $\sum_{i=1}^{k} 1/c_i = \sum_{i=1}^{k} 1/a_i - (1/a_{r+1} - 1/a_{r+2}) - (1/a_{r+3} - 1/a_{r+4}) - \cdots \geq \sum_{i=1}^{k} 1/a_i - \frac{1}{M+1} = t - \frac{1}{M+1}$.

Next, we construct $b := (b_1, \ldots, b_m)$ from c. First, we add all elements in c that are at most M to b. Other elements in c can be divided into pairs of same integers. If c contains the pair (x, x), we add an integer $\lceil x/2 \rceil$ to b. Two agents with intervals (x, x) can work as a single agent with interval $\lceil x/2 \rceil$, so b is also bad. This process reduces the inverse sum by the factor of at most $\frac{M+2}{M+1}$. Thus, we can bound the inverse sum of b as follows: $\sum_{i=1}^{m} 1/b_i \geq \frac{M+1}{M+2} \sum_{i=1}^{k} 1/c_i \geq \frac{M+1}{M+2} t - \frac{1}{M+2}$. $\qquad \square$

Theorem 14. *If $\sum_{i=1}^{k} 1/a_i > 1.546$, $(a_1, \ldots, a_k)$ is good.*

Proof. Suppose that there exists bad $(a_1, \ldots, a_k)$ such that $\sum_{i=1}^{k} 1/a_i > 1.546$ and $a_i \leq 12 \cdot 2^r$ for all i. By using the previous lemma r times, we can prove that there exists $(b_1, \ldots, b_m)$ such that $\sum_{i=1}^{m} 1/b_i > f(r)$ and $b_i \leq 12$ for all i, where $f(r) = (\cdots((1.546 \cdot \frac{12 \cdot 2^{r-1}+1}{12 \cdot 2^{r-1}+2} - \frac{1}{12 \cdot 2^{r-1}+2}) \cdot \frac{12 \cdot 2^{r-2}+1}{12 \cdot 2^{r-2}+2} - \frac{1}{12 \cdot 2^{r-2}+2}) \cdots) \cdot \frac{13}{14} - \frac{1}{14}$. We verified that $f(r) > 1.1822$ for all r, hence $\sum_{i=1}^{m} 1/b_i > 1.1822$. There are finite number of $(b_1, \ldots, b_m)$ that satisfy $b_1 \leq \ldots \leq b_m$ and $\sum_{i=1}^{m} 1/b_i \geq 1.1822 > \sum_{i=1}^{m-1} 1/b_i$ and $b_i \leq 12$ for all i. We verified that all these cases are good. This is a contradiction. $\qquad \square$

On the other hand, we can prove that the constant cannot be smaller than $\sum_{i=0}^{\infty} 1/(2^i + 1) = 1.264 \ldots$ (see the full version [7] for a proof).

Theorem 15. *$(2, 3, 5, \ldots, 2^k + 1)$ is bad.*

We suspect that this cannot be improved:

Conjecture 16. Let $\alpha := \sum_{0}^{\infty} 1/(2^i + 1) \approx 1.264$. If $\sum_{i=1}^{k} 1/a_i > \alpha$, $(a_1, \ldots, a_k)$ is good.

6 Complexity of Problems Related to Point Patrolling

In previous sections, we discussed approximation algorithms of patrolling problems. This is because patrolling problems look unsolvable in polynomial time. In this section, we will try to justify this intuition. Ideally, we should prove NP hardness of patrolling problems, but we failed to prove that. Instead, we will prove NP completeness of problems related to Discretized Point Patrolling Problem.

We conjecture that even in the special case where $\sum_{i=1}^{k} 1/a_i = 1$, there is no pseudo-polynomial time algorithm for the Discretized Point Patrolling Problem. It turns out that this special case is closely related to a well-studied object called *Disjoint Covering Systems*.

A set of pairs of integers (m_i, r_i) is called a *Disjoint Covering System* [9] if for all integer x, there exists unique i such that $x \equiv r_i \pmod{m_i}$.

We define a new decision problem:

Disjoint Covering System Problem. You are given a list of integers $(m_1, \ldots, m_k)$. Determine whether there exists a list of integers $(r_1, \ldots, r_k)$ such that (m_i, r_i) is a disjoint covering system.

This problem is equivalent to the special case of the Discretized Point Patrolling Problem (see the full version [7] for a proof):

Theorem 17. *Suppose that $\sum_{i=1}^{k} 1/m_i = 1$. Then there exists a list of integers $(r_1, \ldots, r_k)$ such that (m_i, r_i) is a disjoint covering system if and only if $(m_1, \ldots, m_k)$ is good.*

Conjecture 18. Disjoint Covering System Problem is strongly NP-complete. In particular, if this conjecture is true, Point Patrolling Problem is strongly NP-hard. Here a problem is called strongly NP-complete if the problem is NP-complete when the integers in the input are given in unary notation.

We will prove that a similar problem is strongly NP-complete.

A set of pairs of integers (m_i, r_i) is called a *Disjoint Residue Class* [8] if for every integer x, there exists at most one i such that $x \equiv r_i \pmod{m_i}$.

We define a new decision problem in a similar way:

Disjoint Residue Class Problem. We are given a list of integers $(m_1, \ldots, m_k)$. Determine whether there exists a list of integers $(r_1, \ldots, r_k)$ such that (m_i, r_i) is a disjoint residue class.

Theorem 19. *The Disjoint Residue Class Problem is strongly NP-complete.*

Proof. The vertex cover problem for triangle-free graphs is known to be NP-complete. We will reduce this problem to the Disjoint Residue Class Problem. Let $G = (V, E)$ be a triangle-free graph, and k be an integer. Let $n = |V|$, and $p_1, \ldots, p_n$ be the smallest n primes greater than n. Label the vertices of G with $p_1, \ldots, p_n$. For each $e \in E$, assign a label $m_e := kp_s p_t$, where p_s and p_t are the primes assigned to the endpoints of e. We claim that G has a vertex cover of size $\leq k$ if and only if each edge $e \in E$ can be assigned an integer r_e such that (m_e, r_e) forms a disjoint residue class.

Suppose that $S = \{v_1, \ldots, v_k\} \subseteq V$ is a vertex cover. It is possible to assign a pair of integers (a_e, b_e) to each $e \in E$ such that:

- One of the endpoints of e is v_{a_e}.
- $0 \leq b_e < n$.
- No two edges are assigned the same pair of integers.

Then, we choose r_e such that $r_e \equiv a_e \pmod{k}$ and $r_e \equiv b_e \pmod{p_{a_e}}$. Let e_1, e_2 be two distinct edges.

- If $a := a_{e_1} = a_{e_2}$, both m_{e_1} and m_{e_2} are multiples of p_a and $r_{e_1} \neq r_{e_2} \pmod{p_a}$.
- If $a_{e_1} \neq a_{e_2}$, both m_{e_1} and m_{e_2} are multiples of k and $r_{e_1} \neq r_{e_2} \pmod{k}$.

Thus, (m_e, r_e) forms a disjoint residue class.

Conversely, let (m_i, r_i) be a disjoint residue class. Suppose that $e_1, e_2 \in E$ doesn't share a vertex. Assume that $r_1 \equiv r_2 \pmod{k}$. Since $\gcd(m_{e_1}, m_{e_2}) = k$, by Chinese remainder theorem, there exists an integer x that satisfies $x \equiv r_1$

(mod m_{e_1}) and $x \equiv r_2$ (mod m_{e_2}). This contradicts the fact that (m_i, r_i) is a disjoint residue class. Therefore $r_1 \neq r_2$ (mod k). This means that if we divide E into k disjoint sets $E_1, \ldots, E_k$ by r_i mod k, any two edges in the same subset shares a vertex. Since G is triangle-free, for each E_i, there must exist a vertex v_i such that all edges in E_i contain v_i. $v_1, \ldots, v_k$ is a vertex cover. $\square$

We also obtain an NP-complete problem if we specify the set of times at which the point must be visited (see the full version [7] for a proof):

Generalized Point Patrolling Problem. We are given a finite set of times $S \subseteq \mathbb{Z}$ and integers $a_1, \ldots, a_k > 0$. For each $t \in S$, at least one agent must visit the point at time t. If the ith agent visits the point at two distinct times t_1 and t_2, we must have $|t_1 - t_2| \geq a_i$. Determine whether this is possible.

Theorem 20. *Generalized Point Patrolling Problem is NP-complete.*

References

1. Chen, K., Dumitrescu, A., Ghosh, A.: On fence patrolling by mobile agents. In: Proc. 25th Canadian Conference on Computational Geometry (CCCG) (2013)
2. Czyzowicz, J., Gąsieniec, L., Kosowski, A., Kranakis, E.: Boundary patrolling by mobile agents with distinct maximal speeds. In: Demetrescu, C., Halldórsson, M.M. (eds.) ESA 2011. LNCS, vol. 6942, pp. 701–712. Springer, Heidelberg (2011)
3. Dumitrescu, A., Ghosh, A., Tóth, C.D.: On fence patrolling by mobile agents. Electronic Journal of Combinatorics **21**, P3.4 (2014)
4. Dumitrescu, A., Tóth, C.D.: Computational Geometry Column 59. ACM SIGACT News **45**(2) (2014)
5. Garey, M., Johnson, D.S.: Computers and Intractability: A Guide to the Theory of NP-Completeness. W. H. Freeman (1979)
6. Kawamura, A., Kobayashi, Y.: Fence patrolling by mobile agents with distinct speeds. Distributed Computing **28**(2), 147–154 (2015)
7. Kawamura, A., Soejima, M.: Simple strategies versus optimal schedules in multi-agent patrolling (2014). Preprint, arXiv:1411.6853
8. Sun, Z.: On disjoint residue classes. Discrete Mathematics **104**(3), 321–326 (1992)
9. Znám, Š.: On exactly covering systems of arithmetic sequences. Mathematische Annalen **180**(3), 227–232 (1969)

Sharing Non-anonymous Costs
of Multiple Resources Optimally

Max Klimm[1] and Daniel Schmand[2(✉)]

[1] Department of Mathematics, Technische Universität Berlin, Berlin, Germany
klimm@math.tu-berlin.de
[2] RWTH Aachen University, Aachen, Germany
daniel.schmand@oms.rwth-aachen.de

Abstract. In cost sharing games, the existence and efficiency of pure Nash equilibria fundamentally depends on the method that is used to share the resources' costs. We consider a general class of resource allocation problems in which a set of resources is used by a heterogeneous set of selfish users. The cost of a resource is a (non-decreasing) function of the *set* of its users. Under the assumption that the costs of the resources are shared by *uniform* cost sharing protocols, i.e., protocols that use only local information of the resource's cost structure and its users to determine the cost shares, we exactly quantify the inefficiency of the resulting pure Nash equilibria. Specifically, we show tight bounds on prices of stability and anarchy for games with only submodular and only supermodular cost functions, respectively, and an asymptotically tight bound for games with arbitrary set-functions. While all our upper bounds are attained for the well-known Shapley cost sharing protocol, our lower bounds hold for arbitrary uniform cost sharing protocols and are even valid for games with anonymous costs, i.e., games in which the cost of each resource only depends on the cardinality of the set of its users.

1 Introduction

Resource allocation problems are omnipresent in many areas of economics, computer science, and operations research with many applications, e.g., in routing, network design, and scheduling. Roughly speaking, when solving these problems the central question is how to allocate a given set of resources to a set of potential users so as to optimize a given social welfare function. In many applications, a major issue is that the users of the system are striving to optimize their own private objective instead of the overall performance of the system. Such systems of users with different objectives are analyzed using the theory of non-cooperative games.

A fundamental model of resource allocation problems with selfish users are congestion games [16]. In a congestion game, each user chooses a subset of a given set of resources out of a set of allowable subsets. The cost of each resource depends on the number of players using the particular resource. The private cost of each

---

Most of the work was done while this author was at Technische Universität Berlin.

© Springer International Publishing Switzerland 2015
V.Th. Paschos and P. Widmayer (Eds.): CIAC 2015, LNCS 9079, pp. 274–287, 2015.
DOI: 10.1007/978-3-319-18173-8_20

user equals the sum of costs given by the cost functions of the resources contained in the chosen subset. Congestion games can be interpreted as cost sharing games with fair cost allocation in which the cost of a resource is a function of the number of its users and each user pays the same cost share, see also Anshelevich et al. [1]. Rosenthal [16] showed that every congestion game has a pure Nash equilibrium, i.e., a strategy vector such that no player can decrease its cost by a unilateral change of her strategy.

This existence result depends severely on the assumption that the players are identical in the sense that they contribute in equal terms to the congestion—and, thus, the cost—on the resources, which is unrealistic in many applications. As a better model for heterogeneous users, Milchtaich [14] introduced *weighted* congestion games, where each player is associated with a positive weight and resource costs are functions of the aggregated weights of their respective users. It is well known that these games may lack a pure Nash equilibrium [4,6,13]. A further generalization of weighted congestion games with even more modeling power are congestion games with set-dependent cost functions introduced by Fabrikant et al. [3]. Here, the cost of each resource is a (usually non-decreasing) function of the *set* of its users. Set-dependent cost function can be used to model multi-dimensional cost structures on the resources that may arise form different technologies at the resources required by different users, such as bandwidth, personal, machines, etc.

For the class of congestion games with set-dependent costs, Gopalakrishnan et al. [7] precisely characterized how the resources' costs can be distributed among its users such that the existence of a pure Nash equilibrium is guaranteed. Specifically, they showed that the class of generalized weighted Shapley protocols is the unique maximal class of cost sharing protocols that guarantees the existence of a pure Nash equilibrium in all induced cost sharing games. This class of protocols is quite rich as it contains, e.g., weighted versions of the Shapley protocol and ordered protocols as considered by Chen et al. [2]. They examined which protocols give rise to good equilibria for cost sharing games where each resource has a fixed cost that has to be paid if the resource is used by at least one player. In subsequent work, von Falkenhausen and Harks [20] gave a complete characterization of the prices of anarchy and stability that is achievable by cost sharing protocols in games with weighted players and matroid strategy spaces. In this paper, we follow this line of research asking which cost sharing protocols guarantee the existence of good equilibria for *arbitrary* (non-decreasing) set-dependent cost functions and *arbitrary* strategy spaces.

Our Results. We study cost sharing protocols for a general resource allocation model with set-dependent costs that guarantee the existence of efficient pure Nash equilibria. Our results are summarized in Table 1. Specifically, we give tight and asymptotically tight bounds on the inefficiency of Nash equilibria in games that use the Shapley protocol both in terms of the price of anarchy and the price of stability and for games with submodular, supermodular and arbitrary non-decreasing cost functions, respectively. The lower bounds that we provide hold for *arbitrary* uniform cost sharing protocols and even in network games with anonymous costs in which the cost of a resource depends only of the cardinality of the set of its users.

Table 1. The inefficiency of n-player cost sharing games

Cost Functions	Price of stability			Price of anarchy		
	Value	Lower bd.	Upper bd.	Value	Lower bd.	Upper bd.
submodular	H_n	[1]	Thm. 3	n	[2]	Thm. 4
supermodular	n	Pro. 3	Thm. 2	∞	[20], Thm. 6	–
arbitrary	$\Theta(nH_n)$	Thm. 5	Thm. 1	∞	[20], Thm. 6	–

Our upper and lower bounds are exactly matching except for the price of stability in games with arbitrary non-decreasing cost functions where they only match asymptotically. Nonetheless, the lower bound of $\Theta(nH_n) = \Theta(n \log n)$ for any uniform protocol is our technically most challenging result and relies on a combination of characterizations of stable protocols for constant cost functions taken from [2] and set-dependent cost functions from [7]. Our results imply that both for submodular and supermodular costs there is no other uniform protocol that gives rise to better pure Nash equilibria than the Shapley protocol, in the worst case. As another interesting side-product of our results, we obtain that moving from anonymous costs (that depend only on the number of users) to set-dependent costs does not deteriorate the quality of pure Nash equilibria in cost sharing games.

Related Work. To measure the inefficiency of equilibria, two notions have evolved. The price of anarchy [12,15] is the worst case ratio of the social cost of an equilibrium and that of a social optimum. The price of stability [1,18] is the ratio of the social cost of the most favorable equilibrium and that of the social optimum.

Chen et al. [2] initiated the study of cost sharing protocols that guarantee the existence of efficient pure Nash equilibria. They considered the case of constant resource costs and characterized the set of linear uniform protocols. For uniform protocols (that solely depend on local information) they showed that proportional cost sharing guarantees a price of stability equal to the n-th harmonic number and a price of anarchy of n for n-player games. Von Falkenhausen and Harks [20] studied cost sharing protocols for weighted congestion games with matroid strategy spaces. They gave various tight bounds for the prices of anarchy and stability that is achievable by uniform and more general protocols. Among other results, they showed that even for a bounded number of players, no uniform protocol can archive a constant price of anarchy. Kollias and Roughgarden [11] studied weighted congestion games with arbitrary strategy spaces. They recalled a result from Hart and Mas-Colell [8] to deduce that every weighted congestion game has a pure Nash equilibrium if the cost of each resource is distributed according to the (weighted) Shapley value. Furthermore, bounds on the price of stability for this cost sharing protocol are given. Gopalakrishnan et al. [7] considered congestion games with arbitrary set-dependent cost functions. They give a full characterization of the cost sharing protocols that guarantee the existence of a pure Nash equilibrium in all such games.

Independently of our work, Roughgarden and Schrijvers [17] showed that the price of stability of Shapley cost sharing games is H_n provided that the cost of each resource is a submodular function of its users. They further showed that the strong price of anarchy is H_n as well. In forthcoming work, Gkatzelis et al. [5] examined optimal cost sharing rules for weighted congestion games with polynomial cost functions. In particular, they showed that among the set of weighted Shapley cost sharing methods, the best price of anarchy can be guaranteed by unweighted Shapley cost sharing. Although in a similar vein, their result is independent from ours since our results hold even for unweighted players and arbitrary (submodular, supermodular, or arbitrarily non-decreasing) costs, while their result holds for weighted players and convex and polynomial costs. Yet, we believe that the combination of the results of Gkatzelis et al. and ours give strong evidence that in multiple scenarios of interest, Shapley cost sharing is the best way to share the cost among selfish users. Both papers thus contribute to the *quantitative* justification of Shapley cost sharing, as opposed to the *axiomatic* justification originally proposed by Shapley [19].

2 Preliminaries

We are given a finite set of *players* $N = \{1, \ldots, n\}$ and a finite and non-empty set of *resources* R. For each player i, the set of *strategies* $\mathcal{P}_i$ is a non-empty set of subsets of R. We set $\mathcal{P} = \mathcal{P}_1 \times \cdots \times \mathcal{P}_n$ and call $P = (P_1, \ldots, P_n) \in \mathcal{P}$ a *strategy vector*. For $P \in \mathcal{P}$ and $r \in R$, let $P^r = \{i : r \in P_i\}$ denote the set of players that use resource r in strategy vector P. For each resource r, we are given a non-negative cost function $C^r : 2^N \to \mathbb{R}_{\geq 0}$ mapping the set of its users to the respective cost value. We assume that all cost functions C^r are non-decreasing, in the sense that $C^r(T) \leq C^r(U)$ for all $T \subseteq U$ with $T, U \in 2^N$, and that $C^r(\emptyset) = 0$. The function C^r is *submodular* if $C^r(X \cup \{i\}) - C^r(X) \geq C^r(Y \cup \{i\}) - C^r(Y)$ for all $X \subseteq Y \subseteq N$, $i \in N \setminus Y$ and *supermodular* if $C^r(X \cup \{i\}) - C^r(X) \leq C^r(Y \cup \{i\}) - C^r(Y)$ for all $X \subseteq Y \subseteq N$, $i \in N \setminus Y$. We call C^r *anonymous* if $C^r(X) = C^r(Y)$ for all $X, Y \subseteq N$ with $|X| = |Y|$. For anonymous cost functions, submodularity and supermodularity are equivalent to concavity and convexity, respectively.

The tuple $\mathcal{M} = (N, R, \mathcal{P}, (C^r)_{r \in R})$ is called a *resource allocation model*. For the special case that R corresponds to the set of edges of a graph and for every i the set $\mathcal{P}_i$ corresponds to the set of (s_i, t_i)-paths for two designated vertices s_i and t_i, we call $\mathcal{M}$ a *network* resource allocation model.

The players' private costs are governed by local cost sharing protocols that decide how the cost of each resource is divided among its users. A cost sharing protocol is a family of functions $(c_i^r)_{i \in N, r \in R} : \mathcal{P} \to \mathbb{R}$ that determines for each resource r and each player i, the cost share $c_i^r(P)$ that player i has to pay for using resource r under strategy vector P. The private cost of player i under strategy vector P is then defined as $C_i(P) = \sum_{r \in P_i} c_i^r(P)$.

A resource allocation model $\mathcal{M} = (N, R, \mathcal{P}, (C^r)_{r \in R})$ together with a cost sharing protocol $(c_i^r)_{i \in N, r \in R}$ thus defines a strategic game G with player set N, strategy space $\mathcal{P}$ and private cost functions $(C_i)_{i \in N}$. For a strategic game G, let

PNE $\subseteq \mathcal{P}$ denote the set of pure Nash equilibria of G. The price of anarchy of G is then defined as $\max_{P \in \text{PNE}(G)} C(P)/C(\hat{P})$, and the price of stability is defined as $\min_{P \in \text{PNE}(G)} C(P)/C(\hat{P})$, where $C(P) = \sum_{r \in R} C^r(P^r)$ is the *social cost* of a strategy vector P and $\hat{P}$ is a strategy vector that minimizes C. The strategy vector $\hat{P}$ is called *socially optimal*.

Throughout this paper, we impose the following assumptions on the cost sharing protocol that have become standard in the literature, cf. [2, 20]:

- *Stability:* There exists at least one pure Nash equilibrium, i.e, there is $P \in \mathcal{P}$ such that $C_i(P) \leq C_i(\tilde{P}_i, P_{-i})$ for all $i \in N$ and $\tilde{P}_i \in \mathcal{P}_i$.
- *Budget-balance:* The cost of all used resources is exactly covered by the cost shares of its users, i.e., $\sum_{i \in P^r} c_i^r(P) = C^r(P^r)$ and $c_i^r(P) = 0$ for all $i \notin P^r$ for all $r \in R$ and $P \in \mathcal{P}$.
- *Uniformity:* The cost shares depend only on the cost structure of the resource and the set of users, i.e., $c_i^r(P) = c_i^{\tilde{r}}(\tilde{P})$ for all $i \in N$ and all resource allocation models $(N, \mathcal{P}, (C^r)_{r \in R})$, $(N, \tilde{\mathcal{P}}, (\tilde{C}^{\tilde{r}})_{\tilde{r} \in \tilde{R}})$ with $C^r \equiv \tilde{C}^{\tilde{r}}$ and all strategy vectors $P \in \mathcal{P}$, $\tilde{P} \in \tilde{\mathcal{P}}$ with $P^r = \tilde{P}^r$.

With some abuse of notation, we henceforth write $c_i^r(P^r)$ instead of $c_i^r(P)$.

In this work, we only consider cost sharing protocols that satisfy these assumptions but it is worth noting that, except for the lower bound of $\Theta(n H_n)$ on the price of stability for arbitrary non-decreasing cost functions, all our results are valid for a slightly weaker notion of uniformity where the cost shares may depend on the identity of the resource.

A protocol that is stable, budget-balanced and uniform is the *Shapley cost sharing* protocol. To give a formal definition, for a set $S \subseteq N$ of players, let us denote by $\Pi(S)$ the set of permutations $\pi : S \to \{1, \ldots, |S|\}$. Let $\pi \in \Pi(P^r)$ be a permutation of the players in P^r and let $P_{i,\pi}^r = \{j \in P^r : \pi(j) < \pi(i)\}$ be the set of players that precede player i in π. The *Shapley value* of player i at resource r with users P^r is then defined as

$$
\phi_i^r(P^r) = \begin{cases} \frac{1}{|P^r|!} \sum_{\pi \in \Pi(P^r)} C^r(P_{i,\pi}^r \cup \{i\}) - C^r(P_{i,\pi}^r), & \text{if } i \in P^r, \\ 0 & \text{otherwise,} \end{cases} \tag{1}
$$

i.e., the Shapley cost share is the marginal increase in the cost due to player i averaged over all possible permutations of the players in P^r.

The following proposition is an immediate consequence of (1) and well known in the literature. For the sake of completeness, we give a proof in the full version [10].

Proposition 1. *The Shapley cost sharing protocol is budget-balanced and uniform.*

To show that Shapley cost sharing is also stable, we follow the road taken by Kollias and Roughgarden [11], who gave a potential function for Shapley cost sharing for weighted congestion games. The following result can be proven simply by verifying that each step in the proof by Kollias applies to set-dependent costs, too. For the sake of completeness, we give a proof in the full version [10].

Proposition 2. *Let $\pi \in \Pi(N)$ be arbitrary and let $\Phi : \mathcal{P} \to \mathbb{R}$ with $P \mapsto \sum_{r \in R} \Phi^r(P)$ and $\Phi^r(P) = \sum_{i \in P^r} \phi_i^r(P_{i,\pi}^r \cup \{i\})$. Then, Φ is an exact potential function for Shapley cost sharing games with set-dependent costs.*

Using that potential games always have a pure Nash equilibrium, we obtain the following immediate corollary.

Corollary 1. *The Shapley cost sharing protocol is stable.*

As a side-product of the proof of Proposition 2, we obtain the following alternative representation of the exact potential function of a Shapley cost sharing game.

Corollary 2. *The exact potential function for Shapley cost sharing games can be written as $\Phi(P) = \sum_{r \in R} \sum_{T \subseteq P^r} \alpha_T \cdot C^r(T)$ where $\alpha_T = \frac{(|P^r| - |T|)! \cdot (|T| - 1)!}{|P^r|!}$, and $\alpha_\emptyset = 0$.*

3 The Efficiency of Shapley Cost Sharing

Having established the existence of pure Nash equilibria in Shapley cost sharing games we proceed to analyze their efficiency. We start to consider the price of stability.

3.1 Price of Stability

We first need the following technical lemma that bounds the coefficients $\alpha_T, T \subseteq N$ that occur when writing the potential function as in Corollary 2.

Lemma 1. *Let $P \in \mathcal{P}$ and $r \in R$. Then, $\sum_{T \subseteq P^r} \alpha_T = H_k$, where $k = |P^r|$ and H_k is the k-th harmonic number.*

Proof. Let us fix $P^r \subseteq N$ with $|P^r| = k$. Recall that $\alpha_T = \frac{(|P^r| - |T|)!(|T| - 1)!}{|P^r|!}$ and $\alpha_\emptyset = 0$, thus, α_T does not depend on the set T but only on its cardinality $|T|$. In particular, $\alpha_S = \alpha_T$ for all $S, T \in 2^N$ with $|S| = |T|$. Defining $\alpha_l = \alpha_T$ where $T \subset N$ with $|T| = l$ is arbitrary, we obtain $\alpha_l = \frac{(|P^r| - l)!(l - 1)!}{|P^r|!}$ and $\alpha_0 = 0$. We calculate

$$\sum_{T \subseteq P^r} \alpha_T = \sum_{l=1}^{k} \sum_{\substack{T \subseteq P^r \\ |T| = l}} \alpha_T = \sum_{l=1}^{k} \sum_{\substack{T \subseteq P^r \\ |T| = l}} \alpha_l = \sum_{l=1}^{k} \alpha_l \binom{k}{l} = \sum_{l=1}^{k} \frac{(l-1)!(k-l)!}{k!} \binom{k}{l}$$

$$= \sum_{l=1}^{k} \frac{(l-1)!(k-l)!}{k!} \cdot \frac{k!}{l!(k-l)!} = \sum_{l=1}^{k} \frac{1}{l} = H_k, \qquad \square$$

We obtain the following upper bound on the price of stability of n-player Shapley cost sharing games.

Theorem 1. *The price of stability for n-player Shapley cost sharing games with set-dependent cost functions is at most $n H_n$.*

Proof. Fix a Shapley cost sharing game and a socially optimal strategy vector $\hat{P}$. We proceed to show that a global minimum P of the potential function given in Proposition 2 has cost no larger than $nH_n \cdot C(\hat{P})$. We first calculate

$$C(P) = \sum_{r \in R} C^r(P^r) \le \sum_{r \in R} |P^r| \cdot \left(\frac{1}{|P^r|} C^r(P^r) + \sum_{T \subset P^r} \alpha_T C^r(T) \right),$$

since $\alpha_T = \frac{(|T|-1)!(|P^r|-|T|)!}{|P^r|!} \ge 0$. We use $\alpha_{P^r} = \frac{1}{|P^r|}$ to obtain

$$C(P) \le \sum_{r \in R} |P^r| \cdot \left(\alpha_{P^r} C^r(P^r) + \sum_{T \subset P^r} \alpha_T C^r(T) \right) = \sum_{r \in R} |P^r| \cdot \sum_{T \subseteq P^r} \alpha_T C^r(T)$$

$$\le n \cdot \sum_{r \in R} \sum_{T \subseteq P^r} \alpha_T C^r(T) = n \cdot \Phi(P). \tag{2}$$

As $\Phi(P) \le \Phi(\hat{P})$ and cost functions C^r are non-decreasing, we obtain

$$C(P) \le n \cdot \Phi(\hat{P}) = n \cdot \sum_{r \in R} \sum_{T \subseteq \hat{P}^r} \alpha_T C^r(T) \le n \cdot \sum_{r \in R} \sum_{T \subseteq \hat{P}^r} \alpha_T C^r(\hat{P}^r)$$

$$= n \cdot \sum_{r \in R} C^r(\hat{P}^r) \sum_{T \subseteq \hat{P}^r} \alpha_T.$$

Using Lemma 1, we obtain $C(P) \le nH_n \cdot C(\hat{P})$, as claimed.   □

For n-player games with supermodular cost functions, we obtain a better upper bound for the price of stability of n.

Theorem 2. *The price of stability for n-player Shapley cost sharing games with supermodular cost functions is at most n.*

Proof. Let us again fix an arbitrary game and let us denote a socially optimal strategy vector and a potential minimum by $\hat{P}$ and P, respectively. Using inequality (2) from the proof of Theorem 1, we obtain $C(P) \le n\Phi(P) \le n\Phi(\hat{P})$, so it suffices to show that $\Phi(\hat{P}) \le C(\hat{P})$ for supermodular cost functions.

To this end, we first remove all players from the game and then add them iteratively to their strategy played in $\hat{P}$. Formally, for $i \in \{1, \dots, n\}$, let $\hat{P}_{(i)}$ denote the (partial) strategy vector in which only the players $j \in \{1, \dots, i\}$ play their strategies $\hat{P}_j$ and all other players are removed from the game (and, thus, have costs 0).

For every i, we get $C(\hat{P}_{(i)}) - C(\hat{P}_{(i-1)}) = \sum_{r \in \hat{P}_i} C^r(\hat{P}^r_{(i-1)} \cup \{i\}) - C^r(\hat{P}^r_{(i-1)})$, where $\hat{P}^r_{(i-1)}$ denotes the set of players using resource r under strategy vector $\hat{P}_{(i-1)}$. We write this expression as

$$C(\hat{P}_{(i)}) - C(\hat{P}_{(i-1)}) = \sum_{r \in \hat{P}_i} \sum_{\pi \in \Pi(\hat{P}^r_{(i-1)} \cup \{i\})} \frac{C^r(\hat{P}^r \cup \{i\}) - C^r(\hat{P}^r)}{|\hat{P}^r_{(i-1)} \cup \{i\}|!}.$$

As C^r is supermodular, the marginal cost $C^r(\hat{P}^r \cup \{i\}) - C^r(\hat{P}^r)$ of player i does not increase when considering only those players that appear before i in the permutation π. Hence,

$$C(\hat{P}_{(i)}) - C(\hat{P}_{(i-1)}) \geq \sum_{r \in \hat{P}_i} \sum_{\pi \in \Pi(\hat{P}^r_{(i-1)} \cup \{i\})} \frac{C^r(\hat{P}^r_{i,\pi} \cup \{i\}) - C^r(\hat{P}^r_{i,\pi})}{|\hat{P}^r_{(i-1)} \cup \{i\}|!}$$

$$= \sum_{r \in \hat{P}_i} \phi^r_i(\hat{P}^r_{(i-1)} \cup \{i\})$$

$$= C_i(\hat{P}_{(i)}) - C_i(\hat{P}_{(i-1)}) = \Phi(\hat{P}_{(i)}) - \Phi(\hat{P}_{(i-1)}),$$

where the last equation is due to the fact that Φ is an exact potential function. Thus, in each step, the cost increases at least as much as the potential, which finally implies $\Phi(\hat{P}) \leq C(\hat{P})$, as claimed. $\square$

We obtain an even better bound on the price of stability of Shapley cost sharing games with submodular costs. This result has been obtained independently by Roughgarden and Schrijvers [17]. The proof is moved to the full version [10].

Theorem 3. *The price of stability of n-player Shapley cost sharing games with submodular costs is at most H_n.*

3.2   Price of Anarchy

We proceed to analyze the price of anarchy of Shapley cost sharing games. It turns out that we obtain a finite bound on the price of anarchy (for a fixed number of players) only for submodular cost functions.

Theorem 4. *The price of anarchy of n-player Shapley cost sharing games with submodular costs is at most n.*

Proof. Let P be a pure Nash equilibrium and let $\hat{P}$ be a socially optimal strategy vector. We calculate

$$C_i(P) \leq C_i(\hat{P}_i, P_{-i}) = \sum_{r \in \hat{P}_i} \sum_{\pi \in \Pi(P^r \cup \{i\})} \frac{C^r(P^r_{i,\pi} \cup \{i\}) - C^r(P^r_{i,\pi})}{|\hat{P}^r \cup \{i\}|!},$$

and, by submodularity,

$$C_i(P) \leq \sum_{r \in \hat{P}_i} \sum_{\pi \in \Pi(P^r \cup \{i\})} \frac{C^r(\{i\}) - C^r(\emptyset)}{|\hat{P}_r \cup \{i\}|!} = \sum_{r \in \hat{P}_i} C^r(\{i\}) \leq C(\hat{P}). \square$$

4   General Uniform Cost Sharing

In Section 3, we showed upper bounds on the inefficiency of pure Nash equilibria for the Shapley cost sharing protocol. In this section, we show that these upper bounds are essentially tight, even for network games with anonymous resource costs that depend only on the cardinality of the set of the resource's users.

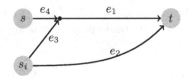

Fig. 1. Cost sharing game with price of stability arbitrarily close to n

4.1 Price of Stability

It is well known [1,2] that no uniform cost sharing protocol can guarantee a price of stability strictly below H_n for all n-player Shapley cost sharing games with submodular costs. In fact, this negative result holds even for anonymous and constant costs. We complement this result with the observation that even for anonymous and convex costs, no uniform cost sharing protocol can guarantee a price of stability strictly below n.

Proposition 3. *For all uniform cost sharing protocols and $\epsilon \in (0,1)$ there is a n-player network resource allocation model with anonymous convex cost functions such that the price of stability is at least $n - \epsilon$.*

Proof. Let us fix a uniform cost sharing protocol and a player set $N = \{1, \ldots, n\}$. Because of the uniformity of the cost sharing protocol, the cost shares for a given edge are completely determined by the local information on that edge and do not depend on the structure of the network.

So let us consider an edge e_1 with the following convex cost function

$$C^{e_1}(P^{e_1}) = \begin{cases} 0, & \text{if } |P^{e_1}| \neq n, \\ n - \epsilon, & \text{if } |P^{e_1}| = n. \end{cases}$$

By budget-balance, $\sum_{i \in N} c_i^{e_1}(N) = n - \epsilon$, implying that there is a player $i \in N$ paying less than 1 on this edge when all users are using it.

Consider the network shown in Figure 1 where the cost of edge e_2 equals the number of its users. The other edges e_3 and e_4 are free. Player i has to route from node s_i to node t and all the other players route from s to t and must use the strategy $\{e_1, e_4\}$. If player i uses the strategy $\{e_1, e_3\}$, the costs equal $C(\{e_1, e_3\}, \{e_1, e_4\}, \ldots, \{e_1, e_4\}) = C^{e_1}(N) = n - \epsilon$. If she chooses $\{e_2\}$ instead, the costs are 1. Clearly, the first strategy vector is the unique pure Nash equilibrium while the latter is the system optimum. Thus, the price of stability is $n - \epsilon$.
□

We proceed to show a lower bound on the price of stability of $\Theta(nH_n)$ for any uniform cost sharing protocol. To prove this, we use a result of Gopalakrishnan et al. [7, Theorem 1] who showed that the set of generalized weighted Shapley cost sharing protocols is the unique maximal set of uniform cost sharing protocols that is stable. To state their result, we first recall the definition of the generalized weighted Shapley cost sharing protocols [9].

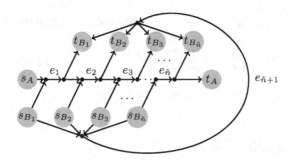

Fig. 2. Cost sharing game with price of stability equal to $\left(\frac{n}{2}+1\right) H_{n/2}$

Definition 1 (Generalized weighted Shapley cost sharing). *Let* $N =$ $\{1,\ldots,n\}$ *be a set of players. A tuple* $w = (\lambda, \Sigma)$ *with* $\lambda = (\lambda_1,\ldots,\lambda_n)$ *and* $\Sigma = (S_1,\ldots,S_m)$ *is called a* weight system *if* $\lambda_i > 0$ *for all* $i \in N$ *and* Σ *is a partition of* N, *i.e.,* $S_1 \cup \cdots \cup S_m = N$ *and* $S_i \cap S_j = \emptyset$ *for all* $i \neq j$.

Given a resource allocation model $\mathcal{M} = (N, R, \mathcal{P}, (C^r)_{r \in R})$ *and a weight system* $w = (\lambda, \Sigma)$, *the* generalized weighted Shapley cost sharing protocol *assigns the cost shares* $(\psi_i^r)_{i \in N, r \in R}$ *defined as*

$$
\psi_i^r(P^r) = \begin{cases} \sum_{T \subseteq P^r : i \in \overline{T}} \frac{\lambda_i}{\sum_{j \in \overline{T}} \lambda_j} \left(\sum_{U \subseteq T} (-1)^{|T| - |U|} C^r(U) \right), & \text{if } i \in P^r \\ 0, & \text{otherwise,} \end{cases}
$$

where $\overline{T} = T \cap S_k$ *and* $k = \min\{j : S_j \cap T \neq \emptyset\}$.

The following proposition is a special case of [7, Theorem 1].

Proposition 4 ((Gopalakrishnan et al. [7]). *For every budget-balanced, uniform and stable cost sharing protocol* ξ *there is a weight system* w *such that* ξ *is equivalent to the generalized weighted Shapley cost sharing protocol with weight system* w.

We use this characterization to obtain the following lower bound.

Theorem 5. *For all uniform cost sharing protocols,* $\epsilon \in \left(0, \frac{1}{2}\right)$ *and* n *even, there is a network* n-*player resource allocation model with anonymous costs for which the price of stability is equal to* $\frac{n/2+1}{1+\epsilon} H_{n/2} \in \Theta(nH_n)$.

Proof. Referring to Proposition 4, it is sufficient to show the claimed result for an arbitrary generalized weighted Shapley cost sharing protocol with weight system w only.

To this end, consider the network resource allocation model with anonymous costs shown in Figure 2. There are n players that we partition into two sets A and

B of cardinality $\tilde{n} := n/2$, i.e., $N = A \cup B$, $A \cap B = \emptyset$, and $|A| = |B| = \frac{n}{2} = \tilde{n}$. For edges $e_1, e_2, \ldots, e_{\tilde{n}}, e_{\tilde{n}+1}$ we assume the following cost functions:

$$C^{e_i}(P^{e_i}) = \begin{cases} 0, & \text{if } |P^{e_i}| < \tilde{n} + 1, \\ \frac{\tilde{n}+1}{i}, & \text{if } |P^{e_i}| = \tilde{n} + 1, \end{cases} \qquad \text{for all } i = 1, \ldots, \tilde{n}.$$

$$C^{e_{\tilde{n}+1}}(P^{e_{\tilde{n}+1}}) = \begin{cases} 0, & \text{if } P^{e_{\tilde{n}+1}} = \emptyset, \\ 1 + \epsilon, & \text{if } P^{e_{\tilde{n}+1}} \neq \emptyset. \end{cases}$$

All other edges are for free.

Note that the cost function of edge $e_{\tilde{n}+1}$ is constant as the cost is equal for every non-empty set of users. In previous work [2, Lemma 5.2] it was shown that in games where all cost functions are constant, all uniform cost sharing protocols have to be monotone in the sense that a player does not pay less if the cost of an edge has to be divided among less players, i.e.,

$$c_i^{e_{\tilde{n}+1}}(P^{e_{\tilde{n}+1}}) \leq c_i^{e_{\tilde{n}+1}}(P^{e_{\tilde{n}+1}} \setminus \{j\}) \qquad \forall i \neq j \in P^{e_{\tilde{n}+1}} \subseteq N.$$

This result applies to our setting as well due to the uniformity of the protocol. It is worth noting, however, that it does not necessarily hold on the other edges in our game, since their cost is not constant.

For a given generalized weighted Shapley cost sharing protocol with weight system $w = (\lambda, \Sigma)$ we assign the players to the sets and start nodes as follows. First, we sort the players ascending to their sets S_i where ties are broken in favor of larger λ_j, i.e., for $a \in S_i$ and $b \in S_j$ we have the property

$$a < b \Leftrightarrow S_i < S_j \vee (S_i = S_j \wedge \lambda_a \geq \lambda_b).$$

Up to renaming we can assume that the obtained order is $(1, \ldots, n)$. We then put the first $n/2$ players into A and the remaining $n/2$ into B.

All players in A have start node s_A and target node t_A. The start and target nodes of the players in B are assigned in a similar fashion as in the proof of Proposition 3. We first pick the player, say $i_{\tilde{n}}$ with largest cost share on edge $e_{\tilde{n}+1}$ among the players in B using $e_{\tilde{n}+1}$. We let $i_{\tilde{n}}$ route from node $s_{\tilde{n}}$ to node $t_{\tilde{n}}$. Then, we consider the remaining players $B \setminus \{i_{\tilde{n}}\}$ and choose a player

$$i_{\tilde{n}-1} \in \underset{b \in B \setminus \{i_{\tilde{n}}\}}{\operatorname{argmax}} c_b^{e_{\tilde{n}+1}}(B \setminus \{i_{\tilde{n}}\}),$$

and let $i_{\tilde{n}-1}$ route from $s_{\tilde{n}-1}$ to $t_{\tilde{n}-1}$. We iterate this process until all players in B are assigned to their respective start and target nodes.

We proceed to show that in the unique pure Nash equilibrium no player uses edge $e_{\tilde{n}+1}$. For a contradiction, suppose P is a pure Nash equilibrium and there is a non-empty set of players $\tilde{B} \subseteq B$ using edge $e_{\tilde{n}+1}$. Let $j = \max\{k \in \{1, \ldots, \tilde{n}\} : i_k \in \tilde{B}\}$ be the player chosen first among the players in $\tilde{B}$ in the assignment procedure described above. First, we show that i_j pays more than $1/j$ for edge $e_{\tilde{n}+1}$ in P. To this end, note that $c_{i_j}^{e_{\tilde{n}+1}}(\tilde{B}) \geq c_{i_j}^{e_{\tilde{n}+1}}(\cup_{k=1}^{j}\{i_k\}) \geq (1+\epsilon)/j > 1/j$, where for the first inequality we used the monotonicity of the cost shares, and for the second inequality, we used that player i_j pays most among the players in $\cup_{k=1}^{j}\{i_k\}$ by construction.

We proceed to argue that player $i_j \in B$ would not pay more than $1/j$ on edge e_j when deviating to her other path, which is then a contradiction to P being a pure Nash equilibrium with a non-empty set of players using edge $e_{\tilde{n}+1}$. We calculate the cost share of player i_j on e_j by the definition of the generalized weighted Shapley cost sharing protocol.

$$c_{i_j}^{e_j}(A \cup \{i_j\}) = \sum_{T \subseteq A \cup \{i_j\}: i_j \in T} \frac{\lambda_{i_j}}{\sum_{k \in \overline{T}} \lambda_k} \left(\sum_{R \subseteq T} (-1)^{|T|-|R|} C^{e_j}(R) \right)$$

$$= \sum_{T = A \cup \{i_j\}: i_j \in T} \frac{\lambda_{i_j}}{\sum_{k \in \overline{T}} \lambda_k} (C^{e_j}(T)),$$

where we used that $C^{e_i}(R) = 0$ for $R \neq A \cup \{i_j\}$. We obtain

$$c_{i_j}^{e_j}(A \cup \{i_j\}) = \begin{cases} \frac{\lambda_{i_j}}{\sum_{k \in S_1} \lambda_k} (C^{e_j}(A \cup \{i_j\})), & \text{if } i_j \in S_1, \\ 0, & \text{otherwise.} \end{cases}$$

In particular, only the players in S_1 have to pay for edge e_j. If $i_j \notin S_1$ we immediately obtain $c_{i_j}^{e_j}(A \cup \{i_j\}) = 0 \leq 1/j$. If, on the other hand, player $i_j \in S_1$ we know from the construction of A and B that $A \subseteq S_1$. Player i_j has the smallest weight of all these players by construction, so i_j pays not than all the other players. We then obtain $c_{i_j}^{e_j}(A \cup \{i_j\}) \leq \frac{(\tilde{n}+1)/j}{\tilde{n}+1} = 1/j$.

We have shown that in no pure Nash equilibrium there is a player using edge $e_{\tilde{n}+1}$. This implies that in all pure Nash equilibria, each player i_j uses edge e_j, so the players in N pay in total $\sum_{j=1}^{\tilde{n}} C^{e_j}(A \cup i_j) = \sum_{j=1}^{\tilde{n}} \frac{\tilde{n}+1}{j} = (\tilde{n}+1)H_{\tilde{n}}$. The price of stability thus amounts to $\frac{1}{1+\epsilon}(n/2+1)H_{n/2}$, as claimed. $\square$

4.2   Price of Anarchy

Von Falkenhausen and Harks [20] showed that any uniform cost sharing protocol leads to unbounded price of anarchy for cost sharing games with three (weighted) players and three parallel arcs.

We complement their result by showing that also for anonymous costs but more complicated networks, no constant price of anarchy can be obtained. The proof can be found in the full version [10].

Theorem 6. *For all uniform cost sharing protocols there is a 2-player network resource allocation model with anonymous convex cost functions such that the price of anarchy is unbounded.*

References

1. Anshelevich, E., Dasgupta, A., Kleinberg, J., Tardos, É., Wexler, T., Roughgarden, T.: The price of stability for network design with fair cost allocation. SIAM J. Comput. **38**(4), 1602–1623 (2008)
2. Chen, H.-L., Roughgarden, T., Valiant, G.: Designing network protocols for good equilibria. SIAM J. Comput. **39**(5), 1799–1832 (2010)
3. Fabrikant, A., Papadimitriou, C.H., Talwar, K.: The complexity of pure nash equilibria. In: Babai, L., (ed.) Proc. 36th Annual ACM Sympos. Theory Comput., pp. 604–612 (2004)
4. Fotakis, D., Kontogiannis, S., Spirakis, P.G.: Selfish unsplittable flows. Theoret. Comput. Sci. **348**(2–3), 226–239 (2005)
5. Gkatzelis, V., Kollias, K., Roughgarden, T.: Optimal cost-sharing in weighted congestion games. In: Liu, T.-Y., Qi, Q., Ye, Y. (eds.) WINE 2014. LNCS, vol. 8877, pp. 72–88. Springer, Heidelberg (2014)
6. Goemans, M.X., Mirrokni, V.S., Vetta, A.: Sink equilibria and convergence. In: Proc. 46th Annual IEEE Sympos. Foundations Comput. Sci, pp. 142–154 (2005)
7. Gopalakrishnan, R., Marden, J.R., Wierman, A.: Potential games are necessary to ensure pure nash equilibria in cost sharing games. In: Proc. 14th ACM Conf. Electronic Commerce, pp. 563–564 (2013)
8. Hart, S., Mas-Colell, A.: Potential, value, and consistency. Econometrica **57**(3), 589–614 (1989)
9. Kalai, E., Samet, D.: On weighted shapley values. Internat. J. Game Theory **16**(3), 205–222 (1987)
10. Klimm, M., Schmand, D.: Sharing non-anonymous costs of multiple resources optimally (2014). arXiv preprint arXiv:1412.4456
11. Kollias, K., Roughgarden, T.: Restoring pure equilibria to weighted congestion games. In: Aceto, L., Henzinger, M., Sgall, J. (eds.) ICALP 2011, Part II. LNCS, vol. 6756, pp. 539–551. Springer, Heidelberg (2011)
12. Koutsoupias, E., Papadimitriou, C.H.: Worst-case equilibria. In: Meinel, C., Tison, S. (eds.) STACS 1999. LNCS, vol. 1563, pp. 404–413. Springer, Heidelberg (1999)
13. Libman, L., Orda, A.: Atomic resource sharing in noncooperative networks. Telecommun. Syst. **17**(4), 385–409 (2001)
14. Milchtaich, I.: Congestion games with player-specific payoff functions. Games Econom. Behav. **13**(1), 111–124 (1996)
15. Papadimitriou, C.H.: Alogithms, games, and the Internet. In: Proc. 33th Annual ACM Sympos. Theory Comput., pp. 749–753 (2001)
16. Rosenthal, R.W.: A class of games possessing pure-strategy Nash equilibria. Internat. J. Game Theory **2**(1), 65–67 (1973)
17. Roughgarden, T., Schrijvers, O.: Network cost-sharing without anonymity. In: Lavi, R. (ed.) SAGT 2014. LNCS, vol. 8768, pp. 134–145. Springer, Heidelberg (2014)

18. Schulz, A.S., Stier-Moses, N.E.: On the performance of user equilibria in traffic networks. In: Proc. 14th Annual ACM-SIAM Sympos. on Discrete Algorithms, pp. 86–87. Society for Industrial and Applied Mathematics (2003)
19. Shapley, L.S.: A value for n-person games. In: Kuhn, H.W., Tucker, A.W. (ed.) Contributions to the Theory of Games, vol. 2, pp. 307–317. Princeton University Press (1953)
20. von Falkenhausen, P., Harks, T.: Optimal cost sharing for resource selection games. Math. Oper. Res. **38**(1), 184–208 (2013)

Algorithms Solving the Matching Cut Problem

Dieter Kratsch[1] and Van Bang Le[2]($\boxtimes$)

[1] Laboratoire d'Informatique Théorique et Appliquée,
Université de Lorraine, 57045 Metz Cedex 01, France
kratsch@univ-metz.fr
[2] Institut für Informatik, Universität Rostock, Rostock, Germany
le@informatik.uni-rostock.de

Abstract. In a graph, a matching cut is an edge cut that is a matching. MATCHING CUT is the problem of deciding whether or not a given graph has a matching cut, which is known to be NP-complete. This paper provides a first branching algorithm solving MATCHING CUT in time $O^*(2^{n/2}) = O^*(1.4143^n)$ for an n-vertex input graph, and shows that MATCHING CUT parameterized by vertex cover number $\tau(G)$ can be solved by a single-exponential algorithm in time $2^{\tau(G)}O(n^2)$. Moreover, the paper also gives a polynomially solvable case for MATCHING CUT which covers previous known results on graphs of maximum degree three, line graphs, and claw-free graphs.

1 Introduction

In a graph $G = (V, E)$, a *cut* is a partition $V = A \cup B$ of the vertex set into disjoint, nonempty sets A and B, written (A, B). The set of all edges in G having an endvertex in A and the other endvertex in B, also written (A, B), is called the *edge cut* of the cut (A, B). A *matching cut* is an edge cut that is a matching. Note that, by our definition, a matching whose removal disconnects the graph need not be a matching cut.

In [13], Farley and Proskurowski studied matching cuts in graphs in the context of network applications. Patrignani and Pizzonia [25] pointed out an application of matching cuts in graph drawing.

Not every graph has a matching cut; the MATCHING CUT problem is the problem of deciding whether or not a given graph has a matching cut:

MATCHING CUT
Instance: A graph $G = (V, E)$.
Question: Does G have a matching cut?

This paper deals with algorithms solving the MATCHING CUT problem.

Previous Results and Related Work. Graphs admitting a matching cut were first discussed by Graham in [16] under the name *decomposable graphs*. The first complexity results for MATCHING CUT have been obtained by Chvátal, who proved in [8] that MATCHING CUT is NP-complete, even when restricted to

© Springer International Publishing Switzerland 2015
V.Th. Paschos and P. Widmayer (Eds.): CIAC 2015, LNCS 9079, pp. 288–299, 2015.
DOI: 10.1007/978-3-319-18173-8_21

graphs of maximum degree four and polynomially solvable for graphs of maximum degree three. In fact, Chvátal's reduction works even for $K_{1,4}$-free graphs of maximum degree four. Being not aware of Chvátal's result, Patrignani and Pizzonia [25] gave another NP-completeness proof for MATCHING CUT. Later, by modifying Chvátal's reduction, Le and Randerath [21] proved that MATCHING CUT remains NP-complete for bipartite graphs with one color class consisting only of vertices of degree three and the other color class consisting only of vertices of degree four. Using a completely different reduction, Bonsma proved the NP-hardness of MATCHING CUT for planar graphs of maximum degree four and for planar graphs of girth five [2].

Besides the case of maximum degree three mentioned above, it has been shown that MATCHING CUT can be solved in polynomial time for line graphs[1] and for graphs without induced cycles of length at least five (Moshi [23]), for claw-free graphs (Bonsma [2]), for cographs and graphs of bounded tree-width or clique-width (Bonsma [2])[2], and for graphs of diameter two (Borowiecki and Jesse-Józefczyk [3]).

A closely related problem to MATCHING CUT is that of deciding if a given graph admits a stable cutset. Here, a *stable cutset* in a graph $G = (V, E)$ is a stable set $S \subseteq V$ such that $G - S$ is disconnected. It can be seen that, for graphs G with minimum degree at least two, G has a matching cut if and only if the line graph $L(G)$ admits a stable cutset. For information on applications and algorithmic results on stable cutsets we refer to [4–7,9,19–22,26].

Our Contributions. First, we provide a new polynomially solvable case for MATCHING CUT, namely for graphs without induced $K_{1,4}$ and $K_{1,4}+e$ (the graph obtained from $K_{1,4}$ by adding an edge). Thus, extending and unifying Chvátal's results for graphs of maximum degree three, Moshi's results for line graphs, and Bonsma's results for claw-free graphs. Second, we provide, for the first time, an exact branching algorithm for MATCHING CUT that has time complexity $O^*(2^{n/2})$ [3]. Third, we initiate the study of matching cuts from the viewpoint of parameterized complexity. We show that MATCHING CUT is fixed-parameter tractable when parameterized by the vertex cover number $\tau(G)$. Much stronger, we establish a single-exponential algorithm running in time $2^{\tau(G)}O(n^2)$.

Notation and Terminology. Let $G = (V, E)$ be a graph with vertex set $V(G) = V$ and edge set $E(G) = E$. We assume that a (input) graph has n vertices and m edges. A *stable set* (a *clique*) in G is a set of pairwise non-adjacent (adjacent) vertices. The neighborhood of a vertex v in G, denoted by $N_G(v)$, is the set of all vertices in G adjacent to v; if the context is clear, we simply write $N(v)$. Set $\deg(v) = |N(v)|$, the degree of the vertex v. For a subset $W \subseteq V$, $G[W]$ is the subgraph of G induced by W, and $G - W$ stands for

---

[1] The line graph of a graph G is the graph whose vertices correspond to the edges of G, and two vertices are adjacent iff the corresponding edges have a common endvertex in G.

[2] We note that MATCHING CUT can be expressed in MSOL; see also [2].

[3] For two functions f and g, we write $f(n) = O^*(g(n))$ if $f(n) = g(n) \cdot \text{poly}(n)$ for a polynomial $\text{poly}(n)$ in n.

$G[V \setminus W]$. We write $N_W(v)$ for $N(v) \cap W$ and call the vertices in $N(v) \cap W$ the W-*neighbors* of v. A *vertex cover* of G is a subset $C \subseteq V$ such that every edge of G has at least one endvertex in C, i.e., $V \setminus C$ is a stable set in G. The vertex cover number of G, denoted by $\tau(G)$, is the smallest size of a vertex cover of G.

The complete graph and the cycle on n vertices is denoted by K_n and C_n, respectively; K_3 is also called a *triangle*. The tree on $t + 1$ vertices with t leaves is denoted by $K_{1,t}$; $K_{1,3}$ is also called a *claw*. The graph obtained from $K_{1,t}$ by adding a new edge is denoted by $K_{1,t} + e$.

When an algorithm branches on the current instance of size n into subproblems of sizes at most $n - t_1, n - t_2, \ldots, n - t_r$, then $(t_1, t_2, \ldots, t_r)$ is called the *branching vector* of this branching, and the unique positive root of $x^n - x^{n-t_1} - x^{n-t_2} - \cdots - x^{n-t_r} = 0$, written $\tau(t_1, t_2, \ldots, t_r)$, is called its *branching number*. The running time of the branching algorithm is $O^*(\alpha^n)$, where $\alpha = \max_i \alpha_i$ and the maximum is taken over all branching rules. Furthermore for every i, α_i is the branching number of branching rule i. We refer to [15] for more details on exact branching algorithms.

Parameterized complexity deals with NP-hard problems whose instances come equipped with an additional integer parameter k. The objective is to design algorithms whose running time is $f(k) \cdot \text{poly}(n)$ for some computable function f. Problems admitting such algorithms are called *fixed-parameter tractable*. See [11,14,24] for more information.

The paper is organized as follows. In section 2 we show that MATCHING CUT can be solved in polynomial time for graphs without induced $K_{1,4}$ and $K_{1,4} + e$. In Section 3 we describe our branching algorithm and point out that MATCHING CUT does not admit a subexponential time algorithm, unless the exponential time hypothesis fails. In Section 4 we describe a single-exponential algorithm for MATCHING CUT when parameterized by vertex cover number.

2   A Polynomially Solvable Case of Matching Cut

In this section we will unify the known polynomially solvable cases for MATCHING CUT on graphs of maximum degree three, on line graphs, and on claw-free graphs, by proving the following theorem.

Theorem 1. *There is an algorithm solving* MATCHING CUT *for* $(K_{1,4}, K_{1,4}+e)$-*free graphs in time* $O(mn)$.

Proof. Let G be a $(K_{1,4}, K_{1,4} + e)$-free graph. Since forests and cycles of length at least 4 have matching cuts, we may assume that G properly contains a cycle. Let C be a shortest cycle in G.

If C is of length at least 5, then every vertex v of $G \setminus C$ has at most one neighbor in C (otherwise $vC[v_i, v_j]v$ or $vC[v_j, v_i]v$ would be a shorter cycle than C, where v_i and v_j are two neighbors of v on C and $C[v_i, v_j]$ and $C[v_j, v_i]$ are the two subpaths of C between and including v_i and v_j) and every vertex $u \in C$ has at most one neighbor in $G \setminus C$ (otherwise u, two neighbors of u in C, and two neighbors of u in $G \setminus C$ would induce a $K_{1,4}$). Note that $G \setminus C \neq \emptyset$ as G is not a cycle. Thus, the edge set $(V(C), V(G) \setminus V(C))$ forms a matching cut.

So, let us assume that $C = K_3$ or $C = C_4$. Construct a subgraph H of G as follows. First, $H := C$. Then apply the two following steps as long as possible:

1. If there exists $v \in V(G) \setminus V(H)$ with (at least) two neighbors in H, set $H := G[V(H) \cup \{v\}]$.
2. If there exists $u \in V(H)$ with two adjacent neighbors v_1, v_2 in $G - V(C)$, set $H := G[V(H) \cup \{v_1, v_2\}]$.

It should be remarked that in case the shortest cycle C is the C_4, step 2 will never be applied. Note that H is connected and has minimum degree 2. Note also that in case $C = K_3$ or H contains a $K_{2,3}$, H has no matching cut. So, in this case, if $H = G$, then G also has no matching cut. Otherwise, by the construction of H, every vertex in $G - V(H)$ has at most one neighbor in H, and every vertex in H has at most one neighbor in $G - V(H)$ (if $u \in V(H)$ has two neighbors $v_1, v_2 \in G - V(H)$, then v_1 and v_2 are non-adjacent, hence v, u_1, u_2 and two neighbors of v in H form a $K_{1,4}$ or a $K_{1,4} + e$). Therefore, the edge set $(V(H), V(G) \setminus V(H))$ forms a matching cut.

Note that a shortest cycle C in G can be computed in time $O(nm)$ (cf. [18]), and that the construction of H can be done in linear time. Hence a matching cut of G, if any, can be found in $O(nm)$ time. □

Corollary 1 ([2,8,23]). MATCHING CUT *is solvable in polynomial time for graphs of maximum degree 3, for line graphs and for claw-free graphs.*

3 An Exact Exponential Algorithm Solving Matching Cut

Our algorithm takes as input a graph $G = (V, E)$ and decides whether or not there is an edge set $M \subseteq E$ such that M is a matching cut of G. If G is disconnected or has a vertex of degree at most one, G clearly has a matching cut. So, we may assume that G is connected and has minimum degree at least two. The idea is to compute a partition of the vertex set into subsets A and B such that A and B are non-empty unions of components of $G - M$ and all M-edges have one extremity in A and the other in B. Our algorithm is a branching algorithm consisting of reduction rules and branching rules. It is worth mentioning that the algorithm labels the vertices of the input graph by either A or B but never changes the graph G. Finally we provide a final lemma stating that if neither a reduction rule nor a branching rule can be applied then there is a matching cut in the graph G, respecting the current partial partition into the sets of A's and the set of B's.

3.1 Description of the Algorithm

The branching algorithm below will be executed for all possible pairs $a, b \in V$, hence $O(n^2)$ times. To do this set $A := \{a\}$, $B := \{b\}$ and $F := V \setminus \{a, b\}$

and call the branching algorithm. At each stage of the branching algorithm, A and/or B will be extended or it will be determined that there is no matching cut that separates A from B by applying a reduction rule or a branching rule, if possible. The algorithm never changes the input graph G. For more details on branching algorithms and their analysis we refer to [15].

Now we describe our branching algorithm by a list of reduction and branching rules given in preference order, i.e. in an execution of the algorithm on any instance of a subproblem one always applies the first rule applicable to the instance, which could be a reduction or a branching rule. A reduction rule produces one instance/subproblem while a branching rule results in at least two instances/subproblems, with different extensions of A and B. If any recursive call outputs that G has a matching cut, then G has indeed a matching cut, otherwise not. Note that G has a matching cut that separates A from B if and only if in at least one recursive branch, extensions A' of A and B' of B are obtained such that G has a matching cut that separates A' from B'. Typically a rule assigns one or more free vertices, vertices of F, either to A or to B; such vertices are automatically removed from F (not explicitly mentioned in our description).

(R1) If an A-vertex has two B-neighbors, or a B-vertex has two A-neighbors then STOP: "G has no matching cut separating A, B".

(R2) If $v \in F$, $|N(v) \cap A| \geq 2$ and $|N(v) \cap B| \geq 2$ then STOP: "G has no matching cut separating A, B".
   If $v \in F$ and $|N(v) \cap A| \geq 2$ then $A := A \cup \{v\}$.
   If $v \in F$ and $|N(v) \cap B| \geq 2$ then $B := B \cup \{v\}$.

(R3) If $v \in A$ has two adjacent F-neighbors w_1, w_2 then $A := A \cup \{w_1, w_2\}$.
   If $v \in B$ has two adjacent F-neighbors w_3, w_4 then $B := B \cup \{w_3, w_4\}$.

(R4) If there is an edge xy in G such that $x \in A$ and $y \in B$ and $N(x) \cap N(y) \cap F \neq \emptyset$ then STOP: "G has no matching cut separating A, B".
   If there is an edge xy in G such that $x \in A$ and $y \in B$ then add $N(x) \cap F$ to A, and add $N(y) \cap F$ to B.

If (R1) is not applicable then the A, B-edges of G form a matching cut in $G[A \cup B] = G - F$. If (R2) is not applicable then every F-vertex is adjacent to at most one A- and at most one B-vertex. If (R3) is not applicable then the F-neighbors of any A-vertex and the F-neighbors of any B-vertex form an independent set. If (R4) is not applicable then every A-vertex adjacent to a B-vertex has no F-neighbor and every B-vertex adjacent to an A-vertex has no F-neighbor.

Our algorithm consists of nine branching rules dealing with small configurations (connected subgraphs with at most eight vertices some of them may already belong to A or B.) See Fig. 1. To determine the branching vectors we set the size of an instance as its number of free vertices.

(B1) On this configuration we branch into a first subproblem by adding v_1 to A and v_2 to B, and into a second subproblem by adding v_1 to B and v_2 to A. The number of F-vertices is decreased by 2 in each branch. Hence the branching vector is $(2, 2)$.

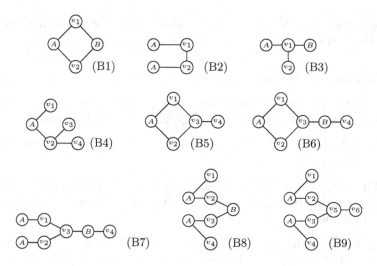

Fig. 1. Branching configurations: Vertices with label A and B belong to A, respectively, to B; vertices with labels v_i are in F

(B2) The only possible choices to label v_1 and v_2 are AA and BB. Hence the branching vector is $(2,2)$.

(B3) The only possible choices to label v_1 and v_2 are AA and BB. Hence the branching vector is $(2,2)$.

(B4) First, add v_2 to B. This implies that v_1 has to be added to A and v_3 and v_4 have be added to B. In the second subproblem we add v_2 to A. Hence the branching vector is $(4,1)$.

(B5) We branch into three subproblems. First, add v_3 to A then v_1 and v_2 have to be added to A too. Second, add v_3 to B then the only possible choices for v_1, v_2 and v_4 are ABB and BAB. Hence the branching vector is $(3,4,4)$.

(B6) On this configuration we branch into three subproblems. First, add v_3 to B then v_1v_2 can be labeled either AB or BA, hence two subproblems. Finally, add v_3 to A then v_4 has to be added to B and v_1, v_2 have to be added to A. Hence the branching vector is $(3,3,4)$.

(B7) In the first subproblem we add v_3 to A. This implies v_1 and v_2 have to be added to A, and v_4 has to be added to B. In the second subproblem we add v_3 to B. Hence the branching vector is $(4,1)$.

(B8) We branch into three subproblems. There are three choices to label v_2 and v_3: AB, BA and BB. In each of the three choices whenever $v_2 \in B$ then we add v_1 to A, and whenever $v_3 \in B$ then add v_4 to A. Hence the branching vector is $(3,3,4)$.

(B9) We branch into four subproblems. First, we add v_5 to A. This implies that v_2, v_3 have to be added to A. Next, add v_5 to B. There are three choices to label v_2 and v_3: AB, BA, BB. In the first two choices, v_6 has to be added to B and v_1 or v_4 has to be added to A. In the last choice, v_1 and v_4 have to be added to A. Hence the branching vector is $(3,5,5,5)$.

Note that the branching vector $(2, 2)$ has the largest branching number $\sqrt{2} \approx$ 1.4143. Hence the running time of our algorithm is $O^*(2^{n/2})$.

3.2   Termination Lemma

Our goal is to show that whenever none of the reduction rules and none of the branching rules can be applied then the graph has a matching cut.

Definition 1. *Let* $A' = \{a \in A : |N(a) \cap F| \geq 2\}$. *A is called final if every vertex in* $N(A') \cap F$ *has degree two.*

Remark 1. Note that if A is not final, then the above definition implies that A has a vertex with a configuration (B3) or (B4), provided none of the reduction rules is applicable.

Proof. If A is not final, there exists a vertex $v \in N(A') \cap F$ with degree at least 3. (Recall that G has minimum degree at least two.) Let $a \in A'$ be the neighbor of v in A. (As (R2) is not applicable, v has exactly one neighbor in A.) Let $v_1, v_2 \in F \cup B$ be two other neighbors of v. Since (R2), (R3) and (R4) are not applicable, a is not adjacent to v_1, v_2 and at most one of v_1 and v_2 may belong to B. In particular, a has another neighbor $u \in F \setminus \{v_1, v_2\}$. Now, if both v_1, v_2 are not in B, then a, u, v, v_1, v_2 form a configuration (B4). If $v_1 \in B$, say, then a, v, v_1, v_2 form a configuration (B3). $\square$

Lemma 1. *Suppose that all reduction and branching rules are not applicable. Then G has a matching cut (and such a matching cut can be computed in polynomial time).*

Proof. It follows from Remark 1 that A is final. Hence every vertex in $F' := N(A') \cap F$ has degree two.

Write $F'' = \{u \in F - F' \mid u$ has two neighbors in $F'\}$, and set

$$X = A \cup F' \cup F'' \text{ and } Y = V(G) - X.$$

Note that F' is an independent set (because of (R3) and (B2)). See also Fig. 2.

We claim that the edge set (X, Y) is a matching cut. First we show that every vertex in X has at most one neighbor in Y. This can be seen as follows:

- Every vertex $x \in A$ has at most one neighbor in Y. This is because (R1) and (R4) are not applicable, and, by definition of X, x has at most one neighbor in $F - (F' \cup F'')$.
- Every vertex $v \in F'$ has at most one neighbor in Y, as $\deg(v) = 2$.
- Every vertex $v \in F''$ has at most one neighbor in Y. By contradiction, assume that v has two neighbors in Y. Then, by (R2), v has a neighbor $u \in Y - B$. Let v_1, v_2 be two neighbors of v in F' and let $a_1, a_2 \in A'$ be the neighbor of v_1 and v_2, respectively. Note that $a_1 \neq a_2$ otherwise a_1, v_1, v_2, v and u would form a configuration (B5). Now, let $v_i' \in F' - \{v_i\}$ be another neighbor of a_i in F', $i = 1, 2$. Then $a_1, a_2, v_1, v_1', v_2, v_2', v$ and u form a configuration (B9), a contradiction. Thus, v has at most one neighbor in Y, as claimed.

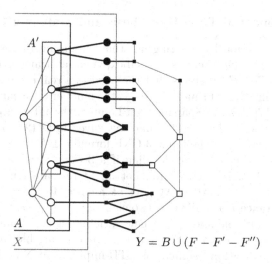

A'

A

X

$$Y = B \cup (F - F' - F'')$$

Fig. 2. When A is final: F'-vertices: black circles; F''-vertices: big black squares, B-vertices: white squares

Next we show that every vertex in Y has at most one neighbor in X. This can be seen as follows:

- Assume to the contrary that $y \in B$ has two neighbors $u_1, u_2 \in X$. Then, by (R1) and (R4), $u_1, u_2 \in F' \cup F''$. First, suppose $u_1, u_2 \in F'$, and let $u_i \in N(a_i)$ for some $a_i \in A'$, $i = 1, 2$. If $a_1 = a_2$, then a_1, u_1, u_2, y form a configuration (B1), a contradiction. If $a_1 \neq a_2$, then $a_1, u_1, u'_1, a_2, u_2, u'_2, y$ form a configuration (B8), where u'_i is a neighbor of a_i in $F' - u_i$, a contradiction again. So let $u_1 \in F''$, say. Let v_1, v_2 be two neighbors of u_1 in F'. By (R3), $u_2 \neq v_1, v_2$. If there is $a \in A'$ such that a, v_1, v_2, u_1 induce a C_4, then a, v_1, v_2, u_1, y and u_2 form a configuration (B6). Thus, v_i is adjacent to $a_i \in A'$ with $a_1 \neq a_2$. But then $a_1, a_2, v_1, v_2, u_1, y$ and u_2 form a configuration (B7). Thus, no vertex in B has two neighbors in X.
- Assume to the contrary that $v \in F - F' - F''$ has two neighbors in X. Consider first the case that a neighbor u of v is in F''. Let v_1, v_2 be two neighbors of u in F' and let $a_i \in A'$ be adjacent to v_i, $i = 1, 2$. Now, if $a_1 = a_2$ then a_1, v_1, v_2, u and v form a configuration (B5), a contradiction. If $a_1 \neq a_2$ then $a_1, a_2, v_1, v'_1, v_2, v'_2, u$ and v form a configuration (B9), a contradiction again. Thus, v cannot have a neighbor in F''. Moreover, by definition of F'', v has at most one neighbor in F', and by (R2), v has at most a neighbor in A. Thus, a neighbor u_1 of v must be in F' and another neighbor u_2 of v must be in A. Then a, u_1, u_2, v form a configuration (B2), where $a \in A'$ is the neighbor of u_1 (note that $a \neq u_2$ by (R3)), a contradiction. Thus, no vertex in $Y - B$ has two neighbors in X.

This implies that (X, Y) is a matching cut of G.   □

3.3   The Exponential Time Hypothesis and Matching Cut

We conclude this section by pointing out that, assuming the exponential time hypothesis (ETH), then there is no subexponential time algorithm solving MATCHING CUT. The ETH states that there exists a real $\epsilon > 0$ such that 3SAT cannot be solved in $O^*(2^{\epsilon n})$ time, where n is the number of variables in the input 3-CNF formula [17]. By the Sparsification Lemma, the ETH implies that there exists a real $\epsilon' > 0$ such that 3SAT cannot be solved in $O^*(2^{\epsilon' m})$ time, where m is the number of clauses in the input 3-CNF formula [17].

NAE 3SAT is the problem of deciding if a 3-CNF formula admits a truth assignment such that, in each clause, not all literals are true. The standard reduction from 3SAT to NAE 3SAT goes as follows: Replace each clause $c = (x, y, z)$ of an instance F for 3SAT by two clauses $c_1 = (x, y, v_c), c_2 = (\neg v_c, z, u)$, where u is a new variable and v_c is a new variable for each clause c. Thus, the obtained instance F' for NAE 3SAT has $n + 2m + 1$ variables and $2m$ clauses. Hence by the Sparsification Lemma, the ETH implies that NAE 3SAT cannot be solved in $O^*(2^{o(m)})$.

Now, the polynomial time reduction from NAE 3SAT to MATCHING CUT given in [25] has the following property: Given a NAE 3SAT formula F with n variables and m clauses, it is possible to construct a graph G with $O(n + m)$ vertices and edges in polynomial time such that F is NAE-satisfiable if and only if G has a matching cut. Thus, we conclude

Lemma 2. *Assuming ETH, there is no subexponential time algorithm for* MATCHING CUT.

Note that the polynomial time reductions from 3-UNIFORM HYPERGRAPH 2-COLORING (the same problem as NAE 3SAT) to MATCHING CUT given in [8, 21] in obtaining graphs of maximum degree 4 yield instances with $O(nm)$ vertices.

4   A Single-Exponential FPT Algorithm for Matching Cut

We consider MATCHING CUT parameterized by vertex cover number.

Lemma 3. *Let I be an independent set and let $U = V(G) - I$. Given a partition (A, B) of $G[U]$, it can be decided in time $O(n^2)$ if G has a matching cut (X, Y) such that $A \subseteq X$ and $B \subseteq Y$.*

Proof. We first consider the case A or B is empty, say $B = \emptyset$. In this case, G has a matching cut (X, Y) such that $A \subseteq X$ if and only if $Y = \{v \in I \mid v$ has at most one neighbor in $A = U\}$ is not empty.

Now assume that A and B are non empty. Write

$$I_A = \{v \in I \mid v \text{ has at least two neighbors in } A\},$$
$$I_B = \{v \in I \mid v \text{ has at least two neighbors in } B\}.$$

Then, we may assume that $I_A \cap I_B = \emptyset$, otherwise G clearly has no matching cut (X, Y) such that $A \subseteq X$ and $B \subseteq Y$.

Consider now the partition $I = I_A \cup I_B \cup I_{01} \cup I_{10} \cup I_{11}$, with

$$I_{01} = \{v \in I : |N(v) \cap A| = 0, |N(v) \cap B| = 1\},$$
$$I_{10} = \{v \in I : |N(v) \cap A| = 1, |N(v) \cap B| = 0\},$$
$$I_{11} = \{v \in I : |N(v) \cap A| = 1, |N(v) \cap B| = 1\},$$

and write $I_{11A} = \{v \in I_{11} \mid$ the neighbor of v in A has a neighbor in $B\}$ and $I_{11B} = \{v \in I_{11} \mid$ the neighbor of v in B has a neighbor in $A\}$.

Then, we may assume that $I_{11A} \cap I_{11B} = \emptyset$, otherwise G clearly has no matching cut (X, Y) such that $A \subseteq X$ and $B \subseteq Y$. Moreover, writing $I_{11C} = I_{11} - I_{11A} - I_{11B}$, we may assume that

$$(A \cup I_A \cup I_{10} \cup I_{11A}, B \cup I_B \cup I_{01} \cup I_{11B}) \text{ is a} \tag{1}$$
$$\text{matching cut of } G - I_{11C},$$

otherwise G has no matching cut (X, Y) such that $A \subseteq X$ and $B \subseteq Y$.

Now, create a boolean formula F as follows: For each $v \in I_{11C}$ we have two boolean variables v_A and v_B (indicating v should be added to A, respectively, to B). The clauses of F are as follows:

– For each $v \in I_{11C}$: $(v_A \vee v_B)$, $(\neg v_A \vee \neg v_B)$. These clauses ensure that v will be moved to A or else to B.
– For each $u, v \in I_{11C}$ having a common neighbor in A: $(\neg u_B \vee \neg v_B)$. This clause ensures that in this case, u or v must go to A.
– For each $u, v \in I_{11C}$ having a common neighbor in B: $(\neg u_A \vee \neg v_A)$. This clause ensures that in this case, u or v must go to B.

Then F is the conjunction of all these clauses over all $v \in I_{11C}$. We claim that G has a matching cut (X, Y) such that $A \subseteq X$ and $B \subseteq Y$ if and only if F is satisfiable.

Indeed, suppose (X, Y) is a matching cut of G such that $A \subseteq X$ and $B \subseteq Y$. Then, for each $v \in I_{11C}$ set $v_A = \text{true}$, $v_B = \text{false}$ if $v \in X$ and $v_A = \text{false}$, $v_B = \text{true}$ if $v \in Y$. It is a routine to check that F is satisfied by this assignment.

Conversely, suppose there is a satisfying assignment for F. Then set $X := A \cup I_A \cup I_{10} \cup I_{11A} \cup \{v \in I_{11C} \mid v_A \text{ is true}\}$, and $Y := B \cup I_B \cup I_{01} \cup I_{11B} \cup \{v \in I_{11C} \mid v_B \text{ is true}\}$. To see that (X, Y) is a matching cut of G, observe that if $u \in X$ has two neighbors v, w in Y, then by (1) and by the fact that I is an independent set, $u \in A$ and $v, w \in I_{11C}$. But this contradicts the assumption that the clause $(\neg v_B \vee \neg w_B)$ is satisfied. Thus, no vertex in X can have two neighbors in Y, and similarly, no vertex in Y can have two neighbors in X.

Obviously, the length of the formula F is $O(n^2)$. Since 2-Sat can be solved in linear time (cf. [1, 10, 12]), the above discussion yields an $O(n^2)$-algorithm for deciding if G has a matching cut (X, Y) such that $A \subseteq X$ and $B \subseteq Y$. □

Running the algorithm of Lemma 3 for all partitions (A, B) of $G[U]$, where U is a minimum vertex cover of the input graph G, one obtains.

Theorem 2. MATCHING CUT *parameterized by the vertex cover number* $\tau(G)$ *can be solved in time* $2^{\tau(G)}O(n^2)$.

Consequently MATCHING CUT can be solved in time $2^{n-\alpha(G)}O(n^2)$, where $\alpha(G)$ is the maximum size of a stable set of G. An obvious consequence of this is another $O^*(2^{n/2})$ time algorithm for bipartite graphs.

References

1. Aspvall, B., Plass, M.F., Tarjan, R.E.: A linear-time algorithm for testing the truth of certain quantified boolean formulas. Information Processing Letters **8**, 121–123 (1979). Erratum: 14 (1982) 195
2. Bonsma, P.: The complexity of the Matching-Cut problem for planar graphs and other graph classes. J. Graph Theory **62**, 109–126 (2009)
3. Borowiecki, M., Jesse-Józefczyk, K.: Matching cutsets in graphs of diameter 2. Theoret. Comp. Sci. **407**, 574–582 (2008)
4. Brandstädt, A., Dragan, F., Le, V.B., Szymczak, T.: On stable cutsets in graphs. Discrete Appl. Math. **105**, 39–50 (2000)
5. Caro, Y., Yuster, R.: Decomposition of slim graphs. Graphs Combinatorics **15**, 5–19 (1999)
6. Chen, G., Faudree, R.J., Jacobson, M.S.: Fragile graphs with small independent cuts. J. Graph Theory **41**, 327–341 (2002)
7. Chen, G., Xingxing, Y.: A note on fragile graphs. Discrete Math. **249**, 41–43 (2002)
8. Chvátal, V.: Recognizing decomposable graphs. J. Graph Theory **8**, 51–53 (1984)
9. Derek, G.: Corneil, Jean Fonlupt, Stable set bonding in perfect graphs and parity graphs. J. Combin. Theory (B) **59**, 1–14 (1993)
10. Davis, M., Putnam, H.: A computing procedure for quantification theory. J. ACM **7**, 201–215 (1960)
11. Downey, R.G., Fellows, M.R.: Fundamentals of Parameterized Complexity. Springer (2013)
12. Even, S., Itai, A., Shamir, A.: On the complexity of timetable and multicommodity flow problems. SIAM J. Computing **5**, 691–703 (1976)
13. Arthur, M.F., Proskurowski, A.: Networks immune to isolated line failures. Networks **12**, 393–403 (1982)
14. Flum, J., Grohe, M.: Parameterized Complexity Theory. Springer (2006)
15. Fomin, F.V., Kratsch, D.: Exact Exponential Algorithms. Springer (2010)
16. Graham, R.L.: On primitive graphs and optimal vertex assignments. Ann. N.Y. Acad. Sci. **175**, 170–186 (1970)
17. Impagliazzo, R., Paturi, R., Zane, F.: Which problems have strongly exponential complexity? Journal of Computer and System Sciences **63**, 512–530 (2001)
18. Itai, A., Rodeh, M.: Finding a minimum circuit in a graph. SIAM J. Comput. **7**, 413–423 (1978)
19. Klein, S., de Figueiredo, C.M.H.: The NP-completeness of multi-partite cutset testing. Congressus Numerantium **119**, 217–222 (1996)
20. Le, V.B., Mosca, R., Müller, H.: On stable cutsets in claw-free graphs and planar graphs. J. Discrete Algorithms **6**, 256–276 (2008)
21. Le, V.B., Randerath, B.: On stable cutsets in line graphs. Theoret. Comput. Sci. **301**, 463–475 (2003)

22. Le, V.B., Pfender, F.: Extremal graphs having no stable cutsets. Electr. J. Comb. **20**, #P35 (2013)
23. Moshi, A.M.: Matching cutsets in graphs. J. Graph Theory **13**, 527–536 (1989)
24. Niedermeier, R.: Invitation to Fixed Parameter Algorithms. Oxford University Press (2006)
25. Patrignani, M., Pizzonia, M.: The complexity of the matching-cut problem. In: Brandstädt, A., Le, V.B. (eds.) WG 2001. LNCS 2204, vol. 2204, pp. 284–295. Springer, Heidelberg (2001)
26. Tucker, A.: Coloring graphs with stable cutsets. J. Combin. Theory (B) **34**, 258–267 (1983)

End-Vertices of Graph Search Algorithms

Dieter Kratsch[1]([⊠]), Mathieu Liedloff[2], and Daniel Meister[3]

[1] Université de Lorraine, LITA, 57045 Metz Cedex 1, France
`dieter.kratsch@univ-lorraine.fr`
[2] Univ. Orléans, INSA Centre Val de Loire, LIFO EA 4022, 45067 Orléans, France
`mathieu.liedloff@univ-orleans.fr`
[3] Theoretical Computer Science, University of Trier, Trier, Germany
`daniel.meister@uni-trier.de`

Abstract. Is it possible to force a graph search algorithm to visit a selected vertex as last? Corneil, Köhler, and Lanlignel showed that this end-vertex decision problem is NP-complete for Lexicographic Breadth-First Search (LexBFS). Charbit, Habib, and Mamcarz extended the intractability result, and showed that the end-vertex problem is hard also for BFS, DFS, and LexDFS. We ask for positive results, and study algorithmic and combinatorial questions. We show that the end-vertex problem for BFS and DFS can be solved in $\mathcal{O}^*(2^n)$ time, hereby improving upon the straightforward and currently best known running-time bound of $\mathcal{O}^*(n!)$. We also determine conditions that preserve end-vertices in subgraphs when extending to larger graphs. Such results are of interest in algorithm design, when applying techniques such as dynamic programming and divide-and-conquer.

1 Introduction

A graph search algorithm is an algorithmic procedure to explore graphs. Corneil and Krueger initialized a systematic study of graph search algorithms [9]. They define GENERIC SEARCH as the algorithm that covers all graph search algorithms which, given as input a connected graph, choose a source vertex as the start, mark it as visited, and examine the vertices of the graph one by one by choosing the next vertex to visit from those adjacent to an already visited vertex. Listing the vertices in their visit order yields a *search ordering*. Concrete graph search algorithms can be formulated as specifications of GENERIC SEARCH, by specifying the choice of the next-to-visit vertex. Famous and broadly known graph search algorithms are Breadth-First Search (BFS) and Depth-First Search (DFS) [5], that store candidate vertices in a queue and in a stack, respectively.

Graph search algorithms have wide-range applications. Specialised graph search algorithms are often applied in efficient algorithms, where properties of their generated orderings are exploited. An easy example is diameter computation by applying BFS. A remarkable example is chordal graph recognition by applying LexBFS [12], or proper interval graph recognition by applying a multisweep LexBFS [6]. A recent and first-time example is an efficient solution of the minimum

© Springer International Publishing Switzerland 2015
V.Th. Paschos and P. Widmayer (Eds.): CIAC 2015, LNCS 9079, pp. 300–312, 2015.
DOI: 10.1007/978-3-319-18173-8_22

path-cover problem on cocomparability graphs by applying LexDFS [7]. The study
of search orderings and their combinatorial properties can often be condensed into
the study of the properties of the last-visited vertex, that is also called the *end-vertex* of the search ordering. Next to their combinatorial study, Corneil, Köhler,
and Lanlignel initiated the algorithmic study of graph search end-vertices [8], more
precisely, they asked for the computational complexity of solving the end-vertex
problem for a specific graph search algorithm. Let Ω be a specific graph search
algorithm.

Ω-END-VERTEX
Given: graph G and vertex t
Question: Does G have an Ω-search that visits t as last vertex?

Corneil, Köhler, and Lanlignel showed that the LEXBFS-END-VERTEX prob-
lem is NP-complete already on weakly chordal graphs [8]. Charbit, Habib, and
Mamcarz continued this research, and showed the NP-completeness of the BFS-,
DFS-, and LEXDFS-END-VERTEX problem on weakly chordal graphs [4].

We are interested in complementary results, namely in fast algorithms to
solve the Ω-END-VERTEX problem. As our main algorithmic results, we give
exact algorithms for the BFS- and DFS-END-VERTEX problem. Our algorithm
for deciding the end-vertices of DFS is a highlight of this paper. We show that a
vertex t is a DFS-end-vertex of a graph G if and only if G has a path containing all
neighbours of t and t itself as an endpoint. In other words, t is a DFS-end-vertex
if and only if G has an induced subgraph that contains all neighbours of t and has
a Hamilton path with t as an endpoint. Thus, solving the DFS-END-VERTEX
problem is equivalent to solving a new type of Hamilton path problem. Our
algorithm will have $\mathcal{O}^*(2^n)$ running time, and no essentially better deterministic
algorithm is known for the HAMILTON PATH problem. It is interesting that
Björklund recently and for the first time in decades improved upon this running
time by constructing a randomised Monte-Carlo algorithm of expected running
time $\mathcal{O}^*(1.67^n)$ for the HAMILTON PATH problem [2].

Next to our chief interest in fast exact algorithms, we also study combinatorial properties of graph search algorithms, with a special interest in partition
properties. As already mentioned above, graph search algorithms are of great
algorithmic importance. Major algorithm design techniques like dynamic programming and divide-and-conquer [5,10] solve problems on small instances and
combine their solutions into solutions for larger instances. We investigate dependencies for such combination steps, and we particularly focus on the much less
understood depth-first search strategies.

2   Preliminaries

Our graphs are simple, finite, and undirected. A graph is an ordered pair $G =
(V, E)$, where $V = V(G)$ is the *vertex set* and $E = E(G)$ is the *edge set* of G. We
adopt the convention of letting n denote $|V(G)|$. Edges are denoted as uv, where
u and v are *adjacent* in G, or *neighbours*. For a vertex u of G, the *neighbourhood*
of u, $N_G(u)$, is the set of neighbours of u in G, the *closed neighbourhood* of

u is $N_G[u] = N_G(u) \cup \{u\}$, and for $U \subseteq V(G)$, $N_G(U) = \bigcup_{u \in U} N_G(u) \setminus U$. For $X \subseteq V(G)$, the *subgraph* of G *induced* by X, $G[X]$, is the graph on vertex set X and with edge set $\{xy \in E(G) : x, y \in X\}$, and by $G \setminus X$, we denote the graph $G[V(G) \setminus X]$.

Let G be a graph. For u, v a vertex pair of G, a u, v-*path* of G is a sequence $(x_0, \ldots, x_r)$ of pairwise different vertices of G such that $x_i x_{i+1} \in E(G)$ for $0 \leq i < r$, $x_0 = u$, $x_r = v$, and r is the *length* of $(x_0, \ldots, x_r)$. A graph is *connected* if it has a u, v-path for every vertex pair u, v of G, and a *connected component* is a maximal connected induced subgraph. The *distance* of u and v, $d_G(u, v)$, is the smallest length of a u, v-path of G.

Let G be a graph. A *vertex ordering* for G is a linear arrangement of the vertices of G, that we often denote as $\tau = \langle x_1, \ldots, x_n \rangle$. We call x_1 and x_n the respectively *start-* and *end-vertex* of τ. For a vertex pair x_i, x_j, we write $x_i \prec_\tau x_j$ if $i < j$, and we write $x_i \preccurlyeq_\tau x_j$ if $i \leq j$. In our applications, vertex orderings will be generated by graph search algorithms, and the end-vertex of the generated vertex ordering is the vertex visited last by the algorithm. The start-vertex of the ordering will also be called the *source vertex* of the search algorithm [5]. Given a vertex x, a vertex y is *left-side* of x in τ if $y \prec_\tau x$. Given a set X of vertices, a vertex x is *leftmost* of X in τ if $x \in X$ and $x \preccurlyeq_\tau y$ for every $y \in X$. Analogously, x is *rightmost* of X in τ if $x \in X$ and $y \preccurlyeq_\tau x$ for every $y \in X$. If X is empty then no leftmost or rightmost vertex exists.

Given two orderings $\tau' = \langle x_1, \ldots, x_{n'} \rangle$ and $\tau'' = \langle y_1, \ldots, y_{n''} \rangle$, the *composition* of τ' and τ'', $\tau' \circ \tau''$, is $\langle x_1, \ldots, x_{n'}, y_1, \ldots, y_{n''} \rangle$. We say that an ordering τ *respects* an ordered partition $\langle X_1, \ldots, X_r \rangle$ if there are orderings $\varrho_1, \ldots, \varrho_r$ for respectively $X_1, \ldots, X_r$ such that $\tau = \varrho_1 \circ \cdots \circ \varrho_r$.

3   End-Vertices of Breadth-First Search

We first study the graph search algorithm *Breadth-First Search* (BFS), which starts from a source vertex s and explores the input graph by visiting unvisited neighbours of already visited vertices. The vertex selection rule follows a preference order that respects the discover order of the vertices. A vertex is *discovered* when a neighbour is visited. This discover preference order selection rule is implemented by using a queue for memorizing the discovered vertices, and a queue is a first-in first-out data structure [5]. Our algorithm for the BFS-END-VERTEX problem has to resolve the dependencies that are caused by the queue, which is based on the following characterisation of BFS-end-vertices.

Lemma 1. *Let G be a connected graph, and let t be a vertex of G.*

Then, t is a BFS-end-vertex of G if and only if there are a vertex s of G, an ordered partition $\langle L_0, \ldots, L_k \rangle$ of $V(G)$ and triples $(P_1, x_1, Q_1), \ldots, (P_k, x_k, Q_k)$ that satisfy the following conditions:

1) $L_0 = \{s\}$, $L_i = \{x \in V(G) : d_G(s, x) = i\}$ for $0 < i \leq k$, and $t \in L_k$.
2) For $1 \leq i < k$, $P_i \cap (Q_i \cup \{x_i\}) = \emptyset$, $P_i \cup \{x_i\} \cup Q_i = L_i$, $x_i x_{i+1} \in E(G)$, $N_G(P_i) \cap L_{i+1} \subseteq P_{i+1} \subseteq N_G(P_i \cup \{x_i\})$, and $(P_k, x_k, Q_k) = ((L_k \setminus \{t\}), t, \emptyset)$.

Proof. Let σ be a visit order of a BFS on G with end-vertex t. Let s be the start vertex of σ. For $i \geq 0$, define $L_i = \{x \in V(G) : d_G(s,x) = i\}$, and choose k to be the largest integer such that $L_k \neq \emptyset$. Note that the BFS strategy implies that σ respects $\langle L_0, \ldots, L_k \rangle$, and $L_0 = \{s\}$, and $t \in L_k$.

For every $x \in V(G)$ where $s \prec_\sigma x$, let $\varphi(x)$ be the leftmost left-side neighbour of x in σ. Recall that each vertex but s has a left-side neighbour in σ. Note that if $x \in L_i$ then $\varphi(x) \in L_{i-1}$. Now, let $x_k = t$, and for $1 \leq i < k$ in decreasing order, let $x_i = \varphi(x_{i+1})$. Furthermore, for all $i \in \{1, 2, \ldots, k\}$, let

$$ P_i = \{u \in L_i : u \prec_\sigma x_i\} \quad \text{and} \quad Q_i = \{u \in L_i : x_i \prec_\sigma u\}. $$

By the construction, we have $t \in L_k$, $(P_k, x_k, Q_k) = ((L_k \setminus \{t\}), t, \emptyset)$, and for all $i \in \{1, 2, \ldots, k\}$, $P_i \cup \{x_i\} \cup Q_i = L_i$ and $x_i x_{i+1} \in E(G)$.

To verify the remaining conditions, let $i \in \{1, 2, \ldots, k-1\}$. Observe: $\varphi(y) \in L_i$ for every $y \in L_{i+1}$, and $\varphi(y) \prec_\sigma \varphi(x_{i+1})$ for every $y \in P_{i+1}$, and $\varphi(x_{i+1}) \prec_\sigma \varphi(y)$ for every $y \in Q_{i+1} \cup \{x_{i+1}\}$. So, $\varphi(y) \in P_i \cup \{x_i\}$ for every $y \in P_{i+1}$, and $N_G(y) \cap L_i \subseteq Q_i \cup \{x_i\}$ for every $y \in Q_{i+1} \cup \{x_{i+1}\}$. This implies $P_{i+1} \subseteq N_G(P_i \cup \{x_i\})$, $N_G(Q_{i+1} \cup \{x_{i+1}\}) \cap L_i \subseteq Q_i \cup \{x_i\}$, and $N_G(P_i) \cap (Q_{i+1} \cup \{x_{i+1}\}) = \emptyset$.

To prove the converse, assume a vertex s, an ordered partition $\langle L_0, \ldots, L_k \rangle$ of $V(G)$, and triples $(P_1, x_1, Q_1), \ldots, (P_k, x_k, Q_k)$ exist such that all conditions of the lemma are satisfied. Define $Q_i' = Q_i \setminus \{x_i\}$ for $1 \leq i \leq k$. We show that G has a BFS-ordering that respects $\langle \{s\}, P_1, \{x_1\}, Q_1', \ldots, P_k, \{x_k\}, Q_k' \rangle$. To do this, let σ be any vertex ordering of G that respects $\langle \{s\}, P_1, \{x_1\}, Q_1', \ldots, P_k, \{x_k\}, Q_k' \rangle$, where $\sigma = \langle z_1, \ldots, z_n \rangle$.

A new (improved) ordering σ^+ can be constructed as follows. For every $x \in V(G)$ where $s \prec_\sigma x$, let $\varphi(x)$ be the leftmost neighbour of x in σ. Assume that there is a vertex pair u, v of G such that $u \prec_\sigma v$ and $\varphi(v) \prec_\sigma \varphi(u)$; we choose u leftmost possible in σ. Assume $u = z_p$ and $v = z_q$, and define $\sigma^+ = \langle z_1, \ldots, z_{p-1} \rangle \circ \langle v, u, z_{p+1}, \ldots, z_{q-1} \rangle \circ \langle z_{q+1}, \ldots, z_n \rangle$. Since $u \in L_i$, $v \in L_j$, $1 \leq i \leq j$, $\varphi(u) \in L_{i-1}$, $\varphi(v) \in L_{j-1}$, and $j - 1 \leq i - 1$, we obtain $i = j$. We verify that σ^+ indeed respects $\langle \{s\}, P_1, \{x_1\}, Q_1', \ldots, P_k, \{x_k\}, Q_k' \rangle$:

- if $\varphi(u) \in P_{i-1}$ then $\varphi(v) \in P_{i-1}$, and thus $u, v \in P_i$;
- if $\varphi(u) = x_{i-1}$ then $\varphi(v) \in P_{i-1}$, and thus $v \in P_i$, implying $u \in P_i$ since $u \prec_\sigma v$;
- if $\varphi(u) \in Q_{i-1}'$ then $N_G(u) \cap L_{i-1} \subseteq Q_{i-1}'$, thus $u \notin N_G(P_{i-1} \cup \{x_{i-1}\})$, so that $u \notin P_i \cup \{x_i\}$, and $u, v \in Q_i'$ follows, also because of $u \prec_\sigma v$.

By repeatedly applying the described exchange step to the current ordering, we obtain a BFS-ordering of G that respects $\langle \{s\}, P_1, \{x_1\}, Q_1', \ldots, P_k, \{x_k\}, Q_k' \rangle$, that has t as its end-vertex, since $x_k = t$ and $Q_k' = \emptyset$. $\square$

Using the above characterisation, we transform the BFS-END-VERTEX problem into a reachability problem on an oriented graph.

Theorem 1. *There is an $\mathcal{O}(2^n \cdot n^5)$-time algorithm deciding on input a graph G and a vertex t whether t is a BFS-end-vertex of G.*

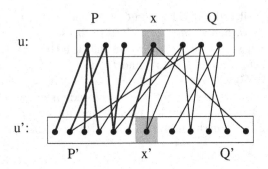

Fig. 1. We consider two vertices u and u' of $\mathfrak{A}(s)$ and their corresponding triples (P, x, Q) and (P', x', Q'). In the depicted situation, (P, x, Q) and (P', x', Q') satisfy the conditions, and $u \to u'$ is an oriented edge of $\mathfrak{A}(s)$. Note that the vertices from $\{x\} \cup Q$ can have neighbours in all of $P' \cup \{x'\} \cup Q'$, and x may be the unique neighbour in $P \cup \{x\} \cup Q$ of a vertex from P'.

Proof. Let G be a connected graph, and let t be a vertex of G. Let S be the set of those vertices of G for which t is a maximum-distance vertex; S will be the set of candidate source vertices. For each vertex in S, we construct an **arena graph**. Let $L_i = \{x \in V(G) : d_G(s, x) = i\}$, and let k be the largest integer such that $L_k \neq \emptyset$. The *arena graph* of s, $\mathfrak{A}(s)$, is defined as follows:

-) the vertices of $\mathfrak{A}(s)$ are partitioned into layers, $\mathcal{L}_1, \ldots, \mathcal{L}_k$;
for $1 \leq i < k$, the vertices of layer $\mathcal{L}_i$ correspond to the triples (P, x, Q) where $P \cup \{x\} \cup Q = L_i$ and $P, \{x\}, Q$ are pairwise disjoint;
the unique vertex of layer $\mathcal{L}_k$ corresponds to $((L_k \setminus \{t\}), t, \emptyset)$
-) the oriented edges of $\mathfrak{A}(s)$ are incident to vertices in consecutive layers;
for $1 \leq i < k$, let u and u' be vertices from respectively $\mathcal{L}_i$ and $\mathcal{L}_{i+1}$, and let (P, x, Q) and (P', x', Q') be the corresponding triples:
$$u \to u' \quad \Longleftrightarrow \quad x_i x_{i+1} \in E(G) \text{ and } N_G(P) \cap L_{i+1} \subseteq P' \subseteq N_G(P \cup \{x\}).$$
(For an illustration of the definition, consider Fig. 1.)

Observe that the definition of the oriented edges of $\mathfrak{A}(s)$ is based on Condition 2 of Lemma 1.

For given G and t, our algorithm computes S, and for each $s \in S$, it computes $\mathfrak{A}(s)$ and checks whether there is an oriented path from a vertex in $\mathcal{L}_1$ to the unique vertex in $\mathcal{L}_k$. Due to Lemma 1, t is a BFS-end-vertex of G if and only if there is an $s \in S$ such that $\mathfrak{A}(s)$ has such a path. The existence of such a path can be verified by checking whether a vertex from $\mathcal{L}_1$ of $\mathfrak{A}(s)$ is reverse-reachable from $\mathcal{L}_k$.

We consider the claimed running time of $\mathcal{O}(2^n \cdot n^5)$. Note that at most n arena graphs are examined. Since the reverse-reachability test can be done in time linear in the size of $\mathfrak{A}(s)$, it remains to determine the size of $\mathfrak{A}(s)$ and its construction time. Let $\langle L_0, \ldots, L_k \rangle$ be the ordered partition of $V(G)$ associated with s. Each layer $\mathcal{L}_i$ of $\mathfrak{A}(s)$ has $|L_i| \cdot 2^{|L_i|-1}$ vertices, which makes a total of at most $2^n \cdot n$ vertices. The number of oriented edges of $\mathfrak{A}(s)$ is bounded from above by:

$$\sum_{i=1}^{k-1} |\mathcal{L}_i| \cdot |\mathcal{L}_{i+1}| = \sum_{i=1}^{k-1} |L_i| \cdot 2^{|L_i|-1} \cdot |L_{i+1}| \cdot 2^{|L_{i+1}|-1} = \sum_{i=1}^{k-1} |L_i| \cdot |L_{i+1}| \cdot 2^{|L_i \cup L_{i+1}|-2},$$

which is at most $n^2 \cdot 2^n$. And the existence of an oriented edge can be decided in $\mathcal{O}(n^2)$ time. $\qquad\square$

4 Mixed Depth-First Searches

We consider graph search strategies that prefer late discovered vertices, in opposition to the preference for early discovered vertices of breadth-first search strategies. Such search strategies are depth-first ones. The most prominent such strategy is Depth-First Search (DFS), and a recently more intensely studied variant is Lexicographic Depth-First Search (LexDFS).

Corneil and Krueger approached graph search strategies in a unified manner by studying their generated vertex visit orderings [9]. This approach is based on the following terminology, that was introduced by Charbit, Habib, and Mamcarz [4], and that we adapt and extend.

Definition 1. *Let G be a graph.*

1) *An* adaptive vertex condition *for G is a vertex labelling $\Pi : V(G) \to \{0, 1, 2\}$.*
2) *Let Π be an adaptive vertex condition and let τ be a vertex ordering for G.*
   a) *A* bridge *of (G, τ) is a vertex triple (a, c, d) of G that satisfies $a \prec_\tau c \prec_\tau d$ and $ac \notin E(G)$ and $ad \in E(G)$.*
   b) *A Π-witness of (G, τ) is a vertex quadruple (a, b, c, d) of G such that (a, c, d) is a bridge of (G, τ) and $a \prec_\tau b \prec_\tau c$ and $bc \in E(G)$, and if $\Pi(d) = 1$ then $bd \notin E(G)$, and if $\Pi(d) = 2$ then $bd \in E(G)$, and if $\Pi(d) = 0$ then there is no restriction.*
   c) *τ is a Π-ordering of G if and only if for every bridge (a, c, d) of (G, τ), there is a Π-witness (a, b, c, d) of (G, τ).*

Corneil and Krueger showed that DFS-orderings are the Π-orderings where Π assigns label 0 to each vertex and LexDFS-orderings are the Π-orderings where Π assigns label 1 to each vertex [9]. Such Π-orderings, where Π assigns the same label to each vertex, are called *uniform*, thus DFS- and LexDFS-orderings are uniform. We will also denote DFS- and LexDFS-orderings as 0- and 1-orderings, respectively, especially when we want to emphasise the bridge-witness condition of Definition 1. A new and so far unconsidered uniform search is coLexicographic Depth-First Search (coLexDFS) that generates the 2-orderings.

Π-orderings for adaptive vertex conditions Π that may assign different vertex labels are called *mixed*. In a mixed ordering, a vertex may "choose" to be selected by regarding or ignoring a lexicographic order. Every graph with every choice of a source vertex has a mixed ordering when only labels 0 and 1 are assigned. When Π may also assign 2, there are graphs without a Π-ordering, and it is an interesting problem to decide whether a graph has a Π-ordering for a given Π. As an example, interval graphs always have Π-orderings and almost all trees do not have 2-orderings.

Proposition 1. *Let G be a connected graph, let Π be an adaptive vertex condition for G, and let τ be a vertex ordering for G. Then, τ is a Π-ordering of G if and only if for every vertex pair x, y of G such that $x \prec_\tau y$, among the vertices in $N(x, y)$, the rightmost left-side vertex of x in τ belongs to $N_G(x)$, where*

$$N(x, y) = \begin{cases} N_G(x) \cup N_G(y) & \text{if } \Pi(y) = 0 \\ N_G(x) \triangle N_G(y) & \text{if } \Pi(y) = 1 \\ N_G(y) & \text{if } \Pi(y) = 2. \end{cases}$$

Proof. Assume that τ is a Π-ordering of G. Consider $x \prec_\tau y$, and observe that always $N(x, y) \subseteq N_G(x) \cup N_G(y)$. Choose $u \in N_G(y) \setminus N_G(x)$ such that $u \prec_\tau x$. Then, $u \in N(x, y)$, and (u, x, y) is a bridge of (G, τ). Thus, (G, τ) has a Π-witness (u, b, x, y), and $b \in N(x, y) \cap N_G(x)$ and $u \prec_\tau b$.

For the converse, assume that every vertex pair of G satisfies the ordering condition of the proposition with respect to τ. Let (a, c, d) be a bridge of (G, τ). Observe that $a \in N_G(d) \setminus N_G(c)$ and $a \prec_\tau c$. Thus, there is a vertex b satisfying $b \in N(c, d) \cap N_G(c)$ and $a \prec_\tau b \prec_\tau c$, and (a, b, c, d) is a Π-witness of (G, τ). $\square$

Now, we would like to better understand which vertices can be end-vertices of arbitrary Π-orderings. For a graph G and an adaptive vertex condition Π, a Π-*end-vertex* of G is the end-vertex of a Π-ordering of G. So, the end-vertices of DFS and LexDFS and coLexDFS are exactly the 0-, 1-, and 2-end-vertices. As a main result, we provide a description of Π-end-vertices. It is worth mentioning that our lemma also implicitly studies the existence of Π-orderings.

Lemma 2. *Let G be a connected graph, let Π be an adaptive vertex condition for G, and let t be a vertex of G. Then, t is a Π-end-vertex of G if and only if there exist an ordered partition $\langle X_1, \ldots, X_r \rangle$ of $V(G)$, vertices $x_1, \ldots, x_r$ such that $x_i \in X_i$ for all $1 \le i \le r$, $X_r = \{x_r\} = \{t\}$, and the following conditions are satisfied for all $i \in \{1, 2, \ldots, r-1\}$, where $X_i^- = X_i \setminus \{x_i\}$:*

1) $x_i x_{i+1} \in E(G)$
2) $N_G(X_i^-) \subseteq \{x_1, \ldots, x_i\}$ and $G[X_i]$ is connected
3) $G[\{x_1, \ldots, x_r\} \cup X_i^-]$ has a Π-ordering τ_i satisfying the following properties
    - τ_i respects $\langle \{x_1\}, \ldots, \{x_i\}, X_i^-, \{x_{i+1}\}, \ldots, \{x_r\} \rangle$
    - for every vertex pair x, y of G where x is the leftmost vertex in τ_i of a connected component of $G[X_i^-]$ and $y \in X_{i+1}^- \cup \cdots \cup X_{r-1}^-$ and $\Pi(y) = 1$, the rightmost vertex from $(N_G(x) \triangle N_G(y)) \cap \{x_1, \ldots, x_i\}$ belongs to $N_G(x)$
    - for every vertex pair x, y of G where $x \in X_i^-$ and $y \in X_{i+1}^- \cup \cdots \cup X_{r-1}^-$ and $\Pi(y) = 2$, the rightmost vertex from $N_G(y) \cap \{x_1, \ldots, x_i\}$ belongs to $N_G(x)$.

Proof. We prove the necessity part.

We assume that G has at least two vertices. Let τ be a Π-ordering of G with end-vertex t. Let $\tau = \langle u_1, \ldots, u_n \rangle$ and with start-vertex s. Let ψ be the *predecessor function* of τ, which assigns to each vertex x the rightmost left-side neighbour of x in τ; also let $\psi(s) = s$. Note that $\psi(x)$ exists, since G is connected and τ is a Π-ordering.

Consider $\mathfrak{S} = \{\psi^{(i)}(t) : i \geq 0\} = \{t, \psi(t), \psi(\psi(t)), \ldots\}$ and observe that $s, t \in \mathfrak{S}$. Let $r = |\mathfrak{S}|$. Choose $g_1, \ldots, g_r$ such that $g_1 < \cdots < g_r$ and $\mathfrak{S} = \{u_{g_1}, \ldots, u_{g_r}\}$. Clearly $u_{g_1} = u_1 = s$ and $u_{g_r} = u_n = t$. Let $x_r = u_{g_r}$, $X_r = \{t\}$, and for $1 \leq i < r$, define $x_i = u_{g_i} = \psi(x_{i+1})$ and

$$X_i = \{u_j : g_i \leq j < g_{i+1}\} \quad \text{and} \quad X_i^- = \{u_j : g_i < j < g_{i+1}\}.$$

Consequently, $\{X_1, \ldots, X_r\}$ is a partition of $V(G)$, and $(x_1, \ldots, x_r)$ is an s, t-path of G. Thus, Condition 1 of the lemma is satisfied.

Let us consider Condition 2. Let $q \in \{1, 2, \ldots, r-1\}$. To consider $N_G(X_q^-)$, let $uv \in E(G)$ such that $u \prec_\tau v$, $u \notin X_q$ and $v \in X_q^-$. Then, $u \in X_1 \cup \cdots \cup X_{q-1}$, and therefore, $u \prec_\tau x_q \prec_\tau v$. Let p be such that $u \in X_p$, hence $p < q$. For a proof by contradiction, suppose $u \notin \mathfrak{S}$, which means $x_p \prec_\tau u \prec_\tau x_{p+1} \prec_\tau v$. The definition of ψ implies $ux_{p+1} \notin E(G)$, and thus, (u, x_{p+1}, v) is a bridge of (G, τ), yielding a vertex b such that (u, b, x_{p+1}, v) is a Π-witness of (G, τ), that satisfies $u \prec_\tau b \preccurlyeq_\tau \psi(x_{p+1})$, which yields the contradiction due to $x_p = \psi(x_{p+1})$.

To show connectedness of $G[X_q]$, it suffices to show that $\psi(x) \in X_q$ for every $x \in X_q^-$. To do this, let $v \in X_q^-$. If $x_q v \in E(G)$ then $x_q \prec_\tau \psi(v) \prec_\tau v$, and $\psi(v) \in X_q$. Otherwise, if $x_q v \notin E(G)$ then (x_q, v, x_{q+1}) is a bridge of (G, τ), and we conclude $x_q \prec_\tau \psi(v) \prec_\tau v$ due to the existence of a Π-witness of (G, τ).

Let us consider Condition 3. Let $p \in \{1, 2, \ldots, r-1\}$, $G_p = G[\{x_1, \ldots, x_r\} \cup X_p^-]$ and $\tau_p = \langle x_1, \ldots, x_p \rangle \circ \langle u_{g_p+1}, \ldots, u_{g_{p+1}-1} \rangle \circ \langle x_{p+1}, \ldots, x_r \rangle$. Note that τ_p respects $\langle \{x_1\}, \ldots, \{x_p\}, X_p^-, \{x_{p+1}, \ldots, x_r\} \rangle$.

We verify the two remaining properties. Let $c \in X_p^-$ and $d \in X_{p+1}^- \cup \cdots \cup X_{r-1}^-$. Then, $x_p \prec_\tau c \prec_\tau x_{p+1} \prec_\tau d$. Assume that $(N_G(d) \setminus N_G(c)) \cap \{x_1, \ldots, x_p\}$ is non-empty. Let $a \in (N_G(d) \setminus N_G(c)) \cap \{x_1, \ldots, x_p\}$. Then, (a, c, d) is a bridge of (G, τ). We continue separately for the two properties.

- $\Pi(d) = 1$

  Assume that c is the leftmost vertex in τ of a connected component of $G[X_p^-]$. Observe that $x_p = \psi(c)$. Thus, $a \prec_\tau x_p$, and there is a vertex b of G such that (a, b, c, d) is a Π-witness of (G, τ), that satisfies $a \prec_\tau b \prec_\tau c$ and $bc \in E(G)$ and $bd \notin E(G)$, so that $b \in N_G(c) \setminus N_G(d)$. Since $b \preccurlyeq_\tau \psi(c)$, we have $b \preccurlyeq_\tau x_p$, and thus, $b \in \{x_1, \ldots, x_p\}$ due to Condition 2.

- $\Pi(d) = 2$

  There is a Π-witness (a, b, c, d), that satisfies $a \prec_\tau b \prec_\tau c$ and $bc, bd \in E(G)$. So, $b \in N_G(c) \cap N_G(d)$ and $b \in N_G(X_{p+1}^- \cup \cdots \cup X_{r-1}^-)$, which implies $b \in \mathfrak{S}$ by Condition 2.

Thus, Condition 3 is satisfied, which completes the proof of the necessity part.

$\square$

5   End-Vertices of Uniform Depth-First Searches

In the previous section, we studied combinatorial properties of depth-first search orderings and obtained Lemma 2 as a main result, that establishes necessary and sufficient conditions for the combination of Π-orderings of induced subgraphs into Π-orderings of the whole graph. Condition 3 of Lemma 2 can be seen as an explanation of the dependencies between different parts of a combination.

Now, we consider consequences of Lemma 2, first combinatorial, then algorithmic. We consider uniform depth-first search orderings, namely 0- and 1-orderings. Recall from Section 4 that 0-orderings are the DFS-orderings and 1-orderings are the LexDFS-orderings [9]. Note that each graph has such orderings for each choice of the source vertex.

Let us start with DFS-orderings and their end-vertices. A path $(x_1, \ldots, x_l)$ of a graph G is a *Hamilton path* if $\{x_1, \ldots, x_l\} = V(G)$, i.e., if it visits each vertex of G exactly once. The *endpoints* of $(x_1, \ldots, x_l)$ are x_1 and x_l. There is a relationship between Hamilton paths and DFS-orderings, that has already been noticed in the NP-completeness proof of the DFS-END-VERTEX problem [4]. We provide an astonishing characterisation of DFS-end-vertices using Hamilton paths.

Proposition 2. *Let G be a connected graph, and let t be a vertex of G.*

Then, t is a DFS-end-vertex of G if and only if there is $X \subseteq V(G)$ such that $N_G[t] \subseteq X$ and $G[X]$ has a Hamilton path with endpoint t.

Proof. If G has a DFS-ordering with end-vertex t then G has a path $(x_1, \ldots, x_r)$ satisfying $x_r = t$ and $N_G(t) \subseteq \{x_1, \ldots, x_{r-1}\}$ due to Lemma 2, and $(x_1, \ldots, x_r)$ is a Hamilton path of $G[\{x_1, \ldots, x_r\}]$.

For the converse, we also apply Lemma 2. Let $X \subseteq V(G)$ such that $N_G[t] \subseteq X$, and $G[X]$ has a Hamilton path, say $(x_1, \ldots, x_r)$, with endpoint t, where $t = x_r$. We may assume $X \neq V(G)$, otherwise we are done.

We construct an ordered partition of $V(G)$. Let C be a connected component of $G \backslash X$. Clearly, $N_G(V(C)) \subseteq \{x_1, \ldots, x_{r-1}\}$ and $N_G(V(C)) \cap X \neq \emptyset$. Let $\mu(C)$ be the largest j such that $1 \leq j < r$ and $x_j \in N_G(V(C))$. For $1 \leq i \leq r$, define

$$X_i = \{x_i\} \cup \bigcup_{\text{connected component } C \text{ of } G \backslash X \text{ s.t. } \mu(C)=i} V(C),$$

and let $X_i^- = X_i \setminus \{x_i\}$ for $1 \leq i < r$. Clearly, $\langle X_1, \ldots, X_r \rangle$ is an ordered partition of $V(G)$ with $X_r = \{x_r\} = \{t\}$. Also $N_G(X_i^-) \subseteq \{x_1, \ldots, x_i\}$, and since each connected component C of $G \setminus X$ satisfying $\mu(C) = i$ contains a neighbour of x_i, $G[X_i]$ is connected. Thus, Conditions 1 and 2 of Lemma 2 are satisfied.

We consider Condition 3 of Lemma 2. Let $p \in \{1, 2, \ldots, r-1\}$ and $G_p = G[\{x_1, \ldots, x_r\} \cup X_p^-]$. Let ϱ_p be a DFS-ordering of $G[X_p]$ with start-vertex x_p; recall that ϱ_p does exist. Let $\tau_p = \langle x_1, \ldots, x_{p-1} \rangle \circ \varrho_p \circ \langle x_{p+1}, \ldots, x_r \rangle$. Note that τ_p respects $\langle \{x_1\}, \ldots, \{x_p\}, X_p^-, \{x_{p+1}, \ldots, x_r\} \rangle$. We show that τ_p is a 0-ordering of G_p. Let (a, c, d) be a bridge of (G_p, τ_p). If $c = x_i$ for some $1 < i < r$ then

$a \neq x_{i-1}$, and (a, x_{i-1}, c, d) is a 0-witness of (G_p, τ_p). If $c \notin \{x_1, \ldots, x_r\}$ then $c \in X_p^-$, and either $a, d \in X_p$ and $(G[X_p], \varrho_p)$ has a 0-witness (a, b, c, d), that is a 0-witness also of (G_p, τ_p), or $a \in \{x_1, \ldots, x_{p-1}\}$ and $d \in X_p^- \cup \{x_{p+1}, \ldots, x_r\}$, and c has a neighbour b' satisfying $x_p \preccurlyeq_{\varrho_p} b' \prec_{\varrho_p} c$, and (a, b', c, d) is a 0-witness of (G_p, τ_p). □

A straightforward consequence of Proposition 2 is that each vertex of a graph with a Hamilton cycle is a DFS-end-vertex. A more important consequence of Proposition 2 is the construction of a simple exponential-time algorithm solving the DFS-END-VERTEX problem. Further algorithmic consequences, e.g. for restricted input graph classes, are to be expected.

Theorem 2. *There is an $\mathcal{O}(2^n \cdot n^2)$-time algorithm deciding on input a graph G and a vertex t whether t is a DFS-end-vertex of G.*

Proof. Our algorithm finds a set X such that $N_G[t] \subseteq X \subseteq V(G)$ and $G[X]$ has a Hamilton path with endpoint t, or it finds out that such X does not exist. By Proposition 2, t is a DFS-end-vertex of G in the first case, and it is not otherwise.

We give a description of our algorithm and analyse its running time. Due to Bellmann [1] and Held-Karp [11], there is a dynamic-programming algorithm generating all pairs (X, u) for $X \subseteq V(G)$ and $u \in X$ satisfying that $G[X]$ has a Hamilton path with endpoint u. The Bellmann-Held-Karp algorithm has running time $\mathcal{O}(2^n \cdot n)$. Additionally, in $\mathcal{O}(n)$ time per pair (X, u), our algorithm checks the condition $N_G(t) \subseteq X$ and $u = t$. □

Next, we consider the combination of LexDFS-orderings, analogously to Proposition 2 for DFS-orderings. We shall see that this case is really more complex.

Proposition 3. *Let G be a connected graph, and let t be a vertex of G.*
Then, t is a LexDFS-end-vertex of G if and only if there exist an ordered partition $\langle X_1, \ldots, X_r \rangle$ of $V(G)$, vertices $x_i \in X_i$ for $1 \leq i \leq r$, $X_r = \{x_r\} = \{t\}$ and for $1 \leq i < r$, the following conditions are satisfied, where $X_i^- = X_i \setminus \{x_i\}$ and $G_i = G[\{x_1, \ldots, x_{i+1}\} \cup X_i^-]$:

1) $x_i x_{i+1} \in E(G)$
2) $N_G(X_i^-) \subseteq \{x_1, \ldots, x_i\}$ and $G[X_i]$ is connected
3) G_i has a LexDFS-ordering ϱ_i that respects $\langle \{x_1\}, \ldots, \{x_i\}, X_i^-, \{x_{i+1}\} \rangle$.

Proof. Necessity is a consequence of Lemma 2. We shall prove sufficiency. Let $\langle X_1, \ldots, X_r \rangle$ be an ordered partition of $V(G)$ and $x_1, \ldots, x_r$ vertices of G such that the assumptions of the lemma are satified. We need to verify Condition 3 of Lemma 2, to make sure that this lemma can be applied.

Let $G_i^+ = G[\{x_1, \ldots, x_r\} \cup X_i^-]$ and $\tau_i = \varrho_i \circ \langle x_{i+2}, \ldots, x_r \rangle$. Note that τ_i respects $\langle \{x_1\}, \ldots, \{x_i\}, X_i^-, \{x_{i+1}, \ldots, x_r\} \rangle$. Also observe, by restricting τ_{r-1} to the vertices of $\{x_1, \ldots, x_r\}$, that $\langle x_1, \ldots, x_r \rangle$ is a 1-ordering of $G[\{x_1, \ldots, x_r\}]$.

Let $p \in \{1, 2, \ldots, r-1\}$. We need to show that τ_p is a 1-ordering of G_p^+ that satisfies Condition 3 of Lemma 2. To do this, let (a, c, d) be a bridge of (G_p^+, τ_p). By our assumptions, we may restrict to consider the situation of $a \preccurlyeq_{\tau_p} x_p \prec_{\tau_p}$

$c \prec_{\tau_p} x_{p+1} \prec_{\tau_p} d$. If c is not the leftmost vertex in τ_p of a connected component of $G[X_p^-]$ then there is a vertex b such that $x_p \prec_{\tau_p} b \prec_{\tau_p} c$ and $bc \in E(G)$, and $bd \notin E(G)$ in particular due to Condition 2, and (a, b, c, d) is a 1-witness of (G_p^+, τ_p). If $c \in N_G(x_p)$ and $x_p d \notin E(G)$ then $a \prec_{\tau_p} x_p$ and (a, x_p, c, d) is a 1-witness of (G_p^+, τ_p).

The case remaining is $c, d \in N_G(x_p)$. Let q be such that $d \in X_q$. It suffices to show $N_G(c) \cap \{x_j : a \prec_{\tau_p} x_j \prec_{\tau_p} c\} \not\subseteq N_G(d)$. This will be done by a (intricated) proof by contradiction. We assume that $N_G(c) \cap \{x_j : a \prec_{\tau_p} x_j \prec_{\tau_p} c\} \subseteq N_G(d)$, and obtain a contradiction by constructing an infinite sequence of bridges of one of the following two types, either

$$\langle\ (a_1, c, x_{p+1}),\ (a_2, x_{p+1}, d),\ (a_3, c, x_{p+1}),\ \ldots\ \rangle\ \text{or}$$
$$\langle\ (a_1, x_{p+1}, d),\ (a_2, c, x_{p+1}),\ (a_3, x_{p+1}, d),\ \ldots\ \rangle.$$

Note that the bridges alternatingly belong to (G_p, ϱ_p) and (G_q, ϱ_q) where $a_1 \prec_{\tau_p} a_2 \prec_{\tau_p} a_3 \prec_{\tau_p} \cdots \prec_{\tau_p} c$. Let us describe the construction of the sequence.

- *First bridge of sequence*
  We choose a_1 as a, and if $ax_{p+1} \in E(G)$ then (a, c, x_{p+1}) is a bridge of (G_p, ϱ_p), and if $ax_{p+1} \notin E(G)$ then (a, x_{p+1}, d) is a bridge of (G_q, ϱ_q).
- *Next bridge of sequence*
  Let (a', x_{p+1}, d) be a bridge of (G_q, ϱ_q). There is a 1-witness (a', b', x_{p+1}, d) of (G_q, ϱ_q). Since $b' \in \{x_1, \ldots, x_p\}$, $b'c \in E(G)$ implies $b'd \in E(G)$ by our assumption, so that $b'c \notin E(G)$, and thus, (b', c, x_{p+1}) is a bridge of (G_p, ϱ_p), where $a' \prec_{\tau_p} b' \prec_{\tau_p} c$.
  Let (a'', c, x_{p+1}) be a bridge of (G_p, ϱ_p). There is a 1-witness (a'', b'', c, x_{p+1}) of (G_p, ϱ_p), and since $b'' \prec_{\varrho_p} x_p$ and $b''c \in E(G)$ and $b''x_{p+1} \notin E(G)$, we conclude that (b'', x_{p+1}, d) is a bridge of (G_q, ϱ_q), where $a'' \prec_{\tau_p} b'' \prec_{\tau_p} c$.

The existence of such an infinite sequence of bridges is impossible in a finite graph, and thus a contradiction. Consequently, τ_p is a 1-ordering of G_p^+ that also satisfies Condition 3 of Lemma 2. Recall here that the leftmost vertex in τ_p of a connected component of $G[X_p^-]$ is adjacent to x_p.   □

Finally, we consider a combination problem for DFS-orderings. Let G be a graph, and let (X, Y) be a partition of $V(G)$. Let τ' and τ'' be DFS-orderings of respectively $G[X]$ and $G[Y]$. We ask, and answer, under which additional assumptions about τ' and τ'', the concatenation $\tau' \circ \tau''$ is a DFS-ordering of G.

Let τ be any vertex ordering for G, and let T and R such that $T \subseteq R \subseteq V(G)$. We say that τ is (R, T)-*augmenting-preferring*, or (R, T)-*aug-pref* for short, if the following conditions are satisfied:

- the rightmost vertex of R in τ also belongs to T
- for each vertex x, the rightmost left-side vertex from $N_G(x) \cup R$ also belongs to $N_G(x)$.

It is worth emphasising that the DFS-END-VERTEX problem is equivalent to our combination problem if $((V(G) \setminus \{t\}), \{t\})$ is the chosen partition of $V(G)$.

We provide a first invariance result for (R, T)-aug-pref DFS-orderings.

Lemma 3. *Let G be a connected graph, let (X, Y) be a partition of $V(G)$, let T and R be such that $T \subseteq R \subseteq V(G)$, and let $s \in V(G)$. Assume that $T \cap Y \neq \emptyset$ and that $G[Y]$ is connected. Let $R' = (R \cup N_G(Y)) \cap X$, let $R'' = R \cap Y$ and $T'' = T \cap Y$.*

Then, G has an (R, T)-aug-pref DFS-ordering with start-vertex s that respects $\langle X, Y \rangle$ if and only if there is a vertex w in $Y \cap N_G(X)$ such that, with $T' = N_G(w) \cap X$, the following two conditions are satisfied:

- *$G[X]$ has an (R', T')-aug-pref DFS-ordering with start-vertex s*
- *$G[Y]$ has an (R'', T'')-aug-pref DFS-ordering with start-vertex w.*

A full description needs a $T \cap Y = \emptyset$ analogue of Lemma 3, that we provide in the future full version of this paper.

6 Conclusions and Future Work

We studied structural properties of graph search strategies and their algorithmic consequences. The major goal of a graph search algorithm is to visit all vertices of a given graph in a special order, and this order is determined by the adjacency relation. Our considered graphs are simple and undirected. It is easy to see that our results directly extend to graphs that may contain loops and multiple edges. It would be interesting to study the extension of our results to directed graphs.

How good are our algorithms? Since HAMILTON PATH reduces to our DFS-END-VERTEX problem [4] and since the currently best deterministic algorithm for finding a Hamilton path has a running time of $\mathcal{O}^*(2^n)$ [1,11], our algorithm for solving the DFS-END-VERTEX problem may be considered optimal at the current state of research. On the other hand, it would be interesting to find out whether Björklund's Monte-Carlo algorithm for HAMILTON PATH [2] can be used to obtain a faster randomised algorithm for the DFS-END-VERTEX problem. It is also interesting to find out if improvements are possible in the case of BFS-end-vertices, for instance by a reduction of the size of the constructed arena graphs $\mathfrak{A}(s)$.

Algorithmic challenges are fast algorithms for the LEXBFS-END-VERTEX and the LEXDFS-END-VERTEX problem. For both problems, no algorithm of running time $O(c^n)$, for some constant c, is known to us. Our approach for BFS-end-vertices of Lemma 1 seems applicable: it suffices to ensure that x_{i+1} of $(P_{i+1}, x_{i+1}, Q_{i+1})$ has lexicographically largest label, and this is pre-determined by (P_i, x_i, Q_i). Such an approach may work if the dependencies inside of the triple $(P_{i+1}, x_{i+1}, Q_{i+1})$ are restricted, such as in bipartite or even k-partite graphs. The tractability status of the LEXBFS-END-VERTEX problem on bipartite graphs is still unknown; Charbit, Habib, and Mamcarz conjecture NP-completeness [4]. We already carried out a detailed analysis of such cross-dependencies for LexDFS in Proposition 3. Is it possible to devise an algorithm with running time of order c^n? Which restrictions of the input graphs will allow polynomial- and subexponential-time algorithms?

References

1. Bellmann, R.: Dynamic programming treatment of the travelling salesman problem. Journal of the ACM **9**, 61–63 (1962)
2. Björklund, A.: Determinant Sums for Undirected Hamiltonicity. SIAM Journal on Computing **43**, 280–299 (2014)
3. Brandstädt, A., Dragan, F.F., Nicolai, F.: LexBFS-orderings and powers of chordal graphs. Discrete Mathematics **171**, 27–42 (1997)
4. Charbit, P., Habib, M., Mamcarz, A.: Influence of the tie-break rule on the end-vertex problem. Discrete Mathematics and Theoretical Computer Science **16.2**, 57–72 (2014)
5. Cormen, T.H., Leiserson, C.E., Rivest, R.L., Stein, C.: Introduction to Algorithms, 3rd edn. MIT Press (2009)
6. Corneil, D.G.: A simple 3-sweep LBFS algorithm for the recognition of unit interval graphs. Discrete Applied Mathematics **138**, 371–379 (2004)
7. Corneil, D.G., Dalton, B., Habib, M.: LDFS-Based Certifying Algorithm for the Minimum Path Cover Problem on Cocomparability Graphs. SIAM Journal on Computing **42**, 792–807 (2013)
8. Corneil, D.G., Köhler, E., Lanlignel, J.-M.: On end-vertices of lexicographic breadth first searches. Discrete Applied Mathematics **158**, 434–443 (2010)
9. Corneil, D.G., Krueger, R.: A Unified View of Graph Searching. SIAM Journal on Discrete Mathematics **22**, 1259–1276 (2008)
10. Fomin, F.V., Kratsch, D.: Exact exponential algorithms. Springer (2010)
11. Held, M., Karp, R.M.: A dynamic programming approach to sequencing problems. Journal of the Society for Industrial and Applied Mathematics **10**, 196–210 (1962)
12. Rose, D.J., Tarjan, R.E., Lueker, G.S.: Algorithmic aspects of vertex elimination on graphs. SIAM Jounal on Computing **5**, 266–283 (1976)

Deciding the On-line Chromatic Number of a Graph with Pre-coloring Is PSPACE-Complete

Christian Kudahl[(✉)]

Department of Mathematics and Computer Science,
University of Southern Denmark, Odense, Denmark
kudahl@imada.sdu.dk

Abstract. In an on-line coloring, the vertices of a graph are revealed one by one. An algorithm assigns a color to each vertex after it is revealed. When a vertex is revealed, it is also revealed which of the previous vertices it is adjacent to. The on-line chromatic number of a graph, G, is the smallest number of colors an algorithm will need when on-line-coloring G. The algorithm may know G, but not the order in which the vertices are revealed. The problem of determining if the on-line chromatic number of a graph is less than or equal to k, given a pre-coloring, is shown to be PSPACE-complete.

1 Introduction

In the on-line graph coloring problem, the vertices of a graph are revealed one by one to an algorithm. When a vertex is revealed the adversary reveals which other of the revealed vertices it is adjacent to. The algorithm gives a color to the vertex. This color has to be different from all colors found on neighboring vertices. The goal is to use as few colors as possible.

We let $\chi(G)$ denote the *chromatic number* of G. This is the number of colors that an optimal off-line algorithm needs to color G. Similarly, we let $\chi^O(G)$ denote the *on-line chromatic number* of G. This is the smallest number of colors that the best on-line algorithm needs to guarantee that for any ordering of the vertices, it will be able to color G using at most $\chi^O(G)$ colors. This algorithm may know the graph in advance but not the vertex ordering. As an example, $\chi^O(P_4) = 3$, since if two isolated vertices are presented first, the algorithm will be unable to decide if it is optimal to give them the same or different colors. Clearly, $\chi(P_4) = 2$.

The traditional measure of performance of an on-line algorithm is *competitive analysis* [10]. Here, the performance of an algorithm is compared to the performance of an optimal off-line algorithm. In the on-line graph coloring problem, an algorithm A is said to be *c-competitive* if it holds, that for any graph

C. Kudahl—Supported in part by the Villum Foundation and the Danish Council for Independent Research, Natural Sciences.

V.Th. Paschos and P. Widmayer (Eds.): CIAC 2015, LNCS 9079, pp. 313–324, 2015.
DOI: 10.1007/978-3-319-18173-8_23

G, and for any ordering of the vertices in G, the number of colors used by A is at most c times the chromatic number of G. For the on-line graph coloring problem, there does not exist c-competitive algorithms for any c even if the class of graphs is restricted to trees [2]. This makes this measure less desirable to use in this context.

As an alternative *on-line competitive* analysis was introduced for on-line graph coloring [3]. The definition is similar to competitive analysis, but instead of comparing with the best off-line algorithm, one compares with the best on-line algorithm. In the case of on-line graph coloring, an algorithm is *on-line c-competitive* if for any graph G, and for any ordering of the vertices, the number of colors it uses is at most c times the on-line chromatic number.

With the definition of on-line competitive analysis, a natural problem arose. How computationally hard is it given a graph G and a $k \in \mathbb{N}$ to decide if $\chi^O(G) \le k$. In [8], it was shown that it is possible in polynomial time to decide if $\chi^O(G) \le 3$ when G is triangle free or connected. They conjectured it NP-complete to decide if $\chi^O(G) \le 4$. In this paper, we consider the generalization of the problem where a part of the graph has already been presented and colored (we refer to this as the pre-coloring). We show that it is PSPACE-complete given a pre-colored graph and a k to decide if the uncolored parts can be colored such that at most k total colors are used.

2 Related Work

Studying pre-coloring extensions is not new. In [4], the author studies the pre-coloring extension problem in an offline setting and shows it to be NP-complete even on bipartite graphs and for three colors. In [9] it is shown to be NP-hard on unit interval graphs. For a survey on offline pre-coloring extensions, see [12]. It is an interesting open question how pre-coloring affects on-line graph coloring problem treated here (see closing remarks).

In [1], the author shows another coloring game to PSPACE-complete. In this version, a graph is known and two players take turns coloring the vertices in a fixed order with a fixed set of colors. The player that first is unable to color a vertex loses the game. In some sense, both players take the role of the painter and the drawer's strategy is given beforehand.

It was recently shown that the type of online coloring, that is analyzed in this paper, can be useful when offline coloring certain classes of geometrical graphs [5]. This gives another motivation for studying the complexity of finding the online chromatic number.

3 Preliminaries

On-line graph coloring can be seen as a game. The two players are known as the drawer and the painter. The two players agree on a graph $G = (V(G), E(G))$ and a $k \in \mathbb{N}$. A move for the drawer is presenting a vertex (sometimes we say it request a vertex). It does not specify which vertex in G the presented vertex corresponds to, but it specifies which of the already presented vertices that this new vertex is adjacent to. The presented graph must always be an induced subgraph of G.

A move for the painter is assigning a color from $\{1, \ldots, k\}$ to the newly presented vertex. The color has to be different from the colors that he previously assigned to its neighbors. If the painter manages to color the entire graph, he wins. If he is ever unable to color a vertex (because all colors are already found on neighbors to this vertex) he loses.

When analyzing games, one is often interested in finding out which player has a winning strategy. A game is said to be *weakly solved* if it is known which player has a winning strategy from the initial position. It is said to be *strongly solved* if it is known which player has a winning strategy from any given position. This definition is the motivation behind the assumption to have a pre-coloring. We prove that to strongly solve the game for a given graph, one must, in some cases, solve positions, where it is PSPACE-hard to determine if the drawer or the painter has a win from that position. Note that it may not be PSPACE-hard to weakly solve the game - see closing remarks.

We consider the state in the game after an even number of moves. This means that the game has not started yet or the painter has just assigned a color to a vertex. Such a *state* can be denoted by (G, k, G', f). Here, G is the graph they are playing on and $k \in \mathbb{N}$ is the number of colors the painter is allowed to use. Furthermore, G' is the induced subgraph that has already been presented and colored and $f : V(G') \rightarrow \{1, \ldots, k\}$ is a function that describes what colors have been assigned to the vertices of G'. Note that the painter does not get information on how to map the vertices of G' into G (in fact, the drawer does not have to decide this yet).

We treat the following problem: Let a game state (G, k, G', f) be given. Does the painter have a winning strategy from this state? We show that this problem is PSPACE-complete. The problem is equivalent to deciding if the on-line chromatic number of G is less than or equal to k given that the vertices in an induced subgraph isomorphic to G' have already been given the colors dictated by f. This is also known as a pre-colored graph.

Note that the proof here also works in the model where the painter gets information on how the vertices in G' are mapped to those in G. In fact, a slightly simpler construction would suffice in that case. The model where this information is not available seems more reasonable though, since the pre-coloring is used to represent a state in the game where this information is indeed not available.

We show a reduction from the totally quantified boolean formula (TQBF) problem. In this problem, we are given a boolean formula:

$$\phi = \forall x_1 \exists x_2 \; \ldots \; \exists x_n F(x_1, x_2, \ldots, x_n)$$

We want to decide if ϕ is true or false. This problem is known to be PSPACE-complete even if F is assumed to be in conjunctive normal form with 3 literals in each clause ([11]). Since the complement to any language in PSPACE is also in PSPACE, this is also PSPACE-complete if F is in disjunctive normal form with 3 literals in each term (3DNF). This is the form we will use here. We let t_i denote the i'th term. For convenience, we will assume the number of variables to be even and that the first quantifier is $\forall$ followed by alternating quantifiers. This is possible since any TQBF in 3DNF can be transformed to such a formula by adding new variables.

One such formula could for example be:

$$\forall x_1 \exists x_2 \forall x_3 \exists x_4 \ (x_1 \wedge x_2 \wedge \bar{x}_4) \vee (\bar{x}_1 \wedge x_2 \wedge x_3) \vee (\bar{x}_1 \wedge \bar{x}_2 \wedge x_3)$$

This formula has four variables, x_1, x_2, x_3, and x_4, and three terms, t_1, t_2, and t_3. The term t_1 contains x_1, x_2, and $\bar{x}_4$ (we also say that they *are in* the first term).

4   PSPACE Completeness

In this section, we show that it is PSPACE-complete to to decide if the painter has a winning strategy from a game state (G, k, G', f). First we note, that the problem is in PSPACE.

Observation 1. *The problem of deciding if the drawer has a winning strategy from state (G, k, G', f) is in* PSPACE.

To see this, we see that the game always ends within at most $2V(G)$ moves. We need to show that from each state, the possible following moves can be enumerated in polynomial space. If the painter is about to move, his possible moves are one of the colors $\{1, \ldots, k\}$. This can be done is polynomial space, and can be enumerated based on the value of the color. If the drawer is about to move, his move consists of presenting a vertex that is adjacent to some of the vertices that have already been presented. If v vertices have been presented already, this means that there are possibly 2^v different moves for him. He can enumerate these but only consider those where the resulting graph is an induced subgraph of G. This problem is NP-complete, but it can be solved in polynomial space.

Using this, we do a post-order search in the game tree. In each vertex, we note who has a winning strategy from that given state. For the leaves, we note who has won the game (done by checking if all vertices have been colored). After traversing the tree, we can read in the root if the painter or the drawer has a winning strategy. This shows that the problem is in PSPACE.

To prove that the problem is PSPACE-hard, we show how to transform a totally quantified boolean formula $\phi = \forall x_1 \exists x_2 \ldots \exists x_n \ F(x_1, \ldots, x_n)$ (F is in 3DNF) into a game state (G, k, G', f) such that ϕ is true if and only if the painter has a winning strategy from (G, k, G', f).

- The number of variables in ϕ is n.
- The number of terms in F is t.

- We define $k = t + 3n/2 + 2$ to be the number of colors that the painter is allowed to use.

We now describe G. It consists of subgraphs X, T, H and A, B, c and m. The relationship between them is sketched in Figure 1.

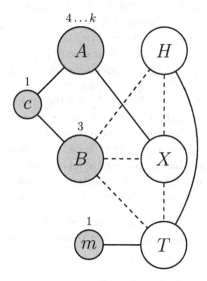

Fig. 1. A sketch of the construction. The small circles represent vertices and the big circles represent parts of the graph containing multiple vertices. The blue circles are subgraphs that have been pre-colored (their color is shown above them). Solid lines are complete connections. Dashed line are connections where not all vertices in both parts are connected. For details on how they are connected, see the description below.

The pre-coloring consists of A, B, c, and m.

- A is a complete graph with $k - 3$ vertices with colors $4 \ldots k$.
- B is an independent set with a large (but polynomial) number of vertices. They all have color 3.
- c is a single vertex with color 1.
- m is a single vertex with color 1.

The vertex c has an edge to each vertex in A and B.

The subgraph X consists of two vertices for each variable, x_i and $\bar{x}_i$, with an edge between them. Each vertex in X is connected to each vertex in A and at least one vertex in B. This means that the only possible colors for vertices in X are 1 and 2 (also called true and false respectively).

The subgraph T corresponds to the terms. It is a complete graph with one vertex, t_j for each term, j. Furthermore, a vertex in T has an edge to x_i if x_i

is in the corresponding term (and similarly one to $\bar{x}_i$ if that is found in the corresponding term).

Since T is a complete graph, each vertex there must be given a different color. However, if one vertex only has neighbors in X with color true, the painter can introduce only $t - 1$ new colors in T instead of t new colors (by reusing the color false in T). This corresponds to one term being satisfied by a truth assignment and it is key in the construction. However, we do not want a color to be saved if one vertex in T only has neighbors in X with color false. To prevent this, we add an edge between m (which has color true) and each vertex in T.

The last subgraph is H. Its purpose is to ensure that the painter requests the existantially quantified vertices in X in ascending order (as they appear in ϕ). It consists of $n/2$ copies of P_4. The copies are named $H_1, \ldots, H_{n/2}$. The vertices in H_i are called $h_i^1, \ldots, h_i^4$ such that h_i^1 and h_i^4 are the endpoints. There are edges between each vertex in H_i and x_{2i-1}. Furthermore, h_i^2 and h_i^4 have edges to x_{2l} for $l \geq i$. Also, each vertex in H has an edge to each vertex in T.

The purpose of B is to give the painter know some information about which vertices are being requested based on the number of edges it has to B. All vertices in the non-pre-colored part (X, T, and H) have at least one edge to a vertex in B. These edges are constructed such that the following holds:

Lemma 1. *When the drawer presents a vertex v, the painter is always able to identify an i and which one of the following statements about v holds*

- *v is x_i and i is even.*
- *v is $\bar{x}_i$ and i is even.*
- *v is either x_i or $\bar{x}_i$ and i is odd (the two cases cannot be distinguished).*
- *v is in H_i (the four cases cannot be distinguished).*
- *v is t_i.*

Clearly, it is possible to use the number of edges to B to encode this information. For the specific construction and the proof of this Lemma, we refer to [7].

Formally, we are only allowed to specify the pre-coloring as a graph (with a function, f, mapping the vertices to colors), but we are not allowed to specify where it is induced in G, it is up to the drawer to decide this. In our case, the pre-colored graph, G' is isomorphic to the graph consisting of A, B, c, and m (with the specified edges between them and the specified colors). In G', we call them A', B', c' and m'.

When inducing G' in G, it is only possible to map c' to the c (we can choose the number of vertices in B large enough that no other vertex in G has as high degree as c). Similarly, the vertices of A' can only be mapped to those of A and those of B' can only be mapped to B. This is because these are the only neighbors of c and they are easily distinguishable since A is a complete graph and B is an independent set. Finally, m' can only be mapped to m since it is the only vertex outside A that has no neighbor in B.

We are now ready for the main proof. We begin with the easier implication.

Lemma 2. *If ϕ is false, then the drawer has a winning strategy from the state (G, k, G', f).*

Proof. We will call color 1 *true* and the color 2 *false*. Since $\phi = \forall x_1 \exists x_2 \ldots \exists x_n$ $F(x_1, \ldots, x_n)$ is false, it holds that

$$\exists x_1 \forall x_2 \ldots \forall x_n \, \neg F(x_1, \ldots, x_n)$$

This means that if two players alternately decide the truth values of $x_1, x_2, \ldots x_n$, there is a strategy S for the player deciding the values of the odd variables which makes F false. The drawer is going to implement S.

The drawer will start by presenting the vertices in X and H. It will do this in rounds. In round i, $1 \leq i \leq n/2$, it first presents x_{2i-1} and $\bar{x}_{2i-1}$ in some order. It then presents the vertices of H_i in some order. Finally, it presents x_{2i} and $\bar{x}_{2i}$ in some order. There are $n/2$ rounds. We want to show that the drawer can ensure that the following holds after round i:

- Each H_j with $j \leq i$ has vertices with at least 3 different colors.
- All x_j and $\bar{x}_j$ with $j \leq i$ have been colored with colors true or false.
- Interpreting coloring as an assignment of truth values to variables $x_1, \ldots, x_i$, the drawer has a winning strategy in the game where the drawer and the painter alternately decide the truth value for the remaining variables.
- Either the colors true and false are not found in H_i or the painter has lost the game.

For $i = 0$, they all hold. Assume that they hold for some i. We show that the drawer can present the vertices in round $i + 1$ in an order that ensures that they hold after it.

The drawer starts by presenting x_{2i-1} and $\bar{x}_{2i-1}$. Among the vertices that have already been presented (including those in the pre-coloring), it holds that each vertex is either adjacent to both x_{2i-1} and $\bar{x}_{2i-1}$ or none of them (note that H_i has not been presented yet). This means that the painter is unable to identify which is which. Since they are both adjacent to all vertices in A and some in B, the only available colors for them are true and false. The painter has to assign true to one of them and false to the other. The drawer now decides which one received the color true according to his winning strategy (we know that he has one from the induction hypothesis). This ensures that he will have a winning strategy independent of whether variable x_{2i} is set to true or false.

Now, the drawer presents two non-adjacent vertices from H_i. The painter cannot identify which, since the vertices in H_i are connected to the same vertices among those that have been presented. If the painter gives these the same color, the drawer decides that they were h_i^1 and h_i^4. Otherwise, the drawer decides that they were h_i^1 and h_i^3. The drawer now presents the remaining two vertices of H_i which results in it containing at least three different colors. Note that the color 3 cannot be used in H_i, since all four vertices are adjacent to some vertices in B.

The drawer now presents x_{2i} and $\bar{x}_{2i}$. According to Lemma 1, the painter can identify which one is x_{2i} and which one is $\bar{x}_{2i}$. Again, the painter must color one true and the other false. This can be interpreted as the painter assigning a truth value to variable x_{2i}. As we argued earlier, the drawer must still have a winning strategy if he decides the truth value of the remaining odd variables.

We need to argue that if a vertex in H_i receives color true or false, the painter will immediately lose. We first consider the case where color true is found on h_i^2 or h_i^4. The painter will color x_{2i} and $\bar{x}_{2i}$. These are adjacent to all vertices in A (and some in B) meaning they can only get color true or false. Since they are adjacent to h_i^2 and h_i^4, they cannot get the color true. Only color false is not available, and after one gets that, the other cannot get any color. If the color true is found on h_i^1 or h_i^3 instead, the drawer changes the positions of h_i^1 and h_i^4 as well as those of h_i^2 and h_i^3. This is possible since it is not at this time possible for the painter to distinguish between h_i^1 and h_i^4 and between h_i^2 and h_i^3. This ensures that the color true does end up on h_i^2 or h_i^4 so we can argue in the same way. The argument is similar if it is the color false is found in H_i.

This concludes the induction. We have now shown that after round $n/2$ all vertices in X have been colored with colors true and false. The truth assignment given to the variables in X makes F false. The drawer now presents all vertices in T in any order. They cannot get the color true, since they are adjacent to m which has that color. Each vertex in T is adjacent to a vertex in X with color false since the truth assignment made F false. Furthermore, there are $3n/2$ colors that cannot be used on T since they were used in H. Also, the color 3 cannot be used, since all vertices in T are adjacent to some in B. This leaves $k - 2 - 3n/2 - 1 = t - 1$ colors. This is not enough to color the t vertices, since they form a clique.   □

We now show the other implication, which completes the proof.

Lemma 3. *If ϕ is true, then the painter has a winning strategy from the state (G, k, G', f).*

Proof. The painter has to color the part of G that is not in G' such that the resulting colored graph has at most k different colors. We notice that all remaining vertices have a least one neighbor in B which means that the color 3 is not available for any vertices. This means that there are $k - 1 = 3n/2 + t + 1$ colors left. Moreover, only the colors true and false are available for vertices in X. We have already defined colors 1 and 2 to be called true and false. We call the next $3n/2$ colors the *H-colors*. The $t - 1$ remaining colors, we call the *T-colors*. The idea is that the H-colors will be used in H, true and false will be used in X and the T colors will be used in T. Since there are only $t - 1$ T-colors, the painter will need to use true, false or an H-color on a vertex in T. This is only possible because ϕ is true.

Before defining the painter strategy, we need a few preliminaries. We start by defining *normal play*. In normal play, when a vertex x_i or $\bar{x}_i$ with i even is requested, the following must hold. In each H_j with $j \leq \frac{i}{2}$, both h_j^1 and h_j^4 have been requested. We also define a *good request*. A good request is a request to a vertex $t_i \in T$, where the following holds: For each neighbor $v \in X$ of t_i, v does not have color false and v's neighbor in X does not have color true.

For example, if t_1 was requested and x_1 was a neighbor, it would be a good request only if x_1 did not have the color false (possibly because it had not been presented yet) and $\bar{x}_1$ did not have the color true (it might also not have been presented yet). Note that when a vertex in T is requested, the painter can identify if it is a good request using Lemma 1 and the fact that it knows for each vertex in T which neighbors in X it has. We call it a good request because it results in the painter being able to use the color false on that vertex, which means that he will have enough colors and win the game.

Since ϕ is true, there must exist a function p, which based on the truth assignment to $x_1, \ldots, x_{i-1}$ computes if variable x_i (i is even) should be true or false if the painter wants to make F true. We define the function p', which computes if variable x_i should be given color true or false if not all variables $x_1, \ldots, x_{i-1}$ have had their truth assignment decided yet. For even i, we let $p'(x_i) = p(p'(x_1), \ldots, p'(x_{i-1}))$. For odd i, we define $p'(x_i) = \text{true}$ if x_i has the color true, if $\bar{x}_i$ has the color false or if none of them have been presented yet. We define $p'(x_i) = \text{false}$ otherwise. It useful to think of it the following way: If x_i is requested before x_j and $\bar{x}_j$, $j < i$, the painter will be able to distinguish between x_j and $\bar{x}_j$ when they get requested. Because of this, the painter just decides that x_j is true and it colors x_j and $\bar{x}_j$ accordingly when they get requested.

We now define a strategy for the painter. There are three phases. The painter starts in Phase 1. Certain events will cause the painter to enter Phase 2 or 3, which in both cases means that the painter from that point can follow a simple strategy to win. Table 1 defines how the painter handles a request to a vertex v when in Phase 1. Phase 2 and 3 will be defined subsequently.

Table 1. Table defining a painter strategy in Phase 1

Case	Subcase	Subsubcase	Color given
$v \in V(H_i)$			Color greedily with H-colors.
$v \in V(X)$	i even	Normal play	Use color $p'(x_i)$
		Not normal play	Color greedily with {true, false} and go to phase 2.
	i odd	No vertex in $H_{\frac{i+1}{2}}$ has been requested	Color greedily with {true, false}.
		At least one vertex in $H_{\frac{i+1}{2}}$ has been requested	The painter can identify if the request is to x_i or $\bar{x}_i$. Use color true for x_i and false for $\bar{x}_i$
$v \in V(T)$	Good request		Use color false and go to phase 3.
	Not good request		Color greedily with T-colors

We show, that under normal play, the drawer will have to eventually make a good request, which makes the painter enter Phase 3. First, we show that the truth assignment that $x_1, \ldots, x_n$ gets will make F true. When an x_i or $\bar{x}_i$ with even i is requested, the painter will color it based on the color of $x_1, \ldots, x_{i-1}$. However, since the drawer decides the order, it may happen that the truth values of these have not already been decided. For the variables with an even index, this is not a problem for the painter, since it can just compute recursively, which color it will apply to it. For a variable with an odd index x_j, we defined that the painter should consider it true (we set $p'(x_j) = \text{true}$ for odd j). This is possible since we are under normal play, which means that $h^1_{\frac{i+1}{2}}$ and $h^4_{\frac{i+1}{2}}$ have already been requested. When x_j and $\bar{x}_j$ are requested, the painter is able to use this to see which one it is. According to Table 1, the painter will give x_j color true and $\bar{x}_j$ color false which is exactly why it is possible for the painter to already consider x_j as true before it has been requested, when x_i is requested under normal play. Note that ϕ is true. Since the painter colors according to p, the resulting truth assignment makes at least one term true. This also gives, that at least one request to a vertex in T will be good (a request to $t_i \in T$ is not good if and only if term t_i cannot be satisfied by the current truth assignment no matter what truth value the undecided variables are given). We have now shown that the drawer must eventually make a good request under normal play. This shows that the game will either deviate from normal play at some point (making the painter enter Phase 2) or make a good request such that the painter enters Phase 3. We now define how the painter behaves in Phase 2 and Phase 3 and show why he will win in both cases.

At the beginning of Phase 2, the drawer has just deviated from normal play. He has presented x_i (or $\bar{x}_i$) with i even, even though there exists a H_j with $j \leq \frac{i}{2}$ where h^1_j and h^4_j have not both been requested. Note that H_j is bipartite (it is a P_4). Since H was colored greedily, and h^1_j and h^4_j have not both been presented, we know that at most one color is already used in each partition and no color is already used in both partitions. For future requests in H_j, the painter will know which partition the requested vertex is in, since h^2_j and h^4_j are connected to the vertex in X that was just requested. Thus, the painter will only have to use 2 colors for H_j. For the remaining requests, the painter colors greedily with H-colors in H. He colors greedily with {true,false} in X and he colors greedily with T-colors in T. When the final vertex in T is requested, there will not be a T-color available (since there are only $t-1$). However, the painter will have one H-color that is not needed (the color saved in H_j). He uses that as a T-color, which ensures that he wins.

At the beginning of phase 3, the painter has just assigned the color false to a vertex in T after a good request. Since the request was good, we know that all adjacent vertices in X have been or can be colored true. Their neighbors in X have been or can be colored false. The remaining vertices in X get colored greedily with {true,false}. The vertices in H will be colored greedily using H-colors. The remaining vertex in T will be colored greedily using T-colors which suffices. This ensures, that the painter wins.

We have now presented a strategy for the painter. We have shown that either Phase 2 or Phase 3 will always be entered and we have shown how the painter wins once such a Phase has been entered. $\qquad\square$

We can now combine Lemmas 2 and 3 and Observation 1 to get the desired theorem.

Theorem 1. *Given a state (G, k, G', f) in the on-line graph coloring game, it is* PSPACE-*complete to decide if the painter has a winning strategy.*

5   Closing Remarks

The complexity of the problem of deciding if $\chi^O(G) \leq k$ is still open. It was shown to be CONP-hard in [6] (unpublished work), and it is certainly in PSPACE using the argument presented here. Adding a pre-coloring ensures that the problem is PSPACE-complete. That result suggests, that it may be harder to do on-line competitive analysis than it is to do competitive analysis, since deciding if $\chi(G) \leq k$ is "only" NP-complete. Note, though, that it is only an indication since it might be possible to do the analysis without computing $\chi^O(G)$ and furthermore, it is not clear if the complexity is changed by the pre-coloring (as it is the case for some offline coloring problems, see [4] and [9]).

Our work with the problem has led to the following conjecture:

Conjecture 1. Let a graph G and a $k \in \mathbb{N}$ be given. The problem of deciding if $\chi^O(G) \leq k$ is PSPACE-complete.

It seems likely to us, that a reduction from totally quantified boolean formula in 3DNF is possible. It may be possible to use a similar construction to the one used here, but special attention has to be given to the case where ϕ is true. It is challenging to allow the painter to implement the winning strategy from the satisfiability game when the drawer is able to request any vertex in the graph without the painter knowing which vertex is being requested.

References

1. Bodlaender, H.L.: On the complexity of some coloring games. Internat. J. Found. Comput. Sci. **2**, 133–147 (1989)
2. Gyarfas, A., Lehel, J.: First fit and on-line chromatic number of families of graphs. Ars Combinatorica **29C**, 168–176 (1990)
3. Gyarfas, A., Kiraly, Z., Lehel, J.: On-line competitive coloring algorithms. Tech. rep., Mathematical Institute of Eotvos Lorand University (1997). http://www.cs.elte.hu/tr97
4. Kratochv, J.: Precoloring extension with fixed color bound. Acta Mathematica Universitatis Comenianae **62**(2), 139–153 (1993). http://eudml.org/doc/118661
5. Krawczyk, T., Walczak, B.: Coloring relatives of interval overlap graphs via on-line games. In: Esparza, J., Fraigniaud, P., Husfeldt, T., Koutsoupias, E. (eds.) ICALP 2014. LNCS, vol. 8572, pp. 738–750. Springer, Heidelberg (2014)
6. Kudahl, C.: On-line Graph Coloring. Master's thesis, University of Southern Denmark (2013)
7. Kudahl, C.: Deciding the on-line chromatic number of a graph with pre-coloring is pspace-complete. CoRR abs/1406.1623 (2014). http://arxiv.org/abs/1406.1623

8. Lehel, A.G.J., Kiraly, Z.: On-line graph coloring and finite basis problems. Combinatorics: Paul Erdos is Eighty, vol. 1, 207–214 (1993)

9. Marx, D.: Precoloring extension on unit interval graphs. Discrete Applied Mathematics **154**(6), 995–1002 (2006). http://www.sciencedirect.com/science/article/pii/S0166218X05003392

10. Sleator, D.D., Tarjan, R.E.: Amortized efficiency of list update and paging rules. Commun. ACM **28**(2), 202–208 (1985)

11. Stockmeyer, L.J.: The polynomial-time hierarchy. Theoretical Computer Science **3**(1), 1–22 (1976)

12. Tuza, Z.: Graph colorings with local constraints - a survey. Math. Graph Theory **17**, 161–228 (1997)

A Lex-BFS-Based Recognition Algorithm
for Robinsonian Matrices

Monique Laurent[1,2] and Matteo Seminaroti[1] ([✉])

[1] Centrum Wiskunde and Informatica (CWI), Amsterdam, The Netherlands
{M.Laurent,M.Seminaroti}@cwi.nl
[2] Tilburg University, Tilburg, The Netherlands

Abstract. Robinsonian matrices arise in the classical seriation problem
and play an important role in many applications where unsorted similarity (or dissimilarity) information must be reordered. We present a
new polynomial time algorithm to recognize Robinsonian matrices based
on a new characterization of Robinsonian matrices in terms of straight
enumerations of unit interval graphs. The algorithm is simple and is
based essentially on lexicographic breadth-first search (Lex-BFS), using
a divide-and-conquer strategy. When applied to a nonnegative symmetric
$n \times n$ matrix with m nonzero entries and given as a weighted adjacency
list, it runs in $O(d(n + m))$ time, where d is the depth of the recursion
tree, which is at most the number of distinct nonzero entries of A.

Keywords: Robinson (dis)similarity · Unit interval graph · Lex-BFS ·
Seriation · Partition refinement · Straight enumeration

1 Introduction

An important question in many classification problems is to find an order of
a collection of objects respecting some given information about their pairwise
(dis)similarities. The classic seriation problem, introduced by Robinson [25] for
chronological dating, asks to order objects in such a way that similar objects are
ordered close to each other, and it has applications in different fields (see [18]).

A symmetric matrix $A = (A_{ij})_{i,j=1}^{n}$ is a *Robinson similarity* matrix if its
entries decrease monotonically in the rows and columns when moving away from
the main diagonal, i.e., if $A_{ik} \leq \min\{A_{ij}, A_{jk}\}$ for all $1 \leq i \leq j \leq k \leq n$. Given
a set of n objects to order and a symmetric matrix $A = (A_{ij})$ which represents
their pairwise correlations, the seriation problem asks to find (if it exists) a permutation π of $[n]$ so that the permuted matrix $A_\pi = (A_{\pi(i)\pi(j)})$ is a Robinson
matrix. If such a permutation exists then A is said to be a *Robinsonian similarity*, otherwise we say that data is affected by noise. The definitions extend to
dissimilarity matrices: A is a Robinson(ian) dissimilarity precisely when $-A$ is
a Robinson(ian) similarity. Hence, results can be directly transferred from one
class to the other. Robinsonian matrices play an important role in several hard
combinatorial optimization problems and recognition algorithms are important

© Springer International Publishing Switzerland 2015
V.Th. Paschos and P. Widmayer (Eds.): CIAC 2015, LNCS 9079, pp. 325–338, 2015.
DOI: 10.1007/978-3-319-18173-8_24

in designing heuristic and approximation algorithms when the Robinsonian property is desired but the data is affected by noise (see e.g. [5],[12],[17]). In the last decades, different characterizations of Robinsonian matrices have appeared in the literature, leading to different polynomial time recognition algorithms. Most characterizations are in terms of interval (hyper)graphs.

A graph $G = (V, E)$ is an *interval graph* if its nodes can be labeled by intervals of the real line so that adjacent nodes correspond to intersecting intervals. Interval graphs arise frequently in applications and have been studied extensively in relation to hard optimization problems (see e.g. [2],[6],[20]). A binary matrix has the *consecutive ones property (C1P)* if its columns can be reordered in such a way that the ones are consecutive in each row. Then, a graph G is an interval graph if and only if its vertex-clique incidence matrix has C1P, where the rows are indexed by the vertices and the columns by the maximal cliques of G [13].

A hypergraph $H = (V, \mathcal{E})$ is a generalization of the notion of graph where elements of $\mathcal{E}$, called *hyperedges*, are subsets of V. The incidence matrix of H is the 0/1 matrix whose rows and columns are labeled, respectively, by the hyperedges and the vertices and with entry 1 when the corresponding hyperedge contains the corresponding vertex. Then, H is an *interval hypergraph* if its incidence matrix has C1P, i.e., its vertices can be ordered so that hyperedges are intervals.

Given a dissimilarity matrix $A \in \mathcal{S}^n$ and a scalar α, the *threshold graph* $G_\alpha = (V, E_\alpha)$ has edge set $E_\alpha = \{\{x, y\} : A_{xy} \leq \alpha\}$ and, for $x \in V$, the ball $B(x, \alpha) := \{y \in V : A_{xy} \leq \alpha\}$ consists of x and its neighbors in G_α. Let $\mathcal{B}$ denote the collection of all the balls of A and $H_\mathcal{B}$ denote the corresponding *ball hypergraph*, with vertex set $V = [n]$ and with $\mathcal{B}$ as set of hyperedges. One can also build the intersection graph $G_\mathcal{B}$ of $\mathcal{B}$, where the balls are the vertices and connecting two vertices if the corresponding balls intersect. Most of the existing algorithms are then based on the fact that a matrix A is Robinsonian if and only if the ball hypergraph $H_\mathcal{B}$ is an interval hypergraph or, equivalently, if the intersection graph $G_\mathcal{B}$ is an interval graph (see [21,22]).

Mirkin and Rodin [21] gave the first polynomial algorithm to recognize Robinsonian matrices, with $O(n^4)$ running time, based on checking whether the ball hypergraph is an interval hypergraph and using the PQ-tree algorithm of Booth and Leuker [3] to check whether the incidence matrix has C1P. Later, Chepoi and Fichet [4] introduced a simpler algorithm that, using a divide-an-conquer strategy and sorting the entries of A, improved the running time to $O(n^3)$. The same sorting preprocessing was used by Seston [27], who improved the algorithm to $O(n^2 \log n)$ by constructing paths in the threshold graphs of A. Very recently, Préa and Fortin [22] presented a more sophisticated $O(n^2)$ algorithm, which uses the fact that the maximal cliques of the graph $G_\mathcal{B}$ are in one-to-one correspondence with the row/column indices of A. Roughly speaking, they use the algorithm from Booth and Leuker [3] to compute a first PQ-tree which they update throughout the algorithm. A numerical spectral algorithm was introduced earlier by Atkins et al. [1] for checking whether a similarity matrix A is Robinsonian, based on reordering the entries of the Fiedler eigenvector of the Laplacian matrix associated to A.

In this paper we introduce a new combinatorial algorithm to recognize Robinsonian matrices. Our approach differs from the existing ones in the sense that it is not directly related to interval (hyper)graphs, but it is based on a new characterization of Robinsonian matrices in terms of *straight enumerations* of unit interval graphs (Section 3). Unit interval graphs are a subclass of interval graphs, where the intervals labeling the vertices are required to have unit length. Several linear time recognition algorithms exist, based in particular on characterizations of unit interval graphs in terms of straight enumerations, that are special orderings of the vertices [7,8].

Our algorithm does not rely on any sophisticated external algorithm such as the Booth and Leuker algorithm for C1P and no preprocessing to order the data is needed. The most difficult task carried out is instead a lexicographic breadth-first search (abbreviated Lex-BFS), which is a variant of the classic breadth-first search (BFS), where the ties in the search are broken by giving preference to those vertices whose neighbors have been visited earliest (see [14, 26]). Following [7], we in fact use the variant Lex-BFS+ introduced by [28] to compute straight enumerations. Our algorithm uses a divide-and-conquer strategy with a merging step, tailored to efficiently exploit the possible sparsity structure of the given similarity matrix A. Assuming the matrix A is given as an adjacency list of an undirected weighted graph, our algorithm runs in $O(d(m + n))$ time, where n is the size of A, m is the number of nonzero entries of A and d is the depth of the recursion tree computed by the algorithm, which is upper bounded by the number L of distinct nonzero entries of A (see Theorem 6). Furthermore, we can return all the permutations reordering A as a Robinson matrix using a PQ-tree data structure on which we perform only a few simple operations (see Section 4.2).

Our algorithm uncovers an interesting link between straight enumerations of unit interval graphs and Robinsonian matrices which, to the best of our knowledge, has not been made before. Moreover it provides an answer to an open question posed by M. Habib at the PRIMA Conference in Shanghai in June 2013, who asked whether it is possible to use Lex-BFS+ to recognize Robinsonian matrices [9]. Alternatively, one could check whether the incidence matrix M of the ball hypergraph of A has C1P, using the Lex-BFS based algorithm of [14], in time $O(r + c + f)$ time if M is $r \times c$ with f ones. As $r \leq nL$, $c = n$ and $f \leq Lm$, the overall time complexity is $O(L(n+m))$. Interestingly, this approach is not mentioned by Habib. In comparison, an advantage of our approach is that it exploits the sparsity structure of the matrix A, as d can be smaller than L.

Contents of the Paper Section 2 contains preliminaries about weak linear orders, straight enumerations and unit interval graphs. In Section 3 we characterize Robinsonian matrices in terms of straight enumerations of unit interval graphs. In Section 4 we introduce our recursive algorithm to recognize Robinsonian matrices, and then we explain how to return all the permutations reordering a given similarity matrix as a Robinson matrix. The final Section 5 contains some questions for possible future work.

2   Preliminaries

Throughout $\mathcal{S}^n$ denotes the set of symmetric $n \times n$ matrices. Given a permutation π of $[n]$ and a matrix $A \in \mathcal{S}^n$, $A_\pi := (A_{\pi(i)\pi(j)})_{i,j=1}^n \in \mathcal{S}^n$ is the matrix obtained by permuting both the rows and columns of A simultaneously according to π. For $U \subseteq [n]$, $A[U] = (A_{ij})_{i,j \in U}$ is the principal submatrix of A indexed by U. As we deal exclusively with Robinson(ian) similarities, when speaking of a Robinson(ian) matrix, we mean a Robinson(ian) similarity matrix.

An ordered partition $(B_1, \ldots, B_p)$ of a finite set V corresponds to a *weak linear order* ψ on V (and vice versa), by setting $x =_\psi y$ if x, y belong to the same class B_i, and $x <_\psi y$ if $x \in B_i$ and $y \in B_j$ with $i < j$. Then we also use the notation $\psi = (B_1, \ldots, B_p)$ and $B_1 <_\psi \ldots <_\psi B_p$. When all classes B_i are singletons then ψ is a linear order (i.e., total order) of V.

The *reversal* of ψ is the weak linear order, denoted $\overleftarrow{\psi}$, of the reversed ordered partition $(B_p, \ldots, B_1)$. For $U \subseteq V$, $\psi[U]$ denotes the *restriction* of the weak linear order ψ to U. Given disjoint subsets $U, W \subseteq V$, we say $U \leq_\psi W$ if $x \leq_\psi y$ for all $x \in U$, $y \in W$. If ψ_1 and ψ_2 are weak linear orders on disjoint sets V_1 and V_2, then $\psi = (\psi_1, \psi_2)$ denotes their *concatenation* which is a weak linear order on $V_1 \cup V_2$.

The following notions of compatibility and refinement will play an important role in our treatment. Two weak linear orders ψ_1 and ψ_2 on the same set V are said to be *compatible* if there do not exist elements $x, y \in V$ such that $x <_{\psi_1} y$ and $y <_{\psi_2} x$. Then their *common refinement* is the weak linear order $\Psi = \psi_1 \wedge \psi_2$ on V defined by $x =_\Psi y$ if $x =_{\psi_\ell} y$ for all $\ell \in \{1, 2\}$, and $x <_\Psi y$ if $x \leq_{\psi_\ell} y$ for all $\ell \in \{1, 2\}$ with at least one strict inequality.

In what follows $V = [n] = \{1, \ldots, n\}$ is the vertex set of a graph $G = (V, E)$, whose edges are pairs $\{x, y\}$ of distinct vertices $x, y \in V$. For $x \in V$, its *closed neighborhood* is the set $N[x] = \{x\} \cup \{y \in V : \{x, y\} \in E\}$. Two vertices $x, y \in V$ are *undistinguishable* if $N[x] = N[y]$. This defines an equivalence relation on V, whose classes are called the *blocks* of G. Clearly, each block is a clique of G. Two distinct blocks B and B' are said to be *adjacent* if there exist two vertices $x \in B$, $y \in B'$ that are adjacent in G or, equivalently, if $B \cup B'$ is a clique of G. A *straight enumeration* of G is a linear order $\phi = (B_1, \ldots, B_p)$ of the blocks of G such that, for any block B_i, the block B_i and the blocks B_j adjacent to it are consecutive in the linear order (see [16]). The blocks B_1 and B_p are called the *end blocks* of ϕ and B_i (with $1 < i < p$) are its *inner blocks*.

The following characterization of unit interval graphs in terms of straight enumerations will play a central role in our paper.

Theorem 1 (Unit interval graphs and straight enumerations). *[10] A graph G is a unit interval graph if and only if it has a straight enumeration. Moreover, if G is connected, then it has a unique (up to reversal) straight enumeration.*

On the other hand, if G is not connected, then any possible linear ordering of the connected components combined with any possible orientation of the straight enumeration of each connected component induces a straight enumeration of G. Several alternative characterizations for unit interval graphs are known (see [7] and references therein), including the following ones.

Theorem 2. *A graph $G = (V, E)$ is a unit interval graph if and only if it satisfies any of the following equivalent conditions:*

(i) **(3-vertex condition)** *[19] There is a linear ordering π of V such that, for all $x, y, z \in V$, $x <_\pi y <_\pi z$ and $\{x, z\} \in E$ implies $\{x, y\}, \{y, z\} \in E$.*
(ii) **(Neighborhood condition)** *[23] There is a linear ordering π of V such that for any $x \in V$ the vertices in $N[x]$ are consecutive with respect to π.*

3   Robinsonian Matrices and Unit Interval Graphs

In this section we characterize Robinsonian matrices in terms of straight enumerations of unit interval graphs. We may view any symmetric binary matrix with all diagonal entries equal to 1 as the *extended* adjacency matrix of a graph. The equivalence between binary Robinsonian matrices and indifference graphs (and thus with unit interval graphs) was first shown by Roberts [23]. Furthermore, as observed, e.g., by Corneil et al. [8], the "neighborhood condition" for a graph is equivalent to its extended adjacency matrix having C1P. Hence we have the following equivalence between Robinsonian binary matrices and unit interval graphs, which also follows as a direct application of Theorem 2(ii).

Lemma 1. *Let $G = (V, E)$ be a graph and A_G be its extended adjacency matrix. Then, A_G is a Robinsonian similarity if and only if G is a unit interval graph.*

The next result characterizes the linear orders that reorder the extended adjacency matrix A_G as a Robinson matrix in terms of the straight enumerations of G. It is simple but will play a central role in our algorithm for recognizing Robinsonian similarities.

Theorem 3. *Let $G = (V, E)$ be a graph. A linear order π of V reorders A_G as a Robinson matrix if and only if there exists a straight enumeration of G whose corresponding weak linear order ψ is compatible with π, i.e., satisfies:*

$$\forall x, y \in V \quad \text{with} \quad x \neq_\psi y \qquad x <_\pi y \iff x <_\psi y. \tag{1}$$

Hence, in order to find the permutations reordering a given binary matrix A as a Robinson matrix, it suffices to find all the possible straight enumerations of the corresponding graph G. As is shown e.g. in [8,10], this is a simple task and can be done in linear time. This is coherent with the fact that C1P can be checked in linear time (see [11] and references therein).

We now consider a general (nonbinary) matrix A. We first introduce its 'level graphs', the analogues for similarity matrices of the threshold graphs for dissimilarities. Let $\alpha_0 < \alpha_1 < \cdots < \alpha_L$ denote the distinct values taken by the entries of A. The graph $G^{(\ell)} = (V, E_\ell)$, whose edges are the pairs $\{x, y\}$

with $A_{xy} \geq \alpha_\ell$, is called the ℓ-th *level graph* of A. Let J be the all ones matrix. Clearly, $\pm J$ is a Robinson matrix. Hence, we may and will assume, without loss of generality, that $\alpha_0 = 0$. Then, A is nonnegative and $G^{(1)}$ is its support graph.

As already observed by Roberts [24], Robinson matrices can be decomposed as conic combinations of binary Robinson matrices (up to a translation by the all-ones matrix). We omit the proof, which is easy.

Lemma 2. *Let $A \in \mathcal{S}^n$ with distinct values $\alpha_0 < \alpha_1 < \cdots < \alpha_L$ and with level graphs $G^{(1)}, \ldots, G^{(L)}$. Then, $A = \alpha_0 J + \sum_{\ell=1}^{L} (\alpha_\ell - \alpha_{\ell-1}) A_{G^{(\ell)}}$.*

Combining the links between binary Robinsonian matrices and unit interval graphs (Lemma 1) and between reorderings of binary Robinsonian matrices and straight enumerations of unit interval graphs (Theorem 3) together with the decomposition result of Lemma 2, we obtain the following characterization of Robinsonian matrices.

Theorem 4. *Let $A \in \mathcal{S}^n$ with level graphs $G^{(1)}, \ldots, G^{(L)}$. Then:*

(i) *A is a Robinsonian matrix if and only if there exist straight enumerations of $G^{(1)}$, ..., $G^{(L)}$ whose corresponding weak linear orders $\psi_1, \ldots, \psi_L$ are pairwise compatible.*
(ii) *A linear order π of V reorders A as a Robinson matrix if and only if there exist straight enumerations of $G^{(1)}, \ldots, G^{(L)}$, whose corresponding common refinement is compatible with π.*

4   The Algorithm

We describe here our algorithm for recognizing whether a given symmetric nonnegative matrix A is Robinsonian. First, we introduce an algorithm which either returns a permutation reordering A as a Robinson matrix or states that A is not a Robinsonian matrix. Then, we show how to modify it in order to return all the permutations reordering A as a Robinson matrix.

4.1   Overview of the Algorithm

The algorithm is based on Theorem 4. The main idea is to find straight enumerations of the level graphs of A that are pairwise compatible and to compute their common refinement. The matrix A is not Robinsonian precisely when these objects cannot be found. One of the main tasks in the algorithm is to find (if it exists) a straight enumeration of a graph G which is compatible with a given weak linear order ψ of V. Roughly speaking, G will correspond to a level graph $G^{(\ell)}$ of A (in fact, to a connected component of it), while ψ will correspond to the common refinement of the previous level graphs $G^{(1)}, \ldots, G^{(\ell-1)}$. Hence, looking for a straight enumeration of G compatible with ψ will correspond to looking for a straight enumeration of $G^{(\ell)}$ compatible with previously selected straight enumerations of the previous level graphs $G^{(1)}, \ldots, G^{(\ell-1)}$.

Our algorithm consists of three main subroutines: *CO-Lex-BFS* (Algorithm 1), a variation of Lex-BFS, which finds and orders the connected components of the level graphs; *Straight_enumeration*, which computes the straight enumeration of a connected graph as in [7]; *Refine* (Algorithm 2), a variation of partition refinement, which finds the common refinement of two weak linear orders. These subroutines are used in the recursive algorithm *Robinson* (Algorithm 3).

Component Ordering Our first subroutine is *CO-Lex-BFS* (where CO stands for 'Component Ordering') in Algorithm 1. Given a graph $G = (V, E)$ and a weak linear order ψ on V, it detects the connected components of G and orders them in a compatible way with respect to ψ (one can show that this is possible if G admits a straight enumeration compatible with ψ).

Straight Enumerations Once the connected components of G are ordered, we need to compute a straight enumeration of each connected component $G[V_\omega]$. We do this with the routine *Straight_enumeration* appplied to $(G[V_\omega], \sigma_\omega)$, where $\sigma_\omega = \sigma[V_\omega]$ and σ is the vertex order returned by *CO-Lex-BFS*(G, ψ, τ).

This routine is essentially the 3-sweep unit interval graph recognition algorithm of Corneil [7] which, briefly, computes three times a Lex-BFS (each is named a *sweep*) and use the vertex ordering coming from the previous sweep to break ties in the search for the next sweep. The only difference with respect to Corneil's algorithm is that we save the first sweep, because we use the order σ_ω given by *CO-Lex-BFS*.

Since the straight enumerations of the level graphs might not be unique, it is important to choose, among all the possible straight enumerations, the ones that lead to a common refinement (if it exists). If G is connected, its straight enumeration ϕ is unique up to reversal and one can show that the 3-sweep Lex-BFS algorithm implicitly returns it correctly oriented with respect to ψ. On the other hand, if G is not connected then any possible ordering of the connected components induces a straight enumeration, obtained by concatenating straight enumerations of its connected components. This freedom in choosing the straight enumerations of the components is crucial in order to return *all* the Robinson orderings of A (see Section 4.2). However, for now we are interested in finding *one* common refinement, and the arbitrary choice made does not affect the correctness of the algorithm.

Refinement of Weak Linear Orders Given two weak linear orders ψ and ϕ on V, our second subroutine *Refine* in Algorithm 2 computes their common refinement $\Phi = \psi \wedge \phi$ (if it exists).

Main Algorithm We can now describe our main algorithm *Robinson*(A, ψ, τ). Given a nonnegative matrix $A \in \mathcal{S}^n$, a weak linear order ψ and an order τ of $V = [n]$ that are compatible, it either returns a weak linear order Φ of V compatible with ψ and with straight enumerations of the level graphs of A, or

---

Algorithm 1. $CO\text{-}Lex\text{-}BFS(G, \psi, \tau)$

---

input: a graph $G = (V, E)$, a weak linear order $\psi = (B_1, \ldots, B_p)$ of V and a
      linear order τ of V compatible with ψ

output: a linear order σ of V and a linear order $(V_1, \ldots, V_c)$ of the connected
        components of G compatible with ψ and σ, or STOP (no such linear
        order of the components exists)

1 mark all the vertices as unvisited
2 u is the first vertex appearing in τ
3 $label(u) = |V|$
4 $\omega = 1$
5 $V_\omega, B_\omega^{\min}, B_\omega^{\max} = \emptyset$
6 **foreach** $v \in V \setminus u$ **do**
7      $label(v) = \emptyset$

8 **for** $i = |V|, \ldots, 1$ **do**
9   let S be the set of unvisited vertices with the lexicographically largest label
10   pick the vertex v in S appearing first in τ and mark it as visited
11    $\sigma(v) = |V| + 1 - i$
12   **if** $label(v) = \emptyset$ **then**
13   **if** $V_\omega \subseteq B_{\omega-1}^{\min}$ **then**
14   swap V_ω and $V_{\omega-1}$ and modify σ accordingly
15   **else**
16   **if** $B_\omega^{\min} <_\psi B_{\omega-1}^{\max}$ *or if there exists a block* B *of* ψ *such that* $B \nsubseteq V_\omega$
                and $B_\omega^{\min} <_\psi B <_\psi B_\omega^{\max}$ **then**
17   **stop** (no ordering of the components compatible with ψ exists)

18        $\omega = \omega + 1$
19        $V_\omega = \emptyset$
20    $V_\omega = V_\omega \cup \{v\}$
21    $B_\omega^{\min}$ is the first block in ψ which meets V_ω
22    $B_\omega^{\max}$ is the last block in ψ which meets V_ω
23   **foreach** *unvisited vertex* w *in* $N(v)$ **do**
24   append i to $label(w)$

25 **return** $(V_1, \ldots, V_c)$ and σ, or STOP

---

it indicates that such Φ does not exist. The idea behind our algorithm is to use the subroutines *CO-Lex-BFS* and *Straight_enumeration* to order the components and compute the straight enumerations of the level graphs of A, and to refine them using the subroutine *Refine*.

However, instead of refining the level graphs one by one on the full set V, we use a recursive algorithm based on a divide-and-conquer strategy, which refines smaller and smaller subgraphs of the level graphs obtained by restricting to the connected components and thus working independently with the corresponding principal submatrices of A. In this way we work with smaller subproblems and one may also skip some level graphs (as some principal submatrices of A may have fewer distinct nonzero entries). This recursive algorithm is Algorithm 3.

Algorithm 2. $Refine(\psi, \phi, \tau)$

input: two weak linear orders $\psi = (B_1, \ldots, B_p)$ and $\phi = (C_1, \ldots, C_q)$ of V, and a linear order τ of V compatible with ψ

output: their common refinement $\Phi = \psi \wedge \phi$, or STOP (ψ and ϕ are not compatible)

1 $B^{\max}$ is the last block of ψ meeting C_1
2 **if** *there exists a block B of ψ such that $B <_\psi B^{\max}$ and $B \not\subseteq C_1$* **then**
3 | **stop**   (ψ and ϕ are not compatible)
4 **else**
5 | $W = V \setminus C_1$
6 | $\Phi = (\psi[C_1], Refine(\psi[W], \phi[W], \tau[W]))$
7 **return** Φ or STOP

Algorithm 3. $Robinson(A, \psi, \tau)$

input: a weak linear order ψ of $V = [n]$, a linear order τ of V compatible with ψ and a nonnegative matrix $A \in \mathcal{S}^n$

output: a weak linear order Φ compatible with ψ and with straight enumerations of all the level graphs of A, or STOP (such an order Φ does not exist)

1 G is the support of A
2 $CO\text{-}Lex\text{-}BFS(G, \psi, \tau)$ returns a linear order $(V_1, \ldots, V_c)$ of the connected components of G compatible with ψ (if it exists) and a vertex order σ
3 $\Phi = \emptyset$
4 **for** $\omega = 1, \ldots, c$ **do**
5 | $a_{\min}$ is the smallest entry of $A[V_\omega]$
6 | **if** $a_{\min} > 0$ **then**
7 | | $A[V_\omega] := A[V_\omega] - a_{\min}J$ and $G[V_\omega]$ is its updated support
8 | $\phi_\omega = Straight\_enumeration(G[V_\omega], \sigma[V_\omega])$ (if $G[V_\omega]$ is a unit interval graph)
9 | $\Phi_\omega = Refine(\psi[V_\omega], \phi_\omega)$ (if $\psi[V_\omega]$ and ϕ_ω are compatible)
10 | $a'_{\min}$ is the smallest nonzero entry of $A[V_\omega]$
11 | $A'[V_\omega]$ is obtained from $A[V_\omega]$ by setting its entries with value $a'_{\min}$ to zero
12 | **if** $A'[V_\omega]$ *is diagonal* **then**
13 | | $\Phi = (\Phi, \Phi_\omega)$
14 | **else**
15 | | τ_ω is a linear order of V_ω compatible with Φ_ω
16 | | $\Phi = (\Phi, Robinson(A'[V_\omega], \Phi_\omega, \tau_\omega))$
17 **return:** Φ or STOP

The final algorithm is Algorithm 4. Roughly speaking, every time we make a recursive call, we are basically passing to the next level graph of A. Hence, each recursive call can be visualized as the node of a recursion tree, whose root is defined by the first recursion in Algorithm 4, and whose leaves (i.e. the pruned nodes) are the subproblems whose corresponding submatrices are diagonal.

Algorithm 4. *Robinsonian*(A)

---

input: a nonnegative matrix $A \in \mathcal{S}^n$
output: a permutation π such that A_π is Robinson or stating that A is not
        Robinsonian

1 $\psi = (V)$
2 $\tau = (1, 2, \ldots, n)$
3 $\Phi = Robinson(A, \psi, \tau)$
4 π is a linear order of V compatible with Φ
5 **return:** π or "A not Robinsonian"

Correctness and Complexity The correctness of Algorithm 4 follows directly from the correctness of Algorithm 3, which is shown by the next theorem.

Theorem 5. *Consider a weak linear order ψ of $V = [n]$ and a nonnegative matrix $A \in \mathcal{S}^n$ ordered compatibly with ψ.*

(i) If Algorithm 3 terminates, then there exist straight enumerations $\phi^{(1)}, \ldots, \phi^{(L)}$ of the level graphs $G^{(1)}, \ldots, G^{(L)}$ of A such that the returned weak linear order Φ is compatible with each of them and with ψ.

(ii) If Algorithm 3 stops then there do not exist straight enumerations of the level graphs of A that are pairwise compatible and compatible with ψ.

For the complexity analysis, we assume that $A \in \mathcal{S}^n$ is nonnegative and is given as weighted adjacency list. One can show that the three subroutines run in linear time in the size of the input. Hence, we have the following result.

Theorem 6. *Let $A \in \mathcal{S}^n$ be a nonnegative matrix and let m be the number of (upper diagonal) nonzero entries of A. Algorithm 4 recognizes whether A is a Robinsonian matrix in time $O(d(m + n))$, where d is the depth of the recursion tree created by Algorithm 4. Moreover, $d \leq L$, where L is the number of distinct nonzero entries of A.*

4.2 Finding All Robinsonian Orderings

In general, there might exist several permutations reordering a given matrix A as a Robinson matrix. We show here how to return all Robinson orderings of a given matrix A, using the PQ-tree data structure of [3].

A PQ-tree $\mathcal{T}$ is a special rooted ordered tree. The leaves are in one-to-one correspondence with the elements of the groundset V and their order gives a linear order of V. The nodes of $\mathcal{T}$ can be of two types, depending on how their children can be ordered. Namely, for a **P-node** (represented by a circle), its children may be arbitrary reordered; for a **Q-node** (represented by a rectangle), only the order of its children may be reversed. Moreover, every node has at least two children. Given a node α of $\mathcal{T}$, $\mathcal{T}_\alpha$ denotes the subtree of $\mathcal{T}$ with root α.

A straight enumeration $\psi = (B_1, \ldots, B_p)$ of a graph $G = (V, E)$ corresponds in a unique way to a PQ-tree $\mathcal{T}$ as follows. If G is connected, then the root of $\mathcal{T}$

is a Q-node, denoted γ, and it has children $\beta_1, \ldots, \beta_p$ (in that order). For $i \in [p]$, the node β_i is a P-node corresponding to the block B_i and its children are the elements of the set B_i, which are the leaves of the subtree T_{β_i}. If a block B_i is a singleton then no node β_i appears and the element of B_i is directly a child of the root γ. If G is not connected, let $V_1, \ldots, V_c$ be its connected components. For each connected component $G[V_\omega]$, T_ω is its PQ-tree (with root γ_ω) as indicated above. Then, the full PQ-tree T is obtained by inserting a P-node α as ancestor, whose children are the subtrees $T_1, \ldots, T_c$.

We now indicate how to modify Algorithms 3 and 4 in order to return a PQ-tree T encoding all the permutations ordering A as a Robinson matrix. We modify Algorithm 3 by taking as input, beside the matrix A, the weak linear order ψ and the linear order τ compatible with ψ, also a node α. Then, the output is a PQ-tree T_α rooted in α, representing all the possible weak linear orders Φ compatible with ψ and with straight enumerations of all the level graphs of A.

It works as follows. Let G be the support of A. The idea is to recursively build a tree T_ω for each connected component V_ω of G and then to merge these trees according to the order of the components found by the routine *CO-Lex-BFS*(G, ψ, τ). To carry out this merging step we classify the components into the following three groups:

1. Θ, which consists of all $\omega \in [c]$ for which the connected component V_ω meets at least two blocks of ψ.
2. Λ, which consists of all $\omega \in [c]$ for which the component V_ω is contained in some block B_i, which contains no other component.
3. $\Omega = \cup_{i=1}^{p} \Omega_i$, where Ω_i consists of all $\omega \in [c]$ for which the component V_ω is contained in the block B_i, which contains at least two components.

Every time we analyze a new connected component $\omega \in [c]$ in Algorithm 3, we create a Q-node γ_ω. After the common refinement Φ_ω (of $\psi[V_\omega]$ and the straight enumeration ϕ_ω of $G[V_\omega]$) have been computed, we have two possibilities. If $A'[V_\omega]$ is diagonal, then we build the tree T_ω rooted in γ_ω and whose children are P-nodes corresponding to the blocks of Φ_ω (and prune the recursion tree at this node). Otherwise, we build the tree T_ω recursively as output of *Robinson*$(A'[V_\omega], \Phi_\omega, \tau_\omega, \gamma_\omega)$.

After all the connected components have been analyzed, we insert the trees T_ω in the final tree T_α in the order they appear according to the routine *CO-Lex-BFS*(G, ψ, τ). The root node is α and is given as input. For each component V_ω, we do the following operation to insert T_ω in T_α, depending on the type of the component V_ω:

1. If $\omega \in \Theta$, then ϕ_ω is the only straight enumeration compatible with $\psi[V_\omega]$. Then we delete the node γ_ω and the children of γ_ω become children of α (in the same order).
2. If $\omega \in \Lambda$, then both ϕ_ω and its reversal $\overline{\phi}_\omega$ are compatible with $\psi[V_\omega]$. Then γ_ω becomes a child of α.
3. If $\omega \in \Omega_i$ for some $i \in [p]$, then both ϕ_ω and $\overline{\phi}_\omega$ are compatible with $\psi[V_\omega]$ and the same holds for any $\omega' \in \Omega_i$. Moreover, arbitrary permuting any

two connected components $V_\omega, V_{\omega'}$ with $\omega, \omega' \in \Omega_i$ will lead to a compatible straight enumeration. Then we insert a new node β_i which is a P-node and becomes a child of α and, for each $\omega' \in \Omega_i$, $\gamma_{\omega'}$ becomes a child of β_i.

Finally, we modify Algorithm 4 by just giving the node $\alpha = \emptyset$ (i.e. undefined) as input to the first recursive call. The overall complexity of the algorithm after the above mentioned modifications is the same as for Algorithm 4. Indeed, determining the type of the connected components can be done in linear time, by just using the information about the initial and final blocks $B_\omega^{\min}$ and $B_\omega^{\max}$ already provided in Algorithm 1. Furthermore, the operations on the PQ-tree are basic operations that do not increase the overall complexity of the algorithm.

5   Conclusions

We introduced a new combinatorial algorithm to recognize Robinsonian matrices, based on a divide-and-conquer strategy and on a new characterization of Robinsonian matrices in terms of straight enumerations of unit interval graphs. The algorithm is simple, rather intuitive and relies only on basic routines like Lex-BFS and partition refinement, and it is well suited for sparse matrices.

The complexity depends on the depth d of the recursion tree. An obvious bound on d is the number L of distinct entries in the matrix. A first natural question is to find other better bounds on the depth d. Is d in the order $O(n)$, where n is the size of the matrix? A possible way to bound the depth is to find criteria to prune recursion nodes. One possibility would be, when a submatrix is found for which the current weak linear order consists only of singletons, to check whether the corresponding permuted matrix is Robinson. Analyzing the complexity implications will be the subject of future work.

Another possible way to improve the complexity might be to compute the straight enumeration of the first level graph and then update it dynamically (in constant time, using a appropriate data structure) without having to compute every time the whole straight enumeration of the next level graphs; this would need to extend the dynamic approach of [16], which considers the case of single edge deletions, to the deletion of sets of edges.

Other possible future work includes investigating how the algorithm could be used to design heuristics or approximation algorithms in the noisy case, when A is not Robinsonian, for example by using (linear) certifying algorithms as in [15] to detect the edges and the nodes of the level graphs which create obstructions to being a unit interval graph.

Acknowledgments. This work was supported by the Marie Curie Initial Training Network "Mixed Integer Nonlinear Optimization" (MINO) grant no. 316647.

References

1. Atkins, J.E., Boman, E.G., Hendrickson, B.: A spectral algorithm for seriation and the consecutive ones problem. SIAM Journal on Computing **28**, 297–310 (1998)
2. Bodlaender, H.L., Kloks, T., Niedermeier, R.: SIMPLE MAX-CUT for unit interval graphs and graphs with few P4s. Electronic Notes in Discrete Mathematics **3**, 19–26 (1999)
3. Booth, K.S., Lueker, G.S.: Testing for the consecutive ones property, interval graphs, and graph planarity using PQ-tree algorithms. Journal of Computer and System Sciences **13**(3), 335–379 (1976)
4. Chepoi, V., Fichet, B.: Recognition of Robinsonian dissimilarities. Journal of Classification **14**(2), 311–325 (1997)
5. Chepoi, V., Seston, M.: Seriation in the presence of errors: A factor 16 approximation algorithm for l_∞-fitting Robinson structures to distances. Algorithmica **59**(4), 521–568 (2011)
6. Cohen, J., Fomin, F.V., Heggernes, P., Kratsch, D., Kucherov, G.: Optimal linear arrangement of interval graphs. In: Královič, R., Urzyczyn, P. (eds.) MFCS 2006. LNCS, vol. 4162, pp. 267–279. Springer, Heidelberg (2006)
7. Corneil, D.G.: A simple 3-sweep LBFS algorithm for the recognition of unit interval graphs. Discrete Applied Mathematics **138**(3), 371–379 (2004)
8. Corneil, D.G., Kim, H., Natarajan, S., Olariu, S., Sprague, A.P.: Simple linear time recognition of unit interval graphs. Information Processing Letters **55**(2), 99–104 (1995)
9. Crescenzi, P., Corneil, D.G., Dusart, J., Habib, M.: New trends for graph search. In: PRIMA Conference in Shanghai, June 2013. http://math.sjtu.edu.cn/conference/Bannai/2013/data/20130629B/slides1.pdf
10. Deng, X., Hell, P., Huang, J.: Linear-time representation algorithms for proper circular-arc graphs and proper interval graphs. SIAM J. Comput. **25**(2), 390–403 (1996)
11. Dom, M.: Algorithmic aspects of the consecutive-ones property. Bulletin of the European Association for Theoretical Computer Science **98**, 27–59 (2009)
12. Fogel, F., Jenatton, R., Bach, F., d'Aspremont, A.: Convex relaxations for permutation problems. In: Advances in Neural Information Processing Systems, pp. 1016–1024 (2013)
13. Fulkerson, D.R., Gross, O.A.: Incidence matrices and interval graphs. Pacific Journal of Mathematics **15**(3), 835–855 (1965)
14. Habib, M., McConnell, R., Paul, C., Viennot, L.: Lex-BFS and partition refinement, with applications to transitive orientation interval graph recognition and consecutive ones testing. Theoretical Computer Science **234**(12), 59–84 (2000)
15. Hell, P., Huang, J.: Certifying LexBFS recognition algorithms for proper interval graphs and proper interval bigraphs. SIAM J. Discret. Math. **18**(3), 554–570 (2005)
16. Hell, P., Shamir, R., Sharan, R.: A fully dynamic algorithm for recognizing and representing proper interval graphs. SIAM J. Comput. **31**(1), 289–305 (2002)
17. Laurent, M., Seminaroti, M.: The quadratic assignment problem is easy for Robinsonian matrices with Toeplitz structure. Operations Research Letters **43**(1), 103–109 (2015)
18. Liiv, I.: Seriation and matrix reordering methods: An historical overview. Statistical Analysis and Data Mining **3**(2), 70–91 (2010)
19. Looges, P.J., Olariu, S.: Optimal greedy algorithms for indifference graphs. Computers & Mathematics with Applications **25**(7), 15–25 (1993)

20. Mahesh, R., Rangan, C.P., Srinivasan, A.: On finding the minimum bandwidth of interval graphs. Information and Computation 95(2), 218–224 (1991)
21. Mirkin, B.G., Rodin, S.N.: Graphs and genes. Biomathematics. Springer, Heidelberg (1984)
22. Préa, P., Fortin, D.: An optimal algorithm to recognize Robinsonian dissimilarities. Journal of Classification 31, 1–35 (2014)
23. Roberts, F.S.: Indifference graphs. In: Harary, F. (ed.) Proof Techniques in Graph Theory: Proceedings of the Second Ann Arbor Graph Theory Conference, pp. 139–146. Academic Press, New York (1969)
24. Roberts, F.S.: Graph theory and its applications to problems of society. Society for Industrial and Applied Mathematics (1978)
25. Robinson, W.S.: A method for chronologically ordering archaeological deposits. American Antiquity 16(4), 293–301 (1951)
26. Rose, D.J., Tarjan, R.E.: Algorithmic aspects of vertex elimination. In: Proceedings of Seventh Annual ACM Symposium on Theory of Computing, STOC 1975, New York, NY, USA, pp. 245–254. ACM (1975)
27. Seston, M.: Dissimilarités de Robinson: algorithmes de reconnaissance et d'approximation. Ph.D. thesis, Université de la Méditerranée (2008)
28. Simon, K.: A new simple linear algorithm to recognize interval graphs. In: Bieri, H., Noltemeier, H. (eds.) CG-WS 1991. LNCS, vol. 553, pp. 289–308. Springer, Heidelberg (1991)

Mixed Map Labeling

Maarten Löffler[1], Martin Nöllenburg[2], and Frank Staals[1]([✉])

[1] Department of Information and Computing Sciences,
Utrecht University, Utrecht, The Netherlands
{m.loffler,f.staals}@uu.nl
[2] Institute of Theoretical Informatics,
Karlsruhe Institute of Technology (KIT), Karlsruhe, Germany
noellenburg@kit.edu

Abstract. Point feature map labeling is a geometric problem, in which a set of input points must be labeled with a set of disjoint rectangles (the bounding boxes of the label texts). Typically, labeling models either use internal labels, which must touch their feature point, or external (boundary) labels, which are placed on one of the four sides of the input points' bounding box and which are connected to their feature points by crossing-free leader lines. In this paper we study polynomial-time algorithms for maximizing the number of internal labels in a mixed labeling model that combines internal and external labels. The model requires that all leaders are parallel to a given orientation $\theta \in [0, 2\pi)$, whose value influences the geometric properties and hence the running times of our algorithms.

1 Introduction

Annotating features of interest in information graphics with textual labels or icons is an important task in information visualization. One classical application, whose principles easily generalize to the labeling of other illustrations, is map labeling, where labels are mostly placed internally in the map. Common cartographic placement guidelines demand that each label is placed in the immediate neighborhood of its feature and that the association between labels and features is unambiguous, while no two labels may overlap each other [12,21]. Point feature labeling has been studied extensively in the computational geometry literature, but also in the application areas. It is known that maximizing the number of non-overlapping labels for a given set of input points is NP-hard, even for very restricted labeling models [9,18]. In terms of labeling algorithms, several approximations, polynomial-time approximation schemes (PTAS), and exact approaches are known [1,7,15,22], as well as many practically effective heuristics, see the bibliography of Wolff and Strijk [23]. If, however, feature points lie too dense in the map or if their labels are relatively large, often only small fractions of the features obtain a label, even in an optimal solution.

An alternative labeling approach using external instead of internal labels is known as *boundary labeling* in the literature. This labeling style is frequently used

© Springer International Publishing Switzerland 2015
V.Th. Paschos and P. Widmayer (Eds.): CIAC 2015, LNCS 9079, pp. 339–351, 2015.
DOI: 10.1007/978-3-319-18173-8_25

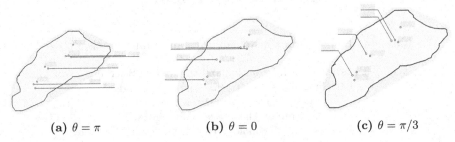

(a) $\theta = \pi$ **(b)** $\theta = 0$ **(c)** $\theta = \pi/3$

Fig. 1. A sample point set with mixed labelings of three different slopes. In (a) five external labels are necessary, whereas (b) and (c) require only four external labels. The slope in (c) yields aesthetically pleasing results.

when annotating anatomical drawings and technical illustrations, where different and often small parts are identified using labels and descriptive texts outside the actual picture, which are connected to their features using leader lines. While the association between points and external labels is often more difficult to see, the big advantages of boundary labeling are that even dense feature sets can be labeled and that larger labels can be accommodated on the margins of the illustration. Many efficient boundary labeling algorithms are known. They can be classified by the leader shapes that are used and by the sides of the picture's bounding box that are used for placing the labels [2,5,6,10,11,14,20].

The combination of internal and external labeling models using internal labels where possible and external labels where necessary seems natural and has been proposed as an open problem by Kaufmann [13]; however, only few results are known in such *hybrid* or *mixed* settings. In a mixed labeling, the final image will be an overlay of the original map or drawing, a collection of internal text labels, and a collection of leaders leading out of the image. Depending on the application, we may wish to forbid intersections between some or all of these layers. When no additional intersection constraints are imposed, the problem reduces to classical internal map labeling. Under the very natural restriction that leaders cannot intersect internal labels, the internal labels need to be placed carefully, as not every set of disjoint internal labels creates sufficient gaps for routing leaders of the prescribed shape from all remaining feature points to the image boundary. Löffler and Nöllenburg [16] studied a restricted case of hybrid labeling, where a partition of the feature points into points with internal fixed position labels and points with external labels to be connected by one-bend orthogonal leaders is given as input. They presented efficient algorithms and hardness results, depending on three different problem parameters. Bekos et al. [3] studied a mixed labeling model with fixed-position internal labels and external labels on one or two opposite sides of the bounding box, connected by two-bend orthogonal leaders. Their goal is to maximize the number of internally labeled points, while labeling all remaining points externally. Polynomial and quasi-polynomial-time algorithms, as well as an approximation algorithm and an ILP formulation were presented.

Contribution. In this paper, we extend the known results on mixed map labeling as follows. We present a mixed labeling model, in which each point is assigned either an axis-aligned fixed-position internal label (e.g., to the top right of the point) or an external label connected with a leader of slope θ, where $\theta \in [0, 2\pi)$ is an input parameter defining the unique leader direction for all external labels, measured clockwise from the negative x-axis (see Fig. 1). In this model, we present a new dynamic-programming algorithm to maximize the number of internally labeled points for any given slope θ, including the left- and right-sided case ($\theta = 0$ or $\theta = \pi$), which was studied by Bekos et al. [3]. While for the right-sided case Bekos et al. provided a faster $O(n \log^2 n)$-time algorithm, where n is the number of input points, our algorithm improves upon their pseudo-polynomial $O(n^{\log n+3})$-time algorithm for the left-sided case. We solve this problem in $O(n^3(\log n + \delta))$ time, where $\delta = \min\{n, 1/d_{\min}\}$ is the inverse of the distance $d_{\min}$ of the closest pair of points in $\mathcal{P}$ and expresses the maximum density of $\mathcal{P}$ (Section 2). In the general case it turns out that the set of slopes can be partitioned into twelve intervals, in each of which the geometric properties of the possible leader-label intersections are similar for all slopes. Depending on the particular slope interval, the amount $\iota(n, \delta, \theta)$ of "interference" between sub-problems varies. This significantly affects the algorithm's performance and leads to running times between $O(n^3 \log n)$ and $O(n^3(\log n + \iota(n, \delta, \theta))) = O(n^7)$ (Section 3). Moreover, we can use our algorithm to optimize the number of internal labels over all slopes θ at an increase in running time by a factor of $O(n^2)$, as is shown in Section 4.1. Omitted proofs can be found in the full version [17].

Problem Statement. We are given a map (or any other illustration) $\mathcal{M}$, which we model for simplicity as a convex polygon (this is easy to relax to larger classes of well-behaved domains), and a set $\mathcal{P}$ of n points in $\mathcal{M}$ that must be labeled by rectangular labels (the bounding boxes of the label texts). In addition, we are given a leader slope $\theta \in [0, 2\pi)$. For simplicity we assume that θ is none of the slopes defined by two points in $\mathcal{P}$. We discuss in Section 4.4 how to remove this restriction. There are two choices for assigning a label to a point $p \in \mathcal{P}$: either we assign an *internal label* λ_p on $\mathcal{M}$ in a one-position model, or an *external label* outside of $\mathcal{M}$ that is connected to p with a leader γ_p. An internal label λ_p is a rectangle that is anchored at p by its lower left corner. A leader γ_p is a line segment of slope θ inside $\mathcal{M}$; it may bend to the horizontal direction outside of $\mathcal{M}$ in order to connect to its horizontally aligned label, see Fig. 1c. So in this model, the labeling is fixed once the choice for an internal or external label has been made for each point $p \in \mathcal{P}$. For a *valid* label assignment we require that (i) the internal labels do not overlap each other or the leaders, and that (ii) the leaders themselves do not intersect each other. Fig. 1 shows valid mixed labelings for three different slopes.

Given a set of points $P \subseteq \mathcal{P}$, let $\Lambda(P) = \{\lambda_p \mid p \in P\}$ denote the set of (candidate) labels corresponding to the points in P and let $\Gamma(P) = \{\gamma_p \mid p \in P\}$ denote the set of (candidate) leaders corresponding to the points in P. A *labeling* of $\mathcal{P}$ is a partition of $\mathcal{P}$ into sets $\mathcal{I}$ and $\mathcal{E}$, the points in $\mathcal{I}$ labeled internally, the points in $\mathcal{E}$ labeled externally, such that no two labels in $\Lambda(\mathcal{I})$ intersect, no two leaders in $\Gamma(\mathcal{E})$ intersect, and no label from $\Lambda(\mathcal{I})$ intersects a leader from $\Gamma(\mathcal{E})$.

For ease of presentation we first assume that all labels have the same size, which, without loss of generality, we assume to be 1×1. Hence, an internal label λ_p is a unit square with its bottom left corner on p. This may be a realistic model in some settings (e.g., unit-size icons as labels), but generally not all labels have the same size. We will sketch how to relax this restriction in Section 4.3.

Each leader γ_p can be split into an *inner* part (or *inner leader*), which is a line segment of slope θ from p to the intersection point with the boundary of $\mathcal{M}$, and an *outer* part (or *outer leader*) from the boundary of $\mathcal{M}$ to the actual label. We focus our attention on the inner leaders as they determine how $\mathcal{P}$ is separated into different subinstances. Hence we can basically think of the leaders as half-lines with slope θ. For completeness, we explain a simple method of routing the outer leaders in Section 4.2.

It is well known that in general not all points in $\mathcal{P}$ can be assigned an internal label. The corresponding label number maximization problem is NP-hard [9,18], even if each label has just one candidate position [16]. If, however, all labels have the same position (e.g., to the top left of the anchor points) and no input point may be covered by any other label, the one-position case can be solved efficiently by first discarding all labels containing an anchor point and then applying a simple greedy algorithm on the resulting staircase patterns [16]. On the other hand, it is also known that any instance can be labeled with external labels using efficient algorithms [4,5]. Mixed labelings combine both label types and sit between the two extremes of purely internal and purely external labeling [3,16]. Here we are interested in the *internal label number maximization problem*, which was first studied for $\theta \in \{0, \pi\}$ by Bekos et al. [3]: Given a map $\mathcal{M}$, a set of points $\mathcal{P}$ in $\mathcal{M}$ and a slope $\theta \in [0, 2\pi)$, we wish to find a valid mixed labeling that maximizes the number $|\mathcal{I}|$ of internally labeled points.

2   Leaders from the Left

We start with the case that $\theta = 0$, i.e., all leaders are horizontal half-lines leading from the points to the left of $\mathcal{M}$. Our approach for maximizing the number of internal labels is to process the points in $\mathcal{P}$ from right to left and to recursively determine the optimal rightmost unprocessed point p to be assigned an external label. Since no leader may cross any internal label, the leader γ_p decomposes the current instance left of p into two (almost) independent parts, one above γ_p and one below. As it turns out, a generic subinstance can be defined by an upper and a lower leader shielding it from the outside and additional information about at most one point outside the subinstance. The problem is then solved using dynamic programming.

2.1   Geometric Properties

Let $p = (p_x, p_y)$ be a point in the plane and let $L_p = \{q \mid q_x < p_x\}$ and $R_p = \{q \mid q_x > p_x\}$ denote the half-planes containing all points strictly to the left and to the right of p, respectively. Analogously, we define the half-planes T_p and

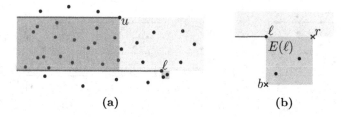

Fig. 2. The slab $\overline{S(\ell,u)}$ in yellow, the region $S(\ell,u)$ and its points from $\mathcal{P}$ in purple, and the region $E(\ell)$ in blue

B_p above and below p, respectively. Let $\overline{S(\ell,u)} = T_\ell \cap B_u$ denote the horizontal slab defined by points ℓ and u (with $\ell_y < u_y$), and let $S(\ell,u) = \overline{S(\ell,u)} \cap L_\ell \cap L_u$ denote the set of points in this slab that lie to the left of both ℓ and u, see Fig. 2. We define $P_{\ell,u}$ as the subset of $\mathcal{P}$ in $S(\ell,u)$ including ℓ and u, i.e., $P_{\ell,u} = \mathcal{P} \cap (S(\ell,u) \cup \{\ell,u\})$. With some abuse of notation we will sometimes also use L_p, R_p, T_p, and B_p to mean the subset of $\mathcal{P}$ that lies in the respective half-plane rather than the entire half-plane.

Recall that $\delta = \min\{n, 1/d_{\min}\}$ is a parameter that captures the maximum density of $\mathcal{P}$ as the inverse of the smallest distance $d_{\min}$ between any two points in $\mathcal{P}$. We can use δ to bound the number of points in a unit square that may be labeled internally.

Lemma 1. *At most $O(\delta)$ points in any unit square have a label λ that does not contain another point in $\mathcal{P}$.*

Next, we characterize which leaders or labels outside of $\overline{S(\ell,u)}$ can interfere with a potential labeling of $P_{\ell,u}$ assuming ℓ and u are labeled externally.

Lemma 2. *Let $\ell, u \in \mathcal{P}$, let $(\mathcal{I}', \mathcal{E}')$, with $\ell, u \in \mathcal{E}'$ be a labeling of $P_{\ell,u}$. There is no point in $T_u \cup B_\ell$ whose leader intersects a label from $\Lambda(\mathcal{I}')$ and there is no point in T_u whose label intersects a label from $\Lambda(\mathcal{I}')$.*

It is not true, however, that labels for points in B_ℓ cannot intersect labels for $P_{\ell,u}$. Still, the influence of B_ℓ is very limited as the next lemma shows. Let $E(\ell)$ denote the open unit square with top-left corner ℓ, i.e., $E(\ell) = R_\ell \cap B_\ell \cap L_r \cap T_b$, where $r = (\ell_x + 1, \ell_y)$ and $b = (\ell_x, \ell_y - 1)$. See Fig. 2b.

Lemma 3. *Let $\ell, u \in \mathcal{P}$, let $(\mathcal{I}', \mathcal{E}')$, with $\ell, u \in \mathcal{E}'$ be a labeling of $P_{\ell,u}$, and let $(\mathcal{I}'', \mathcal{E}'')$ denote a labeling of $\mathcal{P} \cap B_\ell \cup \{\ell\}$ with $\ell \in \mathcal{E}''$. There is at most one point $p \in \mathcal{I}''$ whose label may intersect a label of $\mathcal{I}'$, and $p \in E(\ell)$.*

From Lemma 2 and Lemma 3 it follows that if ℓ and u are labeled externally, there is at most one point r below ℓ that can influence the labeling of the points in $S(\ell,u)$.

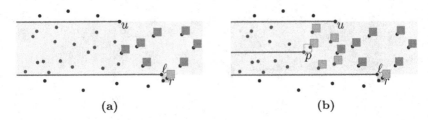

Fig. 3. (a) $\Phi(\ell, u, r)$ expresses the maximum number of points in $S(\ell, u)$ (purple), that can be labeled internally in the depicted situation. (b) The rightmost point p that is labeled with an external label decomposes the problem into two subproblems (the orange and blue points).

2.2 Computing an Optimal Labeling

We define $\Phi(\ell, u, r)$, with $\ell, u \in \mathcal{P}$, and $r \in E(\ell) \cup \{\bot\}$ as the maximum number of points in $S(\ell, u)$ that can be labeled internally, given that

(i) the points ℓ and u are labeled externally,
(ii) all remaining points in $\overline{S(\ell, u)} \setminus S(\ell, u)$ have been labeled internally, and
(iii) point r is labeled internally. If $r = \bot$ then no point in $E(\ell)$ is labeled internally.

See Fig. 3(a) for an illustration. Furthermore, given ℓ, r, and a point $p \in T_\ell$ we define $\varrho(p, \ell, r)$ to be the topmost point in $E(p) \cap (T_\ell \cup \{r\})$ if such a point exists. Otherwise we define $\varrho(p, \ell, r) = \bot$.

Lemma 4. *For any $\ell, u \in \mathcal{P}$, and $r \in E(\ell) \cup \{\bot\}$, we have that $\Phi(\ell, u, r) = |S(\ell, u)|$, or $\Phi(\ell, u, r) = |R_p \cap S(\ell, u)| + \Phi(\ell, p, r) + \Phi(p, u, r')$, where p is the rightmost point in $S(\ell, u)$ with an external label and $r' = \varrho(p, \ell, r)$.*

Proof. et $(\mathcal{I}^*, \mathcal{E}^*)$ be an optimal labeling of $S(\ell, u)$ that satisfies the constraints *(i)–(iii)* on $\Phi(\ell, u, r)$, i.e., $\Phi(\ell, u, r) = |\mathcal{I}^*|$. In case $\mathcal{E}^* = \emptyset$, we have $\Phi(\ell, u, r) = |S(\ell, u)|$ and the lemma trivially holds. Otherwise, there must be a rightmost point $p \in \mathcal{E}^*$ with an external label. Consider the partition of $\mathcal{I}^*$ at point p into the lower left part $B^* = B_p \cap L_p \cap \mathcal{I}^*$, the upper left part $T^* = T_p \cap L_p \cap \mathcal{I}^*$, and the right part $R^* = R_p \cap \mathcal{I}^*$, see Fig. 3b. We show that $|R^*| = |R_p \cap S(\ell, u)|$, $|B^*| = \Phi(\ell, p, r)$, and $|T^*| = \Phi(p, u, r')$, which proves the lemma.

Since p is the rightmost point with an external label it follows that all points in $S(\ell, u)$ right of p are labeled internally. Hence, $R^* = R_p \cap S(\ell, u)$.

Next, we observe that $\mathcal{L}_B = (B^*, S(\ell, p) \setminus B^*)$ as a sub-labeling of $(\mathcal{I}^*, \mathcal{E}^*)$ forms a valid labeling of $S(\ell, p)$, so by Lemma 3 there is at most one point $\hat{r}$ below ℓ that can influence the labeling of $S(\ell, p)$. This point $\hat{r}$, if it exists, lies in $E(\ell)$. By constraint *(iii)* point r lies in $E(\ell)$ or $r = \bot$ and no point in $E(\ell)$ is labeled internally, and thus r can be the only point in $E(\ell)$ labeled internally, i.e., $\hat{r} = r$. So, we have that *(i)* ℓ and p are labeled externally, *(ii)* all points in $\overline{S(\ell, p)} \setminus S(\ell, p)$ are labeled internally, and *(iii)* point r is the only internally labeled point in $E(\ell)$. Thus the definition of Φ applies and we obtain $|B^*| \leq \Phi(\ell, p, r)$.

Lemmas 2 and 3 together imply that any labeling of $S(\ell, p)$ is independent from any labeling of $S(p, u)$. Thus, it follows that $\mathcal{L}_B$ is an optimal labeling of $S(\ell, p)$ (given the constraints), since otherwise $(\mathcal{I}^*, \mathcal{E}^*)$ could also be improved. Thus $|B^*| \geq \Phi(\ell, p, r)$ and we obtain $|B^*| = \Phi(\ell, p, r)$.

Finally, we consider the upper left part T^*. By Lemma 3 there is at most one point r' in B_p with an internal label that can influence the labeling of $S(p, u)$ and we have $r' \in E(p)$. We need to show that $r' = \varrho(p, \ell, r)$. Then the rest of the argument is analogous to the argument for B^*.

We claim that r' is the topmost point in $E(p) \cap (T_\ell \cup \{r\})$. Assume that $r' \notin T_\ell$, which means $r' \in B_\ell$. We know that γ_ℓ does not intersect $\lambda_{r'}$ and hence $r' \in R_\ell$. This means that $r' \in E(p) \cap B_\ell \cap R_\ell =: X$ and since p lies to the top-left of ℓ we have $X \subseteq E(\ell)$. By definition r is the only point with an internal label in $E(\ell)$ and hence $r' = r$. So if $r' \neq r$ we have $r' \in E(p) \cap T_\ell$. Now assume that $r' \neq \varrho(p, \ell, r)$. Then there is another point $q \in E(p) \cap T_\ell$ above r'. This point q must be labeled externally since no two points in $E(p)$ can be labeled internally. This is a contradiction since by definition p is the rightmost externally labeled point in $S(\ell, u)$ and by constraint *(ii)* all points in $\overline{S(\ell, u)} \setminus S(\ell, u)$ are labeled internally. So indeed $r' = \varrho(p, \ell, r)$ and the same arguments as for B^* can be used to obtain $|T^*| = \Phi(p, u, r')$. $\qquad\square$

Let $\ell, u \in \mathcal{P}$, and $p \in S(\ell, u)$. We observe that $|S(\ell, p)|$ and $|S(p, u)|$ are strictly smaller than $|S(\ell, u)|$. Thus, Lemma 4 gives us a proper recursive definition for Φ:

$$\Phi(\ell, u, r) = \max \big\{ \Psi(S(\ell, u)),$$
$$\max_{p \in S(\ell, u)} \{ \Psi(R_p \cap S(\ell, u)) + \Phi(\ell, p, r) + \Phi(p, u, \varrho(p, \ell, r)) \} \big\},$$

where

$$\Psi(P) = \begin{cases} |P| & \text{if all labels in } \Lambda(P \cup \{r\} \cup (\overline{S(\ell, u)} \setminus S(\ell, u))) \text{ are pairwise} \\ & \text{disjoint, and their intersection with } \gamma_\ell \text{ and } \gamma_u \text{ is empty,} \\ -\infty & \text{otherwise.} \end{cases}$$

We can now express the maximum number of points in $\mathcal{P}$ that can be labeled internally using Φ. We add two dummy points to $\mathcal{P}$ that we assume are labeled externally: a point p_∞ that lies sufficiently far above and to the right of all points in $\mathcal{P}$, and a point $p_{-\infty}$ below and to the right of all points in $\mathcal{P}$. The maximum number of points labeled internally is then $\Phi(p_{-\infty}, p_\infty, \perp)$.

We use dynamic programming to compute $\Phi(\ell, u, r)$ for all $\ell, u \in \mathcal{P} \cup \{p_\infty, p_{-\infty}\}$ with $\ell_y < u_y$ and $r \in E(\ell) \cup \{\perp\}$. By finding the maximum in a set of linear size, each value $\Phi(\ell, u, r)$ can be computed in $O(n)$ time, given that the values $\Phi(\ell', u', r')$ for all subproblems have already been computed and stored in a table and the relevant values for the functions ϱ and f have been precomputed. There are $O(n)$ choices for each of ℓ and u; further there are $O(\delta)$ choices for the point r given ℓ

since r is labeled internally and we know from Lemma 1 that there are at most $O(\delta)$ points in $E(\ell)$ as candidates for an internal label. This results in an $O(n^3\delta)$ time and $O(n^2\delta)$ space dynamic-programming algorithm. We show next that the preprocessing of ϱ and f can be done in $O(n^3 \log n)$ time.

To compute $\Phi(\ell, u, r)$ we actually have to compute $\varrho(p, \ell, r)$ and $\Psi'(p, r) := \Psi(R_p \cap S(\ell, u))$ for all points $p \in S(\ell, u)$. We can preprocess all points in $\mathcal{P}$ in $O(n \log n)$ time, such that we can compute each $\varrho(p, \ell, r)$ in $O(1)$ time as follows. First, we compute and store for each point $p \in \mathcal{P}$ the topmost point $q_p \in \mathcal{P}$ in $E(p)$. This requires n standard priority range queries that take $O(n \log n)$ time in total using priority range trees with fractional cascading [8, Chapter5]. To compute $\varrho(p, \ell, r)$ we then check if q_p lies above ℓ. If it does, we have $\varrho(p, \ell, r) = q_p$. Otherwise, the only candidate point for $\varrho(p, \ell, r)$ is r and we can check in $O(1)$ time if r lies in $E(p)$. This takes $O(1)$ time for each triple (p, ℓ, r) and $O(n^2\delta)$ time in total.

Next, we fix ℓ and u, and compute a representation of Ψ' in $O(n \log n)$ time, such that for each $p \in S(\ell, u)$ and $r \in E(\ell) \cup \{\bot\}$ we can obtain $\Psi'(p, r)$ in constant time.

We start by computing the values $\Psi'(p, \bot)$, for all p. We sweep a vertical line from right to left. That is, we sort all points in $\overline{S(\ell, u)}$ by decreasing x-coordinate, and process the points in that order. The status structure of the sweep line contains the number of points N in $S(\ell, u)$ right of the sweep line, and a (semi-)dynamic data structure $\mathcal{T}$, which stores the labels from the points right of the sweep line, and can report all labels intersected by an (axis-parallel) rectangular query window. All labels are unit squares, so λ_r intersects a label λ_q if and only if λ_r contains a corner point of λ_q. Furthermore, we only ever insert new labels (points) into $\mathcal{T}$, thus it suffices if $\mathcal{T}$ supports only insert and query operations. It follows that we can implement $\mathcal{T}$ using a semi-dynamic range tree using dynamic fractional cascading [19]. In this data structure insertions and queries take $O(\log n)$ time.

When we encounter a new point p, $p \notin \{\ell, u\}$ we test if the label of p intersects any of the labels encountered so far. We can test this using a range query in the tree $\mathcal{T}$. If $p \in S(\ell, u)$ we also explicitly test if λ_p intersects γ_ℓ or γ_u. If there are no points in the query range λ_p, and λ_p does not intersect γ_ℓ or γ_u we report $\Psi'(p, \bot) = N$, increment N (if applicable), and insert the corner points of λ_p into $\mathcal{T}$. If the query range λ_p is not empty, it follows that $\Psi'(p', \bot) = -\infty$, for $p' = p$ as well as for any point to the left of p. Hence, we report that and stop the sweep. Our algorithm runs in $O(n \log n)$ time: sorting all points takes $O(n \log n)$ time, and handling each of the $O(n)$ events takes $O(\log n)$ time.

Now consider a point $r \in E(\ell)$. We observe that for all points p right of r, we have that $\Psi'(p, r) = \Psi'(p, \bot) = 0$ since r is right of all points in $S(\ell, u)$. Consider the points left of r ordered by decreasing x-coordinate. There are two options, depending on whether or not λ_r intersects the label λ_p of the current point p. If λ_r intersects λ_p, we have $\Psi'(p, r) = -\infty$ as well as $\Psi'(p', r) = -\infty$ for all points p' left of p. If λ_r does not intersect λ_p we still have $\Psi'(p, r) = \Psi'(p, \bot)$. We can test if λ_r intersects any other label using a range priority query with λ_r

in (the final version of) the range tree $\mathcal{T}$. We need $O(\delta)$ such queries, which take $O(\log n)$ time each. This gives a total running time of $O(n \log n)$.

The above algorithm can also be used when the leaders have a slope $\theta \neq 0$. However, the data structure $\mathcal{T}$ that we use is fairly complicated. In this specific case where $\theta = 0$, we can also use a much easier data structure, and still get a total running time of $O(n \log n)$. Instead of using the semi-dynamic range tree as status structure, we use a simple balanced binary search tree that stores (the end-points of) a set of vertical *forbidden intervals*. When we encounter a new point p, we check if p_y lies in a forbidden interval. If this is the case then λ_p intersects another label. Otherwise we can label p internally. This set of forbidden intervals is easily maintained in $O(\log n)$ time.

We use this algorithm for every pair (ℓ, u). Hence, after a total of $O(n^3 \log n)$ preprocessing time, we can answer $\Psi'(p, r)$ queries for any p and r in constant time. This yields the following result, which improves the previously best known pseudo-polynomial $O(n^{\log n + 3})$-time algorithm of Bekos et al. [3] for the left-sided case $\theta = 0$.

Theorem 1. *Given a set $\mathcal{P}$ of n points, we can compute a labeling of $\mathcal{P}$ that maximizes the number of internal labels for $\theta = 0$ in $O(n^3 \log n + n^3 \delta)$ time and $O(n^2 \delta)$ space, where $\delta = \min\{n, 1/d_{\min}\}$ for the minimum distance $d_{\min}$ in $\mathcal{P}$.*

3   Other Leader Directions

For other leader slopes $\theta \neq 0$ we use a similar approach as before. We consider a sub-problem $S(\ell, u)$ defined by two externally labeled points ℓ and u. We again find the "rightmost" point in the slab labeled externally. This gives us two sub-problems, which we solve recursively using dynamic programming. However, there are four complications:

- The region $E(\ell)$ containing the points "below" the slab $\overline{S(\ell, u)}$ that can influence the labeling of $S(\ell, u)$ is no longer a unit square. Depending on the orientation, it can contain more than one point with an internal label.
- In addition to the region $E(\ell)$, which contains points that can interfere with $S(\ell, u)$ from below, we now also need to consider a second region, which we call $F(u)$, containing points whose labels can interfere with a subproblem from above.
- The labels of points in $S(\ell, u)$ are no longer fully contained in the slab $\overline{S(\ell, u)}$. Hence, we have to check that they do not intersect leaders of points outside $\overline{S(\ell, u)}$.
- The regions $E(p)$ and $F(p)$ are no longer strictly to "the right" of p. Hence, for some sub-problems we may have already decided (by definition of the sub-problem) that a point $q \in E(p)$ that lies "left" of p is labeled internally. Hence, we can no longer choose q to be the rightmost point in $S(\ell, p)$ to be labeled externally.

Details on how to compute a labeling taking these complications into account can be found in the full version [17]. We obtain the following main result, where the running time is at most $O(n^7)$ for the worst case of $\delta = O(n)$:

Theorem 2. *Given a set $\mathcal{P}$ of n points and an angle θ, we can compute a labeling of $\mathcal{P}$ that maximizes the number of internal labels in $O(n^3(\log n + \iota(n, \delta, \theta)))$ time and $O(n^2 \sqrt{\iota(n, \delta, \theta)})$ space, where $\delta = \min\{n, 1/d_{\min}\}$ for the minimum distance $d_{\min}$ in $\mathcal{P}$, and $\iota(n, \delta, \theta)$ models how much subproblems can influence each other. More formally,*

$$\iota(n, \delta, \theta) = n^{2e'(\theta)+2f'(\theta)} \qquad +n^{2e'(\theta)+f'(\theta)}\delta^{f^*(\theta)} +n^{e'(\theta)+2f'(\theta)}\delta^{e^*(\theta)}$$
$$+n^{e'(\theta)+f'(\theta)}\delta^{e^*(\theta)+f^*(\theta)} +n^{2e'(\theta)}\delta^{2f^*(\theta)} \qquad +n^{2f'(\theta)}\delta^{2e^*(\theta)}$$
$$+n^{e'(\theta)}\delta^{e^*(\theta)+2f^*(\theta)} \qquad +n^{f'(\theta)}\delta^{2e^*(\theta)+f^*(\theta)}+\delta^{2e^*(\theta)+2f^*(\theta)},$$

where $e^*(\theta), f^*(\theta), e'(\theta),$ and $f'(\theta)$ are all at most one.

4   Extensions

So far, we have considered a stylized version of the question we set out to solve. In this section we discuss how our solution may be adapted and extended, depending on the exact requirements of the application.

4.1   Optimizing the Direction

Rather than fixing the direction for the leaders in advance, we may be willing to let the algorithm specify the optimal orientation that maximizes the number of points that can be labeled internally. Or, perhaps we wish to compute a chart that plots the maximum number of internally labeled points as a function of the leader orientation θ, leaving the final decision to the judgement of the user.

In both scenarios, we need to efficiently iterate over all possible orientations. We adapt our method straightforwardly. Let Q be the set of all $4n$ corner points of all potential labels. For every pair $p, q \in Q$ consider the slope $\theta_{p,q}$ of the line through p and q. All values $\theta_{p,q}$ partition all possible angles into $O(n^2)$ intervals. For all values θ in the same interval J, any leader γ_p intersects the same set of potential labels, so the optimal set of internal labels is constant throughout J. We compute it separately for each interval.

By applying Theorem 2, we achieve a total of $O(n^2 \cdot n^3(\log n + \iota(n, \delta, \theta))) = O(n^5(\log n + \iota(n, \delta, \theta)))$ time to compute the optimal labelings for all orientations, or to optimize the orientation by performing a simple linear scan.

4.2   Routing the Outer Leaders

Once the core combinatorial problem of deciding which points have to be labeled internally is solved, it remains to route the outer leaders and place the external labels. Since our goal in this paper is to maximize the number of internally

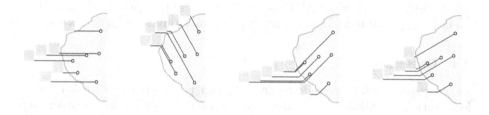

Fig. 4. Examples for routing the outer leaders and placing the external labels for different slopes

labeled points, we are only interested in finding a valid labeling, in which neither labels nor leaders intersect each other. Let us assume that $\theta \in [0, \pi/2] \cup [3\pi/2, 2\pi]$, i.e., all external labels are oriented to the left. The case of labels oriented to the right is symmetric. We consider the leaders in counterclockwise order around the boundary of $\mathcal{M}$ and place them one by one starting with the topmost leader. The first label is placed with its lower right corner anchored at the endpoint of its inner leader. For all subsequent labels we test if the label anchored at the endpoint of the inner leader intersects the previously placed label. If there is no intersection, we use that label position. Otherwise, we draw an outer leader extending horizontally to the left starting from the endpoint of the inner leader until the label can be placed without overlap. Obviously this algorithm takes only linear time. Fig. 4 shows the resulting labelings for four different slopes. We note that depending on the slope θ of the inner leaders other methods for routing the outer leaders might yield more pleasing external labelings. This is, however, beyond the scope of this paper.

4.3   Non-square Labels

Square labels are not very realistic in most map-labeling applications. Their use is justified by the observation that if all labels are homothetic rectangles, we can scale the plane in one dimension to obtain square labels without otherwise changing the problem. Nonetheless, reality is not quite that simple, for two reasons: firstly, the scaling does alter inter-point distances, so if we wish to parametrize our solution by $d_{\min}$ we need to take this into account. Secondly, in real-world applications, labels may arguably have the same height, but not usually the same width.

If all labels are homothetic rectangles with a height of 1 and a width of w, scaling the plane by a factor $1/w$ in the horizontal direction potentially decreases the closest interpoint distance by the same factor. Now, the number of points in a unit-area region that do not contain each other's potential labels is bounded by δw, immediately yielding a result of $O(n^3(\log n + \iota(n, \delta w, \theta)))$ using exactly the same approach.

When all labels have equal heights but may have arbitrary widths, we conjecture that a variation of our approach will still work, but a careful analysis of the intricacies involved is required. If the labels may also have arbitrary heights the problem is open. It is unclear if there is a polynomial time solution in this case.

4.4 Obstacles

In this paper, we have considered only abstract point sets to be labeled, using leaders that are allowed to go anywhere, as long as they do not intersect any internal labels. While this is justified in some applications (e.g., in anatomical drawings, it is common practice to ignore the drawing when placing the leaders, as they are very thin and do not occlude any part of the drawing), in others this may be undesirable (in certain map styles, leaders may be confused for region boundaries or linear features). As a solution, we may identify a set of polygonal *obstacles* in the map, that cannot be intersected by leaders or internal labels.

In this setting, obviously not every input has a valid labeling: a point that lies inside an obstacle can never be labeled, or obstacles may surround points or force points into impossible configurations in more complex ways. Nonetheless, we can test whether an input has a valid labeling and if so, compute the labeling that maximizes the number of internal labels in polynomial time with our approach.

The main idea is to preprocess the input points in a similar way as for the arbitrary leader orientations case (Section 3). Whenever a point has a potential leader that intersects an obstacle, it must be labeled internally; similarly, whenever a point has a potential internal label that intersects an obstacle, it must be labeled externally. If we include such "forced" leaders or labels into our set of obstacles and apply this approach recursively, we will either find a contradiction or be left with a set of points whose potential leaders and potential internal labels do not intersect any obstacle, and we can apply our existing algorithm on this point set.

The same approach may be used for point sets that are not in general position: if we disallow leaders that pass through other points, they are forced to be labeled internally. Note that this again may result in situations where no valid labeling exists.

Acknowledgments. M.L. and F.S. are supported by the Netherlands Organisation for Scientific Research (NWO) under grant 639.021.123 and 612.001.022, respectively.

References

1. Agarwal, P.K., van Kreveld, M., Suri, S.: Label placement by maximum independent set in rectangles. Comput. Geom. Theory Appl. **11**(3–4), 209–218 (1998)
2. Bekos, M., Kaufmann, M., Nöllenburg, M., Symvonis, A.: Boundary labeling with octilinear leaders. Algorithmica **57**, 436–461 (2010)
3. Bekos, M.A., Kaufmann, M., Papadopoulos, D., Symvonis, A.: Combining traditional map labeling with boundary labeling. In: Černá, I., Gyimóthy, T., Hromkovič, J., Jefferey, K., Králović, R., Vukolić, M., Wolf, S. (eds.) SOFSEM 2011. LNCS, vol. 6543, pp. 111–122. Springer, Heidelberg (2011)

4. Bekos, M.A., Kaufmann, M., Symvonis, A.: Efficient labeling of collinear sites. J. Graph Algorithms Appl. **12**(3), 357–380 (2008)
5. Bekos, M.A., Kaufmann, M., Symvonis, A., Wolff, A.: Boundary labeling: Models and efficient algorithms for rectangular maps. Comput. Geom. Theory Appl. **36**(3), 215–236 (2007)
6. Benkert, M., Haverkort, H., Kroll, M., Nöllenburg, M.: Algorithms for multi-criteria boundary labeling. J. Graph Algorithms and Appl. **13**(3), 289–317 (2009)
7. Chalermsook, P., Chuzhoy, J.: Maximum independent set of rectangles. In: Discrete Algorithms (SODA 2009), pp. 892–901 (2009)
8. de Berg, M., van Kreveld, M., Overmars, M., Schwarzkopf, O.: Computational Geometry: Algorithms and Applications, 2nd edn. Springer-Verlag, Berlin, Germany (2000)
9. Formann, M., Wagner, F.: A packing problem with applications to lettering of maps. In: Computational Geometry (SoCG 1991), pp. 281–288. ACM (1991)
10. Gemsa, A., Haunert, J.-H., Nöllenburg, M.: Boundary-labeling algorithms for panorama images. In: Advances in Geographic Information Systems (SIGSPATIAL GIS 2011), pp. 289–298. ACM (2011)
11. Huang, Z.-D., Poon, S.-H., Lin, C.-C.: Boundary labeling with flexible label positions. In: Pal, S.P., Sadakane, K. (eds.) WALCOM 2014. LNCS, vol. 8344, pp. 44–55. Springer, Heidelberg (2014)
12. Imhof, E.: Positioning names on maps. The American Cartographer **2**(2), 128–144 (1975)
13. Kaufmann, M.: On map labeling with leaders. In: Albers, S., Alt, H., Näher, S. (eds.) Efficient Algorithms. LNCS, vol. 5760, pp. 290–304. Springer, Heidelberg (2009)
14. Kindermann, P., Niedermann, B., Rutter, I., Schaefer, M., Schulz, A., Wolff, A.: Two-sided boundary labeling with adjacent sides. In: Dehne, F., Solis-Oba, R., Sack, J.-R. (eds.) WADS 2013. LNCS, vol. 8037, pp. 463–474. Springer, Heidelberg (2013)
15. Klau, G.W., Mutzel, P.: Optimal labeling of point features in rectangular labeling models. Mathematical Programming **94**(2), 435–458 (2003)
16. Löffler, M., Nöllenburg, M.: Shooting bricks with orthogonal laser beams: a first step towards internal/external map labeling. In: Canadian Conf. Computational Geometry (CCCG 2010), pp. 203–206. University of Manitoba (2010)
17. Löffler, M., Nöllenburg, M., Staals, F.: Mixed map labeling. CoRR, abs/1501.06813 (2015)
18. Marks, J., Shieber, S.: The computational complexity of cartographic label placement. Technical report, Harvard University (1991)
19. Mehlhorn, K., Näher, S.: Dynamic fractional cascading. Algorithmica **5**(1–4), 215–241 (1990)
20. Nöllenburg, M., Polishchuk, V., Sysikaski, M.: Dynamic one-sided boundary labeling. In: Advances in Geographic Information Systems (SIGSPATIAL GIS 2010), pp. 310–319, November 2010
21. Reimer, A., Rylov, M.: Point-feature lettering of high cartographic quality: a multi-criteria model with practical implementation. In: EuroCG 2014, Ein-Gedi, Israel (2014)
22. van Kreveld, M., Strijk, T., Wolff, A.: Point labeling with sliding labels. Comput. Geom. Theory Appl. **13**(1), 21–47 (1999)
23. Wolff, A., Strijk, T.: The map labeling bibliography. http://i11www.iti.kit.edu/map-labeling/bibliography/

Optimal Online Edge Coloring of Planar Graphs with Advice

Jesper W. Mikkelsen[✉]

Department of Mathematics and Computer Science,
University of Southern Denmark, Odense, Denmark
jesperwm@imada.sdu.dk

Abstract. Using the framework of advice complexity, we study the amount of knowledge about the future that an online algorithm needs to color the edges of a graph optimally, i.e., using as few colors as possible. For graphs of maximum degree Δ, it follows from Vizing's Theorem that $O(m \log \Delta)$ bits of advice suffice to achieve optimality, where m is the number of edges. We show that for graphs of bounded degeneracy (a class of graphs including e.g. trees and planar graphs), only $O(m)$ bits of advice are needed to compute an optimal solution online, independently of how large Δ is. On the other hand, we show that $\Omega(m)$ bits of advice are necessary just to achieve a competitive ratio better than that of the best deterministic online algorithm without advice. Furthermore, we consider algorithms which use a fixed number of advice bits per edge (our algorithm for graphs of bounded degeneracy belongs to this class of algorithms). We show that for bipartite graphs, any such algorithm must use at least $\Omega(m \log \Delta)$ bits of advice to achieve optimality.

1 Introduction

An *edge coloring* of a graph is an assignment of colors to the edges of the graph such that no two adjacent edges share the same color. Many scheduling and assignment problems can be modeled as edge coloring problems. The *online edge coloring problem*, which we refer to simply as EDGE-COLORING, was introduced by Bar-Noy et al. [3]. In this problem, the edges of a graph arrive one by one. The edges are specified by their endpoints, but the vertices of the graph are not known in advance. Each edge must be assigned a color before the next edge arrives, under the constraint that no two adjacent edges are assigned the same color. The color assigned to an edge cannot be changed later on. The goal is to use as few colors as possible.

Traditionally, worst-case competitive analysis [24,31] is used to measure the performance of an online algorithm. The solution produced by the online algorithm, ALG, is compared to that of an optimal offline algorithm, OPT, which knows the entire input in advance. More precisely, let $\text{ALG}(\sigma)$ ($\text{OPT}(\sigma)$) denote the number of colors used by ALG (OPT) when coloring a sequence, σ, of edges. We say that ALG is *c-competitive* if there exists a constant c_0 such that $\text{ALG}(\sigma) \leq c \cdot \text{OPT}(\sigma) + c_0$

J.W. Mikkelsen—Supported in part by the Villum Foundation and the Danish Council for Independent Research, Natural Sciences.

V.Th. Paschos and P. Widmayer (Eds.): CIAC 2015, LNCS 9079, pp. 352–364, 2015.
DOI: 10.1007/978-3-319-18173-8_26

for any input sequence σ. If the inequality holds with $c_0 = 0$, then ALG is said to be *strictly c-competitive*.

In [3] it is shown that any EDGE-COLORING algorithm, which never introduces a new color unless forced to do so, is strictly 2-competitive and that no online algorithm, even if we allow randomization and restrict the input graph to being a forest, can achieve a better competitive ratio.

The underlying assumption of competitive analysis, that nothing is known about future parts of the input, is sometimes unrealistic. Therefore, for many online problems, various relaxations of this assumption have been suggested, including look-ahead [5], locality of reference [10] and several models where the input is generated from some known probability distribution [25,27,28]. In this paper, we consider the recent idea of *advice complexity* introduced in [18] and further developed in [8,19,22]. Advice complexity provides a quantitative and problem-independent approach for relaxing the online constraint by providing the algorithm partial knowledge of the future. Our main goal in applying the framework of advice complexity to EDGE-COLORING is to better understand the online hardness of the problem. How much (and which kind of) information about the future are we lacking in order to produce an optimal edge coloring in the online setting?

Advice Complexity Models. In this paper, we consider the two most widely used models of advice complexity. In both models, an oracle, which has unlimited computational power and knows the entire input, provides the online algorithm ALG with some advice bits. For EDGE-COLORING, the input is a sequence of m edges $\langle e_1, \ldots, e_m \rangle$. The two models are defined as follows:

Advice-with-request [19]. In this model, ALG receives some fixed number, b, of advice bits along with each request. That is, when the edge e_i arrives, the algorithm receives some advice $b_i \in \{0,1\}^b$ from the oracle. The algorithm then decides which color to assign to e_i based on the edges $e_1, \ldots, e_i$ that have been revealed up until now and the advice $b_1, \ldots, b_i$ received thus far.

Advice-on-tape [8,22]. In this model, the online algorithm ALG is provided access to an infinite advice tape prepared by the oracle. The algorithm may, at any point in time, read some number of advice bits from the tape. When the edge e_i arrives, the algorithm must decide which color to assign to e_i based on the edges $e_1, \ldots, e_i$ that have been revealed up until now and the advice read so far from the tape. The *advice complexity*, $b(m)$, of ALG is the largest number of advice bits read by ALG, over all possible input graphs with at most m edges.

Note that an algorithm in the advice-with-request model receives exactly bm bits of advice in total. Thus, it can be converted into an algorithm with advice complexity $bm + O(\log b)$ in the advice-on-tape model (the $O(\log b)$ bits of advice are used to encode b). Converting in the opposite direction is not always possible in a meaningful way. In particular, an algorithm in the advice-on-tape model is allowed to read only a sublinear number of advice bits. This is not possible in the advice-with-request model.

Preliminaries. All graphs considered are simple. We denote the number of edges in a graph by m, the number of vertices by n and the maximum degree by Δ. A graph G is k-*edge-colorable* if there exists an edge coloring of G with at most k different colors. The *chromatic index* $\chi'(G)$ of G is the smallest integer k such that G is k-edge-colorable. We assume that colors are represented by consecutive positive integers. For a bipartite graph G, we write $G = (L, R)$ if L and R form a bipartition of the vertices of G. We let $K_{a,b}$ denote the complete bipartite graph $G = (L, R)$ where $|L| = a$ and $|R| = b$.

In addition to bipartite graphs, we consider trees, planar graphs and, more generally, d-degenerate graphs. A graph is d-*degenerate* if there is an ordering $v_1, v_2, \ldots, v_n$ of its vertices such that, for $1 \leq i \leq n$, the vertex v_i is adjacent to at most d vertices in $\{v_1, \ldots, v_{i-1}\}$. The *degeneracy* of a graph G is the least integer d such that G is d-degenerate. An edge $e = (v_i, v_k)$ where $i < k$ is said to be a *front-edge* at v_i and a *back-edge* at v_k. Furthermore, $d_f(v_i)$ is the number of front-edges at v_i.

The notion of degeneracy has appeared under other names and many equivalent definitions exist (see e.g. [23]). Note that the degeneracy of a graph is at most Δ. A graph is 1-degenerate if and only if it is a forest. Planar graphs are 5-degenerate. Other graph classes of bounded degeneracy include graphs of bounded genus, bounded tree-width, and graphs excluding a fixed minor.

It is clear that $\Delta \leq \chi'(G)$ for any graph G. The celebrated Vizing's Theorem [33] states that $\chi'(G) \in \{\Delta, \Delta + 1\}$. The following relationship between edge coloring and degeneracy, which is also due to Vizing, will be used extensively in the design of our algorithm.

Theorem 1 (Vizing [32, 34]). *Let G be a d-degenerate graph of maximum degree Δ. If $\Delta \geq 2d$, then Δ colors suffice for edge coloring G.*

Our Contribution. By Vizing's theorem, there is a trivial upper bound on the advice complexity of EDGE-COLORING of $O(m \log \Delta)$ bits. We improve on this upper bound for d-degenerate graphs by showing that $O(m \log d)$ bits of advice suffice to achieve optimality. In particular, only $O(m)$ bits of advice are needed for graphs of bounded degeneracy. The algorithm that we present works in both the advice-on-tape and the advice-with-request model. On the hardness side, we show that $\Omega(m)$ bits of advice are required in order to achieve a competitive ratio better than 2. This lower bound holds even for forests. Finally, we show that in the advice-with-request model, $\Omega(m \log \Delta)$ bits of advice are necessary to achieve optimality for bipartite graphs.

Related Work. While EDGE-COLORING has not previously been studied in the framework of advice complexity, many other online problems have, see e.g. [1, 2, 4, 6–9, 11, 12, 16–20, 22, 29, 30]. We remark that, contrary to EDGE-COLORING, for several of the problems studied in the literature, sublinear advice (in the number of requests) suffice to achieve a competitive ratio better than that of the best deterministic algorithm without advice. This is the case for online problems such as bin-packing [13], list accessing [12], knapsack [9], makespan minimization [2], paging [8] and ski-rental[18].

The computational complexity of (offline) edge coloring is well-studied. In general, deciding if $\chi'(G) = \Delta$ is NP-complete [21], but a $(\Delta + 1)$-edge-coloring can always be found in polynomial time. For planar graphs, fast algorithms exist for most values of Δ (see e.g. [14, 15]). Also, the edge coloring guaranteed to exist by Theorem 1 can be computed efficiently [35].

2   An Algorithm for d-Degenerate Graphs

In this section, we present the algorithm for d-degenerate graphs in the advice-with-request model. As mentioned in the introduction, converting the algorithm to the advice-on-tape model is straightforward.

Theorem 2. *Let $d \in \mathbb{N}$. For the class of d-degenerate graphs, there exists an* EDGE-COLORING *algorithm which always produces an optimal coloring and uses*

$$1 + \lceil \log(2d) \rceil + \lceil \log(d + 1) \rceil = O(\log d)$$

bits of advice per edge and, hence, $O(m \log d)$ bits of advice in total.

Theorem 2 assumes that the degeneracy of the input graph is at most d, where d is a constant hard-wired into both the algorithm and the oracle. In Theorem 3, we show how the assumption that d is constant can be removed by communicating d as part of the advice. Exactly how to do this depends on the model of advice complexity.

In order to prove Theorem 2, we start by assuming that $2d$ divides the maximum degree, Δ, of the input graph. Later on, we will show how to reduce the general case to this special case.

Let $G = (V, E)$ be a d-degenerate input graph of maximum degree Δ and let $a = \frac{\Delta}{2d} \in \mathbb{N}$. We will first give a high-level description of the oracle and the corresponding algorithm. The main idea is to partition the edges E into a disjoint subsets, $E_1, \ldots, E_a$, such that the maximum degree of the graph (V, E_j) is $2d$ for $1 \leq j \leq a$ (this is possible since we are assuming $\Delta = 2d \cdot a$). By Theorem 1, (E, V_j) is $2d$-edge-colorable. Thus, if the algorithm knew how to make such a partition, then, using $O(\log d)$ bits of advice per edge, the algorithm could make an optimal edge coloring of each E_j, and hence all of G.

However, the oracle cannot afford to compute such a partition and then simply encode the index j such that $e \in E_j$ for each edge, since this would require too much advice per edge if a is large. Instead, the oracle finds a specific partition which is based on the arrival time of the edges and the fact that G is d-degenerate. This partition is such that when an edge e is revealed, the algorithm itself can (without advice) compute a sufficiently small set of indices which always contains the correct index j. This makes it possible to reduce the number of advice bits needed for the algorithm to learn the index j.

In order to produce this partition, the oracle orders the vertices of the d-degenerate input graph such that no vertex has more than d back-edges.

Starting with the first vertex in this ordering, the oracle processes the front-edges of each vertex ordered by (increasing) time of arrival. For each edge, the oracle determines the lowest index j' such that the edge can be assigned to $E_{j'}$ while maintaining that $(V, E_{j'})$ has maximum degree at most $2d$. Note that whenever a front-edge, e, at v is being processed, the oracle has already assigned all back-edges of v to some sets in the partition. Since these back-edges may arrive later than e, they may be unknown to the algorithm at the time when e is revealed. Therefore, the advice for the front-edge e will warn the algorithm not to assign e to E_j if this would later on prevent the intended assignment of some back-edge at v to E_j.

The Oracle for the Case Where $2d$ Divides Δ. We now give a formal description of the oracle and the algorithm. To each edge $e \in E$, the oracle associates a bit string, $B(e)$, of length $\lceil \log(2d) \rceil + \lceil \log(d+1) \rceil$ by following Procedure 1.

Procedure 1. Constructing the advice in the case where $2d$ divides Δ.

Input: A d-degenerate graph $G = (V, E)$ of maximum degree Δ where $a \cdot 2d = \Delta$ for some $a \in \mathbb{N}$, together with arrival times of the edges.

Output: A bit string $B(e)$ of length $\lceil \log 2d \rceil + \lceil \log(d+1) \rceil$ for each edge $e \in E$.

1: $E_j \leftarrow \emptyset$ for $1 \le j \le a$
2: Compute an ordering $\{v_1, \ldots, v_n\}$ of the vertices of G such that, for $1 \le i \le n$, the vertex v_i is adjacent to at most d vertices in $\{v_1, \ldots, v_{i-1}\}$.
3: Let $E(v_i)$ denote the edges incident to $v_i \in V$.
4: $\text{Prev}(e, v_i) \leftarrow \{f \in E(v_i): f \text{ arrives before } e\}$ for $e \in E, v_i \in V$.
5: **for** $i = 1$ **to** n **do**
6:     Let $\{e_1, \ldots, e_{d_f(v_i)}\}$ be the front-edges at v_i ordered by time of arrival.
7:     **for** $s = 1$ **to** $d_f(v_i)$ **do**
8:         $e \leftarrow e_s$
9:         $J(e) \leftarrow \{j : |E_j \cap \text{Prev}(e, v_i)| \le 2d - 1\}$
10:        Let j' be the lowest index such that $|E_{j'} \cap E(v_i)| \le 2d - 1$.
11:        $E_{j'} \leftarrow E_{j'} \cup \{e\}$
12:        Use the last $\lceil \log(d+1) \rceil$ bits of $B(e)$ to encode $|\{j \in J(e): j < j'\}|$.
13: Compute $2d$-edge-colorings $\mathcal{C}_j$ of (V, E_j) for all $1 \le j \le a$.
14: For each edge $e \in E$, use the first $\lceil \log(2d) \rceil$ bits of $B(e)$ to encode the color assigned to e in $\mathcal{C}_j$, where j is such that $e \in E_j$.

In order to prove the correctness of Procedure 1, we introduce the following terminology: At any point during the execution of Procedure 1, we say that an edge can *legally* be assigned to a subset E_j if this assignment does not make the maximum degree of (V, E_j) larger than $2d$. Also, we let $\mathcal{P} = \{E_1, \ldots, E_a\}$. We will show in Lemma 1 that the index j' in line 10 is such that e can legally be assigned to $E_{j'}$ and that the number in line 12 can be encoded using $\lceil \log(d+1) \rceil$ bits.

Lemma 1. *Suppose that during the execution of Procedure 1, the second for-loop has just been entered and that $e = e_s$. Let j' be the lowest index such that $|E_{j'} \cap E(v_i)| \leq 2d - 1$. Then, e can legally be assigned to $E_{j'}$. Furthermore, j' is among the $d + 1$ lowest indices in $J(e)$.*

Proof. Assume that $e = (v_i, v_k)$ is a front-edge at v_i and a back-edge at v_k. Because $i < k$, none of the front-edges at v_k has yet been assigned to any subset in $\mathcal{P}$. Since v_k has at most d back-edges (including e), it follows that no subset in $\mathcal{P}$ currently contains more than $d - 1$ edges incident to v_k. Thus, if e cannot legally be assigned to some subset E_j, this can only be because it would violate the degree constraint at v_i. That is, e can be legally assigned to E_j if and only if $|E_j \cap E(v_i)| \leq 2d - 1$. Since at most $\Delta - 1 = a \cdot 2d - 1$ edges incident to v_i have arrived earlier than e, and since there are a subsets in $\mathcal{P}$, this implies that e can legally be assigned to at least one subset in $\mathcal{P}$.

Let j' be the lowest index such that e can legally be assigned to $E_{j'}$. Clearly, $j' \in J(e)$ since $\mathrm{Prev}(e, v_i) \subseteq E(v_i)$. We will show that j' is in fact among the $d + 1$ lowest indices in $J(e)$. Let $j \in J(e)$. By definition of $J(e)$, the number of edges in E_j which are incident to v_i and arrive before e is at most $2d - 1$. Thus, if e cannot legally be assigned to E_j, then there must be an edge $f \in E_j$ which is incident to v_i but arrives later than e. The front-edges at v_i arriving later than e has not yet been assigned to any subset in $\mathcal{P}$, and so f must be a back-edge at v_i. Since v_i has at most d back-edges, there can be at most d indices $j \in J(e)$ such that e cannot legally be assigned to E_j. It follows that j' must be among the $d + 1$ lowest indices in $J(e)$.   □

Combining the assumption that G has maximum degree $\Delta = a2d$ with Lemma 1 shows that Procedure 1 constructs a partition $E_1, \ldots, E_a$ of E such that the maximum degree of (V, E_i) is $2d$, for $1 \leq i \leq n$. Furthermore, the number in line 12 is at most $d + 1$ (and non-negative), and hence it can be encoded in binary using $\lceil \log(d + 1) \rceil$ bits. It follows from Theorem 1 that each of the graphs (V, E_i) can be edge colored using $2d$ colors since they all have maximum degree $2d$ and are d-degenerate (because they are subgraphs of G which is d-degenerate). This proves the correctness of Procedure 1.

The Algorithm for the Case Where $2d$ Divides Δ. We now describe how the algorithm, ALG, uses the advice provided by the oracle. Note that when an edge e arrives, ALG is able to compute the set of indices $J(e)$ as defined in line 9 of Procedure 1, since $J(e)$ only depends on d and the edges that have arrived earlier than e. Thus, ALG can compute the index j' such that e was assigned to $E_{j'}$ by Procedure 1 by learning the number $|\{j \in J(e): j < j'\}|$ from the last $\lceil \log(d+1) \rceil$ bits of $B(e)$. The algorithm (internally) assigns e to $E_{j'}$. Then, ALG reads the integer, c, encoded by the first $\lceil \log(2d) \rceil$ bits of $B(e)$ and colors e with the color $((j' - 1)2d + c)$.

It follows that for all $1 \leq j \leq a$, the algorithm colors the edges of E_j with the colors $((j - 1)2d + 1), \ldots, j2d$ and produces a coloring of E_j which is equivalent to the coloring $\mathcal{C}_j$ computed by the oracle. Thus, ALG produces an optimal edge coloring of G.

The General Case. Using the algorithm for the case where $2d$ divides Δ as a subroutine, we are now ready to prove Theorem 2.

Proof (of Theorem 2). Fix $d \in \mathbb{N}$. We will describe an algorithm, ALG, and an oracle, O, satisfying the conditions of the theorem. Let $G = (V, E)$ be a d-degenerate input graph of maximum degree Δ. To each edge $e \in E$, the oracle associates a bit string, $B(e)$, of length $1 + \lceil\log(2d)\rceil + \lceil\log(d+1)\rceil$. The definition of B falls into two cases.

  Case: $\Delta < 2d$. The oracle computes an optimal edge coloring $\mathcal{C}$ of G. Since $\Delta < 2d$, Vizing's Theorem implies that at most $2d$ different colors are used in $\mathcal{C}$. Let $e \in E$. The first bit of $B(e)$ will be a 0. The next $\lceil\log(2d)\rceil$ bits will encode the color assigned to e in $\mathcal{C}$. The remaining $\lceil\log d\rceil$ bits of $B(e)$ are set arbitrarily.

  Case: $\Delta \geq 2d$. Fix $a, b \in \mathbb{N}$ such that $\Delta = a2d + b$ and $0 \leq b \leq 2d - 1$. By assumption, $a \geq 1$. The oracle computes an optimal edge coloring $\mathcal{C}$ of G. Since $\Delta \geq 2d$, Theorem 1 implies that $\Delta = a2d + b$ colors are used in $\mathcal{C}$. Let E_0 be the edges colored with the colors $1, 2, \ldots, b$. For $e \in E_0$, the bit string $B(e)$ is defined as follows: The first bit is a 0. The next $\lceil\log(2d)\rceil$ bits encode the color assigned to e in $\mathcal{C}$ (this is clearly possible since $b \leq 2d - 1$). The remaining $\lceil\log d\rceil$ bits of $B(e)$ are set arbitrarily.

  Let $G' = (V, E \setminus E_0)$. Since G' is $a2d$-edge-colorable, its maximum degree is at most $a2d$. On the other hand, no vertex in V is incident to more than b edges from E_0, and so the maximum degree is at least $\Delta - b = a2d$. Furthermore, removal of edges cannot increase the degeneracy of a graph. Thus, G' must be d-degenerate. For edges in G', the first bit of $B(e)$ is set to 1. The remaining bits of $B(e)$ are constructed by running Procedure 1 on G'.

  We will now define the algorithm, ALG. For technical reasons, and since the algorithm does not know $\chi'(G)$, we begin by allowing the algorithm to use colors from $\{0, 1\} \times \mathbb{N}$. The algorithm receives the advice $B(e)$ along with each edge $e \in E$. If the first bit of $B(e)$ is a 0, the algorithm learns which color, c, to use for $e \in E_0$ by reading the next $\lceil\log(2d)\rceil$ bits of $B(e)$. It then assigns the color $(0, c)$ to e. If the first bit of $B(e)$ is a 1 then ALG simulates, using the remaining bits of $B(e)$, the algorithm for the case where $2d$ divides Δ with G' as input graph. If that algorithm would assign the color c to e, ALG assigns the color $(1, c)$ to e. One can easily modify ALG to use colors from the set $\{1, \ldots, \chi'(G)\}$ as follows: The first time some color $(i, c), i \in \{0, 1\}$, is supposed to be used, ALG selects the lowest color c' from $\{1, \ldots, \chi'(G)\}$ which has not yet been used. From then on, ALG always uses the color c' instead of (i, c).   □

Improvements of the Algorithm. The family of algorithms from Theorem 2 can be used to create a single algorithm which works even if we do not assume that a constant upper bound on the degeneracy is known a priori. In the advice-with-request model, the oracle starts by computing the degeneracy, $\mathrm{dgn}(G)$, of the input graph G. Then, the oracle finds the largest integer d such that $1 + \lceil\log(2d)\rceil + \lceil\log(d+1)\rceil = 1 + \lceil\log(2\,\mathrm{dgn}(G))\rceil + \lceil\log(\mathrm{dgn}(G)+1)\rceil$. Clearly, $\mathrm{dgn}(G) \leq d$ and hence G is d-degenerate. When the algorithm receives the very first advice string, it determines d from the length of the advice received. From there on, Theorem 2 applies. Note that we do not increase the amount of advice by using d instead of $\text{dgn}(G)$ as an upper bound on the degeneracy. In the advice-on-tape model, the oracle can simply write the value d onto the advice tape in a self-delimiting way using $O(\log d)$ bits (e.g., by writing $\lceil \log d \rceil$ in unary and then d itself in binary). This gives the following theorem.

Theorem 3. *In both the advice-with-request and the advice-on-tape model, there exists an* EDGE-COLORING *algorithm which produces an optimal coloring and uses $O(m \log d)$ bits of advice in total, where d is the degeneracy of the input graph.*

We will show in Theorem 5 that in the advice-with-request model, at least $\Omega(\log d)$ bits per edge are required to achieve optimality. Note that this matches asymptotically the upper bound of Theorems 2 and 3. However, the exact number of bits used by the algorithm presented can be lowered. For example, we would like to mention that the algorithm can rather easily be modified to use only a single bit per edge in the case of 1-degenerate graphs (forests).

3   Lower Bounds

3.1   Sublinear Advice Is No Better Than No Advice

Recall that one very interesting aspect of the advice-on-tape model is that it allows an algorithm to read a sublinear number of advice bits. However, we will now show that linear advice is required to break the lower bound of 2 on the competitive ratio for EDGE-COLORING. We remark that the hard input instances used in Theorem 4 are the same as the ones used in [3] to obtain the lower bound of 2 for the competitive ratio of algorithms without advice. The proof of Theorem 4 essentially shows how the techniques used in [3] can be extended to obtain a lower bound which holds even for algorithms with sublinear advice.

Theorem 4. *Let $\varepsilon > 0$ and let* ALG *be a $(2 - \varepsilon)$-competitive* EDGE-COLORING *algorithm. Then* ALG *must read at least $\Omega(m)$ bits of advice, where m is the number of edges. This lower bound holds even if the input graph is required to be a forest. The constant of $\Omega(m)$ depends on ε.*

Proof. Let ALG be a $(2 - \varepsilon)$ competitive algorithm in the advice-on-tape model (in the advice-with-request model, Theorem 4 follows directly from [3]). By definition, there exists a constant c_0 such that $\text{ALG}(\sigma) \leq (2 - \varepsilon)\text{OPT}(\sigma) + c_0$ for any input sequence σ. Fix $\Delta \geq 2$ such that $\varepsilon\Delta > c_0 + 1$. The adversary graph will be a forest of maximum degree Δ (and therefore Δ-edge-colorable). We introduce some notation which will be used in the proof. Let $\alpha = (\Delta - 1) \cdot \binom{2\Delta - 2}{\Delta - 1} + 1$, let $\beta = \binom{\alpha}{\Delta}$ and let $R \in \mathbb{N}$ be a large integer. A *star* is the complete bipartite graph $K_{1, \Delta - 1}$. We say that a collection of stars are *colored the same* if the edges of all of the stars are colored using the same $\Delta - 1$ colors. The values α and β have been chosen such that the following holds:

Fact 1: Let $\mathcal{C}$ be an edge coloring of α stars using at most $2\Delta - 2$ colors. Then, at least Δ stars must be colored the same in $\mathcal{C}$.

Fact 2: Let $C_1, C_2, \ldots, C_k$ be edge colorings of α stars such that each edge coloring uses at most $2\Delta - 2$ colors. Then, there exist Δ stars such that these stars are colored the same in at least $\lceil k/\beta \rceil$ of the colorings.

Fact 1 follows from the pigeonhole principle. Fact 2 follows since there are β ways to select the Δ stars guaranteed to exist by Fact 1.

The total number of edges in the forest will be $m = (\alpha + \Delta)R$. Let b be the maximum number of advice bits read by ALG on inputs of length m. Note that each of the 2^b possible advice strings read by ALG on inputs of length m corresponds to a deterministic online algorithm without advice. Using this observation, we convert ALG into 2^b deterministic algorithms, $A_1, \ldots, A_{2^b}$, such that $\min_j A_j(\sigma) \leq \text{ALG}(\sigma)$ for any sequence σ of m edges. We say that A_j is *alive* if the number of colors used by A_j so far is at most $2\Delta - 2$.

We will now describe the adversary. The adversary starts by revealing R *rows*, where each row consists of α stars. The remaining edges are revealed in a number of rounds, one for each row.

In round i, where $1 \leq i \leq R$, the adversary uses the following strategy: Let k be the number of algorithms alive just before round i. The adversary selects Δ stars from row i which are colored the same by at least $\lceil k/\beta \rceil$ of the algorithms alive. Since an algorithm which is alive has used at most $2\Delta - 2$ colors, this is always possible by Fact 2. Let $v_1^i, \ldots, v_\Delta^i$ be the vertices of degree $\Delta - 1$ in the stars selected. The adversary reveals Δ edges, $(v, v_1^i), \ldots, (v, v_\Delta^i)$, connecting these vertices to a new vertex, v. An algorithm A_j which have colored the selected stars with the same $\Delta - 1$ colors is forced to use Δ other colors for these new edges and, hence, to use $2\Delta - 1$ colors in total. Thus, at the end of round i, the number of algorithms alive is at most $k - \lceil k/\beta \rceil \leq (1 - 1/\beta)\, k$.

Since $\varepsilon\Delta > c_0 + 1$, we have that $2\Delta - 1 > (2 - \varepsilon)\Delta + c_0$. In particular, at least one algorithm A_j must be alive after round R since the number of colors used by ALG is at most $(2 - \varepsilon)\Delta + c_0$. Before the first round, the number of algorithms alive is at most 2^b. After the last round, the number of algorithms alive is therefore at most $(1 - 1/\beta)^R 2^b$. Thus, it must hold that $1 \leq (1 - 1/\beta)^R 2^b$. But, this implies that

$$b \geq R \log\left(\frac{\beta}{\beta - 1}\right) = \frac{\log\left(\frac{\beta}{\beta - 1}\right)}{\alpha + \Delta} m = \Omega(m). \tag{1}$$

This proves the theorem since the number of rounds R (and therefore also m) can be chosen arbitrarily large. □

We remark that the hidden constant in the $\Omega(m)$ lower bound of Theorem 4 decreases very fast as ε tends to zero. However, the main point of Theorem 4, that sublinear advice does not offer any advantage, is not affected by this.

3.2   Tight Lower Bounds in the Advice-with-Request Model

As we have shown, $O(m)$ bits of advice suffice to achieve optimality for graphs of bounded degeneracy. A natural question is whether this holds for general graphs.

We give a partial negative answer by showing that in the advice-with-request model, this is not the case, not even for bipartite graphs.

It is a well-known result of König [26] that bipartite graphs are Δ-edge-colorable. In the proof of Theorem 5, we will use the following gadget to ensure that two edges cannot be assigned different colors in an optimal edge coloring.

Definition 1. *Let $n \geq 1$. The graph H_n consists of a complete bipartite graph $K_{n,n} = (L, R)$ together with vertices v_l, v_r and edges $\{(v_l, v) : v \in L\}$, $\{(v_r, v) : v \in R\}$. The vertex v_l (v_r) is denoted the leftmost (rightmost) vertex.*

We say that two edges $e_1 = (x_1, y_1)$ and $e_2 = (x_2, y_2)$ are *connected by an H_n* if y_1 is the leftmost vertex and y_2 is the rightmost vertex of the same H_n (and neither x_1 nor x_2 is part of that H_n). See Figure 1.

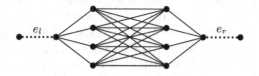

Fig. 1. G_4: Two edges e_l and e_r (dashed lines) connected by an H_4 (solid lines)

Lemma 2. *For $n \geq 1$, let G_n be the graph consisting of two edges, e_l and e_r, connected by an H_n. Then, G_n is $(n+1)$-edge-colorable. On the other hand, an edge coloring of G_n in which e_l and e_r are assigned different colors must use at least $n + 2$ colors.*

Proof. G_n can be edge colored using $n + 1$ colors since it is a bipartite graph of maximum degree $n+1$. Let $\mathcal{C}$ be an edge coloring of G_n such that $C(e_l) \neq C(e_r)$, where $C(e)$ is the color assigned to the edge e. Suppose, by way of contradiction, that only $n+1$ different colors are used in $\mathcal{C}$. Since e_l and e_r are colored differently, the set of colors used for edges between v_l and L cannot be identical to the set of colors used for edges between v_r and R, since this would contradict that $\mathcal{C}$ uses only $n + 1$ colors. Thus, there exists a color, c, such that there is an edge $e = (v_l, v), v \in L$ colored with the color c, while no edge between v_r and R is colored with the color c. It follows that for each $u \in R$, there must be an edge between u and a vertex in L colored with the color c, since u has degree $n + 1$, $\mathcal{C}$ uses $n+1$ colors and the edge (v_r, u) is not colored with the color c. In particular, since $|L| = |R|$, there must be an edge from a vertex in R to v colored with the color c. This is a contradiction, since (v_l, v) is also colored with the color c. $\square$

Theorem 5. *An optimal* EDGE-COLORING *algorithm in the advice-with-request model must use at least $\Omega(\log \Delta)$ bits of advice per edge, even for bipartite graphs, where Δ is the maximum degree of the input graph.*

Proof. Fix $\Delta \geq 2$. At the beginning, the adversary reveals two stars $K_{1,\Delta}$. Let $\{x_1, \ldots, x_\Delta\}$ and $\{y_1, \ldots, y_\Delta\}$ be the vertices of degree 1 in each of these two stars, and let x and y be the center vertices. Furthermore, let t be the time step right after both stars have been revealed. At time t, the adversary picks a permutation π of $\{1, 2, \ldots, \Delta\}$. For each $1 \leq i \leq \Delta$, the edge (x, x_i) is then connected to the edge $(y, y_{\pi(i)})$ by an $H_{\Delta-1}$ through x_i and $y_{\pi(i)}$. Since the resulting graph is bipartite and has maximum degree Δ, it can be colored using Δ colors.

Let ALG be an algorithm in the advice-with-request model such that, at time t, the total number of advice bits received by ALG is strictly less than $\log(\Delta!)$. We claim that ALG cannot be optimal. Note that the adversary has $\Delta!$ different permutations to choose from. This implies that there must exist two different permutations π, π' such that up until time t, the algorithm receives exactly the same bits of advice for both of these permutations. Thus, ALG produces the same coloring, $\mathcal{C}$, of the two stars no matter which of π and π' the adversary chooses to use. Let $C(u, v)$ be the color assigned to the edge $e = (u, v)$ in $\mathcal{C}$. Fix i such that $\pi(i) \neq \pi'(i)$. Because the edges are adjacent, $C(y, y_{\pi(i)}) \neq C(y, y_{\pi'(i)})$. Since $C(x, x_i)$ cannot be the same as both $C(y, y_{\pi(i)})$ and $C(y, y_{\pi'(y)})$, we may assume without loss of generality that $C(x, x_i) \neq C(y, y_{\pi(i)})$. By Lemma 2, this implies that ALG will use at least $\Delta + 1$ colors when the adversary chooses the permutation π. We conclude that an optimal algorithm must have received at least $\log(\Delta!)$ bits of advice at time t. Since only 2Δ edges are revealed before time t, this is only possible if the algorithm receives at least $\frac{\log(\Delta!)}{2\Delta} = \Omega(\log \Delta)$ bits of advice per edge.   □

For the adversary graph used in Theorem 5, the number of edges is $m = O(\Delta^3)$. Thus, we may restate the lower bound of $\Omega(\log \Delta)$ bits per edge in terms of m and get a lower bound of $\Omega(\log(m^{1/3})) = \Omega(\log m)$ bits per edge. This shows that even if we insist on measuring the amount of advice solely as a function of m (and not also Δ), the trivial upper bound of $O(m \log m)$ on the advice complexity is still asymptotically tight. Furthermore, the graph is Δ-regular. It follows that the degeneracy d of the graph is $d = \Delta$. Thus, the lower bound may also be stated in terms of the degeneracy as $\Omega(\log d)$ bits per edge. This matches asymptotically the upper bound of Theorems 2 and 3.

4   Concluding Remarks and Open Problems

As a consequence of Euler's formula, the degeneracy of a planar graph is at most 5. Thus, Theorem 2 implies that 8 bits of advice per edge (and hence $8m = O(m)$ bits in total) suffice to achieve optimality for planar graphs. On the other hand, since a forest is a planar graph, Theorem 4 shows that $\Omega(m)$ bits of advice are necessary just to achieve a competitive ratio better than 2. As mentioned, the greedy algorithm is 2-competitive [3] and uses no advice at all. Thus, Theorems 2 and 4 completely determines (asymptotically) the advice complexity of edge coloring planar graphs, in both models of advice complexity.

For bipartite graphs, the picture is not as clear. The lower bound of Theorem 5 for bipartite graphs relies on the assumption that an algorithm receives a fixed number of advice bits per edge, and so it only holds in the advice-with-request model. The lower bound may be viewed as a worst-case lower bound: We show that there exist some edges for which $\Omega(\log \Delta)$ bits of advice are required. The advice-on-tape model allows us to also study the amortized number of advice bits per edge. Determining the advice complexity of EDGE-COLORING for bipartite graphs in the advice-on-tape model is left as an interesting open problem.

Acknowledgments. The author wishes to thank Joan Boyar and Lene M. Favrholdt for helpful discussions.

References

1. Adamaszek, A., Renault, M.P., Rosén, A., van Stee, R.: Reordering buffer management with advice. In: Kaklamanis, C., Pruhs, K. (eds.) WAOA 2013. LNCS, vol. 8447, pp. 132–143. Springer, Heidelberg (2014)
2. Albers, S., Hellwig, M.: Online makespan minimization with parallel schedules. In: Ravi, R., Gørtz, I.L. (eds.) SWAT 2014. LNCS, vol. 8503, pp. 13–25. Springer, Heidelberg (2014)
3. Bar-Noy, A., Motwani, R., Naor, J.: The greedy algorithm is optimal for on-line edge coloring. Inf. Process. Lett. **44**(5), 251–253 (1992)
4. Barhum, K.: Tight bounds for the advice complexity of the online minimum steiner tree problem. In: Geffert, V., Preneel, B., Rovan, B., Štuller, J., Tjoa, A.M. (eds.) SOFSEM 2014. LNCS, vol. 8327, pp. 77–88. Springer, Heidelberg (2014)
5. Ben-David, S., Borodin, A.: A new measure for the study of on-line algorithms. Algorithmica **11**(1), 73–91 (1994)
6. Bianchi, M.P., Böckenhauer, H.-J., Hromkovič, J., Keller, L.: Online coloring of bipartite graphs with and without advice. Algorithmica **70**(1), 92–111 (2014)
7. Böckenhauer, H.-J., Komm, D., Královič, R., Královič, R.: On the advice complexity of the k-server problem. In: Aceto, L., Henzinger, M., Sgall, J. (eds.) ICALP 2011, Part I. LNCS, vol. 6755, pp. 207–218. Springer, Heidelberg (2011)
8. Böckenhauer, H.-J., Komm, D., Královič, R., Královič, R., Mömke, T.: On the advice complexity of online problems. In: Dong, Y., Du, D.-Z., Ibarra, O. (eds.) ISAAC 2009. LNCS, vol. 5878, pp. 331–340. Springer, Heidelberg (2009)
9. Böckenhauer, H.-J., Komm, D., Královič, R., Rossmanith, P.: The online knapsack problem: Advice and randomization. Theor. Comput. Sci. **527**, 61–72 (2014)
10. Borodin, A., Irani, S., Raghavan, P., Schieber, B.: Competitive paging with locality of reference. J. Comput. Syst. Sci. **50**(2), 244–258 (1995)
11. Boyar, J., Favrholdt, L.M., Kudahl, C., Mikkelsen, J.W.: Advice complexity for a class of online problems. In: STACS. LIPIcs, vol. 30, pp. 116–129 (2015). http://dx.doi.org/10.4230/LIPIcs.STACS.2014.174
12. Boyar, J., Kamali, S., Larsen, K.S., López-Ortiz, A.: On the list update problem with advice. In: Dediu, A.-H., Martín-Vide, C., Sierra-Rodríguez, J.-L., Truthe, B. (eds.) LATA 2014. LNCS, vol. 8370, pp. 210–221. Springer, Heidelberg (2014)
13. Boyar, J., Kamali, S., Larsen, K.S., López-Ortiz, A.: Online bin packing with advice. In: STACS. LIPIcs, vol. 25, pp. 174–186 (2014)
14. Chrobak, M., Nishizeki, T.: Improved edge-coloring algorithms for planar graphs. J. Algorithms **11**(1), 102–116 (1990)
15. Cole, R., Kowalik, L.: New linear-time algorithms for edge-coloring planar graphs. Algorithmica **50**(3), 351–368 (2008)

16. Dobrev, S., Královič, R., Markou, E.: Online graph exploration with advice. In: Even, G., Halldórsson, M.M. (eds.) SIROCCO 2012. LNCS, vol. 7355, pp. 267–278. Springer, Heidelberg (2012)

17. Dobrev, S., Královič, R., Královič, R.: Advice complexity of maximum independent set in sparse and bipartite graphs. Theor. Comput. Syst. **56**(1), 197–219 (2015)

18. Dobrev, S., Královič, R., Pardubská, D.: Measuring the problem-relevant information in input. RAIRO Theor. Inform. Appl. **43**(3), 585–613 (2009)

19. Emek, Y., Fraigniaud, P., Korman, A., Rosén, A.: Online computation with advice. Theor. Comput. Sci. **412**(24), 2642–2656 (2011)

20. Forišek, M., Keller, L., Steinová, M.: Advice complexity of online coloring for paths. In: Dediu, A.-H., Martín-Vide, C. (eds.) LATA 2012. LNCS, vol. 7183, pp. 228–239. Springer, Heidelberg (2012)

21. Holyer, I.: The NP-completeness of edge-coloring. SIAM J. Comput. **10**(4), 718–720 (1981)

22. Hromkovič, J., Královič, R., Královič, R.: Information complexity of online problems. In: Hliněný, P., Kučera, A. (eds.) MFCS 2010. LNCS, vol. 6281, pp. 24–36. Springer, Heidelberg (2010)

23. Jensen, T.R., Toft, B.: Graph Coloring Problems. Wiley (2011)

24. Karlin, A.R., Manasse, M.S., Rudolph, L., Sleator, D.D.: Competitive snoopy caching. Algorithmica **3**, 77–119 (1988)

25. Kenyon, C.: Best-fit bin-packing with random order. In: SODA, pp. 359–364 (1996)

26. König, D.: Über Graphen und ihre Anwendung auf Determinantentheorie und Mengenlehre. Math. Ann. **77**(4), 453–465 (1916)

27. Koutsoupias, E., Papadimitriou, C.H.: Beyond competitive analysis. SIAM J. Comput. **30**(1), 300–317 (2000)

28. Raghavan, P.: A statistical adversary for on-line algorithms. In: On-Line Algorithms, DIMACS Series in Discrete Mathematics and Theoretical Computer Science, pp. 79–83 (1991)

29. Renault, M.P., Rosén, A., van Stee, R.: Online algorithms with advice for bin packing and scheduling problems. CoRR, abs/1311.7589 (2013)

30. Seibert, S., Sprock, A., Unger, W.: Advice complexity of the online coloring problem. In: Spirakis, P.G., Serna, M. (eds.) CIAC 2013. LNCS, vol. 7878, pp. 345–357. Springer, Heidelberg (2013)

31. Sleator, D.D., Tarjan, R.E.: Amortized efficiency of list update and paging rules. Commun. ACM **28**(2), 202–208 (1985)

32. Stiebitz, M., Scheide, D., Toft, B., Favrholdt, L.M.: Graph Edge Coloring: Vizing's Theorem and Goldberg's Conjecture. Wiley (2012)

33. Vizing, V.G.: On an estimate of the chromatic class of a p-graph (in Russian). Metody Diskret. Analiz. **3**, 25–30 (1964)

34. Vizing, V.G.: Critical graphs with given chromatic class (in Russian). Metody Diskret. Analiz. **5**, 9–17 (1965)

35. Zhou, X., Nishizeki, T.: Edge-coloring and f-coloring for various classes of graphs. J. Graph Algorithms Appl. **3**(1) (1999)

Approximability of Two Variants of Multiple Knapsack Problems

Shuichi Miyazaki [1], Naoyuki Morimoto[2](✉), and Yasuo Okabe[1]

[1] Academic Center for Computing and Media Studies,
Kyoto University, Kyoto, Japan
shuichi@media.kyoto-u.ac.jp, okabe@i.kyoto-u.ac.jp
[2] Institute for Integrated Cell-Material Sciences (iCeMS), Kyoto University,
Yoshida-honmachi, Sakyo-ku, Kyoto 606-8501, Japan
nmorimoto@icems.kyoto-u.ac.jp

Abstract. This paper considers two variants of Multiple Knapsack Problems. The first one is the Multiple Knapsack Problem with Assignment Restrictions and Capacity Constraints (MK-AR-CC). In the MK-AR-CC(k) (where k is a positive integer), a subset of knapsacks is associated with each item and the item can be packed into only those knapsacks (Assignment Restrictions). Furthermore, the size of each knapsack is at least k times the largest item assignable to the knapsack (Capacity Constraints). The MK-AR-CC(k) is NP-hard for any constant k. In this paper, we give a polynomial-time $\left(1 + \frac{2}{k+1} + \epsilon\right)$-approximation algorithm for the MK-AR-CC(k), and give a lower bound on the approximation ratio of our algorithm by showing an integrality gap of $\left(1 + \frac{1}{k} - \epsilon\right)$ for the IP formulation we use in our algorithm, where ϵ is an arbitrary small positive constant. The second problem is the Splittable Multiple Knapsack Problem with Assignment Restrictions (S-MK-AR), in which the size of items may exceed the capacity of knapsacks and items can be split and packed into multiple knapsacks. We show that approximating the S-MK-AR with the ratio of $n^{1-\epsilon}$ is NP-hard even when all the items have the same profit, where n is the number of items and ϵ is an arbitrary positive constant.

Keywords: Multiple knapsack problem · Assignment restrictions · Approximation algorithms

1 Introduction

This paper considers two variants of Multiple Knapsack Problems, motivated by efficient power allocation in recent and future power networks. Efficient usage of natural power sources, e.g. solar power or wind power, has been studied actively with the spread of in-home power generations such as solar panels. These kinds of relatively small sources are generally called distributed power sources. Future power networks are supposed to include various distributed generations in addition to conventional commercial power sources, e.g. fossil fuel plants or

© Springer International Publishing Switzerland 2015
V.Th. Paschos and P. Widmayer (Eds.): CIAC 2015, LNCS 9079, pp. 365–376, 2015.
DOI: 10.1007/978-3-319-18173-8_27

nuclear power plants, and it is strongly desired these power sources are utilized effectively. Power sources have various characteristics such as cost, stability, and CO_2 emission. For example, power from fossil fuel plants is stable but relatively costly, while power from in-home solar panels is low-cost but unstable because it depends on weather conditions. Power consuming devices also have characteristics on quality of power that they require. For example, a desktop PC needs stable power, therefore it should be supplied from commercial sources, whereas a laptop with a battery accepts power from solar panels, because it can work with the battery even when the solar panels fails to generate stable power due to weather conditions. Therefore, it is desired to match power sources and power consuming devices in an appropriate manner [13,15].

Power allocation can be naturally formalized as a combinatorial optimization problem as follows: There are power devices and power sources. Each power device d has two values, the profit $p(d)$ and the power consumption $c(d)$, meaning that using the device d requires the power of $c(d)$, and if d can be used we gain the profit of $p(d)$. Each power source s has a capacity, that is a maximum power s can supply. Our goal is to allocate devices to sources so as to maximize the sum of the profits of allocated devices, while keeping the capacity constraint of each source. This problem can be viewed as the *Multiple Knapsack Problem* (*MK*) by regarding power sources as knapsacks and power consuming devices as items. Note that in the above mentioned characteristics-based power allocation, a device can be allocated to only a power source whose power quality matches the requirement of the device. One of the suitable extensions of the MK in this scenario is the *Multiple Knapsack Problem with Assignment Restriction* (*MK-AR*), where a subset of knapsacks is associated with each item and the item can be packed into only those knapsacks. The MK-AR is NP-hard since it is a generalization of the classical Knapsack Problem. As for the approximability, Nutov et al. [12] showed a simple 2-approximation algorithm for the MK-AR. They also proposed an $\frac{e}{e-1}$-approximation algorithm for the fixed-profit Generalized Assignment Problem (GAP), which includes the MK-AR as a special case.

Our Results. In this paper, we consider two extensions of the MK-AR based on the observations on real power networks.

First, the capacity of power sources, such as a commercial power source, is usually much larger than the power consumption of devices. Therefore, it is reasonable to consider instances in which item sizes are small and capacities of knapsacks are large. The problem we propose in this context is the *Multiple Knapsack Problem with Assignment Restrictions and Capacity Constraints* (*MK-AR-CC*). In the MK-AR-CC(k) (where k is a positive integer), the size of each knapsack is at least k times the largest item assignable to the knapsack. It is easy to see that the MK-AR-CC(k) is NP-hard for any constant k (by a straightforward reduction from the classical Knapsack Problem). In this paper we extend Nutov et al.'s 2-approximation algorithm [12] and give a polynomial-time $\left(1 + \frac{2}{k+1} + \epsilon\right)$-approximation algorithm for the MK-AR-CC(k). (In the case of $k = 1$, our algorithm is equivalent to Nutov et al.'s algorithm and hence the approximation ratio is 2 rather than $2+\epsilon$.) We also give a lower bound on the approximation ratio of our algorithm by showing an integrality gap of $\left(1 + \frac{1}{k} - \epsilon\right)$ for the IP formulation we use in our algorithm, where ϵ is an arbitrary positive constant.

The second scenario is that the capacity of distributed power sources is relatively small and one device may need to be allocated to two or more sources. We therefore consider the *Splittable Multiple Knapsack Problem with Assignment Restrictions (S-MK-AR)*, in which the size of items may exceed the capacity of knapsacks and one item can be split and packed into multiple knapsacks. We show that approximating the S-MK-AR with the ratio of $n^{1-\epsilon}$ is NP-hard even when all the items have the same profit, where n is the number of items and ϵ is an arbitrary positive constant.

Related Work. The MK-AR is a special case of the Generalized Assignment Problems (GAP). Approximation algorithms for the GAP and their variants have been studied actively. Shmoys and Tardos [14] presented a 2-approximation algorithm for the GAP. Later, Fleischer et al. [6] derived an $\frac{e}{e-1}$-approximation algorithm for restricted instances of Separable Assignment Problems that includes the GAP as a special case. Feige and Vondrak [5] have broken this barrier using randomization; they presented a randomized $\left(\frac{e}{e-1} - \epsilon\right)$-approximation algorithm for the GAP for some absolute constant $\epsilon > 0$. Cohen et al. [3] showed a combinatorial translation of any algorithm for the single knapsack problem into an approximation algorithm for the GAP, and showed a $(1 + \alpha)$-approximation algorithm for the GAP, where α is an approximation ratio for the single knapsack problem.

Dawande et al. [4] showed a combinatorial 2-approximation algorithm for the restricted case of the MK-AR where the size of an item is equal to its profit. Approximation algorithms for other restricted instances of the MK-AR have been studied as well [1,2].

2 The Multiple Knapsack Problem with Assignment Restrictions and Capacity Constraints

2.1 Problem Formulation

We define the Multiple Knapsack Problem with Assignment Restrictions (MK-AR) as follows. Its input is a bipartite graph $G = (I, J, E)$ with a set of edges E between I and J. Vertices of $I = \{a_1, a_2, \ldots, a_n\}$ correspond to items, and vertices of $J = \{b_1, b_2, \ldots, b_m\}$ correspond to knapsacks. Item $a \in I$ is assignable to knapsack $b \in J$ only if $(a, b) \in E$. For each item $a \in I$, the *profit* and the *size* of a, denoted by $p(a)$ and $\ell(a)$ respectively, are associated. For each knapsack $b \in J$, its *capacity* $c(b)$ is associated. A feasible solution of MK-AR is an assignment of items to knapsacks such that, for each b, the total size of assigned items to knapsack b is at most $c(b)$. The goal of the MK-AR is to maximize the total profit of assigned items. The MK-AR-CC(k) is a restriction of the MK-AR in which any instance satisfies the capacity constraints, namely, $c(b) \geq k\ell(a)$ for any a and b such that $(a, b) \in E$.

For convenience, we define the profit and the size of an edge e as the profit and the size, respectively, of the item incident to e. Formally, the edge $e = (a, b)$ has the profit $p(e) = p(a)$ and the size $\ell(e) = \ell(a)$. For a vertex v of G, $\delta(v)$ denotes the set of edges that are incident to vertex v. Then the MK-AR can be formulated as an integer program (IP) as follows, where x_e is a decision variable;

$$\max \sum_{e \in E} \frac{p(e)}{\ell(e)} x_e$$

$$\text{s.t.} \sum_{e \in \delta(b)} x_e \leq c(b), \forall b \in J$$

$$\sum_{e \in \delta(a)} x_e \leq \ell(a), \forall a \in I$$

$$x_e \in \{0, \ell(e)\}, \forall e \in E$$

The LP-relaxation of the MK-AR is defined by replacing the last constraint of IP formulation by "$x_e \leq \ell(e), \forall e \in E$".

2.2   Algorithm Match-and-FPTAS

The following corollary by Nutov et al. [12] is crucial for constructing and analyzing our algorithm. For a feasible solution x of the relaxation problem of the MK-AR, let $F(x)$ be the graph that consists of the set of *fractional edges* in x, namely, the set of edges e such that $0 < x_e < \ell(e)$, and their endpoint vertices.

Corollary 1. *(Nutov et al. [12]) Given a feasible solution x to the LP relaxation of the MK-AR, we can find in $O(|E(G)|^2)$ time a feasible solution z such that (i) $\sum_{e \in E} \frac{p(e)}{\ell(e)} z_e \geq \sum_{e \in E} \frac{p(e)}{\ell(e)} x_e$, (ii) $F(z)$ is a forest, and (iii) in any connected component of $F(z)$, at most one leaf belongs to I.*

A formal description of our approximation algorithm Match-and-FPTAS, which is based on Nutov et al.'s algorithm for the MK-AR [12], is given in Algorithm 1. It first obtains an optimal solution x^* of an LP relaxation of the MK-AR-CC(k), using a polynomial time algorithm for linear programming [7,11]. It then constructs a solution z from x^* using Corollary 1. In Step 3, for each knapsack b_j, we construct a single knapsack problem I_j consisting of knapsack b_j, *full items*, and at most one *matched item*. Full items are those assigned to b_j in x^* by an integral edge (i.e., an edge e such that $x_e = \ell(e)$). A matched item is defined as follows. We construct a maximum cardinality matching M in $F(z)$ (the graph that consists of the set of fractional edges in the solution z and their endpoint vertices) using the Hungarian method [9]. Then the matched item is the one matched with b_j in M if any. In Step 4, Match-and-FPTAS obtains a near-optimal solution for each single knapsack problem I_j, using an FPTAS [8,10], and finally in Step 5, it outputs the union of selected items for each solution of I_j.

Algorithm 1. Algorithm Match-and-FPTAS for the MK-AR-CC(k)

Step 1. Obtain an optimal solution x^* of a relaxed instance of the MK-AR-CC(k), using a polynomial time algorithm for linear programming.
Step 2. Construct a solution z from x^* as shown in Corollary 1.
Step 3. Construct instances of the single knapsack problem by matching fractional items in $F(z)$ to knapsacks.
Step 4. Using an FPTAS, obtain near-optimal solution for each instance of the single knapsack problem derived in Step 3.
Step 5. Output the union of the solutions obtained in Step 4 as a solution for the original MK-AR-CC(k) instance.

2.3   Analysis of the Approximation Ratio of Match-and-FPTAS

In our analysis, we focus on an instance I_j of the single knapsack problem derived in Step 3, and analyze its optimal solution.

For a set S of items, let $\ell(S)$ and $p(S)$ denote the total size and profit, respectively, of items in S. Let A_j denote the set of items in I_j, and let m_j denote, if any, the matched item of I_j derived in Step 3. Note that, if there does exist the matched item, then $\ell(A_j) < c(b_j) + \ell_{max_j}$ holds where ℓ_{max_j} is the maximum size of the items assignable to knapsack b_j, since the total size of the full items is less than $c(b_j)$ and $\ell(m_j) \leq \ell_{max_j}$ by definition. Let OPT_j be an optimal solution of I_j, and X_j be the set of items packed in the knapsack b_j in OPT_j. Define $Y_j = A_j \setminus X_j$ as the set of items not packed in OPT_j.

First, we show some properties of an optimal solution OPT_j.

Lemma 1. *For any $S \subseteq X_j$ such that $\ell(S) \geq \ell_{max_j}$, $p(S) \geq p(Y_j)$.*

Proof. If we remove S from OPT_j, the vacancy of the knapsack is $\ell(S) + (c(b_j) - \ell(X_j)) \geq \ell_{max_j} + (c(b_j) - \ell(X_j))$. On the other hand, $\ell(Y_j) = \ell(A_j) - \ell(X_j) = (\ell(A_j) - c(b_j)) + (c(b_j) - \ell(X_j)) < \ell_{max_j} + (c(b_j) - \ell(X_j))$. Therefore, we can replace the item set S by Y_j in OPT_j to obtain another feasible solution. If $p(S) < p(Y_j)$, the profit of the new solution is larger than that of OPT_j, contradicting the optimality of OPT_j. □

Lemma 2. *For any $S \subseteq X_j$ and a positive integer k' such that $\ell(S) \geq k'\ell_{max_j}$, $p(S) \geq \left\lceil \frac{k'}{2} \right\rceil p(Y_j)$.*

Proof. Since the size of any item is at most ℓ_{max_j}, we can partition S into at least $\left\lceil \frac{k'\ell_{max_j}}{2\ell_{max_j}} \right\rceil = \left\lceil \frac{k'}{2} \right\rceil$ subsets $S_1, S_2, \ldots, S_z$ ($z \geq \left\lceil \frac{k'}{2} \right\rceil$), such that $\ell_{max_j} \leq \ell(S_i) < 2\ell_{max_j}$ for all i. Since each S_i satisfies the condition of Lemma 1, we have that $p(S_i) \geq p(Y_j)$. Hence, $p(S) = \sum_{i=1}^{z} p(S_i) \geq \left\lceil \frac{k'}{2} \right\rceil p(Y_j)$. □

Lemma 3. *Suppose $|Y_j| = 1$. Then, for any $S \subseteq X_j$ such that $\ell(S) + (c(b_j) - \ell(X_j)) \geq \ell_{max_j}$, $p(S) \geq p(Y_j)$.*

Proof. The proof goes like that of Lemma 1. If we remove S from OPT_j, the vacancy of the knapsack is $\ell(S) + (c(b_j) - \ell(X_j)) \geq \ell_{max_j}$. Since $|Y_j| = 1$ and hence $\ell(Y_j) \leq \ell_{max_j}$, we can replace S by Y_j to obtain a feasible solution. If $p(S) < p(Y_j)$, the new solution is better than OPT_j, a contradiction. $\square$

Lemma 4. *Suppose* $|Y_j| \geq 2$. *Then, for any* $S \subseteq X_j$ *such that* $\ell(S) + (c(b_j) - \ell(X_j)) \geq \ell_{max_j}$, $p(S) \geq \frac{1}{2}p(Y_j)$.

Proof. As before, if we remove all the items in S from OPT_j, we have the vacancy of $\ell(S) + (c(b_j) - \ell(X_j)) \geq \ell_{max_j}$. Let d denote an item with the lowest profit in Y_j. Then $p(d) \leq \frac{p(Y_j)}{|Y_j|} \leq \frac{1}{2}p(Y_j)$ holds. Since X_j does not contain d, $\ell(X_j) + \ell(d) > c(b_j)$ (otherwise, we can add d to OPT_j to get a better solution). Since $\ell(Y_j \setminus \{d\}) = \ell(A_j) - (\ell(X_j) + \ell(d)) < \ell(A_j) - c(b_j) < \ell_{max_j}$, we can pack all the items in $Y_j \setminus \{d\}$ if we remove S from OPT_j. Hence, by the optimality of OPT_j, $p(S) \geq p(Y_j) - p(d) \geq p(Y_j) - \frac{1}{2}p(Y_j) = \frac{1}{2}p(Y_j)$. $\square$

The following lemma is crucial to the analysis.

Lemma 5. $\frac{p(A_j)}{p(X_j)} \leq 1 + \frac{2}{k+1}$.

Proof. Since $\frac{p(A_j)}{p(X_j)} = 1 + \frac{p(Y_j)}{p(X_j)}$, it suffices to show that $p(X_j) \geq \frac{k+1}{2}p(Y_j)$. If $|Y_j| = 0$, then $p(Y_j) = 0$ and the above inequality is satisfied trivially. Hence, from now on, we assume that $|Y_j| \geq 1$. Note that $\ell(X_j) > c(b_j) - \ell_{max_j}$, since otherwise, at least one item in Y_j could be packed in the knapsack b_j, contradicting the optimality of OPT_j.

We will do a case analysis depending on whether $m_j \in X_j$ or not. Let k_1 denote the positive integer such that $k_1 \leq \frac{c(b_j)}{\ell_{max_j}} < k_1 + 1$, i.e. the integer part of the ratio of $c(b_j)$ to ℓ_{max_j}. Note that $k \leq k_1$ holds by the capacity constraint.

Case 1. $m_j \in X_j$. In this case, all the items in Y_j are full items. Note that all the full items in I_j can be packed in the knapsack b_j. Hence, if we remove m_j from OPT_j and add all the items in Y_j, the result is a feasible solution. Therefore, by the optimality of OPT_j, $p(m_j) \geq p(Y_j)$ holds. Hereafter, we will do a case analysis depending on k_1.

Case 1-(i). $k_1 = 1$. By the optimality of OPT_j, $p(X_j) \geq p(m_j) \geq p(Y_j) = \frac{k_1+1}{2}p(Y_j) \geq \frac{k+1}{2}p(Y_j)$.

Case 1-(ii). $k_1 = 2$. Let $L = X_j \setminus \{m_j\}$. Since $|Y_j| \geq 1$ and $\ell(L) + (c(b_j) - \ell(X_j)) = c(b_j) - \ell(m_j) \geq k_1\ell_{max_j} - \ell_{max_j} = \ell_{max_j}$, $p(L) \geq \frac{1}{2}p(Y_j)$ holds by Lemmas 3 and 4. Hence, $p(X_j) = p(m_j) + p(L) \geq p(Y_j) + \frac{1}{2}p(Y_j) = \frac{3}{2}p(Y_j) = \frac{k_1+1}{2}p(Y_j) \geq \frac{k+1}{2}p(Y_j)$.

Case 1-(iii). k_1 **is odd and** $k_1 \geq 3$. Let $L = X_j \setminus \{m_j\}$ as shown in Fig. 1. Since $\ell(m_j) \leq \ell_{max_j}$ and $\ell(m_j) + \ell(L) = \ell(X_j) > c(b_j) - \ell_{max_j}$, $\ell(L) > c(b_j) - \ell_{max_j} - \ell(m_j) \geq (k_1 - 2)\ell_{max_j}$. Therefore, we obtain $p(L) \geq \lceil \frac{k_1-2}{2} \rceil p(Y_j) = \frac{k_1-1}{2}p(Y_j)$ by Lemma 2. Hence, $p(X_j) = p(m_j) + p(L) \geq p(Y_j) + \frac{k_1-1}{2}p(Y_j) \geq \frac{k_1+1}{2}p(Y_j) \geq \frac{k+1}{2}p(Y_j)$.

Case 1-(iv). k_1 **is even and** $k_1 \geq 4$. We partition X_j into three subsets $\{m_j\}$, L, and R as follows: L is a subset of $X_j \setminus \{m_j\}$ that satisfies $(k_1 - 2)\ell_{max_j} \leq$

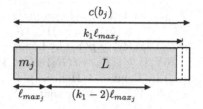

Fig. 1. Case 1-(iii): $m_j \in X_j$ and $k_1 (\geq 3)$ is odd

$\ell(m_j) + \ell(L) < (k_1 - 1)\ell_{max_j}$ and $R = X_j \setminus (L \cup \{m_j\})$ (see Fig. 2). Since $\ell(m_j) \leq \ell_{max_j}$, we have $\ell(L) \geq (k_1 - 3)\ell_{max_j}$. Therefore, $p(L) \geq \lceil \frac{k_1 - 3}{2} \rceil p(Y_j) = \left(\frac{k_1}{2} - 1 \right) p(Y_j)$ by Lemma 2. Also, since $|Y_j| \geq 1$ and $\ell(R) + (c(b_j) - \ell(X_j)) \geq \ell_{max_j}$, $p(R) \geq \frac{1}{2} p(Y_j)$ holds by Lemmas 3 and 4. Hence, $p(X_j) = p(m_j) + p(L) + p(R) \geq p(Y_j) + (\frac{k_1}{2} - 1)p(Y_j) + \frac{1}{2}p(Y_j) = \frac{k_1 + 1}{2}p(Y_j) \geq \frac{k+1}{2}p(Y_j)$.

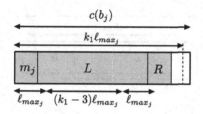

Fig. 2. Case 1-(iv): $m_j \in X_j$ and $k_1 (\geq 4)$ is even

Case 2. $m_j \notin X_j$. In this case, $Y_j = \{m_j\}$ because all the full items can be packed in the knapsack. Again, we will do a case analysis depending on k_1.

Case 2-(i). $k_1 = 1$. By the optimality of OPT_j, $p(X_j) \geq p(m_j) = p(Y_j) = \frac{k_1 + 1}{2}p(Y_j) \geq \frac{k+1}{2}p(Y_j)$.

Case 2-(ii). $k_1 = 2$. We partition X_j into three subsets R_1, R_2, and R_3 as follows (see Fig. 3): First we sort all the items in X_j as $\{a_1, a_2, \ldots, a_{|X_j|}\}$ in descending order of sizes. Then we define $R_1 = \{a_1, a_2, \ldots, a_{q-1}\}$, $R_2 = \{a_q\}$, and $R_3 = \{a_{q+1}, a_{q+2}, \ldots, i_{|OPT'_j|}\}$, where q is the positive integer that satisfies $\ell(R_1) \leq \ell_{max_j} < \ell(R_1) + \ell(R_2)$. Note that $\ell(R_1) \geq \ell(R_2)$ holds.

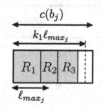

Fig. 3. Case 2-(ii): $m_j \notin X_j$ and $k_1 = 2$

Since $\ell(R_1) + \ell(R_2) > \ell_{max_j}$, we obtain $p(R_1) + p(R_2) \geq p(Y_j)$ by Lemma 1. Since $|Y_j| = 1$ and $\ell(R_2) + \ell(R_3) + (c(b_j) - \ell(X_j)) \geq c(b_j) - \ell(R_1) \geq k_1\ell_{max_j} - \ell_{max_j} = \ell_{max_j}$, we obtain $p(R_2) + p(R_3) \geq p(Y_j)$ by Lemma 3. Also, since $|Y_j| = 1$ and $\ell(R_3) + \ell(R_1) + (c(b_j) - \ell(X_j)) \geq \ell(R_3) + \ell(R_2) + (c(b_j) - \ell(X_j)) \geq \ell_{max_j}$, we obtain $p(R_3) + p(R_1) \geq p(Y_j)$ by Lemma 3. Hence, $(p(R_1) + p(R_2)) + (p(R_2) + p(R_3)) + (p(R_3) + p(R_1)) \geq 3p(Y_j)$, resulting that $p(X_j) = p(R_1) + p(R_2) + p(R_3) \geq \frac{3}{2}p(Y_j) = \frac{k_1+1}{2}p(Y_j) \geq \frac{k+1}{2}p(Y_j)$.

Case 2-(iii). k_1 is odd and $k_1 \geq 3$. We partition X_j into two subsets L and R as follows: L is a subset of X_j that satisfies $(k_1 - 2)\ell_{max_j} \leq \ell(L) < (k_1 - 1)\ell_{max_j}$, and $R = X_j \setminus L$ (see Fig. 4). Since $\ell(L) \geq (k_1 - 2)\ell_{max_j}$, $p(L) \geq \lceil \frac{k_1-2}{2} \rceil p(Y_j) = \frac{k_1-1}{2}p(Y_j)$ holds by Lemma 2. Since $|Y_j| = 1$ and $\ell(R) + (c(b_j) - \ell(X_j)) = c(b_j) - \ell(L) > k_1\ell_{max_j} - (k_1 - 1)\ell_{max_j} = \ell_{max_j}$, we have that $p(R) \geq p(Y_j)$ by Lemma 3. Hence, $p(X_j) = p(L) + p(R) \geq \frac{k_1-1}{2}p(Y_j) + p(Y_j) = \frac{k_1+1}{2}p(Y_j) \geq \frac{k+1}{2}p(Y_j)$.

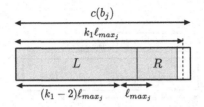

Fig. 4. Case 2-(iii): $m_j \notin X_j$ and $k_1(\geq 3)$ is odd

Case 2-(iv). k_1 is even and $k_1 \geq 4$. We partition X_j into four subsets L, R_1, R_2, and R_3 as follows (see also Fig. 5): First we sort all the items in X_j as $\{a_1, a_2, \ldots, a_{|X_j|}\}$ in descending order of sizes. Then we partition them as $L = \{a_1, a_2, \ldots, a_{q-1}\}$, $R_1 = \{a_q, a_{q+1}, \ldots, a_{r-1}\}$, $R_2 = \{a_r\}$, and $R_3 = \{a_{r+1}, a_{r+2}, \ldots, a_{|X_j|}\}$, where q and r are positive integers that satisfy $(k_1 - 3)\ell_{max_j} \leq \ell(L) < (k_1 - 2)\ell_{max_j}$, $(k_1 - 2)\ell_{max_j} \leq \ell(L) + \ell(R_1) < (k_1 - 1)\ell_{max_j}$, and $(k_1 - 1)\ell_{max_j} \leq \ell(L) + \ell(R_1) + \ell(R_2)$. Note that $\ell(R_1) \geq \ell(R_2)$ holds.

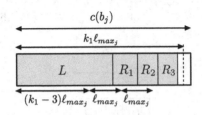

Fig. 5. Case 2-(iv): $m_j \notin X_j$ and $k_1(\geq 4)$ is even

Since $\ell(L) \geq (k_1 - 3)\ell_{max_j}$, we obtain $p(L) \geq \lceil \frac{k_1-3}{2} \rceil p(Y_j) = (\frac{k_1}{2} - 1)p(Y_j)$ by Lemma 2. Since $\ell(R_1) + \ell(R_2) = (\ell(L) + \ell(R_1) + \ell(R_2)) - \ell(L) > (k_1 - 1)\ell_{max_j} -$

$(k_1 - 2)\ell_{max_j} = \ell_{max_j}$, we obtain $p(R_1) + p(R_2) \geq p(Y_j)$ by Lemma 1. Since $|Y_j| = 1$ and $\ell(R_2) + \ell(R_3) + (c(b_j) - \ell(X_j)) \geq c(b_j) - (\ell(L) + \ell(R_1)) > k_1 \ell_{max_j} - (k_1 - 1)\ell_{max_j} = \ell_{max_j}$, we have $p(R_2) + p(R_3) \geq p(Y_j)$ by Lemma 3. Also, since $|Y_j| = 1$ and $\ell(R_3) + \ell(R_1) + (c(b_j) - \ell(X_j)) \geq \ell(R_3) + \ell(R_2) + (c(b_j) - \ell(X_j)) > \ell_{max_j}$, we have that $p(R_3) + p(R_1) \geq p(Y_j)$ by Lemma 3. Therefore, it follows that $(p(R_1) + p(R_2)) + (p(R_2) + p(R_3)) + (p(R_3) + p(R_1)) \geq 3p(Y_j)$, resulting that $p(R_1) + p(R_2) + p(R_3) \geq \frac{3}{2}p(Y_j)$. Hence, $p(X_j) = p(L) + p(R_1) + p(R_2) + p(R_3) \geq (\frac{k_1}{2} - 1)p(Y_j) + \frac{3}{2}p(Y_j) = \frac{k_1+1}{2}p(Y_j) \geq \frac{k+1}{2}p(Y_j)$. ☐

Theorem 1. *Match-and-FPTAS is a $\left(1 + \frac{2}{k+1} + \epsilon\right)$-approximation algorithm for the MK-AR-CC(k), where ϵ is an arbitrary positive constant.*

Proof. Let $p(OPT)$ and $p(LPOPT)$ denote the profits of optimal solutions for the MK-AR-CC(k) and its LP-relaxation, respectively. Note that $p(LPOPT) \leq \sum_j p(A_j)$. Also, let $p(MF)$ denote the profit of the solution obtained by Match-and-FPTAS, and $p(MF_j)$ denote the profit of the solution for I_j obtained in Step 4 of Match-and-FPTAS. Let ϵ' be a positive constant that satisfies $\epsilon' \leq \frac{\epsilon(k+1)}{k+3+\epsilon(k+1)}$. Since we use an FPTAS for the knapsack problem in Step 4, we can have $p(MF_j) \geq (1 - \epsilon')p(X_j)$. By Lemma 5, we obtain

$$\frac{p(OPT)}{p(MF)} \leq \frac{p(LPOPT)}{p(MF)} \leq \frac{\sum_j p(A_j)}{\sum_j p(MF_j)}$$

$$\leq \max_j \left\{\frac{p(A_j)}{p(MF_j)}\right\} \leq \max_j \left\{\frac{p(A_j)}{(1 - \epsilon')p(X_j)}\right\}$$

$$\leq \frac{1}{1 - \epsilon'}\left(1 + \frac{2}{k+1}\right)$$

$$\leq 1 + \frac{2}{k+1} + \epsilon.$$

☐

2.4 Integrality Gap of the IP Formulation Used in Match-and-FPTAS

Theorem 2. *The integrality gap of the IP formulation of the MK-AR-CC(k) used in Match-and-FPTAS is at least $1 + \frac{1}{k} - \epsilon$ for any positive constant ϵ.*

Proof. We consider an instance of the MK-AR-CC(k) including $k+1$ items with profit 1 and size 1, and one knapsack with capacity $k + 1 - \epsilon$. The profit of an optimal solution for LP-relaxation is $k + 1 - \epsilon$, while that for the original MK-AR-CC(k) is k. Hence, the integrality gap is $1 + \frac{1}{k} - \frac{\epsilon}{k} \geq 1 + \frac{1}{k} - \epsilon$. ☐

3 Splittable Multiple Knapsack Problem with Assignment Restrictions

3.1 Problem Definition

We define the Splittable Multiple Knapsack Problem with Assignment Restrictions (S-MK-AR) as follows. We are given a bipartite graph $G = (I, J, E)$, where

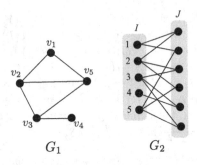

Fig. 6. An example of the reduction from the Maximum Independent Set Problem to the S-MK-AR

I and J correspond to items and knapsacks, respectively. A vertex (item) $a \in I$ has the *size* $\ell(a)$ and the *profit* $p(a)$. A vertex (knapsack) $b \in J$ has the *capacity* $c(b)$. A feasible solution of S-MK-AR is a non-negative weight on each $e \in E$ such that the total weight of edges incident to a vertex $b \in J$ is at most $c(b)$ and the total weight of edges incident to a vertex $a \in I$ is at most $\ell(a)$. If the total weight of incident edges of a equals $\ell(a)$, we say that the item a is *satisfied*. The *profit of a feasible solution* is the sum of the profits of the satisfied items, and the goal of the S-MK-AR is to maximize it.

3.2 Approximation Hardness of the S-MK-AR

Theorem 3. *For any positive constant ϵ, there is no polynomial-time $n^{1-\epsilon}$-approximating algorithm for the S-MK-AR unless P=NP, even if all the items have the same profit, where n is the number of items in an input.*

Proof. The proof is done by a reduction from the Maximum Independent Set Problem (MIS). For a graph $G = (V, E)$, a subset $S \subseteq V$ is called an *independent set* if there is no edge between any pair of vertices in S. MIS is the problem of, given a graph G, finding an independent set of G with the maximum cardinality.

Given an instance $G_1 = (V, E)$ of MIS, we construct an instance L_2 of the S-MK-AR (whose underlying graph is $G_2 = (I, J, E')$) as follows (an example of the reduction is illustrated in Fig. 6).

Without loss of generality, we can assume that G_1 has no isolated vertex. Suppose that G_1 has n vertices $v_1, \ldots, v_n$, and m edges $e_1, \ldots, e_m$. Then, I has n vertices $a_1, \ldots, a_n$, where a_i corresponds to the vertex v_i of G_1. Similarly, J has m vertices $b_1, \ldots, b_m$, where b_j corresponds to the edge e_j of G_1. Finally, we define E' as $(a_i, b_j) \in E'$ if and only if the vertex v_i is incident to the edge e_j in G_1. The capacity of the knapsack b_j is $c(b_j) = 1$ for every j. The profit and the size of an item a_i is $p(a_i) = 1$ and $\ell(a_i) = deg(v_i)$, respectively, where $deg(v_i)$ denotes the degree of v_i. Clearly this reduction can be done in polynomial time.

Suppose that S is an independent set of G_1. We will construct a solution T of L_2 in such a way that if $v_i \in S$ then all the edges in $\delta(a_i)$ have weight 1, and if $v_i \notin S$ then all the edges in $\delta(a_i)$ have weight 0 (recall that $\delta(a_i)$ is the set of edges incident to a_i). Note that this is a feasible solution because, for any $b_j \in J$, at most one edge in $\delta(b_j)$ has weight 1 (and other edge(s) have weight 0), since S is an independent set. Also, it is not hard to see that for any i such that $v_i \in S$, the item a_i is satisfied since all of its incident edges have weight 1. Therefore, the profit of T is $|S|$. This implies that $OPT(L_2) \geq OPT(G_1)$, where $OPT(G_1)$ denotes the size of a maximum independent set of G_1 and $OPT(L_2)$ denotes the profit of an optimal solution of L_2.

Next, consider a feasible solution T of L_2, and denote its profit $p(T)$. It is not hard to see from the above argument that if we choose a vertex v_i for every i such that a_i is satisfied by T, we have an independent set of G_1 whose size is the same as $p(T)$. Therefore, from a feasible solution T of L_2, we can construct in polynomial time an independent set S such that $|S| = p(T)$.

Now, suppose that there is a polynomial time $n^{1-\epsilon}$-approximation algorithm ALG_1 for the S-MK-AR. Consider the following approximation algorithm ALG_2 for MIS: Given an instance G_1 of MIS, ALG_2 first transforms it to L_2 using the above reduction. It then solves L_2 using ALG_1 and obtains a solution T. Finally, ALG_2 transforms T into an independent set S of G_1. By the above arguments, we have that $\frac{OPT(G_1)}{|S|} = \frac{OPT(L_2)}{p(T)} \leq n^{1-\epsilon}$, that is, ALG_2 is an $n^{1-\epsilon}$-approximation algorithm for MIS. It is known that existence of an $n^{1-\epsilon}$-approximation algorithm for MIS implies P=NP [16]. This completes the proof.

$\square$

4 Conclusion

In this paper, we have considered two variants of the Multiple Knapsack Problems, motivated by efficient power allocation in power networks. As future work, it would be challenging to fill the gap between upper and lower bounds on approximation ratios. It would also be interesting to formulate the problem in an online manner, considering the scenario where power consuming devices and power sources appear and disappear dynamically.

Acknowledgements. The authors would like to thank anonymous reviewers for their helpful comments. This work was supported by JSPS KAKENHI Grant Number 24500013, JST Super Cluster Program, and NICT Advanced Telecommunication Research Fund.

References

1. Aerts, J., Korst, J., Spieksma, F.: Approximation of a retrieval problem for parallel disks. In: Proceedings of the 5th Italian Conference on Algorithms and Complexity (CIAC), pp. 178–188 (2003)
2. Aerts, J., Korst, J., Spieksma, F., Verhaegh, W., Woeginger, G.: Random redundant storage in disk arrays: complexity of retrieval problems. IEEE Transactions on Computers **52**(9), 1210–1214 (2003)

3. Cohen, R., Katzir, L., Raz, D.: An efficient approximation for the generalized assignment problem. Information Processing Letters **100**(4), 162–166 (2006)
4. Dawande, M., Kalagnanam, J., Keskinocak, P., Salman, F.S., Ravi, R.: Approximation algorithms for the multiple knapsack problem with assignment restrictions. Journal of Combinatorial Optimization **4**(2), 171–186 (2000)
5. Feige, U., Vondrak, J.: Approximation algorithms for allocation problems: improving the factor of $1 - 1/e$. In: Proceedings of the 47th Annual IEEE Symposium on Foundations of Computer Science (FOCS), pp. 667–676 (2006)
6. Fleischer, L., Goemans, M.X., Mirrokni, V.S., Sviridenko, M.: Tight approximation algorithms for maximum separable assignment problems. Mathematics of Operations Research **36**(3), 416–431 (2011)
7. Karmarkar, N.: A new polynomial-time algorithm for linear programming. In: Proceedings of the 16th Annual ACM Symposium on Theory of Computing (STOC), pp. 302–311 (1984)
8. Kellerer, H., Pferschy, U.: A new fully polynomial time approximation scheme for the knapsack problem. Journal of Combinatorial Optimization **3**(1), 59–71 (1999)
9. Kuhn, H.W.: The hungarian method for the assignment problem. Naval Research Logistics Quarterly **2**(1–2), 83–97 (1955)
10. Magazine, M.J., Oguz, O.: A fully polynomial approximation algorithm for the 0–1 knapsack problem. European Journal of Operational Research **8**(3), 270–273 (1981)
11. Murty, K.G.: Network Programming. Prentice-Hall, Inc (1992)
12. Nutov, Z., Beniaminy, I., Yuster, R.: A $(1 - 1/e)$-approximation algorithm for the generalized assignment problem. Operations Research Letters **34**(3), 283–288 (2006)
13. Sakai, K., Okabe, Y.: Quality-aware energy routing toward on-demand home energy networking. In: Proceeding of the 2011 IEEE Consumer Communications and Networking Conference (CCNC), pp. 1041–1044 (2011)
14. Shmoys, D.B., Tardos, É.: An approximation algorithm for the generalized assignment problem. Mathematical Programming **62**(1–3), 461–474 (1993)
15. Takuno, T., Kitamori, Y., Takahashi, R., Hikihara, T.: AC power routing system in home based on demand and supply utilizing distributed power sources. Energies **4**(5), 717–726 (2011)
16. Zuckerman, D.: Linear degree extractors and the inapproximability of max clique and chromatic number. In: Proceedings of the 38th Annual ACM Symposium on Theory of Computing (STOC), pp. 681–690 (2006)

Block Sorting Is APX-Hard

N.S. Narayanaswamy[1]([⊠]) and Swapnoneel Roy[2]

[1] Department of Computer Science and Engineering,
Indian Institute of Technology Madras, Chennai, TN 600036, India
narayanaswamy@gmail.com
[2] School of Computing, University of North Florida, Jacksonville, Florida 32224, USA
s.roy@unf.edu

Abstract. BLOCK SORTING is an NP-hard combinatorial optimization problem motivated by applications in Computational Biology and Optical Character Recognition (OCR). It has been approximated in P time within a factor of 2 using two different techniques and the complexity of better approximations has been open for close to a decade now. In this work we prove that BLOCK SORTING does not admit a PTAS unless P = NP i.e. it is APX-Hard. The hardness result is based on new properties, that we identify, of the existing NP-hardness reduction from E3-SAT to BLOCK SORTING. In an attempt to obtain an improved approximation for BLOCK SORTING, we consider a generalization of the well-studied BLOCK MERGING, called k-BLOCK MERGING which is defined for each $k \geq 1$, and the 1- BLOCK MERGING problem is the same as the BLOCK MERGING problem. We show that the optimum k-BLOCK MERGING is an $1 + \frac{1}{k}$-approximation to the optimum block sorting. We then show that for each $k \geq 2$, we prove k-BLOCK MERGING to be NP-Hard, thus proving a dichotomy result associated with block sorting.

1 Introduction

A *sorted sequence* in a permutation π is a sequence of consecutive elements which are also consecutive in the identity permutation *id*. In the BLOCK SORTING problem, a block in a permutation π is defined as a maximal sorted sequence. BLOCK SORTING is to find the shortest sequence of blocks to be moved to sort a given permutation π. The number of moves in such a shortest sequence is denoted by $bs(\pi)$. BLOCK SORTING is also motivated by its applications in OCR [2], [7]. In OCR, to measure how good a *zoning algorithm* is, we need to find the minimum number of steps required to transform the string generated by the zoning algorithm to the correct string. This is equivalent to BLOCK SORTING. BLOCK SORTING is also a nontrivial variation of SORTING BY TRANSPOSITIONS. SORTING BY TRANSPOSITIONS arises in the context of genome rearrangements in computational biology. In transpositions, we are allowed to move any substring of π (not necessarily a block) to a different position at each step [1]. SORTING BY TRANSPOSITIONS is to compute the minimum number of such moves to sort π, it has been recently shown to be NP-Hard [4], the current best known algorithm

© Springer International Publishing Switzerland 2015
V.Th. Paschos and P. Widmayer (Eds.): CIAC 2015, LNCS 9079, pp. 377–389, 2015.
DOI: 10.1007/978-3-319-18173-8_28

has an approximation ratio of 1.375 [6], and is not known whether it admits a PTAS. It is not known yet whether BLOCK SORTING approximates SORTING BY TRANSPOSITIONS to any factor better than 3. However, it is known that optimal transpositions never need to *break* existing blocks [5]. This shows how the two problems are closely related. The study of the computational complexity of BLOCK SORTING therefore might provide us with more insight into the complexity of SORTING BY TRANSPOSITIONS. The decision version of BLOCK SORTING has been shown to be NP-Hard [3]. After being known to be 3-approximable in [9], and [3], some results have designed 2-approximation algorithms [2,9–11]. However, no polynomial time approximation hardness results are known.

The BLOCK MERGING problem is a constrained form of the BLOCK SORTING problem in which the most important concept is the *redefinition* of a block by adding additional constraints. The given permutation is first partitioned into a set of *maximal increasing sequences*. At this point it is important to keep in mind that this partitioning is fixed and does not change based on the intermediate permutations obtained during the execution of an algorithm. Given such a partition, a block is defined to be a maximal sorted sequence which occurs in exactly *one* maximal increasing sequence. The BLOCK MERGING problem is the BLOCK SORTING problem in which only the redefined blocks are permitted to be moved. The BLOCK MERGING problem was introduced in [10] and to be polynomial time solvable, and also proved to approximate BLOCK SORTING by a factor of 2. In the subsequent [11], the constraints of BLOCK MERGING have been relaxed to by redefining blocks to be those sorted increasing subsequences which are contained in more than one increasing sequence int he give permutation. Many new heuristics in [11] are shown to work on the tight inputs for BLOCK MERGING, but it remains an open question to prove any improved approximation guarantee.

Our Results. Our main motivation is to either prove hardness of approximation results or to improve the current best approximation ratio for BLOCK SORTING. We prove that BLOCK SORTING does not admit a PTAS unless P = NP, thus settling an open question raised by Mahajan et al. in [10]. In particular, we show that for any $\varepsilon > 0$, there does not exist a polynomial time $\frac{65}{64} - \varepsilon$ approximation algorithm for BLOCK SORTING unless P = NP. This proves BLOCK SORTING to be APX-Hard. We then parameterize BLOCK MERGING by a parameter $k \in \mathbb{Z}^+$. In the generalised k-BLOCK MERGING problem, we redefine a block to be a maximal sorted sequence which is contained in a number of *consecutive* increasing sequences among which at most k are *empty*. Thus in the k-BLOCK MERGING problem, the aim is to find the least number of block moves to transform the given π into the identity permutation. We prove that k-BLOCK MERGING approximates BLOCK SORTING by a factor of $1 + \frac{1}{k}$. While this lower bound looks promising, we show that k-BLOCK MERGING is NP-Hard for $k > 1$, effectively closing out a possible approach to design a better approximation algorithm for BLOCK SORTING by designing polynomial time algorithms for k-BLOCK MERGING. On the other hand, this result is an interesting dichotomy result- k-BLOCK MERGING is P time solvable from $k = 1$, and NP-hard for $k > 1$.

2   Preliminaries

The set $\{1, 2, \cdots, n\}$ is denoted by $[n]$, and let S_n denote the set of all permutations over $[n]$, and id_n the sorted or identity permutation of length n. The given permutation $\pi \in S_n$ is represented as a string $\pi_1 \pi_2 \cdots \pi_n$ without loss of generality. For an element $a \in \pi$, we denote its position in π by $\pi(a)$. For a permutation π, the *kernel permutation* denoted by $ker(\pi)$ is obtained by replacing each block by its rank among the blocks. For example when the permutation 8 2 5 6 3 9 1 4 7 is kernelized, we get the permutation 7 2 5 3 8 1 4 6. The idea of the kernel of a permutation allows us to view blocks as *individual elements* without loss of generality, and use the relations $<$, and $>$ on blocks in their usual senses. In [10] $ker(\pi)$ is referred to as a *reduced* permutation. In the kernel, each block is of length 1, therefore we assume that the permutation has n elements and we use $N(\pi)$ to denote the number of blocks in π. In a permutation π, a pair of blocks (a, b) is a *reversal* if $a > b$ and $\pi(b) = \pi(a) + 1$. Denote the number of reversals in π by $rev(\pi)$. Another important concept in a permutation π is a *run*. The blocks $\{a, a + 1, a + 2, \ldots, a + r\}$ form a run of length r if, r is the largest values such that $\pi(a) < \pi(a + 1) < \pi(a + 2) < \ldots < \pi(a + r)$. Then number of runs in π is denoted by $runs(\pi)$.

A *block move* picks up a block and places it elsewhere in the permutation. For a permutation π, a *block sorting schedule* S is a sequence of block moves such that applying the sequence of block moves to permutation π results in the identity permutation id. The length of a block sorting schedule is the number of block moves in the schedule. Block sorting distance $bs(\pi)$ is the number of a block moves in a minimum length block sorting schedule for π. The BLOCK SORTING PROBLEM can be formally stated as: Given as input a permutation π and an integer m, is $bs(\pi) \leq m$? In [10], it is shown that block sorting π is equivalent to block sorting its kernel $ker(\pi)$. For a block a, we use $a+1$ to denote the successor block in the identity permutation. In [3], it has been shown that $bs(\pi)$ is at least $rev(\pi)$.

---

k-BLOCK MERGING PROBLEM

INPUT: A permutation π represented as an ordered set $\mathbb{S} = \{S_1, S_2, \cdots, S_l\}$ in which each S_i, initially, is a maximal increasing sequence of consecutive elements in π. For each $1 \leq i \leq l - 1$, S_i and S_{i+1} are consecutive sequences in $\mathbb{S}$.

CONSTRAINT: A block b is a maximal sorted sequence if there exists $1 \leq i < j \leq l$ such that b is contained in consecutive increasing sequences in $\{S_i, S_{i+1}, \ldots, S_j\}$, of which at most k are non-empty.

A BLOCK MOVE: When a block b is moved, it is always moved into an increasing sequence in $\mathbb{S}$. After the move, the sequences in $\mathbb{S}$ are changed by removing b from the increasing *sequences* that it was present in and adding it to the increasing sequence that it has been moved into. Note, that in the process some $S_i, 1 \leq i \leq l$ could become empty.

OUTPUT: The minimum number of block moves $k\text{-}bm(\mathbb{S})$ after which id, the identity permutation, is a block in $\mathbb{S} = \{S_1, S_2, \cdots, S_l\}$.

QUESTION: Is $k\text{-}bm(\mathbb{S}) \leq m$?

Note that for $k = 1$, k-BLOCK MERGING is the same problem as BLOCK MERG-ING studied in [10]. If $\pi = 2\ 5\ 6\ 3\ 1\ 4\ 7$, then $\mathbb{S}_\pi = \{(2, 5, 6), (3), (1, 4, 7)\}$ is the instance for BLOCK MERGING. The target is $\mathbb{M}_n = \{(1, 2, 3, 4, 5, 6, 7), \emptyset, \emptyset\}$ where $\emptyset$ is the empty set. If we move the block $\boxed{5\ 6}$ to get $\mathbb{S}'_\pi = \{(2), (3), (1, 4, 5, 6, 7)\}$, the block $\boxed{2\ 3}$ is *fragmented*, and cannot be moved till 2 and 3 are brought in the same increasing subsequence (by another block move). However, for $k = 2$, we are allowed to move block $\boxed{2\ 3}$ in $\mathbb{S}'_\pi$ for k-Block Merging in the above example, since $\boxed{2\ 3}$ is contained in 2 increasing sequences. It must be noted that $\mathbb{S}'_\pi = \{(2), (3), (1, 4, 5, 6, 7)\}$ is different from $\mathbb{S}_\pi = \{(2), (1, 4, 5, 6, 7), (3)\}$.

Examples of Block Sorting and Block Merging
Block sorting and k-block merging schedules with $k = 1$ and $k = 2$ are shown on permutation 8 2 5 6 3 9 1 4 7 in Figure 1, Figure 2 (a) and (b) respectively. The block moves are indicated at each step. Note that e.g. the block $\boxed{8\ 9}$ in Figure 2

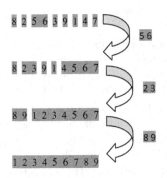

Fig. 1. An example of a block sorting schedule

is allowed to move because it is contained in 2 increasing sequences (and has one or more empty sequences in between).

Finally, G is a graph with V as its vertex set and E as the edge set. For $X \subseteq V$, the edge set $E(X, V \setminus X)$ denotes all those edges with one end point in X and the other in $V \setminus X$.

3   The Red-Blue Graph
In this section, we prove a better lower bound on $bs(\pi)$ based on the number of connected components in the well-studied red-blue graph introduced in [3]. Given a block sorting schedule S for π, in the red-blue graph $G(\pi, S)$, $V(G)$ corresponds to the blocks of π and the edges, which are either red edges or blue edges, are as follows:

1. The pair $\{a, b\}$ is a **blue edge** if (a, b) is a reversal.
2. The pair $\{a, b\}$ is a **red edge** whenever both the following two conditions are satisfied:

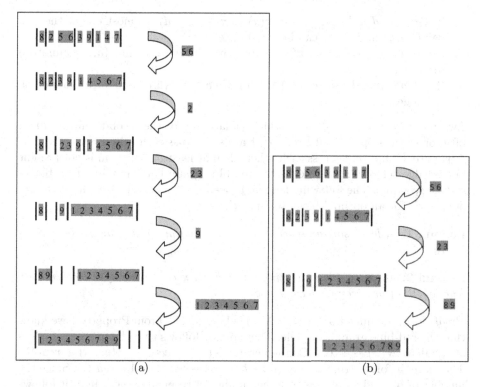

Fig. 2. An example of a k-block merging schedule with (a) $k = 1$ and (b) $k = 2$

- $a < b$ and $\pi(a) < \pi(b)$,
- There is a prefix of blocks moves in S in which a and b are not moved and after this prefix $\pi(a) = \pi(b)$, that is a and b are in the same block. Informally, this is written as a and b are joined in S before either is moved, and for all blocks c such that $\pi(a) < \pi(c) < \pi(b)$, block c is moved in S before a and b are joined.

$G(\pi, S)$ is written as G, and will be written as $G(\pi, S)$ when there could be an ambiguity. A *blue component* is a connected component in the graph formed by removing the red edges in G. Further, it is useful to view each blue component as a substring of π. Let $cc(G)$ denote the number of connected components in G. The number of blue edges, denoted by $blue(G)$ in G is the number of reversals in π, and the number of red edges is denoted by $red(G)$. The following important properties of red edges which have been proved in [3] in Property 1.

Property 1. 1. For each permutation π and any sequence of block sorting moves S, the red-blue graph G is acyclic.
2. If $\pi(a) < \pi(c) < \pi(b) < \pi(d)$, at most one of the two pairs $\{a, b\}$ and $\{c, d\}$ can be a red edge.
3. If $a < c < b < d$, at most one of the two pairs $\{a, b\}$ and $\{c, d\}$ can be a red edge.

4. If $a < c < d < b$ and $\pi(c) < \pi(a) < \pi(b) < \pi(d)$, at most one of the two pairs $\{a,b\}$ and $\{c,d\}$ can be a red edge.
5. If $a < b < c$ and $\pi(a) < \pi(c) < \pi(b)$, then both $\{a,b\}$ and $\{a,c\}$ cannot be red edges.
6. If $a < b < c$ and $\pi(b) < \pi(a) < \pi(c)$, then both $\{b,c\}$ and $\{a,c\}$ cannot be red edges.

Two pairs (a,b) and (c,d) in π which violate any of the second, third, fourth, fifth, or sixth properties of Property 1 are said *cross* each other. For a given π, if pairs (a,b) and (c,d) cross each other, then at least one of them is not present in the edge set of G, where G is the red-blue graph corresponding to a block-sorting schedule. The following lemma, proved in [3], characterizes the structure of $G(\pi,\mathcal{S})$ for an optimal block sorting schedule.

Lemma 1. *A block sorting schedule $\mathcal{S}$ on π is optimal if and only if $G(\pi,\mathcal{S})$ is a tree.*

Lemma 2. *Let $\mathcal{S}$ be a block sorting schedule on π consisting of m block moves. The length of $m \geq rev(\pi) + cc(G(\pi,\mathcal{S})) - 1$.*

Proof. From Lemma 4 of [3] $m = N(\pi) - 1 - red(G)$. From Property 1 we know that the red-blue graph is acyclic. Therefore, it follows that the number of edges is equal to $N(\pi) - cc(G)$. Consequently, $N(\pi) = cc(G) + blue(G) + red(G)$. Therefore, it follows that $m = cc(G) + blue(G) + red(G) - 1 - red(G)$. Since the number of blue edges in G is just the number of reversals $rev(\pi)$ in π, it follows that $m = cc(G) + rev(\pi) - 1$. □

The following theorem now is a corollary and is useful in the proof of our hardness result.

Theorem 1. *Let π be a permutation and let d be an integer such that for each block sorting schedule $\mathcal{S}$, $cc(G(\pi,\mathcal{S})) \geq d+1$. Then $bs(\pi) \geq rev(\pi) + d$.*

4   Hardness of Approximation of BLOCK SORTING

We use the construction from [3] to reduce MAX-E3-SAT to BLOCK SORT-ING. Consider an instance of MAX-E3-SAT consisting of a boolean formula $\phi = \mathcal{C}^1 \mathcal{C}^2 \cdots \mathcal{C}^m$ of n variables, and m clauses $\mathcal{C}^1, \mathcal{C}^2, \cdots, \mathcal{C}^m$. A permutation $\pi(\phi)$ of $8m + 4n + 1$ elements is constructed from ϕ by introducing an ordered alphabet $\Sigma_{n,m}$ with $4nm + 2m + 4n + 1$ elements in [3]. Through out this section we write only π for $\pi(\phi)$.

Alphabet: The alphabet $\Sigma_{n,m}$ consists of the following elements:
1. *Term symbols:* $p_i^j, \bar{p}_i^j, q_i^j, \bar{q}_i^j \; \forall \; 1 \leq i \leq n$, and $1 \leq j \leq m$. $p_i^j, \bar{p}_i^j$ and $q_i^j, \bar{q}_i^j$ are called left and right term symbols respectively.
2. *Clause control symbols:* ℓ^j and $r^j \; \forall \; 1 \leq j \leq m+n$.
3. *Variable control symbols:* u_i and $v_i \; \forall \; 1 \leq i \leq n$.
4. *Separator symbol:* s.

Ordering: The total ordering which is considered to be the identity permutation id on $\Sigma_{n,m}$ is generated by the following rules:

1. $u_i < p_i^k < p_i^j < \bar{p}_i^k < \bar{p}_i^j < q_i^j < q_i^k < \bar{q}_i^j < \bar{q}_i^k < v_i < s \ \forall \ 1 \le i \le n$, and $1 \le j < k \le m$.
2. $v_{i-1} < u_i \ \forall \ 1 \le i \le n$.
3. $s < \ell^k < r^k < \ell^j < r^j \ \forall \ 1 \le j < k \le m+n$.

Reduction Procedure: Given the boolean formula ϕ, let each clause be of the form $(l_i \vee l_j \vee l_k)$, where l_i, l_j and l_k correspond to literals of variables x_i, x_j. and x_k, respectively, and $i > j > k$. We construct the permutation π from ϕ using symbols from $\Sigma_{n,m}$ as follows:

1. The first symbol of π is the starting symbol s.
2. Next we have the clause encodings for every clause $\mathcal{C}^j \in \phi$ in increasing of values of j, from 1 to m. The clause encoding of $\mathcal{C}^l$ is as follows:
   (a) The clause encoding for $\mathcal{C}^j$ starts with symbol ℓ^j and ends with r^j.
   (b) If $x_i \in \mathcal{C}^j$, we have symbols p_i^j and q_i^j, and if $\bar{x}_i \in \mathcal{C}^j$, we have $\bar{p}_i^j$ and $\bar{q}_i^j$ respectively. Hence there are a total of six term symbols (two for each literal) in the clause encoding of $\mathcal{C}^j$.
   (c) The three p symbols come first in decreasing order of subscripts, and are followed by the three q symbols in decreasing order of subscripts. As an example the simple encoding of the clause $(x_5 \vee x_3 \vee \bar{x}_2)$ is $\ell p_5 p_3 \bar{p}_2 q_5 q_3 \bar{q}_2 r$. For the real encoding of a clause, the index of that clause is inserted as its superscript. If $\mathcal{C}^4 = (x_5 \vee x_3 \vee \bar{x}_2)$, then its encoding would be $\ell^4 p_5^4 p_3^4 \bar{p}_2^4 q_5^4 q_3^4 \bar{q}_2^4 r^4$.
3. After the encoding of the clauses, for each $1 \le i \le n$, the control sequences $\ell^{m+i} u_i v_i r^{m+i}$ are added.

That this is indeed a reduction from MAX-E3-SAT to BLOCK SORTING is shown in the following theorem proved in [3].

Theorem 2. *An instance of* MAX-E3-SAT ϕ *is satisfiable if and only if* $\pi(\phi)$ *has a block sorting schedule* $\mathcal{S}$ *of length* $6m + 2n - 1$.

4.1   Maximum Number of Clauses Satisfiable in ϕ and $bs(\pi)$

The blue edges are common to the red-blue graphs of all block sorting schedules of π. There are $6m+2n-1$ blue edges and $2m+2n+2$ components. By Theorem 1 the red-blue graph of an optimal schedule for π is a tree and therefore, there can be at most $2m + 2n + 1$ red edges to connect these $2m + 2n + 2$ blue components. The three q_i^j symbols $1 \le i \le n$, for the components of each clause $\mathcal{C}^j$ for $1 \le j \le m$, contribute to two blue edges, and form a blue component in the red-blue graph $G(\pi, \mathcal{S})$ for any block sorting schedule $\mathcal{S}$. We refer to these components as *blue components containing q symbols*. Therefore out of the $2m + 2n + 2$ blue components, m are blue components containing q symbols. We next consider the set of all pairs that could be red edges. For a clause $\mathcal{C}^j$, consider the occurrence of the variable x_i. If x_i occurs as a positive literal, then the pair of blocks which are in π are $\{p_i^j, q_i^j\}$, and if it occurs a negative literal the pair of blocks are $\{\bar{p}_i^j, \bar{q}_i^j\}$. Note that in π, in the gadget corresponding to clause C^j, all the pairs except for the 3 pairs in which the subscripts are the same, are in the *wrong order* with respect to *id*. Therefore, any red edge among the pairs *can* only be among those with the same subscript, that is, among those pairs that correspond to the same variable. For each clause C^j, let $\mathbf{A}^j$ denote the set of 3 pairs which are not blue edges, and the subscripts in each pair are the same. We now define $\mathbf{A} = \bigcup_{j=1}^{m} \mathbf{A}^j$ to be a set of *potential* red edges. Further, let $\mathbf{B} = \{(u_i, v_i) \mid 1 \leq i \leq n\} \cup \{(\ell^j, r^j) \mid 1 \leq j \leq m+n\} \cup \{(s, \ell^1)\}$ to be another set of *potential* red edges. Note that the set $\mathbf{B}$ is a non-crossing set of red edges. Lemma 3 is proved in Lemma 11 of [3] and it based on the structure of the reduction.

Lemma 3. *[3] Consider the $m + 2n + 2$ blue components that do not contain a q symbol. Any set of non-crossing edges that connects these blue components must contain $\mathbf{B}$.*

Lemma 4. *Let S be a block sorting schedule for π obtained from the reduction such that $G = G(\pi, S)$ is connected, and let $X \subset V$ denote the set of all blue components that contain q. Then, $E(X, V \setminus X) \subseteq \mathbf{A}$.*

Proof. From lemma 3 we know that $\mathbf{B} \subset E(G)$. We also know the the red edges of G are a non-crossing set, as $G = G(\pi, S)$ is a connected graph obtained from S which is a block sorting schedule. Therefore, if a red edge is incident on a blue component containing q which is a subset of the vertices corresponding to a clause gadget corresponding to a clause C^j, then such a red edge must belong to $\mathbf{A}^j$. Therefore, it follows that $E(X, V \setminus X) \subseteq \mathbf{A}$. □

Lemma 5. *Let c be an integer such that the maximum number of satisfied clauses in ϕ under any truth assignment is $m - c$. then $bs(\pi) \geq 6m + 2n - 1 + c$.*

Proof. Let S be a block sorting schedule. Since at most $m - c$ clauses can be satisfied by any truth assignment, and we know from Property 1 that each S gives a truth assignment to the variables of ϕ. Therefore, in $G(\pi, S)$ there exist at least c blue components with q. Therefore, the number of connected components in G, $cc(G)$, is at least $c + 1$. From Theorem 1, it follows that the number of block moves in S is at least $6m + 2n - 1 + c$, and therefore it follows that $bs(\pi) \geq 6m + 2n - 1 + c$. □

Lemma 6. *For any $c > 0$, if there exists an assignment satisfying at least $m - c$ clauses of ϕ, then $bs(\pi) \leq 6m + 2n - 1 + c$.*

Proof. Let ϕ' be a subset of clauses that is satisfiable and let it have $m - c$ clauses for some $c \geq 0$, and involving $n' \leq n$ variables. Let π' be the permutation that is obtained by the reduction from ϕ'. We know from [3] that ϕ' can be sorted in $6(m - c) + 2n' - 1$ block moves. The part of π that involves the remaining c clauses and $n - n'$ variables can be sorted in $7c + 2(n - n')$ block moves, since any permutation obtained from a 3-CNF formula with r clauses and s variables can be sorted in $7r + 2s$ block moves. Therefore, π can be sorted in $6m + 2n - 1 + c$ block moves, and hence the bound on $bs(\pi)$ follows.   □

Theorem 3. *For every $\varepsilon > 0$, it is* NP-*Hard to approximate* BLOCK SORTING *within a factor of* $1 + \frac{1}{64+\varepsilon} - \frac{2\varepsilon}{8+\varepsilon}$. *Therefore,* BLOCK SORTING *does not admit a PTAS unless* P = NP.

Proof. For every $\varepsilon > 0$ and positive integers m, n such that $\frac{m}{n} \geq 1$, let us consider the ratio $\frac{6m+2n-1+(\frac{1}{8}-\varepsilon)m}{6m+2n-1+\varepsilon m}$. This is $1 + \frac{(\frac{1}{8}-2\varepsilon)m}{6m+2n-1+\varepsilon m} = 1 + \frac{(\frac{1}{8}-2\varepsilon)}{6+\varepsilon+\frac{2n-1}{m}}$. Since $\frac{n}{m} \leq 1$, it follows that $\frac{2n-1}{m} < 2$, and therefore the ratio is at least $1 + \frac{(\frac{1}{8}-2\varepsilon)}{8+\varepsilon} = 1 + \frac{1}{64+\varepsilon} - \frac{2\varepsilon}{8+\varepsilon}$. Now for some $\varepsilon > 0$, if we have an approximation algorithm from BLOCK SORTING of approximation ration at most $1 + \frac{1}{64+\varepsilon} - \frac{2\varepsilon}{8+\varepsilon}$, then for this ε, our algorithm is one of approximation ratio at most $\frac{6m+2n-1+(\frac{1}{8}-\varepsilon)m}{6m+2n-1+\varepsilon m}$. Since BLOCK SORTING is NP-Complete, it follows that $\varepsilon < \frac{1}{16}$, otherwise the approximation ratio is at most 1. Let us consider the set of MAX-E3-SAT instances on n variables and m clauses for which $\frac{m}{n} \geq 1$. On the class of such instances, we know from the result of Håstad [8]: For any $\varepsilon > 0$, it is NP-hard to approximate MAX-E3-SAT within a factor of $\frac{8}{7} - \varepsilon$. From lemma 5 and lemma 6 we know that c is the maximum number of clauses that can be satisfied in a MAX-E3-SAT instance ϕ if and only if π can be sorted using $6m + 2n - 1 + c$ block moves. Apply lemma 5 and lemma 6 we can see that for any $\varepsilon > 0$,

- **Completeness.** if there exists an assignment satisfying at least $(1 - \varepsilon)m$ clauses of ϕ, then $bs(\pi) \leq 6m + 2n - 1 + \varepsilon m$.
- **Soundness.** if no assignment satisfies more than $(\frac{7}{8} + \varepsilon)m$ clauses of ϕ, then $bs(\pi) \geq 6m + 2n - 1 + (\frac{1}{8} - \varepsilon)m$.

Given an instance ϕ of MAX-E3-SAT, we run our approximation algorithm for BLOCK SORTING with the approximation ratio of $1 + \frac{1}{64+\varepsilon} - \frac{2\varepsilon}{8+\varepsilon}$ on π. Like we said earlier, this is an approximation ratio of at most $\frac{6m+2n-1+(\frac{1}{8}-\varepsilon)m}{6m+2n-1+\varepsilon m}$. Now, if the output of the algorithm is at most $6m+2n-1+(\frac{1}{8}-\varepsilon)m$, then from Lemma 6, we obtain a truth assignment for ϕ that satisfies at least $m - (\frac{1}{8}-\varepsilon)m = (\frac{7}{8}+\varepsilon)m$ clauses. On the other hand if the algorithm output a solution of value more than $6m + 2n - 1 + (\frac{1}{8} - \varepsilon)m$, then it means that $bs(\pi) > 6m + 2n - 1 + \varepsilon m$. From Lemma 5, we can now conclude that the number of clauses satisfiable in ϕ is less that $(1 - \varepsilon)m$. Now, in P time, we can find a truth assignment that satisfies at least $\frac{7}{8}m$ clauses, since a random truth assignment is expected to satisfy $\frac{7}{8}m$ clauses. By searching over $O(n^3)$ truth assignments from a 3-wise independent sample space on $\{0,1\}^n$, we get a truth assignment that satisfies at least $\frac{7}{8}m$ clauses in P time. The approximation ratio of this algorithm for MAX-E3-SAT is the minimum of $\frac{7}{8} + \varepsilon$ and $\frac{7}{8(1-\varepsilon)}$. For all $\varepsilon < 1$, $\frac{7}{8(1-\varepsilon)} > \frac{7}{8} + \varepsilon$. Therefore, our algorithm for MAX-E3-SAT is a $\frac{7}{8} + \varepsilon$ approximation factor, polynomial time algorithm. However, from the result of Håstad [8], we know that for *every* $\varepsilon > 0$, it is NP-hard to approximate MAX-E3-SAT within a factor of $\frac{7}{8} + \varepsilon$ Consequently, it follows that for any $\varepsilon > 0$, it is NP-hard to approximate BLOCK SORTING with in a ratio of $1 + \frac{1}{64+\varepsilon} - \frac{2\varepsilon}{8+\varepsilon}$. This proves that BLOCK SORTING is APX-Hard.   □

5   Hardness of k-BLOCK MERGING

For each $k \geq 1$, $k\text{-}bm(\mathbb{S})$ is the minimum number k-block merging moves to transform $\mathbb{S}$ to $\mathbb{M}_n$. The following lemma is generalization of the result for block merging from [10].

Lemma 7. *For each $k \geq 1$, $bs(\pi) \leq k\text{-}bm(\mathbb{S}_\pi) \leq bm(\mathbb{S}_\pi)$.*

Proof. For each $k \geq 1$, a k-block merging schedule is a schedule of block moves, and therefore $bs(\pi) \leq k\text{-}bm(\mathbb{S}_\pi)$. Further, a block merging schedule is a k-block merging schedule for $k = 1$. Therefore, for each $k \geq 1$, each block merging schedule is a k-block merging schedule. Therefore, it follows that $k\text{-}bm(\mathbb{S}_\pi) \leq bm(\mathbb{S}_\pi)$. Hence the lemma.   □

The following lemma is a generalization of Lemma 5.3 in [10] in which we show that for each $k \geq 1$, we get a $1 + \frac{1}{k}$ approximation to block sorting. We follow exactly the same proof technique as in [10] based on modified definitions of the potential function to take care of the fact that we consider are k-block merging moves.

Theorem 4. *Let π be a permutation. For each $k \geq 1$, $bs(\pi) \geq \frac{k\text{-}bm(\mathbb{S}_\pi)}{1+\frac{1}{k}}$. Therefore, k-BLOCK MERGING approximates BLOCK SORTING by a factor of $1 + \frac{1}{k}$.*

Proof. Let π be a block sorting schedule $b_1, b_2, \cdots, b_m$ of m moves. We construct a k-block merging schedule of at most $m(1 + \frac{1}{k})$ moves for $\mathbb{S}_\pi$. The way we achieve this is by replacing for each i, the block move b_i by a k-block merging schedule. Let b_i be the move of a block B in the permutation π after the block moves $b_1, \ldots, b_i$. Let us assume that the block merging moves corresponding to $b_1, \ldots, b_{i-1}$ also have been done on $\mathbb{S}_\pi$. If B is contained in at most k contiguous non-empty increasing sequences (Note that there might be zero or more empty sequences in between those k sequences, but we do not count those.) in S_π, then we move B as the move corresponding to b_i in the block merging schedule. On the other hand, if B is contained in $x_i > k$ contiguous non-empty increasing sequences, we keep the first increasing sequence in which the block is contained intact. Then we pick up the next k increasing sequences and put it in the first increasing sequence. We continue doing this until the number of increasing sequences in which the block is contained is at most k. We observe that the number of such defragmentation moves required here is $\lfloor \frac{x_i - 1}{k} \rfloor$ moves (to get B in at most k sequences). Next we move B. This way, each move in the block sorting schedule, can be performed by one or more moves in the k-block merging schedule.

We now perform an amortized analysis of the number of moves in this k-block merging schedule as a function of m. Let π^i denote the permutation after the block moves $b_1, \ldots, b_i$ are applied to π. Let $\mathbb{S}^i_\pi$ be the sequences obtained by applying to $\mathbb{S}$ the k-block merging moves corresponding to $b_1, \ldots, b_i$. For a permutation π, let $Inc(\pi) = \{\pi_i | \pi_i < \pi_{i+1}, \forall 1 \leq i < n\}$, and $Inc(\mathbb{S}_\pi) = \{i \in [n-1] | i$ is not the last element of a sequence $\mathbf{S} \in \mathbb{S}\}$. Clearly, we have $Inc(\pi) = Inc(\mathbb{S}_\pi)$. Observe that for each $i \geq 1$, $Inc(\mathbb{S}^i_\pi) \subseteq Inc(\pi^i)$. For $i \geq 1$, let $p_i = |Inc(\pi^i)|$ and $q_i = |Inc(\mathbb{S}^i_\pi)|$. We define the potential function $\phi_i = \frac{p_i - q_i}{k}$. For each $i \geq 1$, let c_i denote the number of k-block merging moves corresponding to b_i. From the construction of the k-block merging sequence, $c_i \leq 1 + \lfloor \frac{x_i - 1}{k} \rfloor$. We define the amortized cost of the i-th block move as $a_i = c_i + \phi_i - \phi_{i-1}$. Note that for each $i \geq 1$, $a_i \geq c_i$. We now show that for each $i \geq 1$, $a_i \leq 1 + \frac{1}{k}$. For this, we use the fact from [10] that $\phi_i - \phi_{i-1} = \frac{(p_i - q_i) - (p_{i-1} - q_{i-1})}{k} = \frac{(p_i - p_{i-1}) - (q_i - q_{i-1})}{k} = \frac{1}{k} - \frac{x_i - 1}{k}$. Therefore, the total number of k-block merging moves is given by $\sum_{i=1}^{m} c_i = \sum_{i=1}^{m} a_i + \phi_m - \phi_0$. Both, ϕ_0 and ϕ_m are zero. Therefore,

$$\sum_{i=1}^{m} c_i = \sum_{i=1}^{m} a_i \leq (1 + \frac{1}{k})m.$$ Therefore, it follows that $k\text{-}bm(\mathbb{S}_\pi) \leq (1 + \frac{1}{k})bs(\pi)$. Hence the result is proved. $\qquad\square$

From [10], it is well known that 1-BLOCK MERGING is solvable in polynomial time. We now prove that k-BLOCK MERGING is NP-hard for each $k \geq 2$. The way we achieve this by identifying the value of k for which a block sorting schedule is k-block merging schedule. Let $\mathcal{S}$ be a block sorting schedule. A path of red edges in $G(\pi, \mathcal{S})$ is defined to *span* those increasing sequences in $\mathbb{S}_\pi$ from which the red path contains its vertices.

Theorem 5. *Let $\mathcal{S}$ be a block sorting schedule of m moves on π. If every path of red edges in $G(\pi, \mathcal{S})$ spans at most k increasing sequences in $\mathbb{S}_\pi$, then $\mathcal{S}$ is a k-block merging schedule of length m on $\mathbb{S}_\pi$.*

Proof. Let us consider the first block move B in $\mathcal{S}$ such that B is in $k + 1$ increasing sequences. Let $b_1, b_2, \ldots, b_{k+1}$ be the parts of B which are in $k + 1$ increasing sequences, respectively. Clearly, none of the blocks $b_1, \ldots, b_{k+1}$ have been moved in $\mathcal{S}$. The reason is as follows: had they been moved, then they could have been moved to merge with the neighboring block in the neighboring increasing sequence. Therefore, in $G(\pi, \mathcal{S})$ the blocks $b_1, \ldots, b_{k+1}$ form a red path which spans $k + 1$ increasing sequences. This contradicts the premise that every path of red edges in $G(\pi, \mathcal{S})$ spans at most k increasing sequences in $\mathbb{S}_\pi$. Therefore, it follows that $\mathcal{S}$ is a k-block merging sequence on $\mathbb{S}_\pi$. Hence the theorem. $\qquad\square$

Let π be permutation that is output by the reduction in Section 4 on an input instance of MAX-E3-SAT ϕ.

Lemma 8. *Let $\mathcal{S}$ be a block sorting schedule on π. Then each block move in $\mathcal{S}$ is in at most 2 increasing sequences in $\mathbb{S}_\pi$.*

Proof. If a block is in at least 3 increasing sequences, then it means that $G(\pi, \mathcal{S})$ has a red path spanning at least 3 increasing sequences. Recall the potential red edges of Section 4. We know that for each block sorting schedule $\mathcal{S}$ the red edges of $G(\pi, \mathcal{S})$ are a subset of the set of potential red edges. From the structure of the permutation output by the reduction, it is easy to observe that every red path using potential red edges is a path that spans at most two increasing sequences in $\mathbb{S}_{pi}$. The reason is that the only potential red edges that for a path of two edges $\{s, l^1\}$ and $\{l^1, r^1\}$. The remaining potential red edges are all vertex disjoint. Therefore, each red path in $G(\pi, \mathcal{S})$ spans at most two increasing sequences. This contradicts our conclusion that $G(\pi, \mathcal{S})$ has a red path spanning at least 3 increasing sequences. Therefore, every block move in $\mathcal{S}$ is in at most 2 increasing sequences.

Lemma 9. *Let ϕ be an instance of* MAX-E3-SAT. *Let π be the permutation output by the reduction in Section 4. For each $k \geq 2$, $k\text{-}bm(\mathbb{S}_\pi) = 6m + 2n - 1$ if and only if ϕ is satisfiable.*

Proof. We know that ϕ is satisfiable if and only if there is a block sorting schedule of π in $6m + 2n - 1$. Further, from lemma 8 we know that each block sorting schedule is a 2-block merging schedule. Since a 2-block merging schedule is a k-block merging schedule for each $k \geq 2$, it follows that $k\text{-}bm(\mathbb{S}_\pi) \leq 6m + 2n - 1$. However, a k-block merging schedule is a block sorting schedule itself, and therefore $k\text{-}bm(\mathbb{S}_\pi) \geq 6m + 2n - 1$. Hence the lemma follows. $\square$

Finally, Lemma 9 leads to Theorem 6.

Theorem 6. *For each $k \geq 2$, k-BLOCK MERGING is NP-Hard.*

This result can also be seen as an interesting dichotomy in the complexity of k-BLOCK MERGING. For $k = 1$ it is P time solvable [10], and for all $k \geq 2$ it is NP-hard.

References

1. Bafna, V., Pevzner, P.A.: Sorting by Transpositions. SIAM Journal of Discrete Mathematics **11**(2), 224–240 (1998)
2. Bein, W.W., Larmore, L.L., Latifi, S., Sudborough, I.H.: A Quadratic TIme 2-Approximation Algorithm for Block Sorting. Theoretical Computer Science **410**(8–10), 711–717 (2009)
3. Bein, W.W., Larmore, L.L., Latifi, S., Sudborough, I.H.: Block sorting is hard. International Journal of Foundations of Computer Science **14**(3), 425–437 (2003)
4. Bulteau, L., Fertin, G., Rusu, I.: Sorting by Transpositions is Difficult. Automata, Languages and Programming **6755**, 654–665 (2011)
5. Christie, D.A.: Genome Rearrangement Problems. PhD Thesis, University of Glasgow (1999)
6. Elias, I., Hartman, T.: A 1.375-Approximation Algorithm for Sorting by Transpositions. IEEE/ACM Transactions on Computational Biology and Bioinformatics **3**(4), 369–379 (2006). doi:10.1109/TCBB.2006.44
7. Gobi, R., Latifi, S., Bein, W.W.: Adaptive Sorting Algorithms for Evaluation of Automatic Zoning Employed in OCR Devices. In: Proceedings of the 2000 International Conference on Imaging Science, Systems, and Technology, CISST 2000, pp. 253–259. CSREA Press (2000)

8. Håstad, J.: Some Optimal Inapproximability Results. Journal of the ACM **48**(4), 798–859 (2001)
9. Mahajan, M., Rama, R., Raman, V., Vijayakumar, S.: Merging and Sorting ByStrip Moves. In: Pandya, P.K., Radhakrishnan, J. (eds.) FSTTCS 2003. LNCS, vol. 2914, pp. 314–325. Springer, Heidelberg (2003)
10. Mahajan, M., Rama, R., Raman, V., Vijayakumar, S.: lApproximate Block Sorting. International Journal of Foundation of Computer Science, 337–356 (2006)
11. Mahajan, M., Rama, R., Vijayakumar, S.: Block sorting: a characterization and some heuristics. Nordic Journal of Computing **14**(1), 126–150 (2007)
12. Mahajan, M., Rama, R., Vijayakumar, S.: Towards constructing optimal block move sequences. In: Chwa, K.-Y., Munro, J.I. (eds.) COCOON 2004. LNCS, vol. 3106, pp. 33–42. Springer, Heidelberg (2004)

An Opportunistic Text Indexing Structure Based on Run Length Encoding

Yuya Tamakoshi, Keisuke Goto, Shunsuke Inenaga$^{(\boxtimes)}$,
Hideo Bannai, and Masayuki Takeda

Department of Informatics, Kyushu University, Fukuoka, Japan
keisukegotou@gmail.com, {inenaga,bannai,takeda}@inf.kyushu-u.ac.jp

Abstract. We present a new text indexing structure based on the run length encoding (RLE) of a text string T which, given the RLE of a query pattern P, reports all the *occ* occurrences of P in T in $O(m+occ+\log n)$ time, where n and m are the sizes of the RLEs of T and P, respectively. The data structure requires $n(2\log N + \log n + \log \sigma) + O(n)$ bits of space, where N is the length of the uncompressed text string T and σ is the alphabet size. Moreover, using $n(3\log N + \log n + \log \sigma) + 2\sigma \log \frac{N}{\sigma} + O(n\log\log n)$ bits of total space, our data structure can be enhanced to answer the beginning position of the lexicographically ith smallest suffix of T for a given rank i in $O(\log^2 n)$ time. All these data structures can be constructed in $O(n \log n)$ time using $O(n \log N)$ bits of extra space.

1 Introduction

A *maximal character run* of a character c in a string T is a non-extensible, consecutive sequence of c's in T. The *run-length encoding* (RLE) of a string T is a natural compressed representation of T, where each maximal character run of a character a of length p is encoded as a^p. Typical applications of RLE would be bitmap images, black/white fax messages, and musical sequences such as MIDI. Also, RLE strings have quite many interesting features from an algorithmic perspective (e.g., see [1,3,6,11]).

In this paper, we propose a new text indexing structure built on the RLE of text string T which supports (1) an $O(m + occ + \log n)$-time locate query to answer, given the RLE of a pattern P, all the *occ* occurrences of P in T, and (2) an $O(j - i + \log n)$-time substr query to extract, given two positions $1 \le i \le j \le |T|$ in T, the substring of T which starts at position i and ends at position j, where n and m are the sizes of the RLEs of T and P, respectively. The data structure requires $n(2\log N + \log n + \log \sigma) + O(n)$ bits of space, where $N = |T|$ is the length of the uncompressed text string T and σ is the alphabet size. Assuming all characters in the alphabet Σ occur in T, $\sigma \le n \le N$ holds. We remark that our index replaces the original text T thanks to the substr queries. Remark that the time complexity of locate queries is *independent* of the length M of the uncompressed pattern P, but is dependent on the size m of the RLE of P, which can be much smaller than M when P contains long character runs.

© Springer International Publishing Switzerland 2015
V.Th. Paschos and P. Widmayer (Eds.): CIAC 2015, LNCS 9079, pp. 390–402, 2015.
DOI: 10.1007/978-3-319-18173-8_29

Moreover, using $n(3 \log N + \log n + \log \sigma) + 2\sigma \log \frac{N}{\sigma} + O(n \log \log n)$ bits of total space, we can also support (3) an $O(\log^2 n)$-time access_SA query to answer, given an integer $1 \leq i \leq |T|$, the beginning position of the lexicographically ith smallest suffix of T. Given the RLE of text T, all of our data structures can be built in $O(n \log n)$ time and $O(n \log N)$ bits of extra space.

As components of the above index, we propose suffix arrays and longest common prefix (LCP) arrays for strings represented by RLE. To construct these arrays efficiently, we extend the notion of S-type and L-type suffixes [10] to the suffixes of strings represented by RLE. We then show some combinatorial properties of S-type and L-type RLE suffixes, which may be of independent interest. Consequently, we obtain an $O(n \log n)$-time construction algorithms for suffix arrays and LCP arrays for strings represented by RLE.

Related Work. There exist a couple of indexing structures based on RLE. Yamamoto et al. [17] proposed an RLE-based data structure for on-line computation of Lempel-Ziv 77 factorization [18], which requires $O(n \log N)$ bits of space. Given the RLE of size m representing a pattern P, this data structure can find a *single occurrence* of P in T in $O\left(m \cdot \min\left\{ \frac{(\log \log n)(\log \log N)}{\log \log \log N}, \sqrt{\frac{\log n}{\log \log n}} \right\}\right)$ time. However, it does not support locate queries. Eltabakh et al. [4] proposed an external-memory text indexing structure based on RLE, which has an external-memory space complexity of $O(n/B)$ pages, and is able to answer locate query with $O(\log_B n + (m + occ)/B)$ I/O operations, where B is the disk page size. We remark that our data structure is conceptually quite different from the run-length FM-index (RLFM) [12], and from the run-length compressed suffix array (RLCSA) [13], both of which are based on *the RLE of the Burrows-Wheeler transform* [2] of text T. For an alphabet of size $\sigma = O(\mathrm{polylog}(N))$, RLFM requires $N H_k \log \sigma + O(N)$ bits of space, where H_k denotes the kth order empirical entropy of T, and supports locate query in $O(M + \sigma occ \log^{1+\gamma} N)$ time, and substr query in $O(\sigma(j - i + \log^{1+\gamma} N))$ time, for any constant $\gamma > 0$. RLCSA requires $\alpha n' \left(\log \frac{\sigma N}{n'} + \log \frac{N}{n'} + O(\log \log \frac{\sigma N}{n'})\right) + O\left(\frac{N}{d} \log \frac{N}{d} \sigma \log N\right)$ bits of space for some constant $1 < \alpha < 2$, where n' is the number of maximal character runs in the BWT of T and d is the suffix array sample rate, and supports locate query in $O(M \log N + occ \cdot d \log N)$ time, substr query in $O((d + j - i) \log N)$ time, and access_SA query in $O(d \log N)$ time. Since $m \leq M$ and $n \leq N$, our data structure answers locate queries faster than RLFM and RLCSA.

2   Preliminaries

Let $\Sigma = \{1, \ldots, \sigma\}$ be an integer *alphabet*. An element of Σ^* is called a *string*. The length of a string T is denoted by $|T|$. The empty string ε is the string of length 0, namely, $|\varepsilon| = 0$. The i-th character of a string T of length N is denoted by $T[i]$ for $1 \leq i \leq N$. For $1 \leq i \leq j \leq N$, let $T[i..j] = T[i] \cdots T[j]$, i.e., $T[i..j]$ is the *substring* of T starting at position i and ending at position j in T. For convenience, let $T[i..j] = \varepsilon$ if $j < i$. For any $1 \leq i \leq N$, non-empty strings $T[1..i]$ and $T[i..N]$ are called *prefixes* and *suffixes* of T, respectively. Let $suf_T(i) = T[i..N]$. $suf_T(i)$ will be denoted by $suf(i)$ for an arbitrary fixed text string T. Let $lcp(X, Y)$ denote the *longest common prefix* of strings X and Y.

If a string X is lexicographically smaller than another string Y, then we write $X \prec Y$ or $Y \succ X$. Assume any string T terminates with the lexicographically smallest character $\$$ that does not occur anywhere else in T. The *suffix array* $SA[1..N]$ of a string T of length N is an array of length N such that $SA[i] = j$ iff $suf(j)$ is the lexicographically ith suffix of T. For $1 \le i < N$, $suf(i)$ is called *S-type* (resp. *L-type*) if $suf(i) \prec suf(i+1)$ (resp. if $suf(i) \succ suf(i+1)$). For convenience, $suf(n) = \$$ is regarded as both S-type and L-type.

The *run length encoding* (RLE) of string T of length N is a compact representation of T which encodes each maximal character run $T[i..i+p-1]$ by a^p, if $T[j] = a$ for all $i \le j \le i+p-1$, (2) $T[i-1] \ne T[i]$ or $i = 1$, and (3) $T[i+p-1] \ne T[i+p]$ or $i+p-1 = N$. E.g., $RLE(\text{aabbbbcccaaa\$}) = \text{a}^2\text{b}^4\text{c}^3\text{a}^3\$^1$. The *size* of $RLE(T) = a_1^{p_1} \cdots a_n^{p_n}$ is the number n of maximal character runs in T. For any $1 \le i \le j \le n$, let $RLE(T)[i..j] = a_i^{p_i} a_{i+1}^{p_{i+1}} \cdots a_j^{p_j}$. For convenience, let $RLE(T)[i..j] = \varepsilon$ for $i > j$. For any string X, let $val(RLE(X)) = X$. We define the lexicographical order $\prec$ of RLE strings by $RLE(X) \prec RLE(Y) \Leftrightarrow X \prec Y$. Let $RLEsuf_T(i) = RLE(T)[i..n]$ for any $1 \le i \le n$. $RLEsuf_T(i)$ will be denoted by $RLEsuf(i)$ for an arbitrary fixed text string T. Each $RLEsuf(i)$ is called an RLE suffix of $RLE(T)$. For any non-empty strings X and Y, let $lcp(RLE(X), RLE(Y)) = \max\{k \ge 0 \mid RLE(X)[1..k] = RLE(Y)[1..k]\}$.

We will use the following known results in our data structures.

Theorem 1 ([5]). *For an integer array A of length n, there is a data structure which supports range maximum queries (RMQs) in $O(1)$ time, requires $2n+o(n)$ bits of extra space, and can be constructed in $O(n)$ time.*

Lemma 1. *Using an $O(1)$-time RMQ data structure for integer array A, given a range $[i, j]$ and threshold b, we can report all elements in the subarray $A[i..j]$ which are larger than or equal to b in $O(k+1)$ time, where k is the number of output elements.*

Lemma 2 ([15]). *For a string Z of length n over an integer alphabet $\{1, \ldots, \ell\}$ with $\ell \le n$, there is a wavelet tree for Z of $n \log \ell + O(n \log \log n \log \ell / \log n)$ bits of space, which, given a range $[b, e] \subseteq [1, n]$ and an integer $1 \le t \le \ell$, returns (1) the number k of positions i in Z satisfying $b \le i \le e$ and $Z[i] \ge t$ in $O(\log n)$ time and (2) all such k positions in $O(k \log n)$ time. The wavelet tree can be constructed in $O(n \log \ell)$ time and $O(n \log \ell)$ bits of extra space.*

3   Suffix Arrays for Run Length Encoded Strings

In this section, we introduce the *truncated RLE suffix array* (tRLESA) and the *truncated RLE lcp array* (tRLELCP) for the RLE suffixes of T, which are important data structures for our RLE-based text indexing structure. We also introduce two more data structures called the *RLE suffix array* (RLESA) and the *RLE lcp array* (*RLELCP*) for the RLE suffixes of T, which we use as intermediate data structures to efficiently construct $tRLESA$ and $tRLELCP$.

Definition 1. $RLESA[1..n]$ for $RLE(T) = a_1^{p_1} \cdots a_n^{p_n}$ is an array of length n such that $RLESA[i] = j$ iff $RLEsuf(j)$ is the lexicographically ith RLE suffix of $RLE(T)$ for $1 \le i \le n$.

For example, $RLESA$ for $RLE(T) = a^3b^2c^4a^3b^2a^1b^7c^4\$$ is $[9, 4, 1, 6, 5, 7, 2, 8, 3]$.

Definition 2. $RLELCP[1..n]$ for $RLE(T) = a_1^{p_1} \cdots a_n^{p_n}$ is an array of length n such that $RLELCP[i] = lcp(RLEsuf(RLESA[i-1]), RLEsuf(RLESA[i]))$ for $1 < i \le n$, and $RLELCP[1] = nil$.

In the running example, $RLELCP$ is $[-, 0, 2, 0, 0, 0, 0, 0, 1]$.

We classify the RLE suffixes of $RLE(T) = a_1^{p_1} \cdots a_n^{p_n}$ into two types, in a similar way to [10]: $RLEsuf(i)$ is said to be *S-type* if $RLEsuf(i) \prec RLEsuf(i+1)$, and $RLEsuf(i)$ is said to be *L-type* if $RLEsuf(i) \succ RLEsuf(i+1)$. For convenience, let $RLEsuf(n)$ be both S-type and L-type.

Lemma 3. *All RLE suffixes of $RLE(T) = a_1^{p_1} \cdots a_n^{p_n}$ can be classified to either S-type or L-type in $O(n)$ time and $O(\log n)$ bits of extra space.*

Proof. By definition, $a_i \ne a_{i+1}$ for any $1 \le i < n$. Hence all RLE suffixes can be classified to either S-type or L-type by a simple scan on $RLE(T)$. □

The following lemma generalizes the result of [10] regarding the lexicographical order of S-type and L-type suffixes, to that of S-type and L-type RLE suffixes.

Lemma 4. *Let $RLEsuf(i)$ and $RLEsuf(j)$ be any RLE suffixes of $RLE(T)$ such that $1 \le i \ne j \le n$ and $a_i = a_j$. Then the following properties hold: (a) If $RLEsuf(i)$ is L-type and $RLEsuf(j)$ is S-type, then $RLEsuf(i) \prec RLEsuf(j)$; (b) If $RLEsuf(i)$ and $RLEsuf(j)$ are both L-type and $p_i < p_j$, then $RLEsuf(i) \prec RLEsuf(j)$; (c) If $RLEsuf(i)$ and $RLEsuf(j)$ are both S-type and $p_i > p_j$, then $RLEsuf(i) \prec RLEsuf(j)$.*

Proof. Each case can be shown as follows:

Case (a): By assumption, we have $a_{i+1} \prec a_i = a_j \prec a_{j+1}$. If $p_i = p_j$, then $lcp(val(RLEsuf(i)), val(RLEsuf(j))) = p_i$. Since $a_{i+1} \prec a_{j+1}$, $RLEsuf(i) \prec RLEsuf(j)$. If $p_i < p_j$, then $lcp(val(RLEsuf(i)), val(RLEsuf(j))) = p_i$. Since $a_{i+1} \prec a_j$, $RLEsuf(i) \prec RLEsuf(j)$. If $p_i > p_j$, then $lcp(val(RLEsuf(i)), val(RLEsuf(j))) = p_j$. Since $a_i \prec a_{j+1}$, $RLEsuf(i) \prec RLEsuf(j)$.

Case (b): By assumption, we have $a_j = a_i \succ a_{i+1}$. It follows from $p_i < p_j$ that $lcp(val(RLEsuf(i)), val(RLEsuf(j))) = p_i$. Since $a_{i+1} \prec a_j$, $RLEsuf(i) \prec RLEsuf(j)$.

Case (c): By assumption, we have $a_i = a_j \prec a_{j+1}$. It follows from $p_i > p_j$ that $lcp(val(RLEsuf(i)), val(RLEsuf(j))) = p_j$. Since $a_i \prec a_{j+1}$, $RLEsuf(i) \prec RLEsuf(j)$.

Hence the lemma holds. □

We introduce a total order on the positions of $RLE(T)$ based on Lemma 4.

Definition 3. *Let $\lhd$ be the total order on $[1, n]$ such that for any $1 \leq i \neq j \leq n$,*

1. *if $a_i \prec a_j$, then $i \lhd j$;*
2. *if $a_i = a_j$, $RLEsuf(i)$ is L-type, and $RLEsuf(j)$ is S-type, then $i \lhd j$;*
3. *if $a_i = a_j$, $RLEsuf(i)$ and $RLEsuf(j)$ are L-type, and $p_i < p_j$, then $i \lhd j$;*
4. *if $a_i = a_j$, $RLEsuf(i)$ and $RLEsuf(j)$ are S-type, and $p_i > p_j$, then $i \lhd j$;*
5. *if $a_i = a_j$, $RLEsuf(i)$ and $RLEsuf(j)$ are the same type, and $p_i = p_j$, then $i \equiv j$.*

For any $1 \leq i \leq n$, let $RANK_\lhd(i)$ be the rank of position i w.r.t. $\lhd$.

Lemma 5. *Let $R = RANK_\lhd(1) \cdots RANK_\lhd(n)$. Then, the suffix array SA of string R is identical to $RLESA$ of $RLE(T)$.*

Proof. It suffices to show that for any $1 \leq i \neq j \leq n$, $RLEsuf(i) \prec RLEsuf(j)$ iff $R[i..n] \prec R[j..n]$. Let $RLE(T) = a_1^{p_1} \cdots a_n^{p_n}$.

($\Longrightarrow$) Suppose $RLEsuf(i) \prec RLEsuf(j)$. If $a_i \prec a_j$ (case 1 of Definition 3), then clearly $R[i..n] \prec R[j..n]$. Assume $\ell = lcp(RLEsuf(i), RLEsuf(j)) \geq 1$. Since $RLEsuf(i) \prec RLEsuf(j)$, $a_{i+\ell} \preceq a_{j+\ell}$. If $a_{i+\ell} = a_{j+\ell}$, then case 2, 3, or 4 applies to positions $i + \ell$ and $j + \ell$, yielding $i + \ell \lhd j + \ell$ by Lemma 4. If $a_{i+\ell-1} \prec a_{i+\ell} \prec a_{j+\ell}$ or $a_{i+\ell} \prec a_{j+\ell} \prec a_{i+\ell-1}$, then case 1 applies to positions $i + \ell$ and $j + \ell$, yielding $i + \ell \lhd j + \ell$. If $a_{i+\ell} \prec a_{i+\ell-1} \prec a_{j+\ell}$, then case 2 applies to positions $i + \ell - 1$ and $j + \ell - 1$, yielding $i + \ell - 1 \lhd j + \ell - 1$ by Lemma 4. In the former, $RANK_\lhd(i + \ell) < RANK_\lhd(j + \ell)$ and $RANK_\lhd(i + k) = RANK_\lhd(j + k)$ for all $0 \leq k < \ell$. In the latter, $RANK_\lhd(i + \ell - 1) < RANK_\lhd(j + \ell - 1)$ and $RANK_\lhd(i + h) = RANK_\lhd(j + h)$ for all $0 \leq h < \ell - 1$. Hence $R[i..n] \prec R[j..n]$.

($\Longleftarrow$) Suppose $R[i..n] \prec R[j..n]$. If $\ell' = lcp(R[i..n], R[j..n])$, then $R[i + \ell'] \prec R[j + \ell']$. This implies $i + \ell' \lhd j + \ell'$, which falls into one of the cases 1, 2, 3, and 4 of Definition 3. In either case, $RLEsuf(i + \ell') \prec RLEsuf(j + \ell')$ by Lemma 4. Since $RLE(T)[i + k'] = RLE(T)[j + k']$ for all $0 \leq k' < \ell'$, we have $RLEsuf(i) \prec RLEsuf(j)$. $\square$

Theorem 2. *Given $RLE(T) = a_1^{p_1} \cdots a_n^{p_n}$, we can compute $RLESA$ of $RLE(T)$ in $O(n \log n)$ time and $O(n \log n)$ bits of extra space.*

Proof. We classify each $a_i^{p_i}$ to either S-type or L-type in $O(n)$ time by Lemma 3. Since any pair of positions in $RLE(T)$ can be compared w.r.t. $\lhd$ in constant time, $RANK_\lhd(i)$ for all $1 \leq i \leq n$ can be computed in a total of $O(n \log n)$ time and $O(n \log n)$ bits of space, using any suitable comparison-based sorting algorithm. Since $R = RANK_\lhd(1) \cdots RANK_\lhd(n)$ is a string over the integer alphabet $[1, n]$, the suffix array of R can be constructed in $O(n)$ time, using any linear time suffix array construction algorithm (e.g. [10,16]). By Lemma 5, this suffix array is identical to $RLESA$ of $RLE(T)$, the desired output. $\square$

Theorem 3. *Given $RLESA$ for $RLE(T) = a_1^{p_1} \cdots a_n^{p_n}$, we can compute $RLELCP$ for $RLE(T)$ in $O(n)$ time and $O(n \log n)$ bits of extra space.*

Proof. Since $RLESA$ is based on the order $\prec$ between RLE suffixes, and since $RLEsuf(i) \prec RLEsuf(j)$ iff $suf(i) \prec suf(j)$, we can use the linear-time LCP array construction algorithm of Kasai et al. [9], with the following slight modification: When we compare $a_i^{p_i}$ and $a_j^{p_j}$ with $a_i = a_j$ during RLE suffix comparison, we increase the lcp value by $\min\{p_i, p_j\}$. The comparison of two RLE suffixes terminates as soon as we encounter the case where $a_i \neq a_j$, or $a_i = a_j$ and $p_i \neq p_j$. Thus, it takes a total of $O(n)$ time. The total space requirement is $O(n \log n)$ bits due to the use of the reverse array $RLESA^{-1}$ for $RLESA$ that is needed to apply the algorithm of Kasai et al. [9]. □

For any RLE suffix $RLEsuf(i) = a_i^{p_i} a_{i+1}^{p_{i+1}} \cdots a_n^{p_n}$ of $RLE(T)$ with $1 < i \leq n$, let $tRLEsuf(i) = a_{i-1}^1 a_i^{p_i} \cdots a_n^{p_n}$, or equivalently, $tRLEsuf(i) = a_{i-1}^1 RLEsuf(i)$. Namely, $tRLEsuf(i)$ is obtained by replacing the exponent p_{i-1} of $RLEsuf(i-1)[1] = a_{i-1}^{p_{i-1}}$ by 1. Let $tRLEsuf(1) = \$^1$ for convenience. Each $tRLEsuf(i)$ for $1 \leq i \leq n$ is called a *truncated RLE suffix* of $RLE(T)$.

Definition 4. *$tRLESA[1..n]$ for $RLE(T) = a_1^{p_1} \cdots a_n^{p_n}$, is an array of length n such that $tRLESA[i] = j$ iff $tRLEsuf(j)$ is the lexicographically ith truncated RLE suffix of $RLE(T)$ for $1 \leq i \leq n$.*

In the running example, $tRLESA$ is $[1, 5, 7, 2, 6, 8, 3, 9, 4]$.

Definition 5. *$tRLELCP[1..n]$ for $RLE(T) = a_1^{p_1} \cdots a_n^{p_n}$ is an array of length n such that for $1 < i \leq n$,*

$$tRLELCP[i] = \begin{cases} lcp(RLEsuf(tRLESA[i-1]), RLEsuf(tRLESA[i])) & \text{if } a = a', \\ 0 & \text{if } a \neq a', \end{cases}$$

where $a = a_{tRLESA[i-1]}$ and $a' = a_{tRLESA[i]}$, and $tRLELCP[1] = nil$.

In the running example, $tRLELCP$ is $[-, 0, 0, 0, 0, 0, 1, 0, 0]$.

Theorem 4. *Given $RLESA$ for $RLE(T) = a_1^{p_1} \cdots a_n^{p_n}$, we can compute $tRLESA$ for $RLE(T)$ in $O(n)$ time and $O(\sigma \log n)$ bits of extra space.*

Proof. We use the idea of an induced sorting algorithm [16]. For each $a \in \Sigma$, we compute the range of $tRLESA$ corresponding to the truncated RLE suffixes starting at a, in a total of $O(n)$ time and $O(\sigma \log n)$ bits of space by a simple scan on $RLE(T)$. We then scan $RLESA$ from the beginning to the end, and assume we have processed the first $i - 1$ entries of $RLESA$ and have filled $i - 1$ entries of $tRLESA$. Now, $RLEsuf(RLESA[i])$ is induced to the $(k + 1)$th entry in the range of $tRLESA$ for $a_{RLESA[i]-1}$, where k is the number of truncated RLE suffixes that start with $a_{RLESA[i]-1}$ and are lexicographically smaller than $a_{RLESA[i]-1} RLEsuf(RLESA[i])$. Clearly, it takes a total of $O(n)$ time and $O(\sigma \log n)$ bits of extra space. □

Theorem 5. *Given $RLESA$, $tRLESA$ and $RLELCP$ for $RLE(T) = a_1^{p_1} \cdots a_n^{p_n}$, we can compute $tRLELCP$ for $RLE(T)$ in $O(n)$ time and $O(n \log n)$ bits of extra space.*

Proof. We preprocess $RLELCP$ in $O(n)$ time with $2n + o(n)$ bits of extra space so that range minimum queries on $RLELCP$ can be answered in $O(1)$ time, using Theorem 1. We also construct the reverse array $RLESA^{-1}$ for $RLESA$ in $O(n)$ time, using $O(n \log n)$ bits of space. By definition, we can easily determine whether $tRLELCP[i] = 0$ or not in $O(1)$ time. If $tRLELCP[i] \geq 1$, then $tRLELCP[i] = 1 + lcp(RLEsuf(tRLESA[i-1]+1), RLEsuf(tRLESA[i]+1)) = 1 + \min_{u_i < j \leq v_i} RLELCP[j]$, where $u_i = RLESA^{-1}[tRLESA[i-1]+1]$ and $v_i = RLESA^{-1}[tRLESA[i]+1]$. Thus $tRLELCP[1..n]$ can be constructed in $O(n)$ time and $O(n \log n)$ bits of space. □

4   Text Indexing Based on *tRLESA* and *tRLELCP*

This section presents our indexing structure for a string T of length N, which is based on *tRLESA* and *tRLELCP* for $RLE(T)$ and supports the following:

- locate($RLE(P)$): given $RLE(P)$ for a pattern string $P \in \Sigma^*$, return the set $occ(T, P)$ of occurrences of P in T.
- access_SA(i): given an integer $1 \leq i \leq N$, return $SA[i]$.
- substr(i, j): given integers $1 \leq i \leq j \leq N$, return the substring $T[i..j]$ of T.

4.1   Compact Representation for $RLE(T)$ and substr Queries

Lemma 6. *We can represent $RLE(T)$ with $n \log \sigma + n \log \frac{N}{n} + O(n)$ bits of space so that a substr(i, j) query can be answered in $O(j - i + \log n)$ time.*

Proof. We represent the base characters $a_1, \ldots, a_n$ and the exponents $p_1, \ldots, p_n$ of $RLE(T) = a_1^{p_1} \cdots a_n^{p_n}$ separately. The sequence of base characters is stored explicitly using $n \log \sigma$ bits of space. Let Y be the prefix sums of the exponents, i.e., $Y[k] = \sum_{j=1}^{k} p_j$ for any $1 \leq k \leq n$. Recall that $Y[k] \leq N$ for any $1 \leq k \leq n$, and therefore if we store Y explicitly, it takes $n \log N$ bits of space. To represent Y space-efficiently, we use the data structure of [8] which allows $O(1)$-time random access on Y, using $n \log \frac{N}{n} + O(n)$ bits of space. All together, we have a representation of $RLE(T)$ with $n \log \sigma + n \log \frac{N}{n} + O(n)$ bits of space. Using this representation, we can answer substr(i, j) query in $O(j - i + \log n)$ time: We conduct two binary searches on Y in a total of $O(\log n)$ time, to find the maximal character runs $a_b^{p_b}$ and $a_e^{p_e}$ which, respectively, contain the positions i and j when decompressed. Then, $RLE(T[i..j]) = a_b^x a_{b+1}^{p_{b+1}} \cdots a_{e-1}^{p_{e-1}} a_e^w$, where $x = Y[b] - i + 1$ and $w = j - Y[e-1]$. We can then obtain $T[i..j]$ in $O(j - i + 1)$ time. Hence, substr(i, j) query takes a total of $O(j - i + \log n)$ time. □

4.2   Data Structure for locate Queries

For each character $c \in \Sigma$, let $Spos_c$ and $Lpos_c$ be the sets of beginning positions of the S-type and L-type RLE suffixes of $RLE(T)$ starting with c, respectively. Let $SExp_c$ (resp. $LExp_c$) be the set of *distinct* exponents of the maximal character

$RLE(T) = a^5\, b^2\, a^2\, b^3\, a^7\, b^1\, a^2\, b^5\, a^7\, b^5\, a^2\, b^1\, a^8\, \$^1$

types: S L S L S L S L S L S L L S/L

$Spos_a = \{1, 3, 5, 7, 9, 11\}$

$SExp_a = \{2, 5, 7\}$

k	1	2	3	4	5	6
sid_a^{-1}	-	7	5	5	5	2
$SMCR_a$	-	2	7	7	7	19

$sid_a = (7) = 2$
$sid_a = (5) = 3$
$sid_a = (2) = 6$

Fig. 1. $Spos_a$, $SExp_a$, sid_a, sid_a^{-1}, and $SMCR_a$ for $RLE(T) = a^5 b^2 a^2 b^3 a^7 b^1 a^2 b^5 a^7 b^5 a^2 \$^1$. Consider to compute $sid_a(5)$ for a given exponent $5 \in SExp_a$. Since $sid_a(5)$ is equal to the minimum index k on sid_a^{-1} such that $sid_a^{-1}[k] = 5$, and since sid_a^{-1} is non-increasing, $sid_a(5) = 3$ can be obtained in $O(\log n)$ time by a binary search on sid_a^{-1}.

runs of c with which the S-type (resp. L-type) RLE suffixes start, i.e., $SExp_c = \{q \mid j \in Spos_c, RLE(T)[j] = c^q\}$ and $LExp_c = \{p \mid j \in Lpos_c, RLE(T)[j] = c^p\}$. For any $q \in SExp_c$ and $p \in LExp_c$, let $sid_c(q) = |\{j' \mid j' \in Spos_c, RLE(T)[j'] = c^{q'}, q' \geq q\}|$ and $lid_c(p) = |\{i' \mid i' \in Lpos_c, RLE(T)[i'] = c^{p'}, p' \geq p\}|$, which are the numbers of S-type and L-type RLE suffixes starting with c with exponent at least q and p, respectively. Let X be the sequence of length n such that $X[i] = sid_{a_j}(p_j)$ if $RLEsuf(tRLESA[i])$ is S-type, and $X[i] = lid_{a_j}(p_j)$ if $RLEsuf(tRLESA[i])$ is L-type, where $j = tRLESA[i] - 1$. Given $RLE(T) = a_1^{p_1} \cdots a_n^{p_n}$, the sequence X can be computed in $O(n \log n)$ time and $O(n \log n)$ bits of space, by sorting the exponents $p_1, \ldots, p_n$ and using Lemma 3. See Fig. 1 for examples of the above notations.

Theorem 6. *There is a data structure of $n(2 \log N + \log n + \log \sigma) + O(n)$ bits of space which supports $\mathsf{locate}(RLE(P))$ query in $O(m + occ + \log n)$ time, for a given query $RLE(P) = b_1^{q_1} \cdots b_m^{q_m}$. Given $RLE(T)$ of size n, the data structure can be constructed in $O(n \log n)$ time and $O(n \log N)$ bits of extra space.*

Proof. In our data structure, queries are processed in the following 3 steps.

Step (a): Let B_S be a bitmap of length n s.t. $B_S[i] = 1$ if $RLEsuf(tRLESA[i] - 1)$ is S-type and $RLEsuf(tRLESA[i - 1] - 1)[1] \neq RLEsuf(tRLESA[i] - 1)[1]$, and $B[i] = 0$ otherwise. Similarly, let B_L be a bitmap of length n s.t. $B_L[i] = 1$ if $RLEsuf(tRLESA[i] - 1)$ is L-type and $RLEsuf(tRLESA[i - 1] - 1)[1] \neq RLEsuf(tRLESA[i] - 1)[1]$, and $B[i] = 0$ otherwise. Namely, B_S (resp. B_L) indicates the boundary of the first characters of the S-type (resp. L-type) truncated RLE suffixes sorted in $tRLESA$. Using constant-time select dictionaries [7] on two bitmaps B_S and B_L which require a total of $2n + o(n)$ bits, we can find in $O(1)$ time the range of $tRLESA$ which corresponds to the truncated RLE suffixes starting with b_1, which is the first base character of the query pattern $RLE(P)$.

Step (b): We conduct a binary search on the range obtained in Step (a) for $b_1 RLE(P)[2..m - 1] = b_1 b_2^{q_2} \cdots b_{m-1}^{q_{m-1}}$. To speed-up the binary search, we use the data structure of Theorem 1 built on $tRLELCP$, which supports RMQ's on

$tRLELCP$ in $O(1)$ time. Then, we can find an occurrence of $b_1 RLE(P)[2..m-1]$ in the range with only $O(m)$ maximal character run comparisons, by a standard technique for pattern matching with suffix arrays and LCP arrays [14]. Thus, this step takes $O(m + \log n)$ time.

Step (c): Let h be the entry of $tRLESA$ found in Step (b), which corresponds to an occurrence of $b_1 RLE(P)[2..m-1]$, i.e., $aRLE(T)[tRLESA[h]..tRLESA[h] + m-3] = b_1 RLE(P)[2..m-1]$, where $a = a_{tRLESA[h]-1}$. We then find the range $[u, v]$ of $tRLESA$ which corresponds to the truncated RLE suffixes which start with $b_1 RLE(P)[2..m]$, including the last maximal character run $b_m^{p_m}$. This range can be found in $O(\log n)$ time by a binary search based on Lemma 4, starting from the hth entry of $tRLESA$, with a help of $O(1)$-time RMQ's on $tRLELCP$. Now, for any $u \le r \le v$, $RLE(P)$ occurs at position $tRLESA[r]-1$ of $RLE(T)$ iff $p_{tRLESA[r]-1} \le q_1$. Assume that $b_1 \prec b_2$, which implies that $RLE(P)$ corresponds to S-type RLE suffixes of $RLE(T)$. Assume for simplicity that $q_1 \in SExp_{b_1}$ (the case where $q_1 \notin SExp_{b_1}$ can be treated similarly). To evaluate whether the inequality $p_{tRLESA[r]-1} \ge q_1$ holds or not, we can utilize string X and sid_{b_1}, since $p_{tRLESA[r]-1} \ge q_1$ iff $sid_{b_1}(p_{tRLESA[r]-1}) \ge sid_{b_1}(q_1)$. To compute $sid_{b_1}(p)$ for a given exponent $p \in SExp_{b_1}$ efficiently, we do the following: For each character $c \in \Sigma$, we use an array sid_c^{-1} of length $|Spos_c|$ such that $sid_c^{-1}[k] = p$ if p is the smallest element in $SExp_c$ such that $k \ge sid_c(p)$ (see also Fig. 1 for an example). Since sid_c^{-1} is monotonically decreasing, we can retrieve $sid_c(p)$ for a given $p \in SExp_c$ in $O(\log n)$ time by a binary search on sid_c^{-1}. In the case where $b_1 \succ b_2$, we use lid_c^{-1} for L-type truncated RLE suffixes, defined in a similar way to sid_c^{-1}. Now it follows from Lemma 1 that we can locate all the corresponding positions in $RLE(T)$ in $O(1 + occ)$ time. For each of the occ positions in $RLE(T)$ we have found, we can obtain the corresponding position in the original text T in $O(1)$ time, using the prefix sum data structure of Lemma 6. Hence, locate query takes a total of $O(m + occ + \log n)$ time. Let us analyze the space requirement in this step. The RMQ data structure requires $2n + o(n)$ bits of extra space by Theorem 1. We construct sid^{-1} and lid^{-1} by concatenating sid_c^{-1} and lid_c^{-1} respectively, for all $c \in \Sigma$ in the lexicographical order. To access sid_c^{-1} for a given character $c \in \Sigma$ in constant time, we maintain a bitmap B'_S of length $|sid^{-1}|$ such that $B'_S[i] = 1$ iff i is the beginning position of the range of sid_c^{-1} corresponding to some character c, and $B'_S[i] = 0$ otherwise. We construct another array lid^{-1} and its corresponding bitmap B'_L in a similar way. Since $\sum_{c \in \Sigma}(|sid_c^{-1}| + |lid_c^{-1}|) = n$, using the data structure of [8], sid^{-1} and lid^{-1} can be represented with a total of $n \log \frac{N}{n} + O(n)$ bits of space, and the select dictionaries for the bitmaps take a total of $n + o(n)$ bits of space.

Overall, locate queries take $O(m + occ + \log n)$ time for $|RLE(P)| = m \ge 2$. locate queries for $RLE(P) = a_1^{p_1}$ can be answered in $O(occ)$ time, using the RMQ data structure for X. On top of the $n \log \frac{N}{n} + n \log \sigma + O(n)$ bits representation of Lemma 6, we need

- $3n \log n$ bits of space for $tRLESA$, $tRLELCP$, and X,
- $2n + o(n)$ bits for the select dictionaries for B_S and B_L in Step (a),

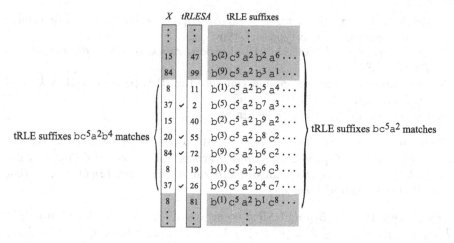

Fig. 2. Illustration of how to search for $RLE(P) = \mathrm{b}^3\mathrm{c}^5\mathrm{a}^2\mathrm{b}^4$ using $tRLESA$ for some $RLE(T)$ such that $sid_b(1) = 8$, $sid_b(2) = 15$, $sid_b(3) = 20$, $sid_b(5) = 37$, and $sid_b(9) = 84$. Since the exponents of the first maximal character runs of the truncated RLE suffixes are ignored in $tRLESA$, they are shown in parentheses. Firstly, we find the range $[u'', v'']$ of $tRLESA$ that corresponds to the truncated RLE suffixes that start with $\mathrm{b} = RLE(P)[1]$. Secondly, we find the subrange $[u', v'] \subseteq [u'', v'']$ of $tRLESA$ that corresponds to $\mathrm{b}RLE(P)[2..3] = \mathrm{bc}^5\mathrm{a}^2$. Thirdly, we find the subrange $[u, v] \subseteq [u', v']$ of $tRLESA$ that correspond to $\mathrm{b}RLE(P)[2..4] = \mathrm{bc}^5\mathrm{a}^2\mathrm{b}^4$. Finally, we find all the elements of $X[u..v]$ which are more than or equal to $sid_b(3) = 20$. The set $\{2, 26, 55, 72\}$ of the checked entries of $tRLESA$ are the positions where $RLE(P)$ occurs in $RLE(T)$.

- $4n + o(n)$ bits for the RMQ data structures for $tRLELCP$ and X in Steps (b) and (c),
- $n \log \frac{N}{n} + O(n)$ bits of space for sid^{-1} and lid^{-1} in Step (c), and
- $n + o(n)$ bits for the select dictionaries for B'_S and B'_L in Step (c).

Hence, the total space requirement for our locate data structure is $n(2 \log N + \log n + \log \sigma) + O(n)$ bits.

It follows from Lemmas 4 and 5 that $tRLESA$ and $tRLELCP$ for $RLE(T)$ of size n can be constructed in $O(n \log n)$ time and $O(n \log n)$ bits of space. The select dictionaries of [7] can be constructed in $O(n)$ time, and the RMQ data structures can be constructed in $O(n)$ time by Theorem 1. The sequences sid^{-1} and lid^{-1} can be computed in $O(n \log n)$ time and $O(n \log N)$ bits of extra space by a suitable comparison based sorting algorithm. □

See Fig. 2 for an example showing how the occurrences of $RLE(P)$ can be found using our data structure of Theorem 6.

4.3 Data Structure for access_SA Queries

In our data structure, an access_SA(i) query is processed in the following 4 steps:

1. Retrieve the first character $a = suf(SA[i])[1]$ of $suf(SA[i])$, and determine whether $suf(SA[i])$ is S-type or L-type.
2. Compute a "small" range $[p', p'']$ of exponents s.t. $p' \leq p \leq p''$, where $a^p = RLE(suf(SA[i]))[1]$.
3. Determine the exponent p, and compute the RLE position k $(1 \leq k \leq n)$ s.t. $RLE(suf(SA[i])) = a^p a_k^{p_k} \cdots a_n^{p_n}$.
4. Compute $SA[i]$ from k obtained in Step 3.

Theorem 7. *There is a data structure of $n(3 \log N + \log n + \log \sigma) + 2\sigma \log \frac{N}{\sigma} + O(n \log \log n)$ bits of space which supports* access_SA(i) *query in $O(\log^2 n)$ time. Given $RLE(T)$ of size n, the data structure can be constructed in $O(n \log n)$ time and $O(n \log N)$ bits of extra space.*

Proof. **Step 1:** Let $LBeg$ and $SBeg$ be arrays of length σ s.t. for each $a \in \Sigma$, $LBeg[a] = |\{j \mid 1 \leq j \leq N, T[j] \prec a\}| + 1$ and $SBeg[a] = LBeg[a] + |\{j \mid 1 \leq j \leq N, T[j] = a, suf(j) \text{ is L-type}\}|$. The suffix array ranges for the L-type and S-type suffixes starting with $b \in \Sigma$ are $[LBeg[b], SBeg[b] - 1]$ and $[SBeg[b], LBeg[c] - 1]$, respectively, where $c \succ b$ is the lexicographical successor of b in Σ. Since both $LBeg$ and $SBeg$ are monotonically non-decreasing, we can binary search these arrays to retrieve the first character $a = T[SA[i]]$, determine whether $suf(SA[i])$ is S-type or L-type, and compute the suffix array range that $suf(SA[i])$ belongs to. Using the data structure of [8] these arrays can be represented by $2\sigma \log \frac{N}{\sigma} + O(\sigma)$ bits of space, so that each binary search takes $O(\log \sigma)$ time.

Step 2: Assume $suf(SA[i])$ is S-type. Let $[s, t]$ be the suffix array range for the S-type suffixes that start with a. Then, $SA[i]$ corresponds to the $(i - s + 1)$th entry of $[s, t]$. For each $c \in \Sigma$ and $1 \leq k \leq |Spos_c|$, let $SMCR_c[k] = |\{j \mid suf(j) \text{ is S-type and starts with } c^{sid_a^{-1}[k]}, 1 \leq j \leq N\}|$. (See also Fig. 1 for an example of $SMCR_c$.) An important observation is that the size of the suffix array range for the S-type suffixes that start with c^q is equal to the number of S-type suffixes of T which start with c^y for $y \geq q$. Since $SA[i]$ corresponds to the $(i - s + 1)$th entry of the suffix array range that is S-type and starts with a, we binary search $SMCR_a$ for $i - s + 1$. There are two cases: (i) If $i - s + 1$ is found in $SMCR_a$, we then conduct another binary search to find the smallest k such that $SMCR_a[k] = i - s + 1$. This gives us the exponent p, since $sid_a(p) = k$. (ii) If $i - s + 1$ is not found in $SMCR_a$, then we terminate the binary search with the largest index r of $SMCR_a$ such that $SMCR_a[r] < i - s + 1$. We then conduct two binary searches on $SMCR_a$ to find the smallest and largest entries r'' and r' such that $SMCR_a[r''] = SMCR_a[r'] = SMCR_a[r]$. This gives us the two exponents $p'', p' \in SExp_a$ such that $p'' \geq p \geq p'$, $sid_a(p'') = r''$ and $sid_a(p') = r' + 1$. In either case, the binary searches take a total of $O(\log |SMCR_a|) = O(\log n)$ time. Hence, the total time complexity of Step 2 is $O(\log n)$. The case where $suf(SA[i])$ is L-type can be handled analogously, using $LMCR_c$ defined for L-type RLE suffixes, in a similar way to $SMCR_c$. Let $SMCR$ be an array of length $\sum_{c \in \Sigma} |Spos_c|$, obtained by concatenating $SMCR_c$ for all $c \in \Sigma$ in the lexicographical order. We construct $LMCR$ and a corresponding bitmap B'_L in a similar way. Since $\sum_{c \in \Sigma}(|SMCR_c| + |LMCR_c|) = n$, using the data structure of [8], $SMCR$ and $LMCR$ can be represented with a total of $n \log \frac{N}{n} + O(n)$ bits of space. We can use the bitmaps B'_S and B'_L of the locate data structure of Theorem 6 to access $SMCR_c$ and/or $LMCR_c$ for any $c \in \Sigma$ in constant time, since $|sid_c| = |SMCR_c| = |Spos_c|$ and $|lid_c| = |LMCR_c| = |Lpos_c|$. Hence, the total space requirement for Step 2 is $n \log \frac{N}{n} + O(n)$ bits (excluding the select dictionaries for the bitmaps B'_S and B'_L).

Step 3: The first goal of Step 3 is to determine the exponent p of the first maximal character run. In what follows we will only consider case (ii) of Step 2, since p is already known in case (i). There is no RLE position $1 \le j \le n$ of $RLE(T) = a_1^{p_1} \cdots a_n^{p_n}$, for which $a_j = a$, $p'' > p_j > p'$, and $RLEsuf(j)$ is the S-type RLE suffix. Thus, for any $p'' \ge x > p'$, there are exactly $sid_a(p'')$ S-type suffixes which start with maximal character run a^x. Also, there are exactly $sid_a(p')$ S-type suffixes which start with maximal character run $a^{p'}$. Let $v = s + SMCR_a[r''] + (p'' - p' - 1)sid_a(p'') - 1$, i.e., v is the largest entry of the suffix array s. t. the first maximal character run of $suf(v)$ is $a^{p'+1}$. If $i \le v$, then $p'' \ge p > p'$, and otherwise $p' = p$ In what follows, we consider the case where $p'' \ge p > p'$, since the other case can be treated similarly. Let $u = s + SMCR_a[r''] - sid_a(p'')$, i.e., u is the smallest entry of the suffix array s.t. the first maximal character run of $suf(u)$ is $a^{p''}$. Let $d = i-u+1$, $g_1 = \lfloor d/sid_a(p'') \rfloor$, and $g_2 = d \mod sid_a(p'')$. Then $p = p'' - g_1$ and $suf(SA[i])$ is the g_2th entry of the suffix array range for the S-type suffixes starting with a^p. We compute the RLE position k s.t. $a^p RLE(suf(SA[i]+p)) = a^p a_k^{p_k} \cdots a_n^{p_n}$, as follows: Let $[\ell, r]$ be the range of $tRLESA$ which corresponds to all the S-type RLE suffixes starting with a. Consider the subset $J \subseteq [\ell, r]$ such that $j \in J$ iff $RLEsuf(tRLESA[j]-1)$ is S-type and $RLEsuf(tRLESA[j] - 1)[1] = a^x$ for some $x \ge p$. Notice that the g_2th element of J corresponds to the entry h of $tRLESA$ such that $tRLESA[h] = k - 1$. Using Lemma 2 with the wavelet tree for X, we can binary search h in $O(\log(r - \ell + 1) \cdot \log n) = O(\log^2 n)$ time. Hence, the overall time complexity of this step is $O(\log^2 n)$. Since each character of X is drawn from $[1, n]$, the wavelet tree for X requires a total of $n \log n + O(n \log \log n)$ bits of space by Lemma 2.

Step 4: Since $RLE(suf(SA[i])) = a^p RLE(T)[k..n]$, $SA[i] = 1 - p + \sum_{j=1}^{k-1} p_j$. This can be calculated in $O(1)$ time from Y of Lemma 6 which represents the prefix sums of the exponents. There is no additional data structure for this step.

Putting all together, $SA[i]$ can be computed in $O(\log^2 n)$ time. On top of the locate data structure of Theorem 6 which requires $n(2 \log N + \log n + \log \sigma) + O(n)$ bits of space we need:

- $2\sigma \log \frac{N}{\sigma} + O(\sigma)$ bits of space for $SBeg$ and $LBeg$ in Step 1;
- $n \log \frac{N}{n} + O(n)$ bits of space for $SMCR$ and $LMCR$ in Step 2;
- $n \log n + O(n \log \log n)$ bits of space for the wavelet tree of X in Step 3.

Hence, the total space requirement for our access_SA data structure is $n(3 \log N + \log n + \log \sigma) + 2\sigma \log \frac{N}{\sigma} + O(n \log \log n)$ bits.

The data structure can be constructed as follows. The $LBeg$ and $SBeg$ arrays of step 1 can be computed in $O(n)$ time by a simple scan on $RLE(T)$, using

$O(\log N)$ bits of extra space. *SMCR* and *LMCR* can be computed in $O(n)$ time with $O(\log N)$ bits of extra space from sid^{-1} and lid^{-1}, respectively, with simple arithmetic. Hence, it follows from Theorem 6 that the whole data structure can be constructed in $O(n \log n)$ time using $O(n \log N)$ bits of extra space.   □

References

1. Apostolico, A., Erdös, P.L., Jüttner, A.: Parameterized searching with mismatches for run-length encoded strings. Theor. Comput., Sci. (2012)
2. Burrows, M., Wheeler, D.J.: A block-sorting lossless data compression algorithm. Tech. Rep. SRC-RR-124, Systems Research Center (1994)
3. Chen, K.Y., Chao, K.M.: A fully compressed algorithm for computing the edit distance of run-length encoded strings. Algorithmica (2011)
4. Eltabakh, M.Y., Hon, W.K., Shah, R., Aref, W.G., Vitter, J.S.: The SBC-tree: an index for run-length compressed sequences. In: Proc. EDBT, pp. 523–534 (2008)
5. Fischer, J., Heun, V.: Space-efficient preprocessing schemes for range minimum queries on static arrays. SIAM J. Comput. **40**(2), 465–492 (2011)
6. Freschi, V., Bogliolo, A.: Longest common subsequence between run-length-encoded strings: a new algorithm with improved parallelism. IPL **90**(4), 167–173 (2004)
7. Golynski, A.: Optimal lower bounds for rank and select indexes. Theor. Comput. Sci. **387**(3), 348–359 (2007)
8. Grossi, R., Vitter, J.S.: Compressed suffix arrays and suffix trees with applications to text indexing and string matching. SIAM J. Comput. **35**(2), 378–407 (2005)
9. Kasai, T., Lee, G., Arimura, H., Arikawa, S., Park, K.: Linear-time longest-common-prefix computation in suffix arrays and its applications. In: Proc. CPM 2001, pp. 181–192 (2001)
10. Ko, P., Aluru, S.: Space efficient linear time construction of suffix arrays. J. Discrete Algorithms **3**(2–4), 143–156 (2005)
11. Lee, S., Park, K.: Dynamic rank/select structures with applications to run-length encoded texts. Theor. Comput. Sci. **410**(43), 4402–4413 (2009)
12. Mäkinen, V., Navarro, G.: Succinct suffix arrays based on run-length encoding. Nord. J. Comput. **12**(1), 40–66 (2005)
13. Mäkinen, V., Navarro, G., Sirén, J., Välimäki, N.: Storage and retrieval of highly repetitive sequence collections. J. Computational Biology **17**(3), 281–308 (2010)
14. Manber, U., Myers, E.W.: Suffix arrays: A new method for on-line string searches. SIAM J. Comput. **22**(5), 935–948 (1993)
15. Navarro, G.: Wavelet trees for all. In: Proc. CPM, pp. 2–26 (2012)
16. Nong, G., Zhang, S., Chan, W.H.: Two efficient algorithms for linear time suffix array construction. IEEE Trans. Computers **60**(10), 1471–1484 (2011)
17. Yamamoto, J., I, T., Bannai, H., Inenaga, S., Takeda, M.: Faster compact on-line Lempel-Ziv factorization. In: Proc. STACS 2014. pp. 675–686 (2014)
18. Ziv, J., Lempel, A.: A universal algorithm for sequential data compression. IEEE Transactions on Information Theory IT-23(3), 337–349 (1977)

PSPACE-Completeness of Bloxorz and of Games with 2-Buttons

Tom C. van der Zanden[1]([⊠]) and Hans L. Bodlaender[1,2]

[1] Department of Computer Science, Utrecht University,
Utrecht, The Netherlands
T.C.vanderZanden@students.uu.nl
[2] Department of Mathematics and Computer Science,
Eindhoven University of Technology, Eindhoven, The Netherlands
H.L.Bodlaender@uu.nl

Abstract. Bloxorz is an online puzzle game where players move a $1 \times 1 \times 2$ block by tilting it on a subset of the two dimensional grid, that also features switches that open and close trapdoors. The puzzle is to move the block from its initial position to an upright position on the goal square. We show that the problem of deciding whether a given Bloxorz level is solvable is PSPACE-complete and that this remains so even when all trapdoors are initially closed or all trapdoors are initially open. We also answer an open question of Viglietta [6], showing that 2-buttons are sufficient for PSPACE-hardness of general puzzle games. We also examine the hardness of some variants of Bloxorz, including variants where the block is a $1 \times 1 \times 1$ cube, and variants with single-use tiles.

1 Introduction

We study the computational complexity of the online puzzle game Bloxorz. We show that deciding whether it is possible to reach the goal square in a level of Bloxorz featuring switches and trapdoors is PSPACE-complete, which we will prove by reduction from Nondeterministic Constraint Logic. We first give a proof that requires us to be able to choose the initial state (open or closed) of each trapdoor. We will then also show that the problem remains PSPACE-complete even when the trapdoors are required to be initially all closed or all open.

Viglietta [6] studied the complexity of general computer games featuring certain common elements such as "destroyable paths, collectible items, doors opened by keys or activated by buttons or pressure plates, etc.". Viglietta established that a game featuring 2-buttons and doors is NP-hard and that a game featuring 3-buttons is PSPACE-hard. Viglietta asked whether this result could be improved to show that a game with 2-buttons is PSPACE-hard. We settle this question, showing that 2-buttons are indeed sufficient for PSPACE-hardness.

We also examine several variants of Bloxorz, and compare the hardness of decision and optimization versions of these. We show that Bloxorz with single-use tiles but without switches or trapdoors is NP-complete. Bloxorz with a $1 \times 1 \times 1$ cube instead of the $1 \times 1 \times 2$ block is NP-complete when both single-use tiles

© Springer International Publishing Switzerland 2015
V.Th. Paschos and P. Widmayer (Eds.): CIAC 2015, LNCS 9079, pp. 403–415, 2015.
DOI: 10.1007/978-3-319-18173-8_30

and switches and trapdoors are included, but becomes polynomially solvable if either is not present in the level.

1.1 Bloxorz

Bloxorz [1] is a "brain twisting puzzle game of rolling blocks and switching bridges" by Damien Clarke. A $1 \times 1 \times 2$ block is tilted around a subset of the two dimensional grid. Games using similar mechanics to Bloxorz are available on multiple platforms, including Android [2] and Apple iOS [3].

Fig. 1. An example Bloxorz level

The block can be in two states: lying down on a rectangular side, or standing up on a square face. From a standing position, the player may make a tilting move (Figure 2a) to place the block on its side. From a lying position, the player can either make a tilting move (Figure 2a) to stand the block up, or a rolling move (Figure 2b) after which the block will still be in a lying state.

(a) Tilting move   (b) Rolling move

Fig. 2. The two types of move available: (a) tilting and (b) rolling

Not all moves are possible: the player may only make a move if after that move, the block is still fully supported by the game level (the online game visualizes this by the block toppling off the stage if an illegal move is attempted).

The goal of the puzzle is to reach some specified goal square (Figure 3b) and get the block to fall through it. This implies the block must be in a standing orientation upon reaching the goal square but this constraint does not contribute to the hardness.

The online version of Bloxorz features various gadgets that make the game more interesting: switches that open and close trapdoors, switches that can only

(a) Trapdoor and switch   (b) Goal square   (c) Reaching the goal square

Fig. 3. More game elements: (a) switches and bridges, (b),(c) goal square

be triggered with the block in a specific orientation, weak tiles that will not support the block standing up, block-splitting portals and more. We will consider the version of Bloxorz with only switches, trapdoors and of course, regular squares. An example of a trapdoor and switch is shown in Figure 3a. Note that the full version of Bloxorz features pressure plates and thus is easily PSPACE-complete (Metatheorem 4c, [6]), we consider a more limited version with switches rather than pressure plates.

A *trapdoor* is a special game square that can be either open or closed. If the trapdoor is open, it is not allowed for the block to be on top of it. Each trapdoor is controlled by exactly one switch and each switch only controls one trapdoor.

A *switch* is a special game square that is associated with exactly one trapdoor. If a move places the block over a switch, the state of the associated trapdoor is toggled (from open to closed and vice-versa). A switch may be triggered more than once. If a move places a block over two switches then they will both be triggered.

If a move places a block over a switch and its associated trapdoor simultaneously, the effect of pressing the switch will be considered before checking the legality of the move. Our results remain valid if we check the legality of the move before activating the switch.

Note that in the original version of the game trapdoors are instead called "bridges" and are of size 1×2 but our constructions will use 1×1 trapdoors. However, our proofs can easily be adapted to work with 1×2 trapdoors instead.

We now formally define BLOXORZ as a decision problem:

BLOXORZ

Instance: A level of Bloxorz, given as a list of game tiles (normal tiles, trapdoors and switches) with their positions, a one-to-one mapping between trapdoors and switches, for each trapdoor an initial state (open/closed) and a start/goal square.

Question: Is there a sequence of legal moves from the start to the goal square?

The precise encoding used for problem instances is not important. We consider the size of a level to be the number of tiles in it.

1.2   Previous Work

Buchin [9] showed that solving rolling block mazes is PSPACE-complete. The block in Bloxorz moves in exactly the same way as a block in a rolling block maze, but a rolling block maze may feature multiple blocks (which may interfere with each other's movement) while Bloxorz only features a single block. Bloxorz also differs from rolling block mazes in that it features switches and trapdoors. These are required for the hardness since a rolling block maze with only a single block is solvable in logarithmic space [12], since it is a connectivity problem.

Viglietta [6] established several metatheorems regarding the complexity of games with buttons[1], doors and pressure plates.

A *door* is a special game square that may be either in a closed or in an open state, and the avatar may pass through the door if and only if it is open. Bloxorz features trapdoors which are similar to doors except that the notions of open and closed are reversed (i.e., the avatar may pass over the trapdoor if and only if it is closed).

The state of a door may be altered by pressure plates and buttons. Viglietta defines a *pressure plate* as a special game square "that is operated whenever the avatar steps on it, and its effect may be either the opening or the closure of a specific door".

In addition, Viglietta defines *buttons* which function similar to pressure plates except that the player has a choice whether to trigger the button or not, while a pressure plate is activated immediately when the player steps on it.

In Bloxorz, a switch instead toggles the state of a trapdoor between open and closed, while a pressure plate or button may either open or close a door, but may not toggle between the two states. Like a pressure plate, a switch is toggled when the block moves over it but due to the $1 \times 1 \times 2$ shape of the block two switches may be triggered simultaneously.

A single door may be controlled by up to two pressure plates: one that opens it, and one that closes it. In Bloxorz, one trapdoor is controlled by exactly one switch.

A k-button may control up to k doors simultaneously, possibly opening some while closing others. Viglietta established that a game featuring doors and k-buttons with $k \geq 2$ is NP-hard, while a game with $k \geq 3$ is PSPACE-hard.

Viglietta poses the question whether his result for $k = 2$ can be improved upon, that is, if a game with 2-buttons is PSPACE-hard. We show that this is the case.

Our reduction is based on the work by Hearn and Demaine [8] on Nondeterministic Constraint Logic.

---

[1] In the conference version of his paper [5], Viglietta uses the term "switch" instead of "button". In the final version of the paper [6], he uses "button" to describe the same object. We use this newer terminology since it better describes the functionality of the object.

2   Nondeterministic Constraint Logic

First, we briefly introduce (restricted) Nondeterministic Constraint Logic (NCL).

A *constraint graph* is an undirected graph $G = (V, E)$ with for each vertex and edge a weight that is a positive integer. We only need to consider weights of 1 and 2. The weight of a vertex is called its *minimum inflow*.

A *configuration* is an assignment of an orientation to each edge, which is *legal* if and only if, for each vertex v, the sum of the weights of the edges pointing into v is at least the minimum inflow of v.

An *edge reversal* is the operation that reverses the orientation of an edge. Thus, its weight starts counting towards the other vertex. A reversal of (v, w) is *legal* if the minimum inflow constraints for v and w remain satisfied.

We consider only constraint graphs featuring two special types of vertices:

- An *AND vertex* is a vertex of degree 3, with minimum inflow 2 and with incident edge weights of 1, 1 and 2. It works like a logical AND gate because either the weight 2 edge must be directed inwards or the two weight 1 edges must be directed inwards.
- An *OR vertex* is a vertex of degree 3, with minimum inflow 2 and with incident edge weights all equal to 2. It works like a logical OR since to satisfy its minimum inflow requirement, at least one of its edges must be directed inwards.

RESTRICTED NCL

Instance: A constraint graph G built exclusively from AND and OR vertices, a legal initial configuration for G and a given target edge e from G.

Question: Is there a sequence of legal edge reversals that, starting from the specified initial configuration, reverses e?

Theorem 1 (Hearn and Demaine [8]). RESTRICTED NCL *is* PSPACE-*complete.*

Note that RESTRICTED NCL remains PSPACE-complete even when restricted to planar graphs, but we do not make use of this property in our proof. This is because the structure of the NCL graph is encoded in the switch-trapdoor correspondence of the Bloxorz level, rather than in its physical structure.

3   PSPACE-Completeness of Standard Bloxorz

Theorem 2. BLOXORZ *is* PSPACE-*complete.*

Proof. By reduction from RESTRICTED NCL. We first establish PSPACE-hardness. We first describe the various constructions used to represent elements from a constraint graph, then show how to put them together into a Bloxorz level. Finally, we show that BLOXORZ is in PSPACE.

3.1   Overview

Figure 4 shows a simplified overview of our construction. The white area represents a section of the game level where the block can move freely. The grey boxes represent copies of the vertex and edge gadgets, which will be described in detail later. The start and goal squares are marked with S and G respectively.

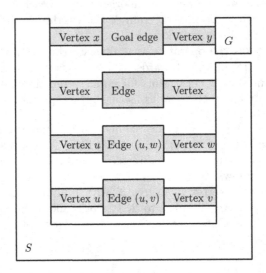

Fig. 4. Overview of the construction

A solution to the Bloxorz level will correspond to a solution to the NCL problem as follows: the player will cross the gadgets from left to right and from right to left a number of times to solve the level. Crossing the edge gadget (u, v) from left to right (resp. from right to left) corresponds to changing the orientation of the edge (u, v) to be towards u (v). To cross the edge gadget, the player will need to activate a number of switches in the edge gadget which also changes the state of the trapdoors in the vertex gadgets. Since reaching the goal square requires crossing the goal edge gadget, the level can thus be solved if and only if the goal edge can be reversed.

The vertex gadgets are constructed so that it is possible to leave the edge via them if and only if their inflow constraint is satisfied. So, if the player crosses the edge (u, v) from left to right, orienting the edge towards u, the vertex v loses inflow while u gains inflow. Since gaining inflow can not cause a constraint to be violated, we only need to check the constraint at vertex v.

Note that the input graph only has degree 3 vertices. For every vertex in the input graph, 3 copies of the vertex gadget are created - one for each of the incident edges. These are related: in our example, crossing the edge (u, v) would also affect the state of the trapdoors in the gadget for vertex u attached to edge (u, w). Also note that the structure of the NCL graph is encoded by the switch-trapdoor correspondence and not in the physical structure of the level.

3.2   Edge Gadget

Figure 5 shows the construction used to represent an edge from an NCL graph. The gray squares represent normal tiles, the squares with fat borders represent trapdoors. Circles represent switches, and the lettering shows the switch-trapdoor correspondence. Trapdoors and switches with white infill correspond to open trapdoors and conversely, grey switches and trapdoors correspond to closed trapdoors. The switches labelled with u, v correspond to trapdoors that are inside vertex gadgets, which will be introduced later.

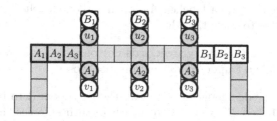

Fig. 5. The edge gadget. The vertex gadget corresponding to vertex v will be attached to the left side of this gadget, and a vertex gadget corresponding to vertex u to the right side.

On either end, the edge gadget will be attached to a corner gadget. The edge gadget is shown pointing to the right vertex, i.e., its weight is counting towards the right vertex.

To reorient the edge gadget, the player would enter it from the left and travel over the trapdoors labelled A. They would then trigger all the switches exactly once, which leaves the trapdoors labelled A open while closing the trapdoors labelled B. The block can then leave the edge via the right side, travelling over the trapdoors labelled B. Note that due to the $1 \times 1 \times 2$ shape of the block it is only possible for the switches to be triggered in pairs (e.g., B_1 is always triggered at the same time as u_1) so this leaves the trapdoors labelled v closed and the ones labelled u open. The edge gadget is now in the opposite orientation (left) and to reverse it again, one would make the same moves but starting from the right.

Note that when we reorient the edge from right to left, the right vertex loses weight while the left vertex gains weight. This means that the only place a constraint could possibly have been violated is the right vertex. Since the trapdoors labelled A are now open, we have to leave via the right side by travelling over the trapdoors labelled B. The vertex gadget will be constructed in such a way that it enforces the constraint: the vertex gadget will prevent us from leaving the edge gadget if its constraint is not satisfied.

Above, we described a "canonical" movement of the block over the edge. Clearly, the player can decide to move the block over the edge gadget in many different ways. For example, they could enter it from the left, activate all the switches on the top side, closing the trapdoors labelled B and opening those labelled u. They could then leave the edge again from the left side, and the edge's weight is essentially removed from the graph: it no longer gives its weight to either vertex. And even more cause for concern: we did this without checking that the constraint of the right vertex was not violated!

However it is never beneficial to only complete a reorientation partially. If at any point we want the weight of the edge to start counting towards the left vertex we have to open (at least one of) the trapdoors labelled A which forces us to exit via the right side and check the constraint for the right vertex. Also, note that the edge will never give its weight to more than one vertex since that would mean both a trapdoor labelled A is open and one labelled B is open which makes it impossible to leave the edge.

Claim. Upon leaving the edge gadget, it is not possible for a trapdoor labelled u to be closed while a trapdoor labelled v is also closed.

Proof. Leaving the edge gadget requires all the trapdoors A to be closed or all the trapdoors labelled B to be closed, which in turn enforces the condition of the lemma.   □

This enables us to define the orientation of the edge gadget: the edge gadget is oriented away from the vertex u if the trapdoors labelled u are open and oriented away from v if all the trapdoors labelled v are open. If both trapdoors u and v are open then we may consider the edge gadget oriented either way.

3.3 Vertices

A vertex is represented by three separate, identical gadgets, that attach to the edge gadgets corresponding to its incident edges. The edge has three switches corresponding to trapdoors in each of the three copies of the vertex gadget. Every vertex has three trapdoors, and every trapdoor is associated with a distinct edge connecting to the vertex.

OR Vertex Gadget. Figure 6a shows one of the three identical components that together form one OR vertex gadget. It is drawn with part of an edge gadget attached. The trapdoors labelled x, y and v_1 each correspond to one of the three edges incident to the vertex. The following claim is self-evident. Note that it enforces the minimum inflow constraint: if one of the trapdoors is closed then the edge associated with that trapdoor must be pointing towards this vertex.

Claim. It is possible for the block to leave the edge via the OR vertex gadget if and only if at least one of the trapdoors is closed.

AND Vertex Gadget. Figure 6b shows one of the three identical components that together form one AND vertex. The following claim regarding the AND vertex gadget is self-evident:

Claim. It is possible for the block to leave the edge via the AND vertex gadget if and only if the trapdoor labelled v_1 is closed or both the trapdoors labelled x and y are closed.

Thus, the edge associated with trapdoor v_1 functions as the weight-2 edge and the edges associated with trapdoors x and y function as weight-1 edges. The vertex gadget enforces that the minimum inflow constraint is satisfied.

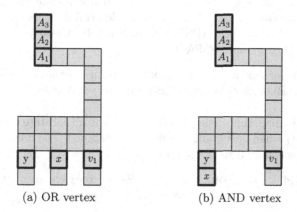

(a) OR vertex   (b) AND vertex

Fig. 6. Vertex gadgets, showing the three identical components that make up: (a) an OR vertex, (b) an AND vertex. The vertex gadgets attach to edge gadgets at the top and to the rest of the game level at the bottom. Note that the trapdoors labelled with A are part of the edge gadget.

3.4   Final Details

We now show how to assemble the component gadgets, forming a Bloxorz level.

For every edge in the constraint graph, one copy of the edge gadget is created. For every vertex, three copies of the appropriate vertex gadget are created and attached to their respective edges. Then, we add some additional blocks so that every vertex can be reached from every other vertex, with the sole exception of the vertex to which the target edge is pointing. It is not attached to any other vertex, instead, it is attached to the goal square. We can place the start square anywhere on the level, except on those squares from which the goal square can be reached directly.

Claim. An instance of BLOXORZ constructed in this way from a RESTRICTED NCL instance has a solution if and only if there exists a sequence of edge reversals that reverses the target edge in the NCL instance.

Proof. If we have a solution to the NCL instance then the sequence of edge reversals translates directly to a solution for the Bloxorz level by traversing the edge gadgets in the same order.

A solution that solves the Bloxorz level can be translated in to a solution for the NCL instance as follows: every time the player traverses an edge gadget associated with edge (u, v) and leaves via the vertex gadget u, this corresponds to reorienting edge (u, v) away from u. Since such a reversal only increases the weight of v, its minimum inflow must remain satisfied. The inflow constraint for u must also remain satisfied since this is enforced by the vertex gadget.   □

Claim. BLOXORZ is in PSPACE.

Proof. Since each state can be represented with polynomial space and we can generate the successor states efficiently, a nondeterministic search of the search space shows that Bloxorz is in NPSPACE. By Savitch's theorem [10], NPSPACE = PSPACE, so Bloxorz is in PSPACE.   □

Since we have shown that BLOXORZ is both contained in PSPACE and PSPACE-hard, BLOXORZ is PSPACE-complete, and we have thus shown Theorem 2.   □

Our proof requires that we are able to choose the initial state of each trapdoor (open or closed) when specifying the problem instance. The following theorems show this is not essential to the hardness:

Theorem 3 ([4]). BLOXORZ *remains* PSPACE-*complete even when limited to the instances where all trapdoors are initially open.*

Theorem 4 ([4]). BLOXORZ *remains* PSPACE-*complete even when limited to the instances where all trapdoors are initially closed.*

4   Hardness of General Puzzles with Buttons

Viglietta [6] established that a game featuring 2-buttons and doors is NP-hard and asked whether this could be improved to show such games are PSPACE-hard. We show that such games are indeed PSPACE-hard.

Note that for a door, the notions of open and closed are inverted from that of a trapdoor.

Theorem 5. *A game where the avatar has to reach an exit location to win that features 2-buttons and doors is* PSPACE-*hard.*

Proof. Our construction to show PSPACE-hardness of BLOXORZ can be adapted to show PSPACE-hardness of a game featuring 2-buttons and doors. Replace every trapdoor by a door. In the edge gadget of Figure 5 replace every pair of switches B_i and u_i by a pair of 2-buttons: one 2-button that opens B_i and closes u_i and another that closes B_i and opens u_i. Do the same for switches labelled A and v. This new construction functions in the exact same way as the edge gadget for Bloxorz.   □

Figure 7 shows an example of this construction. The small white rectangles depict buttons, and are labelled to show which doors they open (marked with a plus sign) and which doors they close (marked with a minus sign).

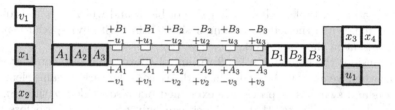

Fig. 7. Edge gadget for 2-button games, depicted with an OR vertex gadget on the left and an AND vertex gadget on the right.

Theorem 6 ([4]). *A game that matches the conditions of Theorem 5 is PSPACE-hard, even when all doors are initially closed. When all doors are initially open the problem can be decided in logarithmic space [12].*

5 Variants of Bloxorz

In order to more closely identify what features make Bloxorz hard, we now look at various simplified variants and determine their complexity. Table 1 shows the complexity for various variants: we list the complexity of the corresponding decision problem (determining whether a solution exists) and optimization problem (determining whether a solution with at most k moves exists) and also the number of moves required in the worst case.

Some of the results in this table are trivial so we will not discuss them.

Table 1. Complexity of various Bloxorz variants, listing the complexity of the decision variant, optimization variant (deciding whether a solution of certain maximum length exists) and the solution length in the worst case.

	$1 \times 1 \times 2$			$1 \times 1 \times 1$		
	Dec.	Opt.	Moves	Dec.	Opt.	Moves
None	P	P	$\Theta(n)$	P	P	$\Theta(n)$
Switches+Trapdoors	PSPACE-C	PSPACE-C	$2^{\Theta(n)^2}$	P	NP-C[3]	$\Theta(n^2)$
Single-use tiles	NP-C	NP-C	$\Theta(n)$	P	P	$\Theta(n)$
S+T+Single-use	PSPACE-C	PSPACE-C	$2^{\Theta(n)^4}$	NP-C[5]	NP-C	$\Theta(n^2)$

---

[2] This follows from our PSPACE-completeness proof. Solving a configuration created in this way from an NCL instance can take exponentially many moves.

[3] Metatheorem 2 of [7]. Collecting items is implemented by forcing a number of trapdoors to be closed before the exit square can be reached.

[4] See footnote 2.

[5] Metatheorem 1 of [6]. Location traversal can be implemented by forcing the player to close a number of trapdoors.

The heading of table 1 shows what size of block and which type of problem is considered, while the leftmost column shows which (if any) special features can be present in the game level.

We consider the regular $1 \times 1 \times 2$ variant of Bloxorz but also consider what happens when we use a $1 \times 1 \times 1$ cube instead. We consider various game elements: trapdoors and switches as previously examined but also single-use tiles. Single-use tiles are game squares that the block can only pass over once before they break and are removed from the game board. Single-use tiles are actually not featured in the original Bloxorz game. We discuss them here since they are an example of a game feature that, when combined with switches and trapdoors will make the decision for the $1 \times 1 \times 1$ case NP-complete.

An interesting observation is that $1 \times 1 \times 2$ Bloxorz with single-use tiles is NP-complete while this is not the case for $1 \times 1 \times 1$ Bloxorz:

Theorem 7 ([4]). $1 \times 1 \times 2$ *Bloxorz with single-use tiles is* NP-*complete, while* $1 \times 1 \times 1$ *Bloxorz with single-use tiles can be solved in polynomial time.*

Furthermore, we can decide Bloxorz with trapdoors and switches in polynomial time when we do not use a rectangular block but a $1 \times 1 \times 1$ cube:

Theorem 8 ([4]). *In the $1 \times 1 \times 1$ case, a Bloxorz instance with switches and trapdoors can be solved in polynomial time, taking $\Theta(n^2)$ moves in the worst case.*

However, when trapdoors, switches and single-use tiles are combined, $1 \times 1 \times 1$ Bloxorz becomes NP-complete again.

6   Conclusion

We have shown by a reduction from RESTRICTED NCL that BLOXORZ is PSPACE-complete, even when the trapdoors are initially fixed either open or closed. We also showed how our proof can be adapted to general games and showed that games with 2-buttons and doors are not only NP-hard, but also PSPACE-hard, answering an open question of Viglietta [6].

We then examined some other variants of Bloxorz, including variants where the $1 \times 1 \times 2$ block is replaced by a $1 \times 1 \times 1$ cube and variants where single-use tiles are included. We showed that Bloxorz with single-use tiles but without switches or trapdoors is NP-complete, and that most variants with a $1 \times 1 \times 1$ block can be polynomially solved. Only the $1 \times 1 \times 1$ variant featuring both single-use tiles and switches and trapdoors is NP-complete.

We think our result clearly illustrates the power of the NCL framework. It is not at all obvious that a universal quantifier may be constructed from the elements in Bloxorz; the work by Demaine and Hearn [8,11] is instrumental for establishing the complexity of this and other puzzle and reconfiguration problems.

Acknowledgement. We thank Damien Clarke for graciously allowing us to reproduce screenshots of Bloxorz and the referees for their useful comments.

References

1. Bloxorz, DX Interactive. http://dxinteractive.com/#!/bloxorz
2. Ice Cube Caveman, Google Play Store.
   https://play.google.com/store/apps/details?id=com.TwistedGames.Caveman
3. Ice Cube Caveman, iTunes App Store.
   https://itunes.apple.com/app/ice-cube-caveman/id456179222?mt=8
4. van der Zanden, T.C., Bodlaender, H.L.: PSPACE-Completeness of Bloxorz and of Games with 2-Buttons arXiv:1411.5951 (2014)
5. Viglietta, G.: Gaming is a hard job, but someone has to do it!. In: Kranakis, E., Krizanc, D., Luccio, F. (eds.) FUN 2012. LNCS, vol. 7288, pp. 357–367. Springer, Heidelberg (2012)
6. Viglietta, G.: Gaming Is a Hard Job, but Someone Has to Do It!. Theory Comput. Syst. **54**(4), 595–621 (2014)
7. Forišek, M.: Computational complexity of two-dimensional platform games. In: Boldi, P. (ed.) FUN 2010. LNCS, vol. 6099, pp. 214–227. Springer, Heidelberg (2010)
8. Hearn, R.A., Demaine, E.D.: PSPACE-completeness of sliding-block puzzles and other problems through the nondeterministic constraint logic model of computation. Theoretical Computer Science **343**, 72–96 (2005)
9. Buchin, K., Buchin, M.: Rolling block mazes are PSPACE-complete. Journal of Information Processing **20**(3), 719–722 (2012)
10. Savitch, W.J.: Relationships between nondeterministic and deterministic tape complexities. Journal of Computer and System Sciences **4**, 177–192 (1970)
11. Hearn, R.A., Demaine, E.D.: Games Puzzles, and Computation. A K Peters/CRC Press (2009)
12. Reingold, O.: Undirected connectivity in log-space. Journal of the ACM **55**(4), article 17 (2008)

Advice Complexity
of Fine-Grained Job Shop Scheduling

David Wehner[(✉)]

ETH Zürich, Rämistrasse 101, Zürich 8092, Switzerland
david@wehner.ch

Abstract. We study the advice complexity, which is a tool to measure the amount of information necessary to achieve a certain output quality, of a specific online scheduling problem. A great deal of research has been carried out in this field; however, this paper studies the problem in a new setting. We redefine the cost function of the considered job shop scheduling problem. Up to now, the makespan was taken as the cost function. Thus, every algorithm has a cost of at least m and at most $2m$ and is, as a consequence, at least 2-competitive, where m denotes the number of jobs. Moreover, Hromkovič et al. [8] constructed an algorithm that has a competitive ratio of at most $1 + 1/\sqrt{m}$. It seems futile to look for better algorithms with respect to this measurement. To allow a more fine-grained analysis, we take the delay of an algorithm as its cost. We prove that with this new cost measure, the best algorithms so far have competitive ratios of $\sqrt{m}$ or $\sqrt{m}/2$ while reading about $\log_2(\sqrt{m})$ advice bits. Then, we describe a deterministic online algorithm with a competitive ratio of at most $0.67\sqrt{m}$. We use this deterministic algorithm to construct an online algorithm with advice that reads $b \geq 2\log_2(m) + 1$ advice bits and has a competitive ratio of at most $\max\{6, \sqrt{8\log_2(m)} \sqrt[4]{m}/\sqrt{b}\}$.

1 Introduction and Preliminaries

How much of the future do we have to know in order to make good decisions? While such a philosophical question is certainly asked by many people in all kinds of situations, making good decisions without knowing their full effect is also an important branch of computer science – namely *online computation*. Here, we consider problems where the input arrives piece by piece, and an *online algorithm* has to make decisions (i.e., compute parts of the final output) without knowing the whole input in advance. To judge the quality of the output generated in such an online manner, we compare it to the solution of an optimal *offline algorithm*, i.e., an algorithm that sees the whole input in advance and that can thus produce an optimal output. The *competitive ratio* is the ratio of the solution computed by the online algorithm and the cost of the optimal offline solution. This approach is called *competitive analysis*; it was introduced by Sleator and Tarjan in 1985

This work was partially supported by the SNF grant 200021-141089.

© Springer International Publishing Switzerland 2015
V.Th. Paschos and P. Widmayer (Eds.): CIAC 2015, LNCS 9079, pp. 416–428, 2015.
DOI: 10.1007/978-3-319-18173-8_31

[11]. We now formally define the above concepts, i.e., online problems (in this paper, we only consider the objective to minimize some cost), online algorithms, and their competitive ratio.

Definition 1 (Online Minimization Problem [9]). *An* online minimization problem *consists of a set $\mathcal{I}$ of inputs; for every $I \in \mathcal{I}$, there is a set of feasible outputs $\mathcal{O}(I)$; and there is a cost function cost for all $I \in \mathcal{I}$. Every input $I \in \mathcal{I}$ is a sequence $I = (x_1, \ldots, x_n)$ of requests. Every output $O \in \mathcal{O}(I)$ is a sequence $O = (y_1, \ldots, y_n)$ of answers to the requests of I. The cost function cost: $\mathcal{O}(I) \to \mathbb{R}$ assigns to every output a real value $cost(I, O)$. For every $I \in \mathcal{I}$, we call an output O an* optimal solution *for I if $cost(I, O)$ is $\min_{O \in \mathcal{O}(I)}\{cost(I, O)\}$.*

We usually omit the instance I, and simply write $cost(O)$ if it is clear from the context to which instance we refer.

Definition 2 (Online Algorithm). *Let $I = (x_1, \ldots, x_n)$ be an input of an online minimization problem. An* online algorithm ALG *calculates a feasible output $\text{ALG}(I) = (y_1, \ldots, y_n) \in \mathcal{O}(I)$ such that for $i \in \{2, \ldots, n\}$, y_i depends solely on $x_1, \ldots, x_i$ and on $y_1, \ldots, y_{i-1}$; y_1 depends solely on x_1. The cost of ALG is defined as $cost(\text{ALG}(I))$.*

Definition 3 (Competitive Ratio). *Let ALG be an online algorithm for an online minimization problem and let OPT be an optimal algorithm for the corresponding offline minimization problem. Let $c \geq 1 \in \mathbb{R}$. ALG has a* competitive ratio *of c and ALG is called c-competitive if $cost(\text{ALG}(I)) \leq c \cdot cost(\text{OPT}(I))$ for every input I.*

Note that this is the usual definition of being strictly c-competitive. In this paper, we do not distinguish between strictly c-competitive and c-competitive. When analyzing online algorithms, one often introduces an adversary ADV [4, 8, 9] that constructs input instances for ALG such that the competitive ratio of ALG is as large as possible.

In 2008, Dobrev, Královič, and Pardubská introduced the concept of *advice complexity* for online problems [5]. Basically, they asked the question we asked at the beginning of this chapter. Maybe knowing some very simple yet crucial characteristic about the future requests helps to design online algorithms that are extremely powerful. Their model was later refined [3,6,7]; in this paper, we use the model that was introduced by Hromkovič et al. [7] and analyzed among others by Komm [9,10]. Let us first describe this model on an intuitive level. We are given an online minimization problem, an online algorithm ALG for the problem, and an adversary ADV. Now an oracle O is introduced that knows the whole input in advance. Before ALG starts its computation, O can write some binary information about the input that is created by ADV on an infinitely large *advice tape*. During its computation, ALG can access this advice tape sequentially. The total number of the advice bits read in the worst case is the *advice complexity* of ALG. We continue with a formal definition.

Definition 4 (Online Algorithm with Advice [9]). *Let I be an input of an online minimization problem, $I = (x_1, \ldots, x_n)$. Let OPT be an optimal algorithm for the corresponding offline problem. An online algorithm ALG with advice calculates a feasible output $\mathrm{ALG}^{\phi}(I) = (y_1, \ldots, y_n) \in \mathcal{O}(I)$ such that for $i \in \{2, \ldots, n\}$, y_i depends solely on $\phi, x_1, \ldots, x_i$ and on $y_1, \ldots, y_{i-1}$ (y_1 depends solely on ϕ, x_1), where ϕ is the content of the advice tape, i.e., an arbitrarily large binary sequence. ALG is c-competitive with advice complexity $b(n)$ if for every $n \in \mathbb{N}$ and for every input of length at most n, there exists some ϕ such that $\mathrm{cost}(\mathrm{ALG}^{\phi}(I)) \leq c \cdot \mathrm{cost}(\mathrm{OPT}(I))$ and at most the first $b(n)$ bits of ϕ have been read during the computation of $\mathrm{ALG}^{\phi}(I)$.*

To get an easier notation, we omit ϕ and write $\mathrm{ALG}(I)$ instead of $\mathrm{ALG}^{\phi}(I)$ as it is always clear from context; moreover, we will write b instead of $b(n)$. We call the first b bits of ϕ *advice bits* and say that ALG *uses b advice bits* when ALG accesses b bits of ϕ during its computation. We can think of "online computation with advice" as the following game between the online algorithm ALG with advice, the adversary ADV, and the oracle O [9, p. 25].

1. ADV knows ALG, O and b. ADV then constructs an instance I such that the competitive ratio of ALG is maximized.
2. O examines I and writes its advice ϕ on the advice tape.
3. ALG computes an output for I using at most the first b bits of ϕ.

The following observation is crucial for our thinking about online algorithms with advice. Let ALG be an online algorithm with advice that reads b advice bits. ADV can treat ALG as a set of 2^b deterministic online algorithms without advice from which the algorithm with the lowest competitive ratio on the given input is chosen [9]. ADV knows each of the 2^b algorithms.

1.1 Job Shop Scheduling with Unit-Length Tasks

In this paper, we study *job shop scheduling* with two jobs and unit-length tasks. Here, we are given a factory with m machines, and two customers that arrive with one job each; each job consists of m pairwise different tasks that each want to use one of the machines. Therefore, each job wants to use all machines in a fixed order. A job can thus be described by a permutation of the numbers $1, \ldots, m$ of the machines in the factory. A machine can process only one task at a time and needs one time unit to process a task (this is why we speak of unit-length tasks). A problem arises if both customers want to use the same machine in some time step. In this case, one of them has to be delayed. The goal is to minimize the cost, which (in the classical model) is the total amount of time needed to process all tasks, which we also call the *makespan*. Thus, an online algorithm for this problem aims at doing as much of the work in parallel as possible.

To get a better intuition, we now explain how we can represent the problem graphically following Akers [1]. Recall that both jobs can be described by a permutation π_1 or π_2, respectively. We take a grid of size $m \times m$, and assign coordinates to every intersection in the grid. The lower left corner has coordinates $(0,0)$, the top right corner has coordinates (m, m); see Figure 1.

We label the x-axis with π_1 and the y-axis with π_2. Each cell (i, j) to $(i+1, j+1)$ of the grid that does not have the same label on both axes, i.e., where $\pi_1(i+1) \neq \pi_2(j+1)$, receives a diagonal edge from (i, j) to $(i+1, j+1)$. Instead of drawing the diagonal edges, we often only draw the *obstacles*, i.e., the squares in the grid where there are no diagonal edges. An algorithm for this problem starts at the lower left corner and moves on the edges of the

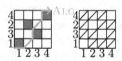

Fig. 1. Grid with obstacles on the left, grid with diagonals on the right

grid to the top right corner. In the classical model, the cost of an algorithm is simply the number of edges it uses. As an example, let $\pi_1 = (1, 2, 3, 4)$ and $\pi_2 = (1, 3, 2, 4)$ be the two jobs. Figure 1 depicts the corresponding grid and a solution computed by some algorithm. We are now ready to give a formal definition.

Definition 5 (Job Shop Scheduling with Two Jobs, JSS). *Let $m > 0$ be a natural number. An input instance I of JSS is of the form (π_1, π_2), where π_1 and π_2 are permutations of $1, \ldots, m$. An output $O \in \mathcal{O}(I)$ consists of two injective functions $f_1, f_2 \colon \{1, \ldots, m\} \to \{1, \ldots, 2m\}$ that map the machines to time steps such that for $r, s \in \{1, \ldots, m\}$:*

1. *$\forall i \in \{1, 2\} \colon$ $f_i(\pi_i(r)) < f_i(\pi_i(s))$ if and only if $r < s$ (the order of the machines defined by the jobs has to be respected), and*
2. *$f_1(r) \neq f_2(r)$ (it is not possible to process two tasks on the same machine at the same time).*

The cost of an algorithm ALG on an instance I is defined as the makespan, i.e., the time when the last task has been executed: $\max_{i \in \{1, 2\}} \{f_i(\pi_i(m))\}$. In the online version, $\pi_1(1)$, $\pi_2(1)$ and m are given first. Then, for $i \in \{1, 2\}$, $\pi_i(j)$ is given as soon as $f_i(\pi_i(j - 1))$ is defined.

Throughout this paper, m denotes the number of machines of the instance at hand. Moreover, the terms "output" and "solution" are used interchangeably. We call the time units a solution needs to schedule the tasks *steps*. We call a step a *diagonal step* if both a task from job π_1 and from π_2 are processed. Similarly, we call it a *horizontal step* (*vertical step*) when only a task from π_1 (π_2) is executed. A first observation is that every solution for JSS needs at least m steps. In other words, the output of every algorithm has a makespan of at least m. The *delay* of a solution $\text{ALG}(I)$ is defined as the number of non-diagonal steps divided by two; the delay of ALG on I is equal to $cost(\text{ALG}(I)) - m$ [8].

If an algorithm ALG for JSS takes a diagonal step whenever possible, we call ALG *smart*; *smart solutions* are defined analogously. We will consider such algorithms later in this paper. For every instance of JSS, there is a smart optimal solution [3]. We will further use *diagonal strategies*, which were introduced by

Hromkovič et al. as a useful combinatorial tool [8, p. 9]. Let I be an instance of JSS. By $Diag_0$ we denote the main diagonal from the lower left corner to the top right corner of the grid. For every $i \in \{1, \ldots, m-1\}$, $Diag_i$ denotes the diagonal from $(i, 0)$ to $(m, m-i)$, which is i cells below $Diag_0$. For $i \in \{-1, \ldots, -(m-1)\}$, $Diag_i$ is the diagonal from $(0, i)$ to $(m-i, m)$. To each such diagonal $Diag_i$, we assign an algorithm D_i. D_i first takes i horizontal or vertical steps towards the starting point of $Diag_i$, i.e., for $i > 0$ towards $(i, 0)$, for $i < 0$ towards $(0, i)$. Then, D_i always takes diagonal steps until it reaches an obstacle. If this happens, D_i takes a horizontal step directly followed by a vertical step – thus avoiding the obstacle – before it continues with diagonal steps. It is not difficult to verify [8, p. 9] that the cost of D_i on an instance I equals the number of obstacles on $Diag_i$ plus m plus $|i|$.

Some proofs in the remainder of this paper are omitted due to space constraints.

2   Implications of New Cost Measure on Former Results

In this section, we are going to change the cost function of JSS since many good results of JSS are due to the fact that the cost of every algorithm is at least m and at most $2m$. This means that we cannot be worse than 2-competitive regardless of how our algorithm behaves, which makes a fine-grained analysis impossible. To foster the search for new algorithms and a more thorough analysis of JSS, we define a new cost measure $newcost$, with which the difference between competitive ratios of different algorithms can be very large. Now let ALG be an algorithm for JSS, let I be an instance for JSS. We define

$$newcost(\text{ALG}(I)) := delay(\text{ALG}(I)) + 1 = cost(\text{ALG}(I)) - m + 1 .$$

We put the $+1$ into the definition of newcost to avoid algorithms with cost 0. Note that before, we had $m \leq cost(\text{ALG}(I)) \leq 2m$; now, we have $1 \leq newcost(\text{ALG}(I)) \leq m + 1$. We insert this new definition of the cost into known results about JSS and obtain the following statements. To enhance readability, we will just write $cost$ instead of $newcost$.

For all $m \in \mathbb{N}$, there is an instance I_m of JSS such that we have for all algorithms ALG $cost(\text{ALG}(I_m)) \geq \lfloor \sqrt{m} \rfloor + 1$ [8]. There is an optimal online algorithm ALG with advice that uses at most $2 \lceil \sqrt{m} \rceil - \frac{1}{4} \log(m)$ advice bits for any instance [10]. This is due to the fact that solutions for the new measure are optimal if and only if they are optimal for the old one. Therefore, bounds on the advice complexity of optimal algorithms carry over immediately. For m large enough, any online algorithm with advice needs to use at least $\sqrt{m/2}$ advice bits to be optimal [10]. Every online algorithm with advice that reads at most b advice bits has a competitive ratio of at least $\sqrt{m}/(3 \cdot 2^b)$ [9].

We are now going to see that the three online algorithms OLR, SURFACE, and ALG_d have competitive ratios of approximately $\sqrt{m}$ or $\sqrt{m}/2$. Such competitive ratios are very high; these results are, therefore, tantamount to a failure of these algorithms in the new cost model. This is surprising considering the fact that these three online algorithms rank among the best online algorithms with advice (for JSS) so far.

We start with the competitive ratio of OLR, an algorithm for JSS presented by Hromkovič et al. [8, p. 5]. OLR chooses a diagonal strategy D_i, $i \in \{-\lfloor\sqrt{m}\rfloor, \ldots, \lfloor\sqrt{m}\rfloor\}$, uniformly at random.

Theorem 6 OLR *has a competitive ratio of* $\sqrt{m} + 1$.

The next algorithm we consider is the deterministic offline algorithm SURFACE from Hromkovič et al. [8] that chooses the best diagonal strategy among the $2\sqrt{m} + 1$ diagonal strategies D_i with $i \in \{-\sqrt{m}, \ldots, \sqrt{m}\}$.

Theorem 7 SURFACE *has a competitive ratio of* $(\lfloor\sqrt{m}\rfloor + 1)/2$.

Komm and Královič [10] suggested an online algorithm ALG_d with advice for any odd $d \in \mathbb{N}$ that reads $\lceil\log d\rceil$ advice bits that tell the algorithm which strategy out of the diagonal strategies D_i, $i \in \{-\frac{d-1}{2}, \ldots, \frac{d-1}{2}\}$, to take. ALG_d achieves a competitive ratio of at most $1 + 1/d$ with respect to the old cost measure. However, with the new measure, we obtain the following theorem, showing a bound that is much weaker.

Theorem 8 ALG_d *has a competitive ratio of at least* $\sqrt{4m + 3}/4 - 1$.[1]

3   New Algorithms

We provide new online algorithms for JSS. We first present BO, a deterministic online algorithm with a fairly good competitive ratio. This is striking since for every deterministic online algorithm ALG, there exists an instance I such that ALG has a delay of at least $m/3$ on I [8, p. 14]. However, the idea of our algorithm is to "move" in such a way that an optimal schedule's delay increases with the delay of the algorithm.

We then present two online algorithms with advice that are based on BO. ADVICEBO only works for $\log(2\sqrt{m} + 1) + 1$ advice bits, which facilitates the comparison with OLR and SURFACE, demonstrating the effectiveness of BO. ADVICEBO2 works for any number $b \geq 2\log(m)+1$ of advice bits and achieves an upper bound on the competitive ratio of $\max\{6, \sqrt{8\log(m)} \sqrt[4]{m}/\sqrt{b}\}$. It combines BO with a *divide-and-conquer* approach.

3.1   A Good Deterministic Online Algorithm

Before we explain the algorithm, we first see how good we can get in the next theorem.

Theorem 9 *No deterministic strategy for JSS has a competitive ratio smaller than* $(m/3 + 1)/(\sqrt{m} + 1) \approx \sqrt{m}/3$.

---

[1] This bound tends to $\frac{\sqrt{m}}{2}$ for large m. It is achieved for $d = \sqrt{4m + 3}$.

We now explain how BO works. To facilitate the description and the calculation of the upper bound on the competitive ratio of BO, we describe BO only on instances of size $m = 3n^2 - 3n + 1$ for an $n \geq 3 \in \mathbb{N}$. The extension to other instances is given later. BO is smart, i.e., BO takes a diagonal step whenever possible, and if BO hits the border of the grid, it takes the necessary amount of horizontal or vertical steps to reach the end.

Let I be an instance of JSS of size $m = 3n^2 - 3n + 1$ for an $n \geq 3 \in \mathbb{N}$. Let OPT be an optimal solution for I. We divide I into at most $2n - 1$ rectangles $B_1, \ldots, B_{2n-1}$, which we call *blocks*, as follows. The number of blocks depends on the number of obstacles BO encounters. In each block B_i, BO encounters a certain number of obstacles. We call a block B_i *complete* if BO encounters (1) $2i - 1$ (if $i \leq n$) or (2) $2(2n - i) - 1$ (if $i > n$) obstacles within B_i.

For $i \in \{2, \ldots, 2n - 1\}$, block B_i exists if block B_{i-1} is complete and BO encounters another obstacle. More constructively, the first block, B_1, starts at the lower left corner of the instance I and it ends when BO has encountered $2 \cdot 1 - 1 = 1$ obstacle and is about to encounter the next obstacle. It ends right before this obstacle. Where B_1 ends, B_2 begins, i.e., B_2 begins right before the second obstacle BO encounters. It ends when BO has encountered $2 \cdot 2 - 1 = 3$ obstacles after B_1 and is about to encounter the next obstacle, which is the 5-th obstacle in total.

We continue iteratively. This way, the

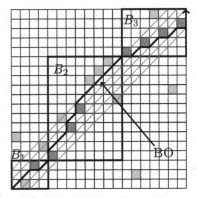

Fig. 2. BO on an instance with three blocks

blocks are aligned as in Figure 2, i.e., the upper right corner of each block B_i touches the lower left corner of B_{i+1}. Moreover, the lower left corner of B_1 is identical with the lower left corner of I, and the upper right corner of the last block is identical with the upper right corner of I. Note that when block B_{2n-1} is complete, BO encounters an obstacle in every second step.

A key point when we prove an upper bound on the competitive ratio of BO is that the delay of OPT depends on the number of blocks, i.e., the more blocks we have, the higher the delay of OPT. We shall now describe how BO behaves.

1. BO takes a diagonal step whenever possible.
2. For all odd i (all even i), BO takes a horizontal (vertical) step upon encountering an obstacle from block B_i.
3. For all $i \geq n$, rule 2 does not apply for the last obstacle of block B_i. There, BO takes a vertical step for odd i and a horizontal step for even i instead.

We have the following upper bound on the competitive ratio for BO, which is close to the best ratio we can expect according to Theorem 9.

Theorem 10 *For $m = 3n^2 - 3n + 1$ and $n \geq 3$, BO has a competitive ratio less than or equal to $(4 - 2\sqrt{2})n \approx (4 - 2\sqrt{2})\sqrt{m}/\sqrt{3} \approx 0.67\sqrt{m}$.*

Proof: Let I be an instance of size $m = 3n^2 - 3n + 1$, $n \geq 3$. We divide I into blocks as we have described above. If there is no obstacle on the main diagonal, BO is optimal because of Rule 1. Therefore, we may assume that we have at least one obstacle on the main diagonal. Moreover, we assume w.l.o.g. that this obstacle touches the lower left corner of I, so BO meets an obstacle right at the beginning. By construction, BO "moves" solely inside the blocks.

The proof has three parts. In the first part, we provide a lower bound on the cost of an optimal solution on an instance with $k \in \mathbb{N}$ complete blocks. In the second part, we calculate the delay of BO on instances with k complete blocks. In the third part, we combine the results to calculate an upper bound on the competitive ratio of BO for instances with k complete blocks. We choose k such that this upper bound on the competitive ratio of BO is maximized.

Part 1: Let ALG be an algorithm for JSS. ALG *leaves block* B_i the moment it touches the right or the upper border of B_i for the first time. If ALG does not touch these borders, we extend the borders of B_i over the whole instance and take the moment it touches these extended borders. We now want to know the cost of an optimal solution on an instance with k complete blocks.

Lemma 11. *Let* $k \leq n$. *If the first* k *blocks are complete, every algorithm has to make at least* k *non-diagonal steps in order to leave* B_k.

This is proven by induction. Note that the non-diagonal steps of Lemma 11 have taken place *when leaving* B_k, i.e., there might be more non-diagonal steps later. We calculate the delay in the next lemma.

Lemma 12. *Let* $k \in \{n+1, \ldots, 2n-1\}$. *If the first* k *blocks are complete, every algorithm has to make at least* k *non-diagonal steps in order to leave* B_k *or it makes* $2n - 1$ *non-diagonal steps in total.*[2]

Proof: Let $k > n$. In B_r, $k \geq r > n$, there is an obstacle on every diagonal $Diag_i$ for $i \in \{-(2n - r - 1), \ldots, 2n - r - 1\}$ since B_r is complete.

Let ALG be any algorithm. ALG leaves B_n on $Diag_i$ and has made at least n non-diagonal steps so far according to Lemma 11. It has to make i non-diagonal steps in order to finish all tasks. Therefore, if $|i| \geq (2n - (n+1) - 1) + 1 = n - 1$, ALG makes at least a total of $n + (n-1) = 2n - 1$ non-diagonal steps and we are done. Else we have $|i| \leq (2n - (n+1) - 1)$ and ALG encounters an obstacle on its way to pass B_{n+1}.

We repeat this reasoning. Either we are done or ALG leaves B_{n+1} on $Diag_i$ and has made at least $n+1$ non-diagonal steps so far. If $|i| > (2n - (n+2) - 1)$, ALG has to make i non-diagonal steps in order to finish all tasks. In this case, ALG makes a total of $n + 1 + (n - 3) + 1 = 2n - 1$ non-diagonal steps and we are done. Else we have $|i| \leq (2n - (n+2) - 1)$ and ALG encounters an obstacle on its way to leave B_{n+1}. We continue iteratively, thus finishing the proof. □

---

[2] Note that if k is even, an optimal algorithm OPT can be on $Diag_0$ after k non-diagonal steps and might not encounter any further obstacle. If k is odd, OPT cannot be on $Diag_0$ after k non-diagonal steps. Therefore, Lemmata 11 and 12 imply a lower bound of $\lfloor (k + 1)/2 \rfloor$ on the delay of OPT.

Part 2: In this part, we calculate the cost of BO. With the following lemma, we know how many non-diagonal steps BO makes in k complete blocks.

Lemma 13. *Let $k \in \{1, \ldots, 2n-1\}$. Let I be an instance where the first k blocks are complete. BO makes exactly (1) k^2 (if $k \leq n$) or (2) $4nk - 2n^2 + 2n - k^2 - 2k$ (if $k > n$) non-diagonal steps in the first k blocks.*

The proof is a simple calculation. The instance that the adversary chooses has k complete blocks and the $(k+1)$-st block may contain some obstacles, but it is not complete. We use this to calculate the delay of BO.

Lemma 14. *Let I be an instance where k blocks are complete. BO has a delay of at most (1) $(k^2 + 3k)/2$ (if $k < n$), (2) $(n^2 + 2n - 4)/2$ (if $k = n$), or (3) $(4nk - 2n^2 + 4n - k^2 - 3k - 2)/2$ (if $k > n$).*

Part 3: We are almost done. No matter what an instance I looks like, it has k complete blocks and some obstacles in the $(k+1)$-st block. The number k and the number of obstacles in the last block completely determine the cost of BO and give a lower bound on the cost of an optimal solution. The adversary chooses k such that the competitive ratio is maximized. The following upper bounds on the competitive ratio for instances with k complete blocks follow immediately from Lemmata 11, 12, and 14.

$$\begin{cases} \left(\frac{k^2+3k}{2} + 1\right) \Big/ \left(\lfloor \frac{k+1}{2} \rfloor + 1\right), & k < n \\ \frac{n^2+2n-4}{2} \Big/ \lfloor \frac{n+1}{2} \rfloor, & k = n \\ \frac{4nk-2n^2+4n-k^2-3k-2}{2} \Big/ \lfloor \frac{k+1}{2} \rfloor, & k > n. \end{cases} \tag{1}$$

Note that we put the $+1$ only in the bound for $k < n$ since we are going to use this bound later and the $+1$ improves it slightly. We can estimate these bounds easily and obtain for $k < n$ an upper bound of $k + 1 \leq n$; for $k = n$ an upper bound of $n + 2$; and for $k > n$ an upper bound of $(4 - 2\sqrt{2})n$, where this last bound is estimated by differentiating, setting to zero and thus finding that the expression is maximal for $k = \sqrt{2}(n+1)$. Therefore, the adversary is going to choose $k = \sqrt{2}(n+1)$, and it cannot force a smaller competitive ratio than $(4 - 2\sqrt{2})n$. This finishes the proof. $\square$

3.2   Two Online Algorithms with Advice

We first describe ADVICEBO and simultaneously prove that it has a competitive ratio of at most $\sqrt{2}\sqrt[4]{m}$. ADVICEBO either runs BO or it takes the best diagonal strategy. When ADVICEBO runs on an instance I of size m, ADVICEBO reads $\log(2\sqrt{m} + 1) + 1$ advice bits. Let OPT be a smart optimal solution for I. The first advice bit is 1 if and only if $delay(\text{OPT})$ is larger than or equal to $\sqrt{m}/c$, where $c \geq 4$ is a number yet to be determined.

In this case, the remaining $\log(2\sqrt{m} + 1)$ advice bit indicate to ADVICEBO which of the $2\sqrt{m} + 1$ diagonal strategies D_i, $i \in \{-\lfloor \sqrt{m} \rfloor, \ldots, \lfloor \sqrt{m} \rfloor\}$ has the smallest delay. ADVICEBO executes this diagonal strategy. There is a diagonal strategy D_i with a delay of at most $\lfloor \sqrt{m} \rfloor$ [12]. It can be shown that this leads to a competitive ratio of at most c. Conversely, if $delay(\text{OPT}) < \sqrt{m}/c$, ADVICEBO simply runs BO. Since for $c \geq 4$, $2\sqrt{m}/c < n$, where n is the smallest number such that $3n^2 - 3n + 1 \geq m$,[3] we are in the case of $k < n$ in (1) and can take the upper bound on the competitive ratio of $k + 1$, where $\lfloor \frac{k+1}{2} \rfloor$ is the delay of OPT. Therefore, $k = 2\,delay(\text{OPT})$ if k is even and $k + 1 = 2\,delay(\text{OPT})$ if k is odd. Since $2\,delay(\text{OPT}) < 2\sqrt{m}/c$, we have a competitive ratio of at most $2\sqrt{m}/c$. As a consequence, we have an overall competitive ratio of at most $\min\{c, 2\sqrt{m}/c\}$. Clearly, a maximized value is obtained by equalizing the ratios, i.e., choosing $c = \sqrt{2}\sqrt[4]{m}$, which finishes the proof.

For ADVICEBO2, we have to do some preliminary work. We are going to split an instance into rectangles, compute the competitive ratio within the rectangles, and then make conclusions on the competitive ratio of the whole instance. We explain how this works. Let $n \in \mathbb{N}$. When we say that we *divide an instance I into n rectangles* $R_1, \ldots, R_n$, we mean that the upper right corner of rectangle R_i, $i < n$, touches the lower left corner of rectangle R_{i+1}. Moreover, the lower left corner of R_1 is identical with the lower left corner of I, and the upper right corner of R_n is identical with the upper right corner of R_n.

For example, the blocks B_i that we constructed for BO in the proof of Theorem 10 divide the instance into rectangles. To define the *delay in a rectangle*, let R_k be an $(i \times j)$-rectangle of a division of I, i.e., a rectangle of size $(i \times j)$. Let ALG be an algorithm that enters R at the lower left corner and leaves it at the upper right corner. The *delay* of ALG is the number of non-diagonal steps it makes divided by two.[4] Now let ALG move diagonally until it encounters the first obstacle. If there is an optimal solution OPT for I that enters the rectangle at the lower left corner and leaves it at the upper right corner, we define the competitive ratio of an algorithm ALG in R as 1 if $delay_R(\text{OPT})$ is zero,[5] else we define it as $delay_R(\text{ALG})/delay_R(\text{OPT})$.

Lemma 15. *We divide an instance I into n rectangles* $R_1, \ldots, R_n$. *Let ALG be an algorithm for I, and let OPT be an optimal solution for I. If we have in all rectangles an upper bound to the competitive ratio of ALG of x, then ALG has a competitive ratio of at most x for I.*

Now we extend BO such that it works for instances I with m tasks where $3(n + 1)^2 - 3(n + 1) + 1 > m > 3n^2 - 3n + 1$. We take BO and act as if it were an instance for $n + 1$. Since we know that the adversary is not going to place all possible blocks, this still works. The competitive ratio is the same since Equation (1) stays the same.

---

[3] See Section "Extension of BO" in this section for the extension of BO to numbers $m \neq 3n^2 - 3n + 1$.

[4] Obviously, in a rectangle, the delay does not have to be a natural number. Moreover, in an $(i \times j)$-rectangle, every algorithm makes at least $|i - j|$ non-diagonal steps, so the delay is larger than or equal to $\frac{|i-j|}{2}$.

[5] If an optimal solution has a delay of zero in an $(i \times j)$-rectangle, then $i = j$ and every algorithm that only moves diagonally is optimal as well.

We extend BO to rectangles as well. Let R be an $(i \times j)$-rectangle. Since a smart solution never passes a coordinate (h, v) with $2h > v$ or $2v > h$ according to Akveld and Bernhard [2], we assume w.l.o.g. that $2j \geq i > j$. In other words, R is at most twice as long as wide. We define $x := (i - j)$. We extend BO as follows. BO runs in the rectangle as if it were an instance of size $(j \times j)$ with a border. Once BO is at the upper right corner of this $(j \times j)$-subinstance, it takes x non-diagonal steps towards the upper right corner of the rectangle. We have the following lemma for the competitive ratio of BO.

Lemma 16. *Let R be an $(i \times j)$-rectangle with k complete blocks, $2j \geq i > j$. BO has a competitive ratio of at most (1) $k + 4 \leq 2\,delay(\text{OPT}) + 4$ (if $k < n$) or (2) $2\sqrt{j/3} \leq 4\,delay(\text{OPT})$ (if $k \geq n$).*

We can now describe ADVICEBO2. ADVICEBO2 is an online algorithm that reads $b \geq 2\log(m) + 1$ advice bits. We are going to prove that ADVICEBO2 has quite a small competitive ratio. We explain the intuition behind the proof first. ADVICEBO2 divides the instance into x rectangles such that there is an optimal solution OPT that passes all the $x - 1$ coordinate points where the rectangles meet. In each rectangle, ADVICEBO2 has an upper bound on its competitive ratio depending on the delay of OPT in this rectangle. We choose the rectangles such that the delay of OPT in each rectangle is $delay(\text{OPT})/x$. Then, we apply Lemma 15.

Theorem 17 ADVICEBO2 *has a competitive ratio of at most the maximum between 6 and $\sqrt{8\log(m)}\,\sqrt[4]{m}/\sqrt{b}$.*

Proof: ADVICEBO2 runs on an instance I of JSS with m tasks. Let OPT be a smart optimal solution of I. As in ADVICEBO, the first advice bit is 1 if and only if $delay(\text{OPT}) \geq \sqrt{m}/c$, where $c \geq 1 \in \mathbb{R}$ is again a number we are going to choose later. On the one hand, in this case, the oracle writes the number of the best diagonal strategy on the advice tape and ADVICEBO2 is c-competitive.

On the other hand, let $x \in \mathbb{N}$ be a natural number which we are going to define later. The oracle indicates to ADVICEBO2 $(x-1)$ coordinates $(h_1, v_1), \ldots,$ (h_{x-1}, v_{x-1}) of the $(m \times m)$-grid. Together with the two known coordinates $(h_0, v_0) := (0, 0)$ and $(h_x, v_x) := (m, m)$, these coordinates are chosen in such a way that they divide the instance I into x rectangles $R_1, \ldots, R_x$. Each rectangle R_i has a lower left corner (h_{i-1}, v_{i-1}) and an upper right corner (h_i, v_i). The oracle chooses the coordinates such that (1) OPT passes these coordinates, and (2) in each of the x rectangles, OPT has a delay of at most $\lceil delay(\text{OPT})/x \rceil$.

Since OPT makes a total of $2\,delay(\text{OPT})$ non-diagonal steps, choosing the coordinate point where OPT is after each $\lceil 2\,delay(\text{OPT})/x \rceil$ non-diagonal steps satisfies these two requirements. Moreover, since OPT is smart, the rectangles we obtain are of the form $(i \times j)$ with $2j \geq i \geq j$ or $2i \geq j \geq i$. ADVICEBO2 runs BO in each of these rectangles. In rectangle R_i, BO has a competitive ratio of at most $\max\{4\,delay_{R_i}(\text{OPT}), 2\,delay_{R_i}(\text{OPT}) + 4\}$ according to Lemma 16 and according to Theorem 10 in the case where R_i is a square. We know that $4\,delay_{R_i}(\text{OPT}) \geq 2\,delay_{R_i}(\text{OPT}) + 4$ is equivalent to $delay_{R_i}(\text{OPT}) \geq 2$.

In the case of $delay_{R_i}(\text{OPT}) = 1$, we cannot guarantee a better upper bound on the competitive ratio than $2 \cdot 1 + 4 = 6$. This is where the 6 in the theorem comes from. We chose the rectangles such that we have for all i, we have $delay_{R_i}(\text{OPT}) \leq \lceil delay(\text{OPT})/x \rceil$. Since we know that $delay(\text{OPT}) \leq \sqrt{m}/c - 1$, BO has a competitive ratio of at most the maximum of 6 and

$$4\, delay_{R_i}(\text{OPT}) \leq 4 \left\lceil \frac{delay(\text{OPT})}{x} \right\rceil \leq \frac{4\sqrt{m}}{cx} \tag{2}$$

in each of the rectangles. We apply Lemma 15 and obtain that (2) holds in I as well. We see that the upper bound in Equation (2) decreases if x increases. We estimate how large x can be. A rough upper bound on the number of advice bits to tell a coordinate (i,j) with $i, j \in \{0, \ldots, m\}$ is $2\log(m)$. The oracle writes $x - 1$ coordinates on the advice tape. For this, it needs at most $(x - 1)2\log(m)$ advice bits. Since there are only $b - 1$ advice bits left to read, we know that $(x - 1)2\log(m) \leq (b - 1)$ holds. We choose x as the largest natural number satisfying the equation above, i.e. with $x \leq \frac{b-1}{2\log(m)}+1$. As a consequence, we have a competitive ratio of at most the maximum between 6 and $\max\{c, 4\sqrt{m}/cx\}$, which is smaller or equal than $\max\{c, 4\sqrt{m}/(cb/2\log m)\}$. This is maximal when the two values in the maximum are equal. We choose $c = \sqrt{8\log(m)}\sqrt[4]{m}/\sqrt{b}$ and obtain a competitive ratio of at most $\max\{c, 6\}$. □

Acknowledgements. I thank Juraj Hromkovič for his useful advice and support; Dennis Komm for his ceaseless effort in giving this work impetus and the many hours he spent with proofreading; and Hans-Joachim Böckenhauer for many helpful discussions.

References

1. Akers, S.B.: A graphical approach to production scheduling problems. Operations Research **4**(2), 244–245 (1956)
2. Akveld, M., Bernhard, R.: Job shop scheduling with unit length tasks. RAIRO - Theoretical Informatics and Applications **46**(3), 329–342 (2012)
3. Böckenhauer, H.-J., Komm, D., Královič, R., Královič, R., Mömke, T.: Online algorithms with advice. Technical Report 614, ETH Zürich (2009)
4. Borodin, A., El-Yaniv, R.: Online Computation and Competitive Analysis. Cambridge University Press (1998)
5. Dobrev, S., Královič, R., Pardubská, D.: How much information about the future is needed? In: Geffert, V., Karhumäki, J., Bertoni, A., Preneel, B., Návrat, P., Bieliková, M. (eds.) SOFSEM 2008. LNCS, vol. 4910, pp. 247–258. Springer, Heidelberg (2008)
6. Emek, Y., Fraigniaud, P., Korman, A., Rosén, A.: Online computation with advice. Theoretical Computer Science **412**(24), 2642–2656 (2011)
7. Hromkovič, J., Královič, R., Královič, R.: Information complexity of online problems. In: Hliněný, P., Kučera, A. (eds.) MFCS 2010. LNCS, vol. 6281, pp. 24–36. Springer, Heidelberg (2010)

8. Hromkovič, J., Steinhöfel, K., Widmayer, P.: Job shop scheduling with unit length tasks: bounds and algorithms. In: Restivo, A., Ronchi Della Rocca, S., Roversi, L. (eds.) ICTCS 2001. LNCS, vol. 2202, pp. 90–106. Springer, Heidelberg (2001)
9. Komm, D.: Advice and randomization in online computation. Dissertation at ETH Zürich No.20164 (2012)
10. Komm, D., Královič, R.: Advice complexity and barely random algorithms. In: Černá, I., Gyimóthy, T., Hromkovič, J., Jefferey, K., Královič, R., Vukolić, M., Wolf, S. (eds.) SOFSEM 2011. LNCS, vol. 6543, pp. 332–343. Springer, Heidelberg (2011)
11. Sleator, D.D., Tarjan, R.E.: Amortized efficiency of list update and paging rules. Communications of the ACM 28(2), 202–208 (1985)
12. Wehner, D.: Job-Shop-Scheduling im Dreidimensionalen. Bachelor's thesis at ETH Zürich (2012)

Author Index

Printed in the United States
By Bookmasters